Game Theory and Exercises

Game Theory and Exercises introduces the main concepts of game theory, along with interactive exercises to aid readers' learning and understanding. Game theory is used to help players understand decision-making, risk-taking and strategy and the impact that the choices they make have on other players; and how the choices of those players, in turn, influence their own behaviour. So, it is not surprising that game theory is used in politics, economics, law and management.

This book covers classic topics of game theory including dominance, Nash equilibrium, backward induction, repeated games, perturbed strategies, beliefs, perfect equilibrium, perfect Bayesian equilibrium and replicator dynamics. It also covers recent topics in game theory such as level-*k* reasoning, best reply matching, regret minimization and quantal responses. This textbook provides many economic applications, namely on auctions and negotiations. It studies original games that are not usually found in other textbooks, including Nim games and traveller's dilemma. The many exercises and the inserts for students throughout the chapters aid the reader's understanding of the concepts.

With more than 20 years' teaching experience, Umbhauer's expertise and classroom experience helps students understand what game theory is and how it can be applied to real life examples. This textbook is suitable for both undergraduate and postgraduate students who study game theory, behavioural economics and microeconomics.

Gisèle Umbhauer is Associate Professor of Economics at the University of Strasbourg, France.

Routledge Advanced Texts in Economics and Finance

1. Financial Econometrics
   Peijie Wang

2. Macroeconomics for Developing Countries, 2nd edition
   Raghbendra Jha

3. Advanced Mathematical Economics
   Rakesh Vohra

4. Advanced Econometric Theory
   John S. Chipman

5. Understanding Macroeconomic Theory
   John M. Barron, Bradley T. Ewing and Gerald J. Lynch

6. Regional Economics
   Roberta Capello

7. Mathematical Finance: Core Theory, Problems and Statistical Algorithms
   Nikolai Dokuchaev

8. Applied Health Economics
   Andrew M. Jones, Nigel Rice, Teresa Bago d'Uva and Silvia Balia

9. Information Economics
   Urs Birchler and Monika Bütler

10. Financial Econometrics (Second Edition)
    Peijie Wang

11. Development Finance
    Debates, Dogmas and New Directions
    Stephen Spratt

12. Culture and Economics
    On Values, Economics and International Business
    Eelke de Jong

13. Modern Public Economics, Second Edition
    Raghbendra Jha

14. Introduction to Estimating Economic Models
    Atsushi Maki

15. Advanced Econometric Theory
    John Chipman

16. Behavioral Economics
    Edward Cartwright

17. Essentials of Advanced Macroeconomic Theory
    Ola Olsson

18. Behavioral Economics and Finance
    Michelle Baddeley

19. Applied Health Economics, Second Edition
    Andrew M. Jones, Nigel Rice, Teresa Bago d'Uva and Silvia Balia

20. Real Estate Economics
   A Point to Point Handbook
   Nicholas G. Pirounakis

21. Finance in Asia
   Institutions, Regulation and Policy
   *Qiao Liu, Paul Lejot and
   Douglas Arner*

22. Behavioral Economics, Second Edition
   Edward Cartwright

23. Understanding Financial Risk Management
   Angelo Corelli

24. Empirical Development Economics
   *Måns Söderbom and Francis Teal with
   Markus Eberhardt, Simon Quinn and Andrew
   Zeitlin*

25. Strategic Entrepreneurial Finance
   From Value Creation to Realization
   Darek Klonowski

26. Computational Economics
   A concise introduction
   *Oscar Afonso and
   Paulo Vasconcelos*

27. Regional Economics, Second Edition
   Roberta Capello

28. Game Theory and Exercises
   Gisèle Umbhauer

Game Theory and Exercises

Gisèle Umbhauer

Routledge
Taylor & Francis Group
LONDON AND NEW YORK

First published 2016
by Routledge
2 Park Square, Milton Park, Abingdon, Oxon OX14 4RN

by Routledge
711 Third Avenue, New York, NY 10017

Routledge is an imprint of the Taylor & Francis Group, an informa business

© 2016 Gisèle Umbhauer

The right of Gisèle Umbhauer to be identified as author of this work has been asserted by her in accordance with the Copyright, Designs and Patent Act 1988.

All rights reserved. No part of this book may be reprinted or reproduced or utilized in any form or by any electronic, mechanical, or other means, now known or hereafter invented, including photocopying and recording, or in any information storage or retrieval system, without permission in writing from the publishers.

Trademark notice: Product or corporate names may be trademarks or registered trademarks, and are used only for identification and explanation without intent to infringe.

British Library Cataloguing in Publication Data
A catalogue record for this book is available from the British Library

Library of Congress Cataloging in Publication Data
Umbhauer, Gisele.
Game theory and exercises / Gisele Umbhauer.
1. Game theory. I. Title.
HB144.U43 2015
519.3--dc23
2015020124

ISBN: 978-0-415-60421-5 (hbk)
ISBN: 978-0-415-60422-2 (pbk)
ISBN: 978-1-315-66906-9 (ebk)

Typeset in Times New Roman and Bell Gothic
by Saxon Graphics Ltd, Derby

Dedication

To my son Victor

Contents

Acknowledgements xix
Introduction 1

1 HOW TO BUILD A GAME 5
INTRODUCTION 5
1 STRATEGIC OR EXTENSIVE FORM GAMES? 5
   1.1 Strategic/normal form games 6
      1.1.1 Definition 6
      1.1.2 Story strategic/normal form games and behavioural comments 6
      1.1.3 All pay auction 12
   1.2 Extensive form games 18
      1.2.1 Definition 18
      1.2.2 Story extensive form games and behavioural comments 20
      1.2.3 Subgames 27
2 STRATEGIES 28
   2.1 Strategies in strategic/normal form games 28
      2.1.1 Pure strategies 28
      2.1.2 Mixed strategies 28
   2.2 Strategies in extensive form games 30
      2.2.1 Pure strategies: a complete description of the behaviour 30
      2.2.2 The Fort Boyard Sticks game and the Envelope game 31
      2.2.3 Behavioural strategies 35
   2.3. Strategic/normal form games and extensive form games: is there a difference? 36
3 INFORMATION AND UTILITIES 40
   3.1 Perfect/imperfect information 40
      3.1.1 A concept linked to the information sets 40
      3.1.2 The prisoners' disks game 41
   3.2 Complete/incomplete information 43
      3.2.1 Common knowledge: does it make sense? 43
      3.2.2 Signalling games and screening games 45
   3.3 Utilities 47
      3.3.1 Taking risks into account 47
      3.3.2 Which is the game you have in mind? 49
      3.3.3 Fairness and reciprocity 49
      3.3.4 Strategic feelings and paranoia 51
      3.3.5 How to bypass utilities 52
CONCLUSION 54

Exercise 1	Easy strategy sets	56
Exercise 2	Children game, how to share a pie equally	57
Exercise 3	Children game, the bazooka game, an endless game	57
Exercise 4	Syndicate game: who will be the free rider?	57
Exercise 5	Rosenthal's centipede (pre-emption) game, reduced normal form game	58
Exercise 6	Duel guessing game, a zero sum game	58
Exercise 7	Akerlof's lemon car, experience good model, switching from an extensive form signalling game to its normal form	59
Exercise 8	Behavioural strategies and mixed strategies, how to switch from the first to the second and vice versa?	60
Exercise 9	Dutch auction and first price sealed bid auction, strategic equivalence of different games	60

2 DOMINANCE 62

INTRODUCTION 62

1 NON ITERATED DOMINANCE 63
   1.1 Definitions, strict and weak dominance 63
   1.2 Dominance in normal form games 63
      1.2.1 Existence of dominated strategies 63
      1.2.2 Normal form games with dominated strategies 64
   1.3 Dominance in extensive form games 67
      1.3.1 Strict and weak dominance in the ascending all pay auction/war of attrition game 67
      1.3.2 Weak dominance and the Fort Boyard sticks game 68

2 ITERATED DOMINANCE 72
   2.1 Iterated dominance, the order matters 72
   2.2 Iterated dominance and first doubts 73

3 CROSSED RATIONALITY AND LIMITS OF ITERATED DOMINANCE 75
   3.1 Envelope game: K–1 iterations for a strange result 75
   3.2 Levels of crossed rationality in theory and reality 77
   3.3 Crossed rationality in extensive form games, a logical inconsistency 78

4 DOMINANCE AND STRUCTURE OF A GAME, A COME BACK TO THE PRISONER'S DISKS GAME 79
   4.1 Solving the game by iterative elimination of strictly dominated strategies 79
   4.2 Dominance and easy rule of behaviour 84

CONCLUSION 87

Exercise 1	Dominance in a game in normal form	89
Exercise 2	Dominance by a mixed strategy	89
Exercise 3	Iterated dominance, the order matters	90
Exercise 4	Iterated dominance in asymmetric all pay auctions	90
Exercise 5	Dominance and value of information	91
Exercise 6	Stackelberg first price all pay auction	92
Exercise 7	Burning money	93
Exercise 8	Pre-emption game in extensive form and normal form, and crossed rationality	93
Exercise 9	Guessing game and crossed rationality	94
Exercise 10	Bertrand duopoly	94
Exercise 11	Traveller's dilemma (Basu's version)	95

Exercise 12	Traveller's dilemma, the students' version	95
Exercise 13	Duel guessing game, the cowboy story	95
Exercise 14	Second price sealed bid auction	96
Exercise 15	Almost common value auction, Bikhchandani and Klemperer's result	97

3 NASH EQUILIBRIUM 98

INTRODUCTION 98

1 NASH EQUILIBRIUM, A FIRST APPROACH 99
  1.1 Definition and existence of Nash equilibrium 99
  1.2 Pure strategy Nash equilibria in normal form games and dominated strategies 100
  1.3 Mixed strategy Nash equilibria in normal form games 102

2 NASH EQUILIBRIA IN EXTENSIVE FORM GAMES 106
  2.1 Nash equilibria in the ascending all pay auction/war of attrition game 107
  2.2 Same Nash equilibria in normal form games and extensive form games 110

3 NASH EQUILIBRIA DO A GOOD JOB 113
  3.1 Nash equilibrium, a good concept in many games 113
    3.1.1 All pay auctions with incomplete information 113
    3.1.2 First price sealed bid auctions 115
    3.1.3 Second price sealed bid auctions, Nash equilibria and the marginal approach 116
  3.2 Nash equilibrium and simple rules of behaviour 118

4 MULTIPLICITY OF NASH EQUILIBRIA 122
  4.1 Multiplicity in normal form games, focal point and talking 122
  4.2 Talking in extensive form games 123
  4.3 Strange out of equilibrium behaviour, a game with a reduced field of vision 124

5 TO PLAY OR NOT TO PLAY A NASH EQUILIBRIUM, CAUTIOUS BEHAVIOUR AND RISK DOMINANCE 127
  5.1 A logical concept, but not always helpful 127
  5.2 Cautious behaviour and risk dominance 130
    5.2.1 Cautious behaviour 130
    5.2.2 Ordinal differences, risk dominance 133

CONCLUSION 136

Exercise 1	Nash equilibria in a normal form game	138
Exercise 2	Story normal form games	138
Exercise 3	Burning money	138
Exercise 4	Mixed Nash equilibria and weak dominance	139
Exercise 5	A unique mixed Nash equilibrium	139
Exercise 6	French variant of the rock paper scissors game	139
Exercise 7	Bazooka game	140
Exercise 8	Pure strategy Nash equilibria in an extensive form game	141
Exercise 9	Gift exchange game	141
Exercise 10	Behavioural Nash equilibria in an extensive form game	142
Exercise 11	Duel guessing game	142
Exercise 12	Pre-emption game (in extensive and normal forms)	142
Exercise 13	Bertrand duopoly	143
Exercise 14	Guessing game	143
Exercise 15	Traveller's dilemma (Basu)	143
Exercise 16	Traveller's dilemma, students' version, P<49	143

CONTENTS

Exercise 17	Traveller's dilemma, students' version, P>49, and cautious behaviour	144
Exercise 18	Focal point in the traveller's dilemma	145
Exercise 19	Asymmetric all pay auctions	145
Exercise 20	Wallet game, first price auction, winner's curse and a robust equilibrium	146
Exercise 21	Wallet game, first price auction, Nash equilibrium and new stories	146
Exercise 22	Two player wallet game, second price auction, a robust symmetric equilibrium	147
Exercise 23	Two player wallet game, second price auction, asymmetric equilibria	147
Exercise 24	N player wallet game, second price auction, marginal approach	147
Exercise 25	Second price all pay auctions	148
Exercise 26	Single crossing in a first price sealed bid auction	148
Exercise 27	Single crossing and Akerlof's lemon car	148
Exercise 28	Dutch auction and first price sealed bid auction	149

4 BACKWARD INDUCTION AND REPEATED GAMES   **150**
INTRODUCTION   150
1 SUBGAME PERFECT NASH EQUILIBRIUM AND BACKWARD INDUCTION   151
   1.1 Subgame Perfect Nash equilibrium and Nash equilibrium   151
   1.2 Backward induction   153
2 BACKWARD INDUCTION, DOMINANCE AND THE GOOD JOB OF BACKWARD INDUCTION/SUBGAME PERFECTION   154
   2.1 Backward induction and dominance   154
   2.2 The good job of backward induction/subgame perfection   156
      2.2.1 Backward induction and the Fort Boyard sticks game   156
      2.2.2 Backward induction and negotiation games   158
3 WHEN THE JOB OF BACKWARD INDUCTION/SUBGAME PERFECTION BECOMES LESS GOOD   160
   3.1 Backward induction, forward induction or thresholds?   160
   3.2 Inconsistency of backward induction, forward induction or momentary insanity?   163
   3.3 When backward induction leads to very strange results.   165
4 FINITELY REPEATED GAMES   169
   4.1 Subgame Perfection in finitely repeated normal form games   169
      4.1.1 New behaviour in finitely repeated normal form games   169
      4.1.2 New behaviour, which links with the facts?   173
      4.1.3 Repetition and backward induction's inconsistency   174
   4.2 Subgame perfection in finitely repeated extensive form games   175
      4.2.1 A forbidden transformation   176
      4.2.2 New behaviour in repeated extensive form games   177
5 INFINITELY REPEATED GAMES   179
   5.1 Adapted backward induction   179
   5.2 Infinitely repeated normal form games   180
      5.2.1 New behaviour in infinitely repeated normal form games, a first approach   180
      5.2.2. Minmax values, individually rational payoffs, folk theorem   182
      5.2.3 Building new behaviour with the folk theorem   184
      5.2.4 Punishing and rewarding in practice   189
   5.3 Infinitely repeated extensive form games   193
CONCLUSION   195

Exercise 1	Stackelberg all pay auction, backward induction and dominance	198
Exercise 2	Duel guessing game	198
Exercise 3	Centipede pre-emption game, backward induction and the students' way of playing	199
Exercise 4	How to share a shrinking pie	200
Exercise 5	English all pay auction	201
Exercise 6	General sequential all pay auction, invest a max if you can	201
Exercise 7	Gift exchange game	202
Exercise 8	Repeated games and strict dominance	203
Exercise 9	Three repetitions are better than two	203
Exercise 10	Alternate rewards in repeated games	204
Exercise 11	Subgame perfection in Rubinstein's finite bargaining game	204
Exercise 12	Subgame perfection in Rubinstein's infinite bargaining game	204
Exercise 13	Gradualism and endogenous offers in a bargaining game, Li's insights	205
Exercise 14	Infinite repetition of the traveller's dilemma game	205
Exercise 15	Infinite repetition of the gift exchange game	205

5 TREMBLES IN A GAME 207
INTRODUCTION 207
1 SELTEN'S PERFECT EQUILIBRIUM 208
   1.1 Selten's horse, what's the impact of perturbing strategies? 208
   1.2 Selten's perfect/trembling hand equilibrium 209
   1.3 Applications and properties of the perfect equilibrium 210
      1.3.1 Selten's horse, trembles and strictly dominated strategies 210
      1.3.2 Trembles and completely mixed behavioural strategies 213
      1.3.3 Trembles and weakly dominated strategies 214
2 SELTEN'S CLOSEST RELATIVES, KREPS, WILSON, HARSANY AND MYERSON: SEQUENTIAL EQUILIBRIUM, PERFECT BAYESIAN EQUILIBRIUM AND PROPER EQUILIBRIUM 216
   2.1 Kreps and Wilson's sequential equilibrium: the introduction of beliefs 216
      2.1.1 Beliefs and strategies: consistency and sequential rationality 216
      2.1.2 Applications of the sequential equilibrium 218
   2.2 Harsanyi's perfect Bayesian equilibrium 221
      2.2.1 Definition, first application and links with the sequential equilibrium 221
      2.2.2 French plea-bargaining and perfect Bayesian equilibria 224
   2.3 Any perturbations or only a selection of some of them? Myerson's proper equilibrium 225
3 SHAKING OF THE GAME STRUCTURE 227
   3.1 When only a strong structural change matters, Myerson's carrier pigeon game 227
   3.2 Small change, a lack of upper hemicontinuity in equilibrium strategies 229
   3.3 Large changes, a lack of upper hemicontinuity in equilibrium payoffs 229
   3.4 Very large changes, a lack of lower hemicontinuity in equilibrium behaviours and payoffs, Rubinstein's e-mail game 232
4 PERTURBING PAYOFFS AND BEST RESPONSES IN A GIVEN WAY 235
   4.1 Trembling*-hand perfection 236
   4.2 Quantal responses 238
   4.3 Replicator equations 240
CONCLUSION 241

Exercise 1	Perfect and proper equilibrium	243
Exercise 2	Perfect equilibrium with weakly dominated strategies	243
Exercise 3	Perfect equilibrium and incompatible perturbations	244
Exercise 4	Strict sequential equiliubrium	245
Exercise 5	Sequential equilibrium, inertia in the players' beliefs	245
Exercise 6	Construct the set of sequential equilibria	246
Exercise 7	Perfect equilibrium, why the normal form is inadequate, a link to the trembling*-hand equilibrium	246
Exercise 8	Perfect Bayesian equilibrium	247
Exercise 9	Perfect Bayesian equilibrium, a complex semi separating equilibrium	247
Exercise 10	Lemon cars, experience good market with two qualities	248
Exercise 11	Lemon cars, experience good market with n qualities	248
Exercise 12	Perfect Bayesian equilibria in alternate negotiation with incomplete information	249

6 SOLVING A GAME DIFFERENTLY 250
INTRODUCTION 250
1 BEST REPLY MATCHING 250
1.1 Definitions and links with correlation and the Nash equilibrium 251
   1.1.1 Best reply matching, a new way to work with mixed strategies 251
   1.1.2 Link between best reply matching equilibria and Nash equilibria 253
   1.1.3 Link between best reply matching equilibria and correlated equilibria 253
1.2 Auction games, best reply matching fits with intuition 254
1.3 Best reply matching in ascending all pay auctions 257
1.4 Limits of best reply matching 259
2 HALPERN AND PASS'S REGRET MINIMIZATION 260
2.1 A new story game, the traveller's dilemma 260
2.2 Regret minimization 263
   2.2.1 Definition 263
   2.2.2 Regret minimization and the traveller's dilemma 264
   2.2.3 Regret minimization and the envelope game 265
2.3 Regret minimization in the ascending all pay auction/war of attrition: a switch from normal form to extensive form 266
2.4 The limits of regret minimization 268
   2.4.1 A deeper look into all pay auctions/wars of attrition 268
   2.4.2 Regret minimization: a too limited rationality in easy games 270
3 LEVEL–K REASONING 271
3.1 Basic level-k reasoning in guessing games and other games 271
   3.1.1 Basic level-k reasoning in guessing games 271
   3.1.2 Basic level-k reasoning in the envelope game and in the traveller's dilemma game 273
3.2 A more sophisticated level-k reasoning in guessing games 274
3.3 Level-k reasoning versus iterated dominance, some limits of level-k reasoning 276
   3.3.1 Risky behaviour and decreasing wins 276
   3.3.2 Level-k reasoning is riskier than iterated dominance 277
   3.3.3 Something odd with basic level-k reasoning? 278
4 FORWARD INDUCTION 278
4.1 Kohlberg and Mertens' stable equilibrium set concept 279

4.2	The large family of forward induction criteria with starting points	282
	4.2.1 Selten's horse and Kohlberg's self-enforcing concept	282
	4.2.2 Local versus global interpretations of actions	283
4.3	The power of forward induction through applications	287
	4.3.1 French plea-bargaining and forward induction	287
	4.3.2 The repeated battle of the sexes or another version of burning money	288
CONCLUSION		289
Exercise 1	Best reply matching in a normal form game	291
Exercise 2	Best reply matching in the traveller's dilemma	291
Exercise 3	Best reply matching in the pre-emption game	292
Exercise 4	Minimizing regret in a normal form game	293
Exercise 5	Minimizing regret in Bertrand's duopoly	293
Exercise 6	Minimizing regret in the pre-emption game	293
Exercise 7	Minimizing regret in the traveller's dilemma	294
Exercise 8	Level-k reasoning in an asymmetric normal form game	295
Exercise 9	Level-1 and level-k reasoning in Basu's traveller's dilemma	295
Exercise 10	level-1 and level-k reasoning in the students' traveller's dilemma	295
Exercise 11	Stable equilibrium, perfect equilibrium and perturbations	295
Exercise 12	Four different forward induction criteria and some dynamics	296
Exercise 13	Forward induction and the experience good market	296
Exercise 14	Forward induction and alternate negotiation	297

ANSWERS TO EXERCISES   299

1 HOW TO BUILD A GAME   301

Answers 1	Easy strategy sets	301
Answers 2	Children game, how to share a pie equally	301
Answers 3	Children game, the bazooka game, an endless game	301
Answers 4	Syndicate game: who will be the free rider?	302
Answers 5	Rosenthal's centipede (pre-emption) game, reduced normal form game	303
Answers 6	Duel guessing game, a zero sum game	305
Answers 7	Akerlof's lemon car, experience good model, switching from an extensive form signalling game to its normal form	307
Answers 8	Behavioural strategies and mixed strategies, how to switch from the first to the second and vice versa?	309
Answers 9	Dutch auction and first price sealed bid auction, strategic equivalence of different games	312

2 DOMINANCE   316

Answers 1	Dominance in a game in normal form	316
Answers 2	Dominance by a mixed strategy	316
Answers 3	Iterated dominance, the order matters	317
Answers 4	Iterated dominance in asymmetric all pay auctions	317
Answers 5	Dominance and value of information	319
Answers 6	Stackelberg first price all pay auction	320
Answers 7	Burning money	321
Answers 8	Pre-emption game in extensive form and normal form, and crossed rationality	322
Answers 9	Guessing game and crossed rationality	323
Answers 10	Bertrand duopoly	324

CONTENTS

Answers 11	Traveller's dilemma (Basu's version)	325
Answers 12	Traveller's dilemma, the students' version	326
Answers 13	Duel guessing game, the cowboy story	327
Answers 14	Second price sealed bid auction	331
Answers 15	Almost common value auction, Bikhchandani and Klemperer's result	333

3 NASH EQUILIBRIUM 336

Answers 1	Nash equilibria in a normal form game	336
Answers 2	Story normal form games	337
Answers 3	Burning money	338
Answers 4	Mixed Nash equilibria and weak dominance	339
Answers 5	A unique mixed Nash equilibrium	339
Answers 6	French variant of the rock paper scissors game	340
Answers 7	Bazooka game	341
Answers 8	Pure strategy Nash equilibria in an extensive form game	342
Answers 9	Gift exchange game	342
Answers 10	Behavioural Nash equilibria in an extensive form game	343
Answers 11	Duel guessing game	344
Answers 12	Pre-emption game (in extensive and normal forms)	345
Answers 13	Bertrand duopoly	347
Answers 14	Guessing game	348
Answers 15	Traveller's dilemma (Basu)	348
Answers 16	Traveller's dilemma, students' version, P<49	350
Answers 17	Traveller's dilemma, students' version, P>49, and cautious behaviour	351
Answers 18	Focal point in the traveller's dilemma	352
Answers 19	Asymmetric all pay auctions	352
Answers 20	Wallet game, first price auction, winner's curse and a robust equilibrium	354
Answers 21	Wallet game, first price auction, Nash equilibrium and new stories	355
Answers 22	Two player wallet game, second price auction, a robust symmetric equilibrium	356
Answers 23	Two player wallet game, second price auction, asymmetric equilibria	357
Answers 24	N player wallet game, second price auction, marginal approach	359
Answers 25	Second price all pay auctions	360
Answers 26	Single crossing in a first price sealed bid auction	361
Answers 27	Single crossing and Akerlof's lemon car	362
Answers 28	Dutch auction and first price sealed bid auction	363

4 BACKWARD INDUCTION AND REPEATED GAMES 366

Answers 1	Stackelberg all pay auction, backward induction and dominance	366
Answers 2	Duel guessing game	367
Answers 3	Centipede pre-emption game, backward induction and the students' way of playing	368
Answers 4	How to share a shrinking pie	370
Answers 5	English all pay auction	371
Answers 6	General sequential all pay auction, invest a max if you can	373
Answers 7	Gift exchange game	375
Answers 8	Repeated games and strict dominance	376
Answers 9	Three repetitions are better than two	378
Answers 10	Alternate rewards in repeated games	379

	Answers 11	Subgame perfection in Rubinstein's finite bargaining game	380
	Answers 12	Subgame perfection in Rubinstein's infinite bargaining game	383
	Answers 13	Gradualism and endogenous offers in a bargaining game, Li's insights	384
	Answers 14	Infinite repetition of the traveller's dilemma game	385
	Answers 15	Infinite repetition of the gift exchange game	393
5	TREMBLES IN A GAME		395
	Answers 1	Perfect and proper equilibrium	395
	Answers 2	Perfect equilibrium with weakly dominated strategies	396
	Answers 3	Perfect equilibrium and incompatible perturbations	398
	Answers 4	Strict sequential equilibrium	399
	Answers 5	Sequential equilibrium, inertia in the players' beliefs	399
	Answers 6	Construct the set of sequential equilibria	401
	Answers 7	Perfect equilibrium, why the normal form is inadequate, a link to the trembling*-hand equilibrium	402
	Answers 8	Perfect Bayesian equilibrium	404
	Answers 9	Perfect Bayesian equilibrium, a complex semi separating equilibrium	406
	Answers 10	Lemon cars, experience good market with two qualities	407
	Answers 11	Lemon cars, experience good market with n qualities	408
	Answers 12	Perfect Bayesian equlibria in alternate negotiation with incomplete information	410
6	SOLVING A GAME DIFFERENTLY		414
	Answers 1	Best reply matching in a normal form game	414
	Answers 2	Best-reply matching in the traveller's dilemma	416
	Answers 3	Best-reply matching in the pre-emption game	419
	Answers 4	Minimizing regret in a normal form game	421
	Answers 5	Minimizing regret in Bertrand's duopoly	422
	Answers 6	Minimizing regret in the pre-emption game	423
	Answers 7	Minimizing regret in the traveller's dilemma	424
	Answers 8	Level–k reasoning in an asymmetric normal form game	426
	Answers 9	Level–1 and level–k reasoning in Basu's traveller's dilemma	426
	Answers 10	Level–1 and level–k reasoning in the students' traveller's dilemma	429
	Answers 11	Stable equilibrium, perfect equilibrium and perturbations	431
	Answers 12	Four different forward induction criteria and some dynamics	433
	Answers 13	Forward induction and the experience good market	434
	Answers 14	Forward induction and alternate negotiation	435

Index 437

Acknowledgements

I wish to thank Paul Pezanis-Christou for the rich and fruitful discussions we had while I was writing the book.

I also thank Linda and the Routledge editors, who helped edit the English wording of the text, given that English is not my native language.

And I am particularly grateful to my students at the Faculté des Sciences Economiques et de Gestion (Faculty of Economic and Management Sciences) of the University of Strasbourg, who regularly play my games with an endless supply of good humour. Their remarks, their sometimes original way to play, always thought-provoking, helped in the writing of this book.

Introduction

"Rock is way too obvious and scissors beat paper. Since they are beginners, scissors are definitely the safest".[1] And the young sisters added: "If they also choose scissors and another round is required, the correct play is to stick to scissors – because everybody expects you to choose rock."

Well, thanks to these young girls, Christie's auction house won the right to sell a Japanese electronic company's art collection, worth more than $20 million. This is not a joke. The president of the company, because he was indifferent between Christie's and Sotheby's, but had to choose an auction house to sell the paintings, simply asked them to play the rock paper scissors game!

You know: rock beats scissors, scissors beat paper, paper beats rock. You also know, because of the strategic symmetry of the weapons (each weapon beats another and is beaten by the remaining one) – or because you have played this game many times in the schoolyard – that there is no winning strategy!

But the young girls won, because they played scissors and Sotheby's players played paper. By luck of course, you will say, but game theory focal point approaches and level-k reasoning would nuance your point of view. "Rock is way too obvious": what did the girls exactly mean by that? That rock is a too primitive weapon, too much played, we may be reluctant to use, but that the opponent may expect you to play? If such a point of view is focal, the young girls are right beginning with scissors, because the symmetry is broken and scissors beat paper. And what about the second part of their reasoning? They said that if both players play scissors so that a second round is needed, then the opponent expects you to play rock in this new round, because rock beats scissors – to say things more theoretically, the opponent expects you to be a level-1 player, a player that just best reacts to a first given action (here scissors). So, because he is a level-2 player – that is to say he is one level more clever than you – he will best react by playing paper, because paper beats rock. Yet, because in fact you are a level-3 player – one more level clever than your opponent – you stick to scissors because scissors beat paper!

Fine, isn't it? But why did the girls stop at "level-3"? What happens if the opponent is a level-4 player, so best replies to the expected level-3 behaviour by playing rock (because rock beats scissors)? Well I may answer in two ways. First, level-4 players do perhaps not grow on trees – every fellow does not run a four-step reasoning – so the girls are surely right not expecting to meet such an opponent. Second, more funny, imagine you are a level-4 player and you meet an opponent, who is only a level-2 player (remember he plays paper): your opponent will win, and worse, he may consider you as a level-1 player (because such a player also plays rock!). That's frustrating, isn't it? More seriously, we perhaps stop a reasoning before it begins cycling, which clearly means, in this game, that we will never run more than a level-3 reasoning.

1

INTRODUCTION

Of course, the Nash equilibrium, which is the central concept of game theory, just says that the only way to win is to play each of the three weapons in a random way. But more recent game theory focuses more and more on the way players play in reality, like the young girls above.

Well, what is a game exactly? I could say that it is a set of players, a set of strategies by players, and vectors of payoffs that depend on the played strategies. That is right, but I prefer saying that a game is a way of structuring the interactions between agents to find strategies with specific properties. So a game has to be built, and building is not so easy: you namely have to delete all what is not necessary to find out the strategies with the specific properties. To give an example, consider match sprint, the cycling event where two riders on a velodrome try to first cross the finish line. The riders, at the beginning, pedal very slowly, observe each other a lot, they even bring their bicycles to a stop, for example to make the other rider take the lead. They seldom ride at very high speed before a given threshold moment, because the follower would benefit from aerodynamics phenomena. In fact, a major event is the moment when they decide to "attack", i.e. to accelerate very quickly, the follower aiming to overtake the leader before the line, and the leader aiming to establish a sufficiently large gap between both riders to cross the line in first position. That is why, in a first approach of this game, despite the riders using plenty of tactics, you can limit the strategy set to the moment of acceleration, depending on physical aptitudes of both riders and the distance that remains to be covered.

So game theory structures a given interactive context to solve it, or more modestly, to highlight strategies with specific properties, for example strategies that best answer to one another. But you may also like some strategies (for example a fair way to share benefits) and look into building a new game so that these nice strategies are spontaneously played by the players in the game. This is the other aim, surely the most stimulating side of game theory. A kind example is again the rock paper scissors game. Even if this game, which dates back to the time of the Chinese Han Dynasty, has not only been played by children, we could say that it is has been written to make each child happy, because each strategy is as winning as the other ones. So each child is happy to play this game, because she has the same probability to win, whether she is logical or not, whether she is patient or impatient. But you surely know less kind examples: when you choose an insurance contract, you in fact play a game written by the insurer, built in such a way that you spontaneously choose the contract that fits your risky or safe lifestyle (you perhaps wanted to hide from the insurer). Even more seriously, it is well known that the electoral procedures, i.e. the game we play to choose the elected person, have a direct impact on the winning candidate. Changing this game may lead to another elected person.

Well, this book on game theory[2] begins with a cover page on (sailboat) match racing, because match racing is of course a stimulating game but it is also ... beautiful, from an aesthetic point of view. The book covers classic topics but also more recent topics of game theory. Chapter 1 focuses on the way to build a game: it displays the notion of strategies, information, knowledge, representation forms, utilities. Chapter 2 is on dominance, weak, strict, iterative. Chapter 3 is on Nash equilibria, risk dominance and cautious behaviour. Chapter 4 is on subgame perfection, backward induction, finitely and infinitely repeated games. Chapter 5 adds perturbations to strategies, payoffs or to the whole structure of the game. It namely presents concepts in the spirit of the perfect and sequential equilibria, but also quantal behaviour and replicator equations. Chapter 6 is on recent topics in game theory, best reply matching, regret minimization, level-k reasoning, and also on forward induction. The book also contains many economic applications, namely on auctions. See the table of contents for more details.

INTRODUCTION

In fact, the book could have a subtitle: *with and for the students*. Surely one of the main originalities of the book is the omnipresence of the students.

First, the book proposes, linked to each chapter, numerous exercises with detailed solutions. These exercises allow one to practise the concepts studied in each chapter, but they also give the opportunity to approach topics not developed in the linked chapter. The difficulty of the exercises (increasing with the number of clovers preceding them) is variable so that both undergraduate as well as postgraduate students will find exercises adapted to their needs.

Second, the chapters contain many "students' inserts"; they generally propose a first application, or a first illustration, of a new concept, so they allow one to easily understand it. Their aim is clearly pedagogical.

Third, and this is the main originality of the book, the students animate the book. As a matter of fact, I have been teaching game theory for more than 20 years and I like studying how my undergraduate, postgraduate and post-doctoral students[3] play games. So I let them play many games during the lectures (like war of attrition games, centipede games, first price and second price all pay auction games, first price and second price wallet games, the traveller's dilemma, the gift exchange game, the ultimatum game…) and I expose their way of playing in the book. As you will see, my students often behave like in experimental game theory, yet not always, and their behaviour is always stimulating from a game theoretical point of view. More, my students are very creative. So for example, because they don't understand the rule of a game, or because they don't like it, they invent and play a new game, which sometimes turns out to be even more interesting than the game I asked them to play. And of course, I couldn't resist exposing in the book one of their inventions, a new version of the traveller's dilemma, which proves to be pedagogically very stimulating. So I continuously interact with the students and this gives a particular dynamic to the book.

Fourth, there are "fils rouges" in both the chapters and in the exercises of the chapters, which means that some games cross all the chapters to successively benefit from the light shed by the different developed concepts. For example two auctions games and the envelope game cross all the chapters to benefit from both classic and recent game theory approaches. The traveller's dilemma crosses the exercises of all the chapters.

To start the book,[4] let me just say that game theory is a very open, very stimulating research field. Remember, we look for strategies with specific properties: by defining new, well-founded, specific properties, you may shed new light on interactive behaviours. It's a kind of magic; I'm sure you will like game theory.

NOTES

1 Vogel, C., April 29, 2005. Rock, paper, payoff: child's play wins auction house an art sale, *The New York Times*.
2 These are three classic books on game theory and one of my previous books:
Binmore, K.G. 1992. *Fun and games*, D.C. Heath, Lexington, Massachusetts.
Fudenberg, D., Tirole, J. 1991. *Game theory*, MIT Press, Massachusetts.
Myerson, R.B. 1991. *Game theory*, Harvard University Press, Cambridge, Massachusetts.
Umbhauer, G. 2004. *Théorie des jeux*, Editions Vuibert, Paris.
3 In the book I mainly talk about the way of playing of my L3 students, who are undergraduate students in their third year of training, because there are always more than 100 students playing the same game, which makes the results more significant. But I also refer to my M1 students (postgraduate students in

INTRODUCTION

their first year of training, 30 to 50 persons playing the same game) and to my postdoctoral students (12 to 16 persons playing the same game).

4   Let me call attention to the facts that, in the book:

x is positive, respectively negative, means $x>0$, respectively $x<0$.

x is preferred, respectively weakly preferred to y, means $x \succ y$, respectively $x \succsim y$.

Chapter 1

How to build a game

INTRODUCTION

What is game theory and what do we expect from it? One way to present game theory is to say that this theory aims to study the interactions between agents that are conscious of their interaction. But of course, this is not specific to game theory given that most social sciences follow the same aim. So, what distinguishes game theory from other social sciences? Rationality of the players? Not so sure. A large part of game theory studies interaction in evolutionary contexts where players only partly understand what is good for them. I would say that the specificity of game theory is *the way it structures a context, that is to say the way a game is written*. Game theory structures an interactive context in a way that helps to find actions – strategies – with specific properties. It automatically follows that different economic, political, management and social contexts may have the same game theoretic structure and therefore will be studied in a similar way. That is why the beginning of this book looks at the structure of a game, at the way to build it. Structuring an interactive context as a game does not mean solving it – this will be discussed in the following chapters. But building and solving are linked activities. First, the structure often "talks a lot", in that it helps underline the specificities of a game. Second, you structure with the aim of solving the game, so you eliminate all that is not necessary to find the solutions. This means that you have a good idea of what is and isn't important for solving a game.

Consequently, building a context as a game is both a fruitful and critical activity. In this chapter we aim to highlight these facts by giving all the elements of the structure of a game. In the first section, we talk about the two main representations of a game: the strategic form game and the extensive form game. In section 2, we turn to a central concept of game theory, the notion of strategy: we develop pure, mixed and behavioural strategies. In section 3 we discuss the concept of information and the way utilities are assigned to the different ways to play the game. We conclude with what can be omitted in a game.

1 STRATEGIC OR EXTENSIVE FORM GAMES?

If you ask a student about what game theory is, s/he will usually suggest a matrix with a small number of rows and columns, usually 2! Well, this follows from the fact that very often, game theory books and lectures start with strategic form games that can be represented by a matrix with only two rows and two columns. Very often, students only discover extensive form games very late. We will not proceed in the same way. Throughout the book we study strategic form and extensive form games together and highlight the links but also the differences between these two ways to represent a game.

HOW TO BUILD A GAME

1.1 Strategic/normal form games

1.1.1 Definition

> **Definition 1.** A *strategic form game or normal form game*[1] is defined by three elements:
>
> $\mathcal{N}$, the set of players, with Card $\mathcal{N}$=N>1
>
> S=XS$_i$, where S$_i$ is the strategy set of player i, i from 1 to N.
>
> N preference relations, one for each player, defined on S. These relations are supposed to be Von Neumann Morgenstern (VNM), and are therefore usually replaced by VNM utility functions.

What is a strategy and a VNM utility function? For now, we say that a strategy set is the set of all possible actions a player can choose and we define a VNM utility function as a function that not only orders the different outcomes of a game (in terms of preferences) but that also takes into account the way people cope with risk (we will be more precise in section 3).

The strategic form game, if there are less than four players and if the strategy sets are finite, is often represented in a matrix or a set of matrices, where the rows are the first player's strategies, the columns the second player's strategies and the matrices the third player's strategies. It usually follows that very easy – often 2×2 – matrices focus the students' attention, because they are not difficult and because they can express some nice story games, often quoted in the literature. Well, these *story games*, that perhaps unfortunately characterize what people usually know from game theory, express some interesting features of interaction that deserve attention and help to illustrate the notion of a strategic form (or normal form) game. So let us present some of them.

1.1.2 Story strategic/normal form games and behavioural comments

> **Prisoner's dilemma**
>
> In the **prisoner's dilemma game** two criminals in separate rooms can choose to deny the implication of both in a crime (they choose to cooperate), or they can say that only the other is implied in the crime (they choose to denounce). The interesting point of this story lies in the pronounced sentences: one year jail for each if both cooperate (the fact that both cooperate is doubtful), no punishment for the denouncer and 15 years for the cooperator if one player cooperates and the other denounces, 10 years for both if they both denounce. Of course these sentences are a little strange (we are not sure that having two doing a crime diminishes the sentence (!), and normally doubts are not enough to convict someone) but they make the story interesting.
>
> Let us write the game in strategic form: Card $\mathcal{N}$=2, S$_i$={C, D}, i=1, 2, where C and D respectively mean that the player cooperates and denounces. Given the sentences, player 1's preferences are given by: (D, C)≻(C, C)≻(D, D)≻(C, D) (by convention, the first and second coordinates of a couple are respectively player 1's and player 2's actions); player 2's preferences are symmetric.

HOW TO BUILD A GAME

By translating these preferences in utilities, the strategic form can be summarized in the following 2×2 Matrix 1.1:

		Player 2	
		C	D
Player 1	C	(u,u')	(0,1)
	D	(1,0)	(v,v')

Matrix 1.1

with 1>u>v>0 and 1>u'>v'>0.

Let the game talk: this game raises a dilemma. Both players perfectly realize that they would be better off both cooperating than both denouncing (u>v and u'>v' – one year jail is better than 10) but they also realize that, whatever is done by the other player – *even if he cooperates* – one is better off denouncing him (because 1>u and v>0, and 1>u' and v'>0). As a matter of fact, if your accomplice denounces you, it is quite logical that it is better for you to denounce him also (10 years' jail is better than 15), but, even if he cooperates, it is better for you to denounce him because no punishment is better than one year jail! So what will happen?

If the story seems rather amazing with respect to actual legal sentences, it is highly interesting because it corresponds to many economic situations. For example, the prisoner's dilemma is used to illustrate *overexploitation of resources* (both players are better off not overexploiting a resource (strategy C), but individually it is always better to exploit a resource more rather than less). This game also illustrates the difficulty of *contributing to a public investment*: it is beneficial for both players that both invest in a public good, but it is always better individually to invest in a private good (even if the other invests in the public good). These situations are called *free rider contexts*. *Bertrand price games* also belong to this category. Both sellers are better off coordinating on a high price (C), but each is better off proposing a lower price (D) than the price of the other seller, in order to get the whole market. And so on…

Chicken game, hawk-dove game, and syndicate game

The chicken and the hawk-dove games are strategically equivalent (see section 3 for a more nuanced point of view) in that they (at least seemingly) lead to the same strategic form.

In the **chicken game** two – stupid – drivers drive straight ahead in opposite directions and rush at the opponent. If nobody swerves, there is a dramatic accident and both die. If one driver swerves (action A) whereas the other goes straight on (action B), the one who swerves is the big loser (the chicken) whereas the other is the winner. If both swerve there is no loser and no winner.

In the **hawk-dove game**, one player (for example a country) can attack another one (Hawk strategy B), but this strategy is beneficial only if the other country does not react (Dove strategy A), i.e. is invaded without any resistance. Both countries are supposed to focus on economic and human losses. If both countries fight, both are very badly off (many human

HOW TO BUILD A GAME

> and financial losses) and they would have preferred not to fight. Furthermore, an attacked country is better off not reacting (the idea is that if it reacts, the battle will be terrible and most of the inhabitants will die).

The strategic form for both games is the same: Card $\mathcal{N}$=2, S_i={A, B}, i=1, 2, where A–B – means that the player swerves or is a dove – goes straight on or is a hawk. Player i's preferences are given by: (B, A)≻(A, A)≻(A, B)≻(B, B); player 2's preferences are symmetric. By translating these preferences in utilities, the strategic form can be summarized in the following 2×2 Matrix 1.2:

		Player 2 A	B
Player 1	A	(u,u')	(v,1)
	B	(1,v')	(0,0)

Matrix 1.2

with 1>u>v>0 and 1>u'>v'>0

Let the game talk: clearly, there is a reasonable situation (A, A), where nobody has a bad utility (if both drivers swerve, nobody is a chicken; if no country fights, they are both well), but this situation is strategically unstable because if one player plays A, it is much better for the other to play B (you are the winner in the chicken game, you get a new country without any fight in the hawk-dove game (1>u and 1>u')). By contrast, the two situations where one player plays A and the other plays B seem strategically stable; we have just observed that is better playing B if the other plays A, but it is also better to play A if the other plays B (v>0, and v'>0). Indeed, it is better to be the chicken than to die, and it is better not to fight if the other fights, given that most of the inhabitants of your country would die if you fought too. The problem is that both couples (A, B) and (B, A) are equally stable, and that each player wishes to be the B player. So clearly, the true interest of the game is to anticipate who will be the A player and who will be the B player.

But is that really so? In the chicken game, yes, because if you play such a silly game, you don't want the outcome (B, B), but you don't want the outcome (A, A) either! Such games want a loser and a winner and it is almost unconceivable to end in a situation where both swerve. In the hawk-dove game, things are different. As a matter of fact, the strategic form game reveals a risky situation – if you play B, you may end up with the worst outcome (both play B) – and a cautious but unstable situation; if you play A, you will never end up in the worst situation but you will never be a happy "winner" either. A cautious behaviour – inconceivable in the chicken game – is not impossible in the hawk-dove game: so, what will the players play?

What is more, given that it seems difficult to conceive that both drivers swerve in the chicken game, is the strategic game as outlined above the true drivers' game? We come back to this point in section 3.

HOW TO BUILD A GAME

What about the syndicate game?

> In a **syndicate game**, a certain number (a threshold number) of players have to join the syndicate (action A) for the syndicate to be created. Each worker can choose to join (action A) or not to join (action B) the syndicate. The work of the members of the syndicate is equally useful for all the workers both inside and outside the syndicate. The worst situation is the one where the syndicate is not created because the threshold number of members is not reached. But the best situation for a worker is to benefit from the work of the syndicate without joining it. Conversely, each member of a syndicate prefers a situation with more syndicated workers than less (namely to share the work). We can represent this game with two players (see the exercises for a game with three players), and a threshold number equal to one. The strategic form of the game is given by: Card $\mathcal{N}=2$, $S_i=\{A, B\}$, $i=1, 2$, where A, respectively B, means that a worker joins the syndicate, does not join the syndicate. Player 1's preferences are given by: (B, A)≻(A, A)≻(A, B)≻(B, B); player 2's preferences are symmetric.

Translating the preferences in utilities leads to exactly the same matrix as both the chicken and the hawk-dove games! As a matter of fact (A, A) – the situation where both join the syndicate – is not a bad situation but it is not strategically stable given that you prefer staying out when the syndicate does not need you to be created! Both situations (A, B) and (B, A) are stable because the worker who plays B (does not join the syndicate) gets his highest payoff (he benefits from the work of the syndicate without contributing) and the worker who plays A (joins the syndicate) cannot be better off leaving the syndicate, because the syndicate would collapse, which is worse for him! And of course, each worker prefers being the unique B player, and both fear the situation (B, B), which is the worst situation for both.

So, it is always nice observing that completely different games (the syndicate game and the chicken game) may have the same strategic structure.

Congestion games, coordination games, and battle of the sexes

> In a **congestion game**, two players have to choose to drive on road A or B, but if both choose the same road there will be a traffic jam. The strategic form of the game is given by: Card $\mathcal{N}=2$, $S_i=\{A, B\}$, $i=1, 2$, where A, respectively B, means that the player chooses road A, respectively road B. Player 1 and player 2's preferences are given by: (B, A)∼(A, B)≻(A, A)∼(B, B).

This game can be represented in Matrix 1.3:

		Player 2 A	Player 2 B
Player 1	A	(0,0)	(1,1)
Player 1	B	(1,1)	(0,0)

Matrix 1.3

HOW TO BUILD A GAME

Let the game talk: what is new in this game, by contrast to the previous ones, is that *there is no conflict at all*. Both players can be equally happy in the same context. The only problem is the coordination of the actions. If players could talk before the game, clearly they both would agree on one of the two situations (A, B) or (B, A) but, given that drivers (they don't know each other) usually do not talk together before choosing a route, a traffic jam is always possible.

> In a **coordination game**, people are best off if they play the same strategy. The strategic form of the game is given by: Card $\mathcal{N}$=2, S_i={A, B}, i=1, 2, and player 1 and player 2's preferences are given by: (A, A)~(B, B)≻(A, B)~(B, A).

This game is represented in Matrix 1.4:

		Player 2	
		A	B
Player 1	A	(1,1)	(0,0)
	B	(0,0)	(1,1)

Matrix 1.4

Let the game talk: this game is the same as the previous one, if one changes the name of the strategy of one of the players (A becomes B and B becomes A). The only problem is to coordinate on a same action. Many stories are linked to this game: the *meeting game* (what matters is that we meet, wherever the place we meet), the choice of a same date (we both win a million, provided we quote the same date on the calendar), but also network games (for example if two complementary producers are better off investing in the same (hence compatible) technology, regardless of this technology).

> The **battle of the sexes** is a variant of a coordination game. Both players still prefer to coordinate on a same action but they have different preferences with regard to the common chosen action. The strategic form of the game is given by: Card $\mathcal{N}$=2, S_i={A, B}, i=1, 2. Player 1's preferences are given by: (A, A)≻(B, B)≻(A, B)~(B, A), and player 2's preferences are given by: (B, B)≻(A, A)≻(A, B)~(B, A).

Translating the preferences in utilities leads to Matrix 1.5:

		Player 2	
		A	B
Player 1	A	(1,u')	(0,0)
	B	(0,0)	(u,1)

Matrix 1.5

with 1>u>0 and 1>u'>0

10

HOW TO BUILD A GAME

Let the game talk: the most famous story around this game is about a man and a woman who like to spend the evening together, but the woman prefers attending a show of wrestling whereas the man prefers spending the evening at the opera. More seriously, network games also belong to this type of game. Complementary producers may prefer to invest in the same technology, but may differ with regard to the most preferred common technology. Battle of the sexes games seem more complicated than pure coordination games. If the aim is still to coordinate on a common action, it is no longer sufficient to meet before the game to choose that action. There is a partial conflict because both players disagree on the most preferred common action, and one easily understands the role that a lottery may play in the coordination: if both players can't agree on A or B, they may agree on a lottery that selects A and B with the same probability.

Conflict games: zero sum games (matching pennies, rock paper scissors game)
As already mentioned, a game is not necessarily a conflict. But some games are pure conflicts.

> **Zero sum games:** in these games, what is won by one player is necessarily lost by the other. There is only one winner and one loser.
>
> The **matching pennies game** is a zero sum game. There are two players. Each puts a penny on the table simultaneously, either on heads or tails. If both pennies are on heads or if both are on tails, player 1 gives her penny to player 2. If not, player 2 gives his penny to player 1. The strategic form of the game is given by: Card $\mathcal{N}=2$, $S_i=\{A, B\}$, $i=1, 2$, where A, respectively B, means that the player chooses heads, respectively tails. Player 1's preferences are given by: (A, B)~(B, A)≻(A, A)~(B, B), and player 2's preferences are given by: (B, B)~(A, A)≻(A, B)~(B, A).

Usually, one translates these preferences into the following utilities, in order to get Matrix 1.6a:

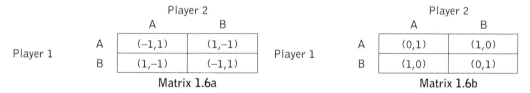

Let the game talk: first observe that the chosen utilities perfectly express that what is won by one player is lost by the other. Yet one can also keep the utilities between 0 and 1: this does not change the philosophy of the solutions (Matrix 1.6b).

By contrast to all the previous games there is no longer any situation that is stabilized: given that there is always a loser, this loser prefers changing his strategy. How should we play in such a game? A and B are both good and bad strategies. *What about playing both with a positive probability?*

For fun, let me just also mention the rock paper scissors game.

> **The rock paper scissors game**: in this game two players – usually children – simultaneously have to mime with their hands either a rock, paper or scissors. The winner is determined by:

11

HOW TO BUILD A GAME

> the scissors cut the paper, the paper covers the rock and the rock destroys the scissors. If both players mime the same object, there is no loser and no winner. The strategic form of the game is given by: Card $\mathscr{N}=2$, $S_i=\{R, Sc, P\}$, i=1, 2, where R, Sc, P mean that the player chooses Rock, Scissors or Paper. Player 1's preferences are given by: (R, Sc)~(Sc, P)~(P, R)≻(R, R)~(Sc, Sc)~(P, P)≻(Sc, R)~(P, Sc)~(R, P). Player 2's preferences are symmetric.

Usually these preferences are translated in order to highlight the zero sum game philosophy – what one player gets is lost by the other – as in Matrix 1.7a:

		Player 2 rock	scissors	paper			
	rock	(0,0)	(1,−1)	(−1,1)	(0.5,0.5)	(1,0)	(0,1)
Player 1	scissors	(−1,1)	(0,0)	(1,−1)	(0,1)	(0.5,0.5)	(1,0)
	paper	(1,−1)	(−1,1)	(0,0)	(1,0)	(0,1)	(0.5,0.5)

Matrix 1.7a   Matrix 1.7b

Let the game talk: as above we can change the utilities in order to keep them in the range [0, 1] (Matrix 1.7b), but if so, we have to slightly change the definition of zero sum games (because of the (0.5, 0.5)).

Despite the fact that there are now situations where both players don't lose and don't win, instability is still at work: whatever situation you consider, there is always a player who can do better by deviating. And children perfectly know that. This is surely why they regularly change the action they mime with their hands!

But do not think that this game is only played by children. It is because no situation seems more strategically stable than another that the president of a Japanese electronics company called on this game for help. He wanted to sell the company's art collection, worth more than $20 million, but was unable to choose the auction house, whether Christie's or Sotheby's, that should sell the collection. So he asked both auction houses to play the rock paper scissors game: the winner got the right to sell the paintings![2]

Just a general remark: in all these highly stylized games, we didn't need to better specify the utilities. We took utilities between 0 and 1 (VNM normalization) that just respected the ranking of the preferences. In other words, in these simple games, all the strategic content is in the ranking and not in the values of the utilities, a property which is highly appreciable (see section 3).

1.1.3 All pay auction

Let us now switch from story games to a two-player *first price sealed bid all pay auction game* which we will often mention in the book.

HOW TO BUILD A GAME

> **First price sealed bid all-pay auction**: there are two players. An object of known value V is sold through a first price sealed bid all-pay auction. This means that each player makes a bid (unknown to the other player), and the object is given to the bidder who makes the highest bid; *the originality of the game is that each player pays his bid*. All-pay auctions for example illustrate the race for a licence: two research teams may engage huge amounts of money to discover a new drug, but only the first who discovers the drug will get the licence. This game is an all-pay auction if the firm who discovers the drug is the one that makes the biggest investment (the highest bid).
>
> Let us suppose in this introductory example that the two players have a revenue M, with V<M. Each player i, i=1, 2 proposes a bid b_i, which is an integer between 0 and M. M and V are known by each player (we later say that they are common knowledge). If $b_1>b_2$, player 1 gets the object, has the payoff M−b_1+V, whereas player 2 has the payoff M−b_2. If b_1=b_2, both players get the object with probability ½. So, for V=3 and M=5, the strategic form of the game is given by Card $\mathcal{N}$=2, S_i={0, 1, 2, 3, 4, 5}, i=1, 2, and each player prefers an outcome with a higher payoff to an outcome with a lower payoff.

If we fix the utilities equal to the payoffs (so they respect the ranking of the preferences), the strategic form is represented by Matrix 1.8a:

				Player 2			
		0	1	2	3	4	5
	0	(6.5,6.5)	(5,7)	(5,6)	(5,5)	(5,4)	(5,3)
	1	(7,5)	(5.5,5.5)	*(4,6)*	(4,5)	(4,4)	(4,3)
Player 1	2	(6,5)	(6,4)	**(4.5,4.5)**	(3,5)	(3,4)	(3,3)
	3	(5,5)	(5,4)	(5,3)	(3.5,3.5)	(2,4)	(2,3)
	4	(4,5)	(4,4)	(4,3)	(4,2)	(2.5,2.5)	(1,3)
	5	(3,5)	(3,4)	(3,3)	(3,2)	(3,1)	(1.5,1.5)

Matrix 1.8a

For example, the bold payoffs in italics are obtained as follows: if player 1 bids 1 and player 2 bids 2, player 2 wins the auction (object) and gets M−b_2+V=5−2+3=6, whereas player 1 loses the auction and gets M−b_1=5−1=4. And the bold underlined payoffs are obtained as follows: if both players bid 2, each player wins the auction (object) half the time and gets M−b_i+V/2=5−2+3/2=4.5 (i=1, 2).

But we could propose a more general matrix with $u_1(7)$=1>$u_1(6.5)$=a>$u_1(6)$=b>$u_1(5.5)$=c>$u_1(5)$=d>$u_1(4.5)$=e>$u_1(4)$=f>$u_1(3.5)$=g>$u_1(3)$=h>$u_1(2.5)$=i>$u_1(2)$=j>$u_1(1.5)$=k>$u_1(1)$=0, and $u_2(7)$=1>$u_2(6.5)$=a'>$u_2(6)$=b'>$u_2(5.5)$=c'>$u_2(5)$=d'>$u_2(4.5)$=e'>$u_2(4)$=f'>$u_2(3.5)$=g'>$u_2(3)$=h'>$u_2(2.5)$=i'>$u_2(2)$=j'>$u_2(1.5)$=k'>$u_2(1)$=0. So we would get the more general Matrix 1.8b:

13

HOW TO BUILD A GAME

		Player 2					
		0	1	2	3	4	5
Player 1	0	(a,a′)	(d,1)	(d,b′)	(d,d′)	(d,f′)	(d,h′)
	1	(1,d′)	(c,c′)	(f,b′)	(f,d′)	(f,f′)	(f,h′)
	2	(b,d′)	(b,f′)	(e,e′)	(h,d′)	(h,f′)	(h,h′)
	3	(d,d′)	(d,f′)	(d,h′)	(g,g′)	(j,f′)	(j,h′)
	4	(f,d′)	(f,f′)	(f,h′)	(f,j′)	(i,i′)	(0,h′)
	5	(h,d′)	(h,f′)	(h,h′)	(h,j′)	(h,0)	(k,k′)

Matrix 1.8b

Let the game talk: first, we can observe that the game shares with the rock paper scissors game the fact that there are no strategies which fit everybody. For example, if player 1 bids 0, then player 2 is best off bidding 1 but, in that case, player 1 is best off switching to bid 2. But if so, player 2 is best off switching to bids 0 or 3. If he bids 0, a new cycle begins, if he bids 3, player 1 switches to bid 0 and a new cycle begins again. So, like in the rock paper scissors game, if we could play several times, we would surely not play the same bid every time.

Yet by contrast to the rock paper scissors game, some strategies seem inappropriate: there is no reason to bid 4, because at best you get an object of value 3 (but you pay 4), and, for the same reasons, it seems inappropriate to bid 5 (at best you pay 5 for an object of value 3).

Second, it seems that M is not important, because you should not invest more than V: in other words, it seems unnecessary to know the common value M (we later say that M does not need to be common knowledge), it is enough that everybody knows that each player is able to invest up to V.

A third comment is about the payoffs. Matrix 1.8a is just the payoff matrix. Matrix 1.8b is the (right) utility matrix. *But in both matrices, the payoffs do not clearly show to a player that he is losing money when he bids more than 3. Yet this may have an impact.*

In other words, the first time I proposed to my third year (L3) students to play the game, I gave them Matrix 1.8a, and many of my students played 4 and 5 (about 1/3 of them!). I knew that the reason for this strange way of playing did not only lie in the fact that the losses do not clearly appear in Matrix 1.8a, in that *many students explained me that they wanted the object regardless of their gains and losses*. But I still wondered if my students would have played in the same way if they had faced the Matrix 1.8c (in this matrix we subtract the revenue M=5 in order to clearly see the net wins or losses). This is why I proposed to another class of L3 students both the game in Matrix 1.8a and the game in Matrix 1.8c. To put it more precisely, *I proposed the game in Matrix 1.8a at the end of an introductory game theory lecture (the students had no idea about Nash equilibrium or dominance), and I proposed the game in Matrix 1.8c at the beginning of the next lecture in game theory, a week later. So the students learned nothing on game theory between the two playing sessions and the week between the two sessions prevented them making a direct link between the two games.*

The result is edifying as can be seen in Figures 1.1a and 1.1b, which give the percentages of students proposing the different bids, in Matrices 1.8a and 1.8c: whereas 29.7% of them played 4 and 5 in Matrix game 1.8a, only 7.2% played 4 and 5 in Matrix game 1.8c. Whereas 20.3% played 3 in Matrix game 1.8a, only 9.7% played 3 in Matrix game 1.8c. By contrast only 8.5% played 2 in Matrix game 1.8a, whereas 15.7% played 2 in Matrix game 1.8c. And only 12.7% played 1 in Matrix game 1.8a, whereas 31.3% played 1 in Matrix game 1.8c (strongest contrast). And whereas 28.8% played 0 in Matrix game 1.8a, 36.1% played 0 in Matrix game 1.8c. So clearly, when the players see their losses *with a negative sign*, they are much more careful than if they do not see the negative sign.

HOW TO BUILD A GAME

				Player 2			
		0	1	2	3	4	5
	0	(1.5,1.5)	(0,2)	(0,1)	(0,0)	(0,−1)	(0,−2)
	1	(2,0)	(0.5,0.5)	(−1,1)	(−1,0)	(−1,−1)	(−1,−2)
Player 1	2	(1,0)	(1,−1)	(−0.5,−0.5)	(−2,0)	(−2,−1)	(−2,−2)
	3	(0,0)	(0,−1)	(0,−2)	(−1.5,−1.5)	(−3,−1)	(−3,−2)
	4	(−1,0)	(−1,−1)	(−1,−2)	(−1,−3)	(−2.5,−2.5)	(−4,−2)
	5	(−2,0)	(−2,−1)	(−2,−2)	(−2,−3)	(−2,−4)	(−3.5,−3.5)

Matrix 1.8c

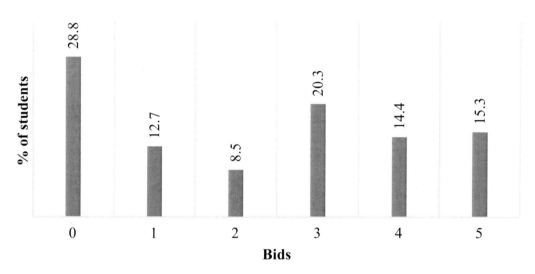

Figure 1.1a Bids played by the students with Matrix 1.8a

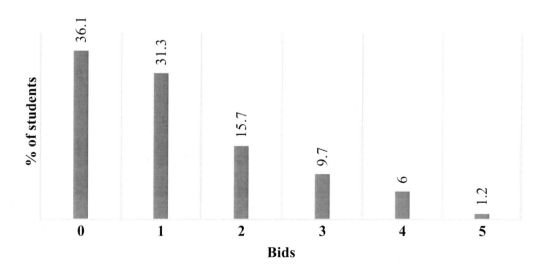

Figure 1.1b Bids played by the students with Matrix 1.8c

15

HOW TO BUILD A GAME

Let us now propose a stranger game, the *second price sealed bid all pay auction*.

> **Second price sealed bid all pay auction**: this game is identical to the first price sealed bid all pay auction except that both bidders, instead of paying their bid, only pay the lowest bid (the bid of the loser). Nothing changes if both make the same bid.

If the utilities are the payoffs, the matrix that represents the strategic form game is Matrix 1.9a:

Player 2

	0	1	2	3	4	5
0	(6.5,6.5)	(5,8)	(5,8)	**(5,8)**	**(5,8)**	**(5,8)**
1	(8,5)	(5.5,5.5)	**(4,7)**	(4,7)	(4,7)	(4,7)
2	(8,5)	(7,4)	(4.5,4.5)	(3,6)	(3,6)	(3,6)
3	(8,5)	(7,4)	(6,3)	(3.5,3.5)	(2,5)	(2,5)
4	(8,5)	(7,4)	(6,3)	(5,2)	(2.5,2.5)	(1,4)
5	(8,5)	(7,4)	(6,3)	(5,2)	(4,1)	(1.5,1.5)

Player 1 (rows)

Matrix 1.9a

For example, the bold underlined payoffs are obtained as follows: if player 1 bids 1 and player 2 bids 2, player 2 wins the auction (object) and gets $M-b_1+V=5-1+3=7$, whereas player 1 loses the auction and gets $M-b_1=5-1=4$.

We can switch to the utilities (Matrix 1.9b): $u_1(8)=1>u_1(7)=a>u_1(6.5)=b>u_1(6)=c>u_1(5.5)=d>u_1(5)=e>u_1(4.5)=f>u_1(4)=g>u_1(3.5)=h>u_1(3)=i>u_1(2.5)=j>u_1(2)=k>u_1(1.5)=m>u_1(1)=0$, and $u_2(8)=1>u_2(7)=a'>u_2(6.5)=b'>u_2(6)=c'>u_2(5.5)=d'>u_2(5)=e'>u_2(4.5)=f'>u_2(4)=g'>u_2(3.5)=h'>u_2(3)=i'>u_2(2.5)=j'>u_2(2)=k'>u_2(1.5)=m'>u_2(1)=0$.

Player 2

	0	1	2	3	4	5
0	(b,b')	(e,1)	(e,1)	(e,1)	(e,1)	(e,1)
1	(1,e')	(d,d')	(g,a')	(g,a')	(g,a')	(g,a')
2	(1,e')	(a,g')	(f,f')	(i,c')	(i,c')	(i,c')
3	(1,e')	(a,g')	(c,i')	(h,h')	(k,e')	(k,e')
4	(1,e')	(a,g')	(c,i')	(e,k')	(j,j')	(0,g')
5	(1,e')	(a,g')	(c,i')	(e,k')	(g,0)	(m,m')

Player 1

Matrix 1.9b

And we can switch to the net benefits (Matrix 1.9c):

Player 2

	0	1	2	3	4	5
0	(1.5,1.5)	(0,3)	(0,3)	(0,3)	(0,3)	(0,3)
1	(3,0)	(0.5,0.5)	(−1,2)	(−1,2)	(−1,2)	(−1,2)
2	(3,0)	(2,−1)	(−0.5,−0.5)	(−2,1)	(−2,1)	(−2,1)
3	(3,0)	(2,−1)	(1,−2)	(−1.5,−1.5)	(−3,0)	(−3,0)
4	(3,0)	(2,−1)	(1,−2)	(0,−3)	(−2.5,−2.5)	(−4,−1)
5	(3,0)	(2,−1)	(1,−2)	(0,−3)	(−1,−4)	(−3.5,−3.5)

Player 1

Matrix 1.9c

16

HOW TO BUILD A GAME

Let the game talk: the fact that both bidders pay the lowest bid has some link with the *English auction*, where the winner just has to bid ε more than the loser's bid (but in our auction, by contrast to the English auction, the loser also pays).

This new game is all but intuitive. By comparison with the previous game, only one thing changed – everybody pays the lowest bid, instead of his own bid – but because of this change, *the game is strategically completely different from the previous one.*

First, there is no longer any inappropriate action, given that bidding 4 or 5 no longer means that you have to pay these (too high) amounts. On the contrary, given that playing a high bid rises the probability to get the object (because you can expect that the other player bids less than you), playing 4 and 5 may even be the best strategies, especially if the opponent plays a bid lower than 3 (because you get the object at a price lower than its value). And you may also observe that it seems cleverer to bid 5 than 4, because bid 5 does better than bid 4 against the bids 4 and 5, and does as well against other bids.

Second, contrary to the previous game, M has to be known by each player, because M may now be a good strategy.

Third, contrary to the previous game, assigning a different bid to both players seems to be a good way to stabilize the game. In some way, there is a common point between this game and the hawk-dove game; as a matter of fact, if we restrict the game to two bids, 0 and 5, we get the hawk-dove game, as is shown by the reduced Matrix 1.9d (out of Matrix 1.9b):

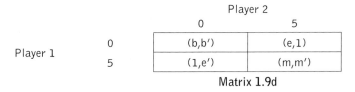

Matrix 1.9d

with 1>b>e>m and 1>b'>e'>m' (we can fix m=m'=0)

So, as in the hawk-dove game, we get asymmetric stable situations: if one of the players bids 0 (and does not get the object), and the other plays 3, 4 or 5 (and gets the object by paying nothing), nobody is incited to choose another action. The player who plays 3, 4 and 5 is most happy given that he gets the object without paying anything. And the other cannot do better than keeping his money (by bidding 0), because winning the object would require to overbid the other player, which means bidding at least 4 or 5, and hence losing money (or getting nothing) because the other player plays at least 3. But of course, as in the hawk-dove game, the question is: who will be the one who bids the large amount, and who will accept bidding 0?

I asked my L3 students to play this game (after having game theory training and therefore knowing the concepts of dominated strategy and Nash equilibrium); 38% (of 116 exploitable data) of them played 0, 20.5% played 3, 16% played 4 and 15% played 5, consequently more than 50% played 3, 4 or 5. And these percentages clearly show that almost 4 out of 10 students played the dove role, which is a rather strong percentage. But this is surely due to the fact that the true played game (Matrix 1.9b and not Matrix 1.9d!) is not exactly a hawk dove game. As a matter of fact, there are more than two actions and 0 is the only bid that never leads to losing money. This surely played a strong role in the students' choice.

17

HOW TO BUILD A GAME

Finally, let me tell you that all my students have difficulties with second price all pay auctions. In some way, these games are very complicated because they are not natural. First, you have to pay even if you do not win and second, if you are the winner, you do not pay your price, but the price of the loser; but if you are the loser, you pay your own price. You don't play such a game every day. From an experimental point of view, I think that a huge amount of time should be devoted to explanations and training before letting the members of an experimental session play the game.

1.2 Extensive form games

Given the strategic/normal form games developed up to now, you may have the impression that strategic form games only allow to study *simultaneous games*, i.e. games where each player plays simultaneously with all the other players, which amounts to saying that he plays without knowing anything about the others' decisions. Further, you may have the impression that each player takes only one decision. This is wrong, but to show it, we first switch to (the surely more intuitive) extensive form games.

1.2.1 Definition

> **Definition 2 (see Selten 1975[3] for a complete definition)**
> An extensive form game is a tree with nodes, branches and payoffs at the terminal (or end) nodes. The **nodes** are partially ordered, with one node (the initial one) preceding all the other ones, and each node having only one immediate predecessor. Each non terminal node is associated with a player, the player who plays at this node. The **branches** that start at a node are the possible actions of the player playing at the node. A **vector of payoffs**, one payoff per player, is associated with each terminal node of the tree.
>
> There is a partition on the non terminal nodes, which is linked to the information each player has on the game. Especially, two nodes x and x' belong to a same **information set** if the same player i plays at x and x' and if player i is unable to distinguish x from x', given his information on the game (and namely the previous actions chosen by the opponents).

To give an example, let's return to our first auction game (first price sealed bid all pay auction). The extensive form game is given in Figure 1.2.

At the initial node, y, player 1 chooses a bid (one of the six branches, each branch being a different bid). Given that the second player bids without knowing player 1's bid, he is unable to know at which of the six nodes x_0, x_1, x_2, x_3, x_4, x_5 he plays. So the 6 nodes belong to the same information set. By convention, *nodes that are in a same information set are linked with a dashed line*.

So the partition of the non terminal nodes is: H={{y}, {x_0, x_1, x_2, x_3, x_4, x_5}}. The partition is compounded of two sub-partitions, one for each player, H_1 for player 1 and H_2 for player 2. H={H_1, H_2}, with H_1={{y}} and H_2={{x_0, x_1, x_2, x_3, x_4, x_5}}.

A direct consequence of player 2's lack of information is that the actions available at each node of a same information set are necessarily the same, given that the player does not know at which node he plays (in other words, if he had different actions at two nodes of a same information set, then he would be able to distinguish the two nodes, just by observing the actions available at the nodes).

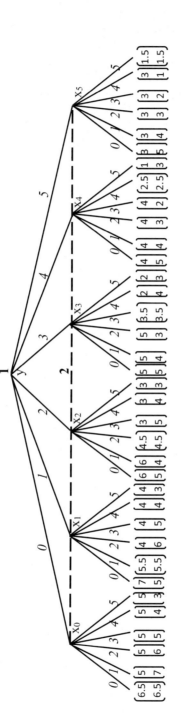

Figure 1.2 By convention, the ith coordinate of each payoff vector is the payoff of player i (here i=1, 2).

HOW TO BUILD A GAME

Let us observe that *we can interchange the roles of players 1 and 2, putting player 2 at the first node, and player 1 at the non singleton information set*. This is due to the fact that we work on a *simultaneous game*, i.e. a game in which no player observes the action played by the other player when he takes his own decision. There is *no sequence of events*, and so the places of players 1 and 2 in the game tree are completely arbitrary.

The above example shows that *we can represent a strategic form game in an extensive form way*. Hence, following the same lines, we could represent any of our story games in an extensive form way, and we would get a game tree with a structure similar to Figure 1.2 (one singleton information set for player 1, one non singleton information set for player 2).

But of course, we can get much richer game trees. For example, we can change the game in Figure 1.2 into *a Stackelberg first price all pay auction*, just by changing player 2's partition of information sets. In the game in Figure 1.3, player 2 observes player 1's bid before playing. So we are in a Stackelberg structure, player 1 being the leader and player 2 the follower.

Now $H_2=\{\{x_0\}, \{x_1\}, \{x_2\}, \{x_3\}, \{x_4\}, \{x_5\}\}$ in that player 2 perfectly knows at which node he is, given that he observes player 1's bid before playing.

Let the game talk: of course, the change in the information sets will affect the way of playing, but how? Economists know that the Stackelberg structure is sometimes profitable to the leader (namely in quantity duopoly models), sometimes to the follower (namely in price duopoly models).

One thing is certain: player 2 will not lose money, because he adapts his bid to player 1's, and if player 1 plays 0 and 3, she will not lose money either (because player 2 will not overbid a bid 3). But what plays player 1? 0, 3 or even 2 (hoping that player 2 does not overbid)?

1.2.2 Story extensive form games and behavioural comments

Strategic form story games are quite well known. But there are also quite interesting extensive form story games. One of the most well-known is the ultimatum game.

> **Ultimatum game:** There are two players that try to share a given amount of money (let us fix it to 10). Player 1 makes an offer x between 0 and 10. Player 2 observes the offer and can accept or refuse it. If he accepts it, player 1 gets x and player 2 gets 10–x, if he refuses, both get 0. So, if x is an integer, and if the utilities are (at least proportional to) the payoffs, the extensive form is given in Figure 1.4.

Let the game talk: *a lot has been written about this game, which is one of the most played games in experimental sessions*.[4] Often this game has been used to criticize game theory. As a matter of fact, game theory – at least some criteria linked to backward induction – will exploit the asymmetric situation of players 1 and 2. Given that player 2 can only accept the offer – and get a positive utility by doing so – or refuse it – and get an associated null utility – it seems that player 1 is in a more powerful position. So she should propose a high amount to herself, given that player 2 prefers getting a small positive utility as opposed to getting a null utility. But, as is usually observed in experimental sessions, player 1 does not play in this way; she usually makes more fair offers because she fears that high offers will not be accepted (and high offers are indeed often refused). How can this happen? Well, *simply because the ultimatum game is badly written*. As we will see in section 3, the utilities in a real offer game are not necessarily proportional to the payoffs. And this largely explains the behaviour in the experimental sessions.

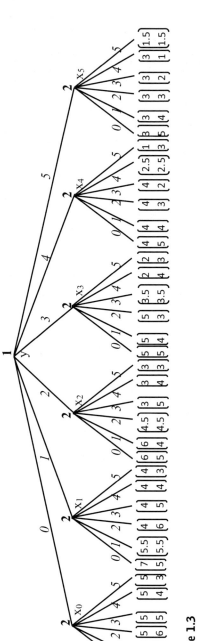

Figure 1.3

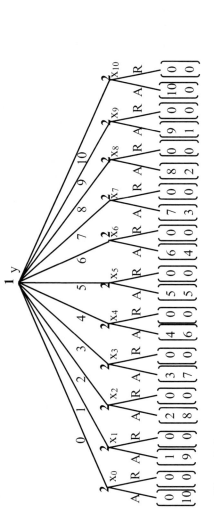

Figure 1.4

HOW TO BUILD A GAME

Another well-known game is the gift exchange game:

> **Gift exchange game**: there are two players, an employer and an employee. The employer plays first and offers a salary, w. The employee observes the salary and then offers a certain level of effort, e. The employer's payoff is, for a given effort, decreasing in the offered salary. The employee's payoff is, for a given salary, decreasing in the provided effort. The interest of this game is that, despite the fact that the employee is better off working less for a given wage and the employer is better off paying less for a given effort, both prefer a high wage associated with a high level of effort to a low wage associated with a low level of effort. The extensive form of this game is given in Figure 1.5, with two levels of salary, w_1 and w_2, $w_1<w_2$, and with two levels of effort, e_1 and e_2, $e_1<e_2$.
>
> This game has a companion game very similar in structure, Berg's (1995)[5] investment game.

Let the game talk: given that the employee plays after the employer, his salary is given when he has to choose an effort, so the employee always offers the lowest effort because working less for a given salary is less tiring than working more (v'>0 and 1>u'). Yet the employer knows this fact, so she knows that the employee chooses e_1 regardless of the wage. So, given that paying less is better than paying more, she offers w_1 (v>0), and the game ends in the sub optimal situation where the players get (v, v'), whereas they could get the higher payoffs (u, u') when switching to a high wage and a high effort. This is why the game is sometimes compared to the prisoner's dilemma, where the players are also unable to switch to the common optimal situation; the structure of the games is however very different (in the gift exchange game, by contrast to the prisoner's dilemma, the second player knows the first player's action before playing).

Yet, in experimental sessions,[6] the employer often does not offer the lowest salary and the employee often reacts to this higher offer with an effort that is higher than the lowest effort. For example, as you will see in section 3.3 my students clearly do not provide a low effort regardless of the offered salary. Why is that? To my mind, there are at least two explanations. First, as in the ultimatum game, the utilities of real players are different from the ones given in Figure 1.5 (see section 3). Second, *real players may be influenced by their real life:* employees suffer from the competitive labour market and they know that if they do not provide the highest effort, they may

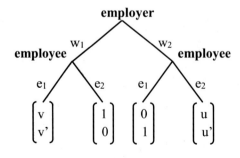

with 1>u>v>0 and 1>u'>v'>0.

Figure 1.5 The first, respectively the second coordinate of each payoff vector, is the employer's, respectively the employee's payoff.

HOW TO BUILD A GAME

lose their job, so they spontaneously offer high levels of effort, and this fact may influence them in experimental sessions. And this is all the more true for students, who often fear the labour market.

War of attrition games and variants
Another story game is the war of attrition game.

> **War of attrition game:** We approach war of attrition games with an ***ascending all pay auction***. So, let us imagine the following auction on an object of value V=3. Two players play the game; both have a revenue M=5 (as in the above all pay auctions) but the game is played as follows: first, player 1 chooses to bid 1 or to quit (stop) the auction, in which case the object is given to the second player for a null price. If she bids 1, player 2 is called on to either overbid (by bidding 2), or to stop the game, in which case the object is given to player 1 for a price equal to 1 (her last bid). If he bids 2, player 1 is again called on to play. Either she overbids – she bids 3 – or she stops; in the latter, player 2 gets the object at price 2, the last bid he proposed, but player 1 pays 1 (her last bid). This continues until player 1 stops or overbids by bidding 5. Each time, when a player stops, the object goes to the other at the last bid s/he proposed, and the player who stops also pays the last bid s/he proposed. When the utilities are given by the payoffs (or are at least proportional to them), the extensive form game is given in Figure 1.6.

Let the game talk: this game is *a war of attrition game* because it has three properties:

- All payoffs decrease when the game goes on.
- A player prefers continuing if the opponent stops at the next round. In our game this is quite natural. If you stop at time t, you lose the amount already offered, say x. If you overbid – bidding x+2 – and if your opponent stops at the next round, then you pay x+2 but you get the value V. As V–x–2>–x (because V>2) you prefer overbidding.
- But a player prefers stopping the game if the other goes on at the next round. As a matter of fact, in our game, if you stop, you pay x, the amount you offered up to now, so you get M–x. If you overbid the opponent and he goes on, you pay at least x+2 if you stop at your next decision round, so you get M–x–2, or, if you overbid at this round (you bid x+4), you get at best M+V–x–4 (if your opponent stops after your second overbid). Well, given that V is equal to 3, we have M–x>M–x+V–4, which explains the wish to stop.

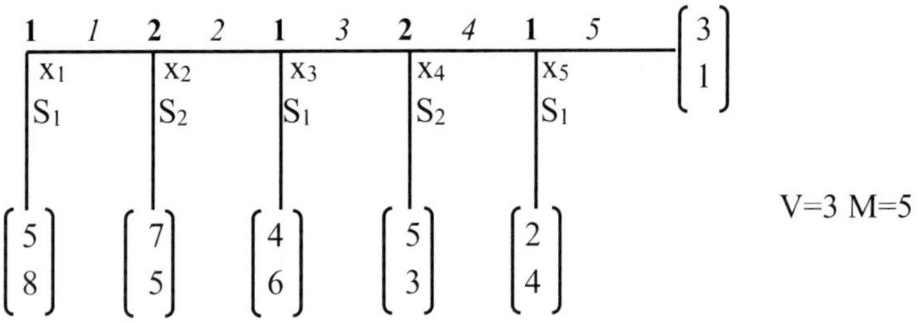

V=3 M=5

■ **Figure 1.6** For example, the payoffs (5, 3) are obtained as follows: Player 2 stops the game at x_4, so the object goes to player 1, who pays her last bid, 3. So player 1 gets M–3+V=5–3+3=5. Player 2 loses the auction but pays his last bid, 2. So he gets M–2=3.

23

HOW TO BUILD A GAME

What happens in such a game, for any values of M and V? Who stops first? Can you reasonably expect that the opponent stops before you, or should you be cautious and stop the game as fast as possible?

Well, if V is equal to 3, then it may be interesting to go on even if the other player stops at the next round with probability p lower than 1: as a matter of fact, going on (and stopping at the next decision round) is best if $(V-x-2)p+(1-p)(-x-2)>-x$, i.e. $p>2/V$, i.e. if the probability that the opponent stops at the next round is higher than 2/3. Things are even more complicated in a modified war of attrition game. If V is higher than 4, say 5 for example, then you may be induced to go on even if your opponent does not stop at his next decision round, but stops at his second next decision round because $M-x<M+V-x-4$. So you are even more induced to go on. Yet, given that the opponent faces the same incentives as you, he may also be induced to go on, and so the game continues but the payoffs inevitably reduce with time.

What can you learn from the past decisions in such a game? The interest to go on sits with the hope that the opponent throws in the towel before you do. But the beliefs on what the opponent does in the next round should depend on the bids he has already made. How should you expect that your opponent stops at the following round if he never stopped before, even when he lost money, i.e. proposed bids larger than V? Previous behaviour necessarily should have an impact on the way you judge your opponent and the way you expect him to play (this way of reasoning is called *forward induction*). A consequence is that, if the game went on for a long time (i.e. the bids become high) it is difficult for you to expect that the opponent will stop at the next round because he didn't do it before, so it is perhaps better to stop now. But if you understand this fact you regret not having stopped at the beginning; but if you don't go on at the start of the game you don't have the opportunity to observe that the opponent is anxious and stops the game.

And what is the impact of a loss, i.e. a payoff lower than M? Do you behave in the same way when your payoff may still be higher than M and when it is definitively lower than M (you are already losing money)? Do you reason in the same way? This leads me to an experimental remark. *In order to test a behaviour in a war of attrition game, we should not only let players play the whole game, but we should also let them play a game which starts at a level where b_1 and b_2 are high, sometimes lower and sometimes higher than V, in order to test their reactions when they lose money.* Let me mention here Shubik[7] (1971)'s paper on his "grab the dollar game", where both players bid for a dollar in a sequential way, as in our ascending all pay auction (V=1 dollar, 5 cents for the increment between two bids). Shubik observed that total payments higher than 3 and 5 dollars are not seldom and he mentioned two thresholds, bids higher than V/2 – so that the player who offers the dollar necessarily makes money – and bids higher than V, where each bidder loses money.

Finally, given that the game seems symmetric, why should you imagine that the opponent is different from you? As a matter of fact, the motivation to pursue lies in the hope that the opponent is different in that he prefers stopping. The expected difference is not in the payoffs, not even in the way people evaluate them (the utilities may be the payoffs), but just in the way you expect others will deal with a difficult decision (stopping and getting on, without knowing which action will be the best). So, for example, you decide to go on because you hope that the opponent is anxious and fears that you are courageous (and hence continue), so that he stops as soon as possible.

And is the game really symmetric? There is a player who begins the game and there is a player who ends the game (in Figure 1.6, both are player 1). Does a player have extra power when he starts or when he ends the game? This is quite a difficult question. Surely, if player 1 reaches point x₅, she will bid 5 – which leads to a higher payoff – but what about the implications of this hypothetical action?

Clearly, this is a very stimulating game; that is why we will study it throughout the book with different tools in order to see how game theory helps us to approach it.

But I can already give you here the way my (97) L3 students played this game, with M=100 and V=60 (so the last player is player 2). I asked them to answer two questions:

- If you were player 1, up to which amount would you bid? If you stop immediately, then say 0.
- If you were player 2, up to which amount would you bid? If you stop immediately, then say 0.

The histograms in Figures 1.7a and 1.7b show their answers.

It is easy to observe that most students either advise the players to stop very fast, either to bid up to the highest possible amount, or to bid up to amounts in [20, 40] or [55, 65], i.e. amounts around

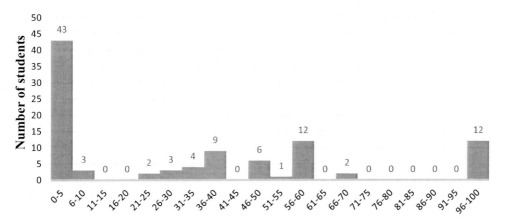

Maximal bids for player 1

▨ **Figure 1.7a** Maximal bids for player 1 in the war of attrition game/ascending all pay auction with M=100, V=60

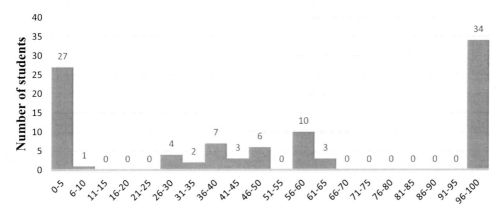

Maximal bids for player 2

▨ **Figure 1.7b** Maximal bids for player 2 in the war of attrition game/ascending all pay auction with M=100, V=60

HOW TO BUILD A GAME

V/2 and V. And we can also observe that player 2 is more encouraged to go on, so there is an asymmetry between the players, linked to the fact that player 2 is the last player in the game tree (see chapter 4, section 3.1 for details on the students' behaviour).

> **Rosenthal's[8] centipede game and pre-emption games**: another story game, Rosenthal's centipede game, and more generally *pre-emption games*, are very stimulating. These games are the *companion to war of attrition games*. There are again two players who play in a sequential way, with only two actions at each information set, i.e. stop or go on. *But this time, payoffs grow as the game goes on*; moreover, it is better to stop before the other stops the game but it is better to go on if the opponent goes on.
> We propose the study of this game as an exercise.

> To close this section let us just add an additional sequential all pay auction game, the **non incremental ascending all pay auction**. In the ascending all pay auction, additional bids were incremental (one unit per period), hence the end player was determined by M, which may have an impact on the result. Let us now suppose that the first player can make a bid b between 0 and M. Then player 2 can leave the auction or overbid, by proposing a bid b' between b+1 and M. Then player 1 can leave the auction, by paying her last bid b, or she overbids with a bid b" between b'+1 and M. And so on. Once the bid is M, the game stops. The game is represented in Figure 1.8, for V=3 and M=5.

Let the game talk: first this procedure fits with the English ascending auction, except that everybody pays. Second, the first player can bid V, which dissuades player 2 from overbidding, so it seems that player 1 has some extra power. But which one? How does M influence the result? As a matter of fact, even if nobody is happy to pay more than V, this game has common features with the

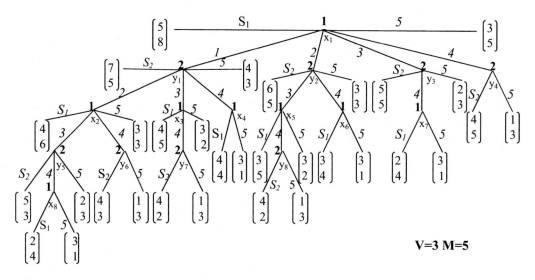

■ Figure 1.8

HOW TO BUILD A GAME

ascending all pay auction, especially because you lose your previous bid: so you may be ready to overbid more than V, if this helps you to win the object. Hence it may be clever to bid up to M, which induces that the value of M may have an impact.

Clearly intuition is not obvious. And we will show that game theory confirms the original link between initial bids, V and M, a link which is neither obvious, nor intuitive.

Just a general remark: later we will present other games we can also call story games, like *the envelope game, the Fort Boyard sticks game, or the prisoners' disks game* (see next section). A general observation will be that, as soon as a game becomes rich, intuition is not enough to solve it, so one of the aims of this book is to show how game theory gives insights on the way to play these games.

1.2.3 Subgames

Just before turning to strategies let us introduce the concept of subgame we use in later chapters.

> **Definition 3 (see Selten 1975 for a complete definition)**
> A subgame K of a game G in extensive form is a node x with all the nodes, branches and payoffs after this node. K has to not cut an information set, that is to say: if a node y belongs to K, then all the nodes in the information set containing y also belong to K.
> G is a subgame of G: we sometimes call *proper subgame* a subgame K different from G.

For example, the first price all pay auction in Figure 1.2 has no (proper) subgame, but the ultimatum game in Figure 1.4 has as many (proper) subgames as nodes x_i, i from 0 to 10.

Selten's horse (a game we study in chapter 5), in Figure 1.9, well illustrates the condition on the information sets.

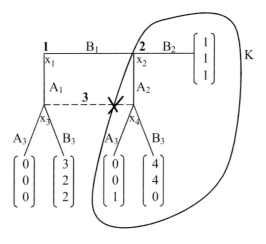

Figure 1.9

HOW TO BUILD A GAME

Contrary to what you might expect, there is no subgame starting at x_2 because the set K with x_2 and the nodes, branches and payoffs following x_2 cuts player 3's information set. In other words, x_4 belongs to K. But x_3, which belongs to the same information set as x_4, does not belong to K (because x_3 is not a successor of x_2).

2 STRATEGIES

Up to now it was as if strategic form games are simultaneous games (with one decision per actor), whereas extensive form games allow people to address games with real sequences of actions. We will see that things are not as simple and that the term strategy helps to bring the two representative forms of a game closer.

2.1 Strategies in strategic/normal form games

Strategic form games lead to pure strategies and mixed strategies.

2.1.1 Pure strategies

> **Definition 4: a pure strategy** of player i, i from 1 to N, gives the action he plays at each of his potential decision rounds. We write S_i for player i's pure strategy set and s_i for an element of this set. A game is *finite* when the set of players and the sets of pure strategies are finite.

> ➤ **STUDENTS' INSERT**
>
> For example, in the first price sealed bid all pay auction game (matrices 1.8a, 1.8b and 1.8c), $S_1=S_2=\{0, 1, 2, 3, 4, 5\}$, given that each player plays only one time and has to choose an integer bid between 0 and 5. Hence any integer between 0 and 5, for example 2, is a pure strategy of player i, i=1, 2.

What comes to mind is that a *strategy is not necessarily clever*. As we observed earlier, it seems irrational to play 4 or 5 in the first price sealed bid all pay auction. So a strategy just expresses a *possible* action (more generally a possible profile of actions). *A strategy may be clever or completely silly by itself (playing 0 seems cleverer than playing 5) but in general, its cleverness depends on what others do.*

2.1.2 Mixed strategies

> **Definition 5: a mixed strategy** of player i, i from 1 to N, is a probability measure on his pure strategy set. We write P_i for player i's mixed strategy set and p_i for an element of this set. Player i's strategy is **completely mixed** if it assigns a positive probability to each of his pure strategies.

> **STUDENTS' INSERT**
>
> For example, in the first price sealed bid all pay auction game (matrices 1.8a, 1.8b and 1.8c), bidding 0 with probability 0.3, 1 with probability 0.2 and 5 with probability 0.5 is a mixed strategy for player i, i=1, 2.

Mixed strategies: what for? We discussed above that it is not easy to advise somebody on the best way to play the first price sealed bid all pay auction. For example playing 0 seems best if the opponent plays 2 or more, but if he plays 0 as well, then it is better to bid 1 and so on. This largely explains why, by playing this game with L3 students over several years, I always got contrasting results. I already mentioned the results for one class of L3 students (118 exploitable data) in section 1.1.3 (28.8% played 0, 12.7% played 1, 8.5% played 2, 20.3% played 3, 14.4% played 4 and 15.3% played 5). I give here the (close) results of another class of L3 students (146 exploitable data): 31% played 0, 13% played 1, 10.5% played 2, 13% played 3, 18% played 4 and 14.5% played 5.

Clearly, the variety of answers means that it is not best playing one strategy with probability 1. In some way, if we replace the last L3 students by a representative one, this player plays a mixed strategy: he plays 0 with probability 31%, 1 with probability 13%, 2 with probability 10.5%, 3 with probability 13%, 4 with probability 18% and 5 with probability 14.5%. And playing in a mixed way seems the best way to play, at least with regard to the positive probabilities assigned to 0, 1, 2 and 3 (as a matter of fact 1 is the best response to 0, 2 is the best response to 1, 3 and 0 are the best responses to 2, 0 is the best response to 3, 4 and 5). So, the introduction of the concept of mixed strategy is not due to the fact that *mathematicians appreciate the compact and convex nature of the set of mixed strategies (which is true)*, but it is due to the fact that mixed strategies are essential, namely if no pure strategy seems to be a unique best reply.

But let me make an observation: *students – and perhaps people in general – do not spontaneously play a true mixed strategy. When called on to play the same game several times, they may change their pure strategy over time, so that the mean strategy looks like a mixed strategy. But, when called on to play only once and authorized to play a lottery (a distribution of probabilities on their actions), they do not spontaneously play in this way.* For example, when I asked my students to play the second price sealed bid all pay auction game (Matrix 1.9a), I allowed them to play in a mixed way. My students played this game after having learned the concepts of pure and mixed strategies, dominated strategies and Nash equilibria. They were able to observe the stabilized (Nash) situations where one of the players bids 0 and the other bids 3 or more. They were able to see that it is nice to be the player who plays 3, 4 or 5 and pays 0, but they also feared that the opponent plays in this way, which induces to play 0. So I hoped that many students would mix on 0 and 5 for example. But only 13 among the 116 exploitable data were mixed strategies! 103 students among 116 still played a pure strategy.

HOW TO BUILD A GAME

2.2 Strategies in extensive form games

2.2.1 Pure strategies: a complete description of the behaviour

In extensive form games, the appropriate strategies are pure and behavioural.

> **Definition 6:** a player i's **pure strategy**, i from 1 to N, is an application $s_i(.)$ that assigns an action to each information set of player i. We call S_i the set of pure strategies and s_i an element of S_i.

> ➤ **STUDENTS' INSERT**
>
> Let us illustrate this concept on the ascending all pay auction in Figure 1.6. Player 1 has 3 information sets $\{x_1\}$, $\{x_3\}$ and $\{x_5\}$. A pure strategy for player 1 is an application that assigns an action to $\{x_1\}$, for example S_1, an action to $\{x_3\}$, for example bid 3, and an action to $\{x_5\}$, for example S_1. For the sake of simplicity this strategy is usually written: $(S_1/\{x_1\}, 3/\{x_3\}, S_1/\{x_5\})$. It expresses an action at each possible turn of play. It follows that player 1 has $8=2^3$ strategies, in that she can choose two actions at each of her three information sets. By extension, the set of pure strategies becomes the set of possible outcomes of the application, i.e. the set (for player 1) becomes:
> $\{(S_1/\{x_1\}, S_1/\{x_3\}, S_1/\{x_5\}), (S_1/\{x_1\}, S_1/\{x_3\}, 5/\{x_5\}), \quad (S_1/\{x_1\}, 3/\{x_3\}, S_1/\{x_5\}),$
> $(S_1/\{x_1\}, 3/\{x_3\}, 5/\{x_5\}), \quad (1/\{x_1\}, S_1/\{x_3\}, S_1/\{x_5\}), \quad (1/\{x_1\}, S_1/\{x_3\}, 5/\{x_5\}),$
> $(1/\{x_1\}, 3/\{x_3\}, S_1/\{x_5\}), \quad (1/\{x_1\}, 3/\{x_3\}, 5/\{x_5\})\}.$

This leads us to three remarks:

First, giving an action at each possible turn of play is also the definition of a pure strategy in the strategic form game. This automatically links the strategic form and the extensive form games (see next paragraph).

Second, the concept of strategy is complete, that is to say it gives an action at each information set. But why do we give player 1's action at nodes x_3 and x_5 when she stops at x_1? More generally why do we give an action at each information set, even if some chosen actions at earlier information sets clearly prevent players from reaching these sets? We do this *because it is necessary to solve a game*. Well, player 1's decision at x_1 clearly depends on player 2's decision at x_2 (it is fine bidding 1 if player 2 stops at x_2, but not so fine in the other case), but player 2's decision at x_2 depends on *player 1's decision at x_3* (it is fine bidding 2 if player 1 stops at x_3 but it is better to stop in the other case), yet player 1's decision at x_3 depends on player 2's decision at x_4 which itself depends on player 1's decision at x_5. *It follows that player 1 is unable to decide what she does at x_1 if she does not say what she does at x_3 and x_5.*

Let me make an experimental observation. In an experimental setting, asking a player to give a "true strategy"; i.e. a profile of actions (an action at each information set) may influence the way he plays, namely because it constrains him to study the whole game, something he would perhaps not have done, if not asked to give such a profile. So asking for a true strategy may not be neutral: for example some authors, like Maximiano, Sloof and Sonnemans (2013)[9] in an experimental gift exchange game, asked the employee players to give a level of effort for each possible salary, albeit the employer players only propose one salary. If the authors had just asked the employee players

HOW TO BUILD A GAME

to react to the unique proposed salary, these players would perhaps not have played in the same way. This claim stems from experiments made by Johnson et al. (2002),[10] where the authors established that when a player plays at a given stage in a sequential game, he sometimes doesn't try to understand the structure of the game further on.

Third, given the completion of the notion of strategy, the strategy sets often become very large. For example, in the simple Stackelberg all pay auction game in Figure 1.3, player 2, who has 6 information sets and 6 possible actions at each of these sets, has $6^6=46,656$ different possible strategies. More generally, as soon as a player has several information sets in a game tree, the set of pure strategies (outcomes of the applications) becomes quite large even for simple games. Well *is that a problem? The answer is clearly no.* As a matter of fact, our aim is not to write the set of strategies but to discover which strategies have some good strategic properties. This usually leads to a focus on a smaller set of strategies.

2.2.2 The Fort Boyard Sticks game and the Envelope game

To illustrate the concept of strategy, let us present the Fort Boyard sticks game (which is played in the French TV summer show Fort Boyard) and the envelope game, two games which we will often talk about throughout the book.

The Fort Boyard Sticks game: two players are in front of K sticks lined up on a table (K is a finite integer). They play in turn. Each player takes 1, 2 or 3 sticks. The game ends when no sticks remain on the table. The loser is the player who takes the last stick.

This game is easy, yet it allows us to show, first that the set of pure strategies is very large, and second that *one should not always try to represent a game in extensive form.* To see why, suppose there are only 6 sticks (usually there are around 20): the game tree is given in Figure 1.10.

The two players are C and M *(C is the candidate in the French TV show and M is the "master of time")* and the candidate is always the first player to take sticks. As you can observe in this reduced game, the candidate has 14 information sets. Given that she has 3 possible actions (take 1, 2 or 3 sticks) at 4 information sets, 2 possible actions (take 1 or 2 sticks) at 4 information sets, and only 1 possible action (take the last stick) at the remaining information sets, her number of pure strategies is: $3^4 \times 2^4 = 1,296$ strategies. M has also 14 information sets; he has 3 possible actions at 4 information sets and 2 possible actions at 3 information sets. Hence his number of pure strategies is $3^4 \times 2^3 = 648$ strategies. Imagine what happens when we switch to the true game with about 20 sticks!

Well, is that a problem? No. In the next chapter we will solve this game regardless of the number of sticks, hence regardless of the large number of strategies and the large structure of the game tree. In other words, everybody is just interested in the best way to play, *and this does not require that someone writes down all the strategies, or draws the (whole) game tree!*

31

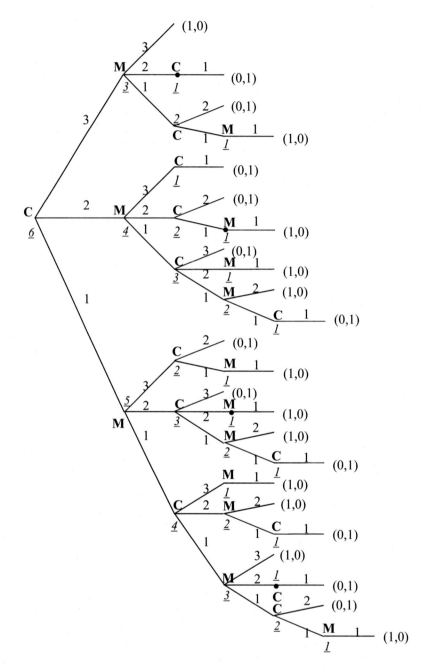

Figure 1.10 The underlined numbers are the numbers of remaining sticks. The first, respectively second coordinate of each payoff vector is the payoff of player C, respectively M.

HOW TO BUILD A GAME

> **The envelope game:** the game is a two player game. Each player is in front of an envelope in which there is an integer between 1 and K. Both players know K (we will say that K is common knowledge). Each player observes the number in his envelope but ignores the number in the opponent's envelope. He only knows that this number is between 1 and K (1 and K included). Then each player pushes on one of two buttons, exchange or not exchange (a player does not see the button chosen by the opponent). If both players push on the button marked "exchange", they exchange their envelope. If not, each keeps his envelope. The game is now ended: each player gets an amount of dollars equal to the number in the envelope he has in hand.

Clearly, in such a game, we usually wonder up to which number the opponent wishes to exchange his envelope. For example, if K=100,000, up to which amount will he exchange – 2,000, 10,000, 50,000, more? This of course influences our own threshold of exchange. If we fear that the opponent is not ready to exchange high amounts, we only wish to exchange very low amounts, but if the opponent loves playing, and is ready to exchange amounts up to 80,000 for example, then of course we can afford to try to exchange higher amounts. So, usually, the only way to play is to try to guess the maximal amount the opponent is ready to exchange, and given this guess and our own fear of losing money, to deduce up to what threshold we are ready to exchange.

In some way, this is much easier (at least it is shorter) than writing down all the pure strategies and drawing the game tree. Let us justify this claim.

Imagine that K is just equal to 3 (a very uninteresting envelope game!). A strategy has to give an action for each information set. What is an information set in this game for player 1? Well, she may be in front of the number 1, 2 or 3, depending on the outcome of the lottery that assigns a number to each envelope. So player 1 has to tell what she does in front of number 1, in front of 2 and in front of 3. Once again, the completeness of the strategy may seem strange: if player 1 has the number 2 in her envelope, why should she bother about what to do in front of 1 and 3? Once again, the completeness is necessary to establish the optimal strategies of both players: player 1, to establish her best strategy, has to know the behaviour of player 2 in front of any possible number, which depends on player 1's behaviour in front of all the possible numbers.

Given that in front of each number, 1, 2, 3, player 1 can choose from two actions (exchange or not exchange) player 1 has $2^3=8$ strategies; for example, one of these strategies is ($\bar{E}$/1, E/2, E/3) – E means 'exchange' and $\bar{E}$ means 'not exchange': once again a strategy is only a possible profile of actions (the chosen one is completely silly).

This game has *originality: a new player plays in this game, Nature.* As a matter of fact, if you try to represent the game, you have to introduce the lottery that assigns a number to each envelope. This lottery is usually called Nature in game theory. Nature has 9 strategies, given that she chooses one among the three numbers 1, 2 and 3 for both players. It follows that the game tree becomes the one given in Figure 1.11.

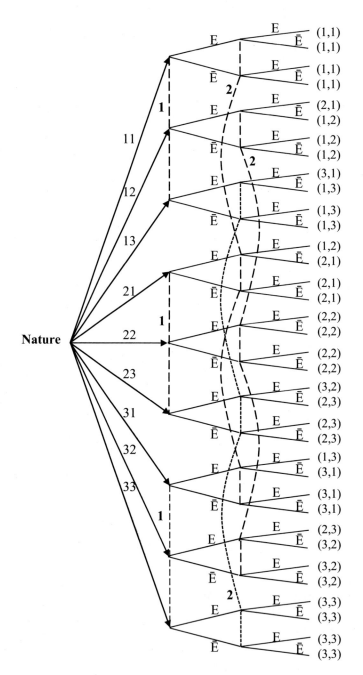

Figure 1.11 12, for example, means that Nature puts number 1 and number 2, respectively in player 1's and player 2's envelope.

HOW TO BUILD A GAME

> **STUDENTS' INSERT**

Let us summarize the richness of the envelope game:
First, it introduces a new player, Nature. Observe that we do not represent Nature's payoffs, in that Nature is supposed to be completely indifferent to the outcomes of the game (she is not a true player, just a lottery). She just ensures that each number can be in each envelope. So Nature plays each of her actions with positive probability. That is why we put an arrow on each action. Observe that in this special game, *we do not know with which probability Nature selects each of the 9 actions*. And *we do not need to know these probabilities because they will not be necessary to solve the game* (see chapters 2 and 3).
Second, this game nicely illustrates the notion of information sets. So for example, player 2 has three information sets. In the first one, the number in his envelope is 1. There are six nodes in this set because player 2 does not observe the number in player 1's envelope, and because he does not observe the button on which player 1 pushes for each number (exchange or not exchange).
Third, the definition of a player's strategy immediately derives from the game tree (each player has 3 information sets with two actions at each of these sets, hence 8 pure strategies).
Fourth, the game tree is very large for K=3. What happens if K=100,000? Well, simply don't try to represent the game, because it is useless: you will not learn more from it than you learn from the game tree for K=3.

2.2.3 Behavioural strategies

Definition 7: a **local strategy** for player i, i from 1 to N, at his information set h is a measure (distribution) of probabilities on the available actions at this set h. We write it π_{ih}; $\pi_{ih} \in \Pi_{ih}$, player i's set of local strategies at h. A **behavioural strategy** for player i, i from 1 to N, is a profile of local strategies at the information sets of player i, one local strategy at each of his information sets. We write it π_i; $\pi_i \in \Pi_i$, player i's set of behavioural strategies.

> **STUDENTS' INSERT**

For example, in the ascending all pay auction in Figure 1.6, a local strategy of player 1 at $\{x_1\}$ consists in bidding 1 with probability 0.1 and stopping with probability 0.9. A local strategy of player 1 at $\{x_3\}$ consists in bidding 3 with probability 0.8 and stopping with probability 0.2, and a local strategy of player 1 at $\{x_5\}$ consists in bidding 5 with probability 0 and stopping with probability 1.

And a behavioural strategy of player 1 may be the profile of the three local strategies listed above.

Hence a player i's behavioural strategy is simply a profile of probability distributions, one at each of player i's information sets.

HOW TO BUILD A GAME

2.3 Strategic/normal form games and extensive form games: is there a difference?

It is now time to clear up a misunderstanding. Normal form games are not necessarily one shot simultaneous games. ***Strategic/normal and extensive forms exist for each game.***

> ### ➤ STUDENTS' INSERT
>
> Let us return to the ascending all pay auction in Figure 1.6. We spontaneously represented this game in extensive form *but we could have represented it in normal form*. So let us recall that a strategy for player i in a normal form game tells what player i does each time he is potentially called on to play. Player 1 is potentially called on to play three times, at the beginning of the play (the node x_1 in the extensive form game) – she can choose 1 or S_1 – after her own bid 1 and the overbid 2 by player 2 (the node x_3 in the extensive form) – she can play 3 or S_1 – and after her bids 1 and 3, and player 2's overbids 2 and 4 (the node x_5 in the extensive form) – she can play 5 or S_1. So, clearly, a pure strategy for player 1 may be to stop at the beginning, to go on after bids 1 and 2, and to stop once asked if she wants to overbid player 2's bid 4. This strategy is nothing other than: ($S_1/\{x_1\}$, $3/\{x_3\}$, $S_1/\{x_5\}$).*
>
> So the concept of pure strategy in the normal form game is equivalent to the concept of pure strategy in the extensive form game.
>
> Given that there are only two players, the normal form game can be summarized in Matrix 1.10a.
>
> * We later write A/x_i instead of A/{x_i} for ease of notations.

		Player 2			
		($S_2/x_2 S_2/x_4$)	($S_2/x_2 4/x_4$)	($2/x_2 S_2/x_4$)	($2/x_2 4/x_4$)
Player 1	($S_1/x_1 S_1/x_3 S_1/x_5$)	(5,8)	(5,8)	(5,8)	(5,8)
	($S_1/x_1 S_1/x_3 5/x_5$)	(5,8)	(5,8)	(5,8)	(5,8)
	($S_1/x_1 3/x_3 S_1/x_5$)	(5,8)	(5,8)	(5,8)	(5,8)
	($S_1/x_1 3/x_3 5/x_5$)	(5,8)	(5,8)	(5,8)	(5,8)
	($1/x_1 S_1/x_3 S_1/x_5$)	(7,5)	(7,5)	**(4,6)**	(4,6)
	($1/x_1 S_1/x_3 5/x_5$)	(7,5)	(7,5)	(4,6)	(4,6)
	($1/x_1 3/x_3 S_1/x_5$)	(7,5)	(7,5)	(5,3)	(2,4)
	($1/x_1 3/x_3 5/x_5$)	(7,5)	(7,5)	(5,3)	(3,1)

Matrix 1.10a

What is the difference between the matrix and the game tree? The answer to this question is not minor and has multiple layers.

With regard to the pure strategies, there is no difference. As long as we know that x_1 means the beginning of the game, x_2 the stage of the game where player 1 already proposed 1 and player 2 has to take a decision to stop or overbid, more generally as long as we know the meaning of each x, the matrix talks as much as the game tree. For example the bold couple in the matrix are the payoffs when player 1 starts by bidding 1 at the beginning, player 2 overbids and player 1 stops after player 2's overbid; moreover we know that if player 1 would overbid at x_3, player 2 would stop the game, and we also know that if player 2 would overbid after bid 3, player 1 would stop thereafter.

HOW TO BUILD A GAME

In the same way, a matrix with 1,296 rows and 648 columns would perfectly represent the six sticks Bort Boyard game. And a matrix with 6 rows and 46,656 columns would perfectly represent the Stackelberg all pay auction.

Let us just call attention to three points:

- First, what matters *is the understanding of the game*; this knowledge means that you *are able* to write the pure strategies of the players, *but you don't have to write them.* And these strategies are the same in both the extensive and the normal form of the game. For example, what matters in the Stackelberg auction game, is that you know how player 2 can react to the different possible bids by player 1, not the writing down of the 46,656 possible strategies in 46,656 columns!
- Second, the representation of the normal form by a matrix sometimes leads to *simplifications that change the nature of the game.* So, for example, game theorists introduce the notion of equivalent strategy.

Definition 8: a pure strategy s_i by player i is (payoff) **equivalent** to another player i pure strategy s_i' if and only if both lead to the same payoffs for all players, regardless of the strategies played by the other players: $\forall j$ from 1 to N, $\forall s_{-i} \in XS_k$, $k \neq i$, k from 1 to N, $u_j(s_i, s_{-i}) = u_j(s_i', s_{-i})$

If, for each player, one suppresses all the equivalent strategies to a given strategy, one gets the **reduced normal/strategic form game.**

So, in the ascending all pay auction, the four strategies where player 1 stops immediately lead to the same outcomes for both players, whatever player 2 does. Let us replace them by one strategy, simply called S_1/x_1. In the same way, the two strategies where player 1 bids 1 but stops if asked to bid 3 lead to the same outcomes for both players, whatever player 2 does, so we can replace them by a unique strategy we call $(1/x_1 S_1/x_3)$. Finally, in the same way, the two strategies where player 2 stops immediately lead to the same outcomes for both players, and we can replace them by a unique strategy we call S_2/x_2. So, we get the reduced normal form (reduced strategic form) game, given in Matrix 1.10b.

		Player 2 S_2/x_2	$(2/x_2 S_2/x_4)$	$(2/x_2 4/x_4)$
Player 1	S_1/x_1	(5,8)	(5,8)	(5,8)
	$(1/x_1 S_1/x_3)$	(7,5)	(4,6)	(4,6)
	$(1/x_1 3/x_3 S_1/x_5)$	(7,5)	(5,3)	(2,4)
	$(1/x_1 3/x_3 5/x_5)$	(7,5)	(5,3)	(3,1)

Matrix 1.10b

We will see later that, with regard to the Nash equilibria, the normal form game and the reduced normal form game are equivalent (they lead to the same equilibrium behaviour); yet, *with regard to the information conveyed, both games are not equivalent.* To justify this remark, imagine that both players push on a button to choose their strategy. In the first game, suppose that player 1 plans to stop at x_1, to bid 3 at x_3 and 5 at x_5: this for example means that player 1 does not dare to bid at

37

HOW TO BUILD A GAME

the start of the game, but, if she bids all the same, she stays in the game until the end. In the second game, player 1 just plans to stop at x_1. Imagine now that player 1, at x_1, pushes on the wrong button (because of a trembling hand), so that she bids 1 instead of stopping. In the first game, we know what she will do next, in the second game, we do not!

- Third, even if we do not switch to the reduced normal form, there is a big difference between normal form games and extensive form games: *behavioural strategies are not mixed strategies.*

To illustrate this point, examine somebody who just knows Matrix 1.10a without knowing the meaning of the x; this person may represent the extensive form of the game as follows (Figure 1.12), a game tree that does not fit the ascending all pay auction game.

In some way, this figure illustrates the difference between mixed and behavioural strategies. A mixed strategy for player 1 is a unique probability distribution on her eight pure strategies whereas a behavioural strategy consists in three distributions of probabilities, one at each of her decision nodes. Intuitively, both strategies do not tell the same story. If each overbid takes one unit of time, player 1's behavioural strategy first gives her decision at time 1 (first distribution of probabilities), then, if she is still in the game at time 3, it gives her behaviour at that time (second distribution of probabilities), and if she is still in the game after player 2's overbid, it gives her behaviour at time 5 (third distribution of probabilities). We clearly have a sequential process of decisions. This is not the case for the mixed strategy. A mixed strategy compresses player 1's 3 steps of decisions into a unique decision at the beginning of the game. For example player 1 plays $(S_1/x_1, 3/x_3, 5/x_5)$ with probability 0.7 and $(1/x_1, S_1/x_3, 5/x_5)$ with probability 0.3. A same number, chosen at time 1, for example 0.7 on $(S_1/x_1, 3/x_3, 5/x_5)$, links the three actions at x_1, x_3 and x_5. It is as if a player decides at only one moment, at the beginning. This does not mean that the true game is a simultaneous game but players treat it as if it were such a game (like in Figure 1.12). So intuitively, the concepts of behavioural and mixed strategies are not the same.

But what about mathematics?
A natural question is: does there exist for each behavioural strategy, a mixed strategy that is equivalent (leads to the same payoffs with the same probabilities), and reciprocally, does there

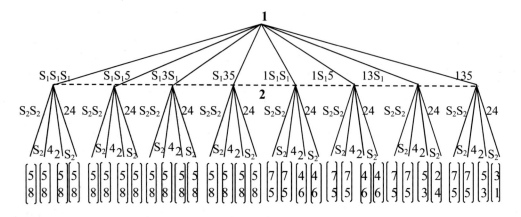

■ Figure 1.12

HOW TO BUILD A GAME

exist for each mixed strategy, a behavioural strategy that is equivalent (leads to the same payoffs with the same probabilities)? There is a link, given in Kuhn's theorem.[11] This theorem says that in each game with perfect recall, [12] on the one hand, for any behavioural strategy of player i, there exists a mixed strategy of player i that ensures him the same payoff, whatever the strategies played by the other players, and, on the other hand, for any mixed strategy of player i, there exists a behavioural strategy of player i that ensures him the same payoff, whatever the strategies played by the other players. *This equivalence is shown in the exercises.*

Yet this doesn't mean that the link between both concepts is perfect. In particular both concepts do not deal with *errors* in the same way. Let us illustrate it on the ascending all pay auction game.

Let us suppose that player 1 is expected to play $(S_1/x_1, 3/x_3, 5/x_5)$, but that she can make a small error. In the normal form game, this means that she plays each of the seven other strategies with a small error e_i, i from 1 to 7, and that she plays $(S_1/x_1, 3/x_3, 5/x_5)$ with probability $1 - \sum_{i=1}^{7} e_i$. In the extensive form game, this means that she plays 1 at x_1 with a small probability ε_1 (and S_1 with probability $(1-\varepsilon_1)$), S_1 at x_3 with a small probability ε_3 (and 3 with probability $(1-\varepsilon_3)$), S_1 at x_5 with a small probability ε_5 (and 5 with probability $(1-\varepsilon_5)$). So she plays the eight pure strategies with different probabilities in the normal form game and in the extensive form game (see Table 1.1).

Table 1.1

	Strategic form	Extensive form
$(S_1/x_1 S_1/x_3 S_1/x_5)$	e_1	$(1-\varepsilon_1)\varepsilon_3 \varepsilon_5$
$(S_1/x_1 S_1/x_3 5/x_5)$	e_2	$(1-\varepsilon_1)\varepsilon_3(1-\varepsilon_5)$
$(S_1/x_1 3/x_3 S_1/x_5)$	e_3	$(1-\varepsilon_1)(1-\varepsilon_3)\varepsilon_5$
$(S_1/x_1 3/x_3 5/x_5)$	$1 - \sum_{i=1}^{7} e_i$	$(1-\varepsilon_1)(1-\varepsilon_3)(1-\varepsilon_5)$
$(1/x_1 S_1/x_3 S_1/x_5)$	e_4	$\varepsilon_1 \varepsilon_3 \varepsilon_5$
$(1/x_1 S_1/x_3 5/x_5)$	e_5	$\varepsilon_1 \varepsilon_3 (1-\varepsilon_5)$
$(1/x_1 3/x_3 S_1/x_5)$	e_6	$\varepsilon_1 (1-\varepsilon_3)\varepsilon_5$
$(1/x_1 3/x_3 5/x_5)$	e_7	$\varepsilon_1(1-\varepsilon_3)(1-\varepsilon_5)$

It derives that in the extensive form game, the seven unplanned strategies cannot be played with the same probabilities, given that they do not include the same *number of errors*: for example, by playing $(S_1/x_1, S_1/x_3, S_1/x_5)$, player 1 makes two errors (at x_3 and x_5), by playing $(S_1/x_1, S_1/x_3, 5/x_5)$ she only makes 1 error (at x_3). It follows that the probability to play $(S_1/x_1, S_1/x_3, 5/x_5)$ is necessarily higher than the probability to play $(S_1/x_1, S_1/x_3, S_1/x_5)$ $((1-\varepsilon_1)\varepsilon_3(1-\varepsilon_5) > (1-\varepsilon_1)\varepsilon_3 \varepsilon_5)$.

This not true in the normal form game, where the probability to play $(S_1/x_1, S_1/x_3, S_1/x_5)$ may be higher, lower or equal to the probability to play $(S_1/x_1, S_1/x_3, 5/x_5)$ (each strategy is just one error). In other words, to play $(S_1/x_1, S_1/x_3, S_1/x_5)$ in the strategic form game, you just tremble one time (you push on the button $(S_1/x_1, S_1/x_3, S_1/x_5)$ instead of the button $(S_1/x_1, 3/x_3, 5/x_5)$) whereas, to play $(S_1/x_1, S_1/x_3, S_1/x_5)$ in the extensive form game, you have to tremble two times (you push on the button S_1 at x_3 instead of the button 3, and you push on the button S_1 at x_5 instead of the button 5).

HOW TO BUILD A GAME

Mathematically, it is not possible for each profile of errors (e_1, ..., e_7) to find three values ε_1, ε_2 and ε_3 such as the probabilities in the normal form game are equal to the probabilities in the extensive form game (seven equations with three unknowns).

> So, despite normal form games and extensive form games having the same pure strategies, *they are not equivalent, as soon as one introduces the notion of error.*

3 INFORMATION AND UTILITIES

3.1 Perfect/imperfect information

3.1.1 A concept linked to the information sets

> **Definition 9**: a game is of **perfect information** if each player, each time he is called on to play, knows exactly what has been played before, otherwise the game is of imperfect information. When the game is in extensive form, a game is of perfect information if all the information sets are singletons.

For example, the Stackelberg auction game, the ascending auction game, the ultimatum game and the Fort Boyard sticks game are all perfect information games. Each player, when called on to play, knows exactly what has been played before.

Let us observe here that in a perfect information game, a player knows exactly what he has played in earlier steps of the game. We say that the game is of perfect recall.

> **Definition 10**: a game is of **perfect recall** if each player, each time he is called on to play, knows exactly what he played before.

We will always work with games of perfect recall.

The normal form story games and the envelope game are imperfect information games.

Let us observe that the term "before" in the definition is not necessarily adequate; in a simultaneous game both players play simultaneously but the game is of imperfect information because the game, in extensive form, *has an information set not reduced to a singleton.*

Let us also focus on a particularity of the envelope game and a companion game, the prisoners' disks game, we describe below: in many games, the game is of imperfect information because of the unknown action played by the special actor Nature. So, in the envelope game, player 1 chooses to exchange her envelope without knowing the number in the opponent's envelope, so she does not know part of the action of Nature. And this is the case in many games.

3.1.2 The prisoners' disks game

Let us turn to one of the most famous imperfect information games, where Nature plays at the beginning of the game, the prisoners' disks game (a version of the dirty faces game).

> **The prisoners' disks game (a version of the dirty faces game):** N prisoners are in a room without a mirror. They all have a white or a black disk on their back. They can only observe the colours of the disks of the other prisoners and they are not allowed to talk. An outside person announces: "At least one person has a black disk. If you discover the colour of your disk, come in ten minutes and tell me the colour. If you are right, you will be released, if you are wrong, you will stay in jail forever. If you do not discover the colour, stay in the room. I will come back every ten minutes." If a prisoner doesn't go out, he spends one additional year in jail (his normal sentence).

Let the game talk: first, this game is rather complicated because it is a priori endless, and the number of pure strategies is infinite.

So let us represent in Figure 1.13 the game tree when N=2. As you can see, even for N=2, this game tree is a nightmare and necessarily incomplete (we have just represented two periods of decisions, three points at a node mean that the game goes on at the node).

Second, the game tree is complicated because it has to represent all possible actions including the inappropriate ones. The complicated game tree contrasts with the easy optimal behaviour of the players. When there are only two players, everybody goes out at the latest at period 2. If a player, say player 1, observes a white disk, she knows her disk is black and goes out, announcing the right colour. If she observes a black disk, and if the opponent goes out at period 1, she deduces that her disk is white (because the opponent could deduce that his disk is black only if hers is white); and if the opponent doesn't go out at period 1, this means that he was unable to deduce that his disk is black, which is only possible if hers is black too! So, without a complicated reasoning, everybody leaves the room latest at period 2. And so it would be silly to try to represent this infinite game in extensive form!

Third, this game, for any number of players, has a very special structure that will be exploited in the resolution of the game. To put it more precisely, even if the game is an imperfect information game, each player has, at the first period of play, an information set where he knows the colour of his disk. This set corresponds to the context where he sees only white disks, and so knows that he has a black disk. And so, if he is rational, he goes out announcing the right colour. And what is more, the information sets, when coupled with rational behaviour (namely going out when one knows the colour of the disk and staying in if not) lead in the following period to new information sets where some players will again go out, and so on, until everybody goes out!

Additionally, this special structure will allow us to propose a simple algorithm of behaviour so that even non clever players will go out in finite time, announcing the right colour.

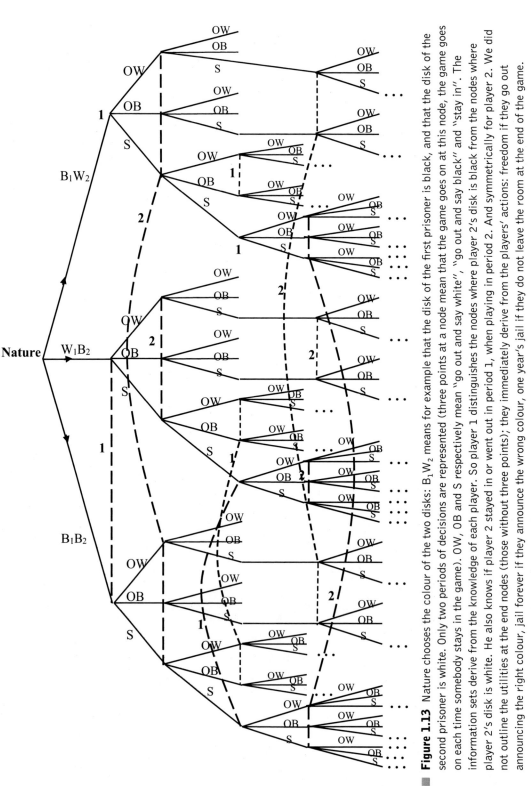

Figure 1.13 Nature chooses the colour of the two disks: B_1W_2 means for example that the disk of the first prisoner is black, and that the disk of the second prisoner is white. Only two periods of decisions are represented (three points at a node mean that the game goes on at this node, the game goes on each time somebody stays in the game). OW, OB and S respectively mean "go out and say white", "go out and say black" and "stay in". The information sets derive from the knowledge of each player. So player 1 distinguishes the nodes where player 2's disk is black from the nodes where player 2's disk is white. He also knows if player 2 stayed in or went out in period 1, when playing in period 2. And symmetrically for player 2. We did not outline the utilities at the end nodes (those without three points); they immediately derive from the players' actions: freedom if they go out announcing the right colour, one year's jail if they announce the wrong colour, jail forever if they do not leave the room at the end of the game.

3.2 Complete/incomplete information

> **Definition 11**: a game is of **incomplete information** if at least one player does not know some elements of the structure of the game. Otherwise the game is of complete information.

So if you ignore the range of possible strategies of your opponent, if you ignore the way an opponent judges the different profiles of strategies (you ignore his utilities), and if you ignore that actions may be repeated then you ignore part of the structure of the game, and the game has incomplete information.

Reality is filled with situations where you ignore part of the game you play. *I even wonder if this fact does not influence players in experimental sessions. For example, Johnson et al. (2002) have shown that some players do not examine the whole game tree (where each player plays three times) when taking their first decision. This is perhaps due to the fact that, in reality, we are used to making decisions without knowing the whole game. So some players just play as they do every day.*

Let us first illustrate incomplete information with an ascending all pay auction, with V=3 and M_2 (player 2's revenue)=2, but where player 2 ignores if player 1's revenue, M_1, is equal to 2 or 3.

In such a context, player 2 suffers from incomplete information in that he does not know player 1's strategy set (nor her payoffs). He doesn't know if player 1, after player 2's overbid 2, is able or not to overbid. So player 2 imperfectly knows the structure of the game. We represent in Figure 1.14a the game he tries to play.

Usually, it is quite impossible to solve a game if you don't know parts of its structure. Yet this is not always true, in that in some contexts players may find interesting strategies without knowing whole parts of the game, especially if they focus on some concepts of equilibria. For example, in the gift exchange game, you don't need to know the exact payoff functions to find the played actions. As soon as the employee's utility decreases in effort for each salary, and as soon as the employer's payoff decreases in salary for each effort, the only stable situation is such that the employer offers a low salary and the employee offers a low level of effort.

3.2.1 Common knowledge: does it make sense?

Nevertheless, usually, players try to escape from incomplete information. *They usually hurry to change incomplete information games into imperfect information games.* So, for example in the studied game, we suppose that player 2, at the beginning of the game, has prior information on M_1, given by a prior probability distribution, that assigns probability ρ to M_1=3 and 1–ρ to M_1=2. This

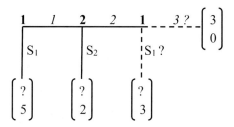

Figure 1.14a The branches are dashed because player 2 does not know if they exist. Player 2 also ignores the payoffs with a ?.

is of course a strong assumption. First, it means that player 2 knows the range of possible values M_1: in our case this means that the true M_1 is necessarily 2 or 3; it can't be 4 or 1 for example. So we exclude games such that player 2 works on completely wrong assumptions as regards M_1. If player 2 is badly informed, he automatically works on a large range of values, to be sure that the true value of M_1 is in the range. Technically, in many cases, this is not problematic, in that players can introduce large ranges of values as necessary. And this does not even necessarily complicate the solving of the game given that players may adopt the same behaviour for a large range of values. What is more problematic to my mind is the existence of a prior probability distribution on the range of possible values M_1, and, even more heroic, the fact that it is common knowledge of both players. What does common knowledge mean?

> **Definition 12**: a fact F is **common knowledge** (CK) of two players 1 and 2 if and only if:
> 1 knows F and 2 knows F
> 1 knows that 2 knows F and 2 knows that 1 knows F
> 1 knows that 2 knows that 1 knows F and 2 knows that 1 knows that 2 knows F
> ... and so on, up to an infinite chain of crossed levels of knowledge.

Why do we make such an assumption? Well, we mainly introduce it to be sure that the new well-defined game we construct is common knowledge for both players. As a matter of fact, if both players don't have the same game in mind, if they are not sure to play the same game, things become necessarily more complicated and they usually try to avoid this.

Does the assumption of a prior distribution CK of all the players make sense? In some contexts it is not stupid to suppose that the prior distribution is known by both players, namely if player 2's probability distribution is based on past play and on past experiences shared by both players for example. But this is of course not always the case. So does it make sense to work with such a new game? The answer depends on what you do with the game. When you look for perfect revelation Nash equilibria (studied in chapter 5), you do not utilize the prior distribution and so this strong assumption is of no impact. Other concepts, like pooling Nash equilibria, usually only require ranges of probabilities: for example, a given equilibrium may exist for ρ higher than ¼. So it is enough that it is common knowledge that this condition is fulfilled; the exact values of the different probabilities need not to be CK. So the assumption of CK of the prior distribution is strong, but, in many cases we do not need it to solve the game: in these cases, we don't have to bother about it. But in the other cases, we should not work with such a game.

Well, what is this new game? The game introduces the additional neutral player we already encountered in the past, Nature. What is its role? Nature begins the game: she plays a lottery on M_1 in accordance with the prior probabilities; she plays M_1=2 with probability 1–ρ and M_1=3 with probability ρ. *And then the true game begins, player 1 knowing the draw of Nature, whereas player 2 is uninformed of this draw.* The new game is given in Figure 1.14b.

As you can see, *the incomplete information has been replaced by imperfect information: player 2, when he plays, does not know the play of Nature.*

HOW TO BUILD A GAME

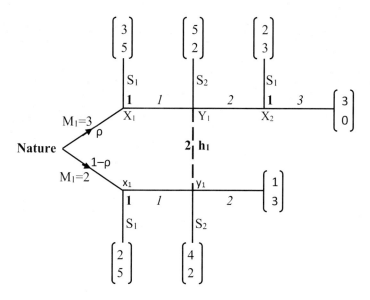

Figure 1.14b

3.2.2 Signalling games and screening games

The game we talked of above is similar to a special class of games called *signalling games*:

Definition 13: a **signalling game** is a two player game. The first player has private information on a parameter t, which can take a finite number of values t_i, $t_i \in T$, where T is a finite set.

t_i is usually called the *type of player 1*. Player 2 does not know player 1's private information but can assign a prior probability distribution on the elements of T, $\rho(.)$, which is CK of both players. Player 1 plays before player 2. She chooses a message m, which may depend on t_i. Player 2 observes player 1's message. This message may allow him to revise his probability distribution on T, which now becomes the posterior probability distribution $\mu(./m)$. Then player 2 chooses a response r that depends on m and the game stops.

Comments
- In a signalling game, the first player to play is Nature, who plays the lottery on the private value t, according the CK prior distribution $\rho(.)$.
- The game is called a signalling game in that the informed player plays before the uninformed one; with her message m, player 1 can try to reveal (or hide) player 2 information on her private information.

Signalling games have companion games: *screening games* where the uninformed player (player 2) plays before the informed one (player 1). Insurance games, where an insurer proposes a contract to a driver without knowing if he is a good or bad driver, are screening games. The insurer just has a prior probability distribution on the possible types of driver (based on statistics linked to the age

HOW TO BUILD A GAME

and experience of the driver, but also on the possible previous accidents of the driver and the nature of these accidents), and makes a decision (she proposes a type of contract to the driver); then the driver, who knows how he drives, accepts or refuses the contract. More generally, *principal agent problems* have links with signalling and screening games.

■ The game in Figure 1.14b is almost a signalling game. Let us suppose that, if $M_1=3$, player 1 overbids (it is the best way to play at X_2). So the game becomes the one in Figure 1.14c: this is a usual signalling game.

Let us discuss another signalling game, a game which we'll come back to in the next section and refer to in further chapters: the French plea bargaining game.[13]

> **The French plea bargaining game:** the game is a two player game, a defendant and a jury (of trial). The defendant can be of two types, innocent (I) or guilty (G); the defendant knows her type but the jury does not. The defendant (player 1) plays first. Either she pleads guilty (action P) and accepts the prosecutor's sentence (this sentence is fixed, it is codified, the prosecutor has no strategic (hence no) role in this game); or she chooses to go to trial (action $\bar{P}$). At trial, the jury (player 2) either releases the defendant or convicts her to a higher sentence than the one pronounced by the prosecutor.

We suppose that the jury, at the beginning of the game, has prior information on the defendant's type, given by a prior probability distribution that assigns probability ρ to guilt and $1-\rho$ to innocence. And we also suppose that this distribution is common knowledge. *But here this assumption is based on true facts*: the prior distribution is given by the information included in the defendant's record when she is placed under arrest. And it is CK that this record is known by both players, the

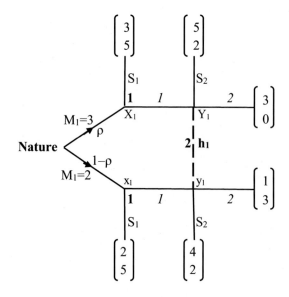

■ **Figure 1.14c**

HOW TO BUILD A GAME

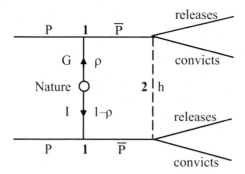

Figure 1.15 The initial node of the game tree is Nature's decision. The new disposition of the game tree is the standard one for signalling games.

defendant and the jury: even if the defendant does not exactly know the jury's prior distribution, she may have a good approximation of this distribution (she may be helped in this by a lawyer or by the prosecutor). So Nature begins the game by playing a lottery on guilt and innocence, which respects the prior distribution. Then the defendant chooses pleading guilty (P) or going to trial ($\bar{P}$). Then, if the defendant goes to trial, the jury decides to convict or to release the defendant. So we get the game in Figure 1.15.

You may observe that the utilities of this game are not given. This is not an omission; we discuss them in the next section.

3.3 Utilities

Let me say first that, to my mind, nothing is more complicated when writing a game than expressing the utilities of the players.

3.3.1 Taking risks into account

First of all there is the *VNM (Von Neumann Morgenstern) definition*. In order to work with the mean expected utilities (what we do as soon as we look for mixed strategies or behavioural strategies), we have to include the notion of risk in the utility. And this is usually a complicated task.

Usually the following story is told to explain how to write utilities:

Suppose that a player can get three payoffs, 100, 500 and 900. Clearly, he prefers 900 to 500 and 500 to 100. Let us write, w.l.o.g. u(100)=0 and u(900)=1. The player's preferences are VNM preferences if there exists a lottery on the worst and the best outcomes such that the player is indifferent between getting 500 and playing the lottery in Table 1.2.

Table 1.2 Lottery

Lottery	100	900
probability	1−p	p

HOW TO BUILD A GAME

The player is indifferent as soon as u(500)=Expected utility of the lottery=E(u (lottery))=(1–p) u(100)+pu(900)=p. Namely, if the player is risk adverse, he requires a rather high p. Given that 500=(100+900)/2, we say that the player is risk adverse if p>½; by contrast, he is a "player", i.e. a risk taker if p<½, and he is risk neutral if p=½.

Another way to say that a player is risk adverse is to say that he prefers getting the mean expected monetary value of the lottery rather than playing the lottery. So the utility of the expected monetary value of the lottery, i.e. u((1–p)100+p900)=u(100+800p), is higher than the expected utility of the lottery, i.e. u(500) (given the way the lottery is defined). It automatically follows that u(500)<u(100+800p), so, given that the utility function is growing in its argument, we get 500<100+800p, hence p>½ (which is consistent with the previous way to define risk adverse behaviour).

Let us come back to the envelope game.

After presenting the envelope game to my L3 students, I asked them to anonymously give me the maximal amount they would like to exchange if K=100€ and if K=100,000€. The mean amount (146 exploitable data) for K=100€, was 48€, and the mean amount for K=100,000€ was 33,256€. For K=100, only 37.7% proposed an amount lower than 50, 33.6% proposed the amount 50, and 28.7% proposed more than 50. For K=100,000, 65.7% proposed an amount lower than 50,000, 19.2% proposed the amount 50,000 and 15.1% an amount higher than 50,000.

It is difficult to exploit these data in terms of utilities, in that the students had to take into account the possible behaviour of the opponent. Yet be that as it may, because it was the first time my students played the game, many of them confessed to behaving as if the other player would always exchange his envelope, and almost all of them supposed that the values of the lotteries are uniformly distributed on [0, K], something I did not say.

Hence let us suppose, just in this setting, that all the students imagined that their opponent always accepts exchanging his envelope. So, when a student exchanges her envelope, she in fact agrees to play the lottery that fixes the amount of the other envelope. And given that the students imagined that the lottery is the uniform distribution on [0, K], they agreed to exchange the amount of the envelope against a lottery whose mean is K/2.

In other terms, for K=100, 37.7% prefer the mean expected payoff to the lottery, given that they are indifferent between playing the lottery and an amount lower than the mean expected value (50) (so they prefer the mean expected value of the lottery to the lottery): hence 37.7% are risk adverse. 33.6% are indifferent between the mean expected value of the lottery and playing the lottery, so they are risk neutral. And 28.7% are risk takers given that they are ready to pay more for playing the lottery than its mean expected monetary value.

For K=100,000, the percentages are completely different. 65.7% are risk adverse, 19.2% are risk neutral and only 15.1% are risk takers.

So clearly, risk aversion rises with the amount to lose (and to win). Of course the above interpretations are not right in that many students (about a 1/3 of them) rightly understood that the opponent may not exchange his envelope when it contains a high number. But nevertheless, even in that case, the results translate a higher risk aversion for higher values of K. So, if you are ready to exchange at most 33,256 ($\approx$K/3) when K=100,000 whereas you were ready to exchange 48 ($\approx$K/2) for K=100, this means that, either you are more risk adverse or you fear that the opponent is more risk adverse.

So, by and large, one expects more risk aversion when K grows and this has to be taken into account when writing the utilities.

3.3.2 Which is the game you have in mind?

Let us now suppose that VNM's requirements are checked. We are still far from knowing exactly how to write utilities, namely *because players usually play a different game than the one you think they play.*

Let us explain this claim:

Recall that my L3 students, by playing the first price sealed bid all pay auction, very often played 4 and 5 (one class played these actions with probability 29.7%, the other with probability 32.5%) albeit these strategies are bad from a payment point of view (the object they can get is only worth 3). So why did almost 1/3 of the students play in this way? *They explained to me that they wanted to win the auction (i.e. get the object): given that they perfectly understood that a reasonable opponent would not invest more than 3 in the object, they hoped to get it by playing 4 or 5, because they hoped that the opponent would play reasonably.* When I explained to them that they lose money by winning the auction with bid 4 or 5, they simply said that *their first aim was to win the object*. What does it mean? Simply that my students did not play the game I proposed to play: the utilities in Matrix 1.8a (or 1.8b) were not the utilities they had in mind. For many of my students it was more important to win the auction than to win money. Everything happens as if they value the object more than its real worth: and you can check that if you replace V=3 by V=6 in the calculi that lead to Matrix 1.8a, bidding 4 and 5 are no longer bad actions.

In other words, I am quite sure that if I proposed the Matrix 1.8a to my students without talking about an auction (just saying that the players win what is written in the matrix), almost no student would have played 4 or 5. It is the pleasure of winning an auction that drove the behaviour.

3.3.3 Fairness and reciprocity

My students, during subsequent lectures, learned to play games without adding personal feelings not written in the payoffs. They also learned the concepts of dominated strategies and Nash equilibrium. Then I asked them to play the second price sealed-bid all pay auction (Matrix 1.9a). Most of my students focused on actions that may appear in a pure strategy Nash equilibrium (0, 3, 4 and 5), but around 11% of them played 1 and 2. When I asked them why, they explained that they knew they had few chances to win the auction by playing 1 and 2, especially if many players play 3, 4 and 5, but that they wanted, *in case they lost the auction, that the winner had something to pay!* It was not possible for them to play 0 because, if they lost the auction, the winner would pay 0, something they felt was completely unfair! So, clearly, in many situations, *the utility of a player also depends on the payoffs of the others*.

And this is a strong social characteristic. Generally the utility we assign to a context does not only depend on the payoff we get, but also on the payoff obtained by the others. The most well-known game that highlights this fact is the ultimatum game. In this game, player 1 (who makes the offer) hesitates over asking too much for herself, because she fears that player 2 will not accept getting much less than she does. Clearly player 2 may be happier getting nothing rather than getting a small part of the sum. And even if this is not the case, player 1 may fear this fact and react accordingly; i.e. she may split the sum in a fair way. Many experimental papers worked on this topic. Some of them conclude on *a utility function that adds to the obtained amount a function that decreases in the difference between the amounts obtained by both players.*

Similar observations were observed for the gift exchange game. *Fairness is often associated with reciprocity.* By offering a higher salary, the employer increases the employee's utility, and so, by reciprocity, the employee offers a higher effort, which increases the employer's utility. Everything happens as if the utility of player i also depends on the utility of player j and vice versa. This has often been commented on in experimental papers on the ultimatum game, the gift exchange game and other games, like Berg et al.'s investment game.

It is interesting here to mention the way my L3 students played the gift exchange game given in Figure 1.16, during one of their first courses in game theory.

We talked about the payoffs for a while, and I insisted on the fact that each employee is systematically better off offering the lowest effort, regardless of the wage he gets, and then I asked them to take a paper and successively answer my three questions: what effort do you offer if I give you W_2? W_3? W_1? The students answered the first question before I asked the second, and they answered the second question before I asked the third. I proceeded in this way in order to avoid the students thinking of the three wages at the same time, which may influence and even induce the reciprocity; it is for the same reason that I chose the order W_2, W_3, W_1 and not the order W_1, W_2 and W_3 (see my remark in section 2.2.1). It is funny here to mention a question asked by the students several times during the play: they wanted to be sure of *keeping* their wage regardless of the effort they offered! This – funny – fear both *shows that they were ready to offer the lowest effort regardless of the wage*, but also that they felt uncomfortable, *even guilty*, by doing so. This shows the strength of true life in a play: in true life, if you work badly, your wage will shrink, and even reach the level 0, given that you will be fired. People cannot forget true life when they play a game; so for example some of my students systematically offered the highest effort and explained that they wanted to get a better job in the firm in the future. Yet, nevertheless, the results (114 students) – summarized in Table 1.3 – are contrasted. Clearly, reciprocity is a strong feature of behaviour given that the efforts clearly rise with the offered wage. More precisely, 45% of the students played a reciprocal strategy – 13.6% played the perfect reciprocal strategy (E_1/W_1, E_2/W_2, E_3/W_3), 31.4% played the partial reciprocal strategies (E_1/W_1, E_1/W_2, E_2/W_3), (E_1/W_1, E_1/W_2, E_3/W_3) or (E_1/W_1, E_2/W_2, E_2/W_3). 8.6% of the students offered even a higher than reciprocal effort – they played the strategies (E_2/W_1, E_2/W_2, E_3/W_3), (E_1/W_1, E_3/W_2, E_3/W_3) or (E_2/W_1, E_3/W_2, E_3/W_3). Yet, 30.7% played the "optimal" strategy (E_1/W_1, E_1/W_2, E_1/W_3), which shows that reciprocity is not systematic!

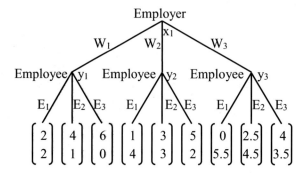

■ **Figure 1.16**

HOW TO BUILD A GAME

Table 1.3 % of students offering different efforts for different wages

	E_1	E_2	E_3
W_1	81.5%	12%	6.5%
W_2	52%	40%	8%
W_3	35%	34.3%	30.7%

3.3.4 Strategic feelings and paranoia

Well, *the need of reciprocity, more generally feelings, are not a problem if we know them:* in that case, we simply include them in the utilities in order to solve the right game. But *feelings are a problem if we do not know them.*

What is more, *feelings may be part of the strategic behaviour of a player*. Let us illustrate it by going back to the chicken game. The chicken game in reality generally involves more than the two drivers. It involves other players who bet on the winner before the race. So, if you swerve, you are not only the chicken (and it is not easy to be the chicken), but you have to suffer the angry reaction of the people who lost their bet, which can make your situation highly uncomfortable. And it may be so uncomfortable that some people prefer risking the accident rather than being the only driver that swerves.

If that is the case for player 1, the played game becomes the one in Matrix 1.11:

		Player 2 swerve	Player 2 straight
Player 1	swerve	(u,u')	(0,1)
	straight	(1,v')	(v,0)

Matrix 1.11

with 1>u>v>0 and 1>u'>v'>0.

It derives that player 2 can deduce that player 1 will never swerve, and this compels him (because he wants to live) to swerve. Yet this is highly profitable to player 1. So, even if player 1 does not prefer dying to swerving, it is interesting for her to give the impression that she prefers dying to swerving, in order to win the game and make herself and the people who bet on her (which may be friends) happy. So, in order to win, *you have to cheat with real feelings*: the friends that bet on you threaten to punish you if you lose, so you can pretend to be a person that prefers death to dishonour, even if your friends will never hurt you and if you have no intention to die. So this silly game will surely be won by the person who best gives the impression, before the game begins, of being compelled to win or to die.

But things go even further. As I asked my students how they would play in the game given in Matrix 1.12, where the payoffs are millions of dollars, I just expected them to say to me:

		Player 2 A_2	Player 2 B_2
Player 1	A_1	(9,9)	(9,0)
	B_1	(0,9)	(10,10)

Matrix 1.12

51

"I prefer playing A, because I get 9 million for sure whereas, if I play B, I get 10 million but only if the other guy plays B too; if he plays A, I get 0." I even expected that some of them would say: "even if I am not cautious myself, I fear that the other player is cautious, so he plays A. It follows that I have to play A, because it is the best answer to a cautious behaviour." But many of my students gave me a quite strange, unexpected answer. They said: "I play A because we live in a competitive world and I fear that the other guy is happy to see me lose a lot of money. So he may hope that I play B, which leads him to play A, so that I get 0." As we will see later, this completely paranoiac view of the game fits with max min strategies, but I was quite astonished to see young people hold such a pessimistic reasoning (my students are about 21 years old). *So many of my students clearly presume that a player's utility can rise with the loss of the opponent* (even in this coordination game with no conflict in the payoffs!); in other words, player 2 is supposed to prefer (0, 9) to (10, 10) because he enjoys seeing that player 1's payoff switches from 10 to 0. The loss he incurs in this switch (from 10 to 9) is not sufficient to counterbalance this pleasure.

A nice example that illustrates the danger of such utility functions is the prisoners' disks game (with two players). Normally, when you see a white disk, you immediately go out and say that you have a black disk. That is why the other player, on the one hand, when you go out, deduces that he has a white disk and goes out a period later announcing the right colour. On the other hand, if you do not go out, he deduces that you observed a black disk, so he goes out in the next period too, announcing the right colour too. *Well, that is right if everybody is only interested in their own utility. But what about a not so kind prisoner who would be happy to see the other in jail forever?* When she observes a white disk, she perfectly deduces that her disk is black, but stays one period longer in the room! At the next period, she goes out announcing the right colour. But the other prisoner, given that she only goes out at the second period, deduces that his disk is black and goes out, announcing the wrong colour! And even if the opponent is fair – i.e. has a utility that only depends on the number of years in jail, hence goes out as soon as she is able to discover the colour of her disk – *you may not be sure that she is fair*. So, either you are lucky because, either you observe a white disk (so you can go out, deducing that your disk is black) or you observe a black disk prisoner who immediately goes out (if so, provided your opponent is not ready to stay in jail forever only to mislead you, this means that she was able to discover that she has a black disk, which is only possible if you have a white disk). Or you stay in forever (because if the opponent only goes out at the second period, this does not mean that she was not able to do so in the first period). *Things are even worse if your opponent is ready to stay in jail forever only to mislead you*: then you can no longer go out at period 2 if you see your black disk opponent going out in period 1! So, when you make your deductions, you always have to guess up to what degree the other is just happy to see you in jail!

3.3.5 How to bypass utilities

Let us make an additional remark. Even if you may be unable to write the right utilities of a player, sometimes *you do not need to write them*. For example, in our story games in section 1.1.2, we were able to highlight the particularities of the games just by knowing the ranking of the payoffs.

More generally, it is not always necessary to write the exact utility functions of the players to solve the game. What matters is that *the obtained equilibria only depend on characteristics of the true sometimes complex and imperfectly known utility functions of the players.*

To illustrate this point let us come back to the plea-bargaining game.

HOW TO BUILD A GAME

Let us outline the payoffs of each player. We restrict the utilities to the interval [0, 1].

We assume that justice is fair in that we assume the jury is happy to release an innocent person and to convict a guilty one. So we say that the jury gets 1 when it releases an innocent person and convicts a guilty one, and that it gets 0 if it convicts an innocent person and if it releases a guilty one. Well, the jury may perhaps prefer releasing a guilty person rather than convicting an innocent one, or the contrary (they prefer convicting an innocent person rather than releasing a guilty one). Our utilities do not express these facts but, on the one hand, we do not know which among the two following relations is the jury's preference relation (Convict/Guilty, Release/Innocent)>(R/G, R/I)>(C/G, C/I)>(R/G, C/I) or (C/G, R/I)>(C/G, C/I)>(R/G, R/I)>(R/G, C/I)), and, on the other hand, these facts have no impact on the kind of equilibria we obtain. In other words, taking them into account would not change the nature of the obtained equilibria, so we can omit them.

We do not represent the prosecutor's payoff when a defendant pleads guilty, because this payoff has no impact on the actions played at equilibrium. A jurist may feel ill-at-ease with such a way of acting: as a matter of fact, many papers focus on the fact that the prosecutor's payoff grows with the prosecutor's sentence whereas some other papers claim that the prosecutor should appreciate that a guilty person pleads guilty because a trial requires a lot of time and sometimes does not allow, for lack of evidence, to convict the guilty person. Well, all these facts may be right, but the point is that they have no impact on the equilibria, in that the defendant's choice only depends on her own payoffs (which only depend on the level of the prosecutor's codified sentence and the decision at trial). In other words, it is not necessary to write a utility for the justice in case of pleading guilty.

But we have to focus our attention on the writing of the defendant's payoffs. These payoffs are difficult to establish. The defendant is a unique player, who is either guilty or innocent. So we have to rank the six payoffs that correspond to six contexts: guilty and plead guilty, innocent and plead guilty, guilty trial and released, guilty trial and convicted, innocent trial and released, and innocent trial and convicted. This is far from obvious. What is the best situation for the defendant? Is a person at trial happier if released and innocent, or if she is released and guilty? Clearly, many things depend on the defendant under study and her morality: a well-intentioned person is better off being released and innocent rather than being released but guilty of an involuntary act. By contrast, an ill-intentioned person, who wants to begin a new life in the tropics, is better off being released but having robbed the bank, rather than being released and innocent. So you may spend a lot of time trying to write a convincing utility function – and usually jurists do – without being sure that the chosen function is the good one.

That is why we prefer proceeding in another way. One more time, we *only focus on what has an impact on the equilibria*. So we only require that the chosen utilities are such that their impact on the equilibria is the right one; in other words, the equilibria we get should only depend on characteristics that correspond to the true unknown utility function of the defendant. *So only the strategic content of our utilities has to fit with the true partly unknown utilities.*

That is why we build the utilities as follows: we say that the defendant, either innocent or guilty, assigns the highest utility 1 to the fact of being released, and assigns 0 to the fact of being convicted at trial, but she judges differently, according to her nature, the sentence given by the prosecutor, in that we assume u>v, where u, respectively v, are the defendant's utility when she pleads guilty and is guilty, respectively innocent. This way of ranking utilities gives rise to the property expressed in Figure 1.17.

The fact that v is lower than u, when a defendant gets 1 if released at trial, whether innocent or guilty, and gets 0 if convicted at trial, whether innocent or guilty, implies that, for given probabilities

HOW TO BUILD A GAME

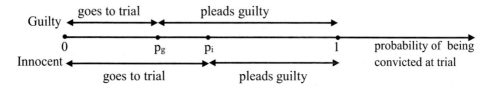

Figure 1.17

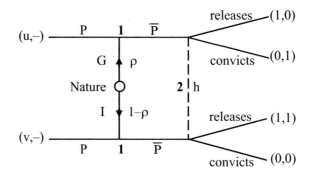

Figure 1.18

of being convicted at trial, the lowest probability such that the guilty individual prefers pleading guilty than going to trial is lower than the lowest probability for which an innocent individual prefers pleading guilty. In other words, all else being equal, it is easier for a guilty person to plead guilty than for an innocent one. This is the strategic fact induced by our way of writing the defendant's utilities. And we believe that this fact corresponds to the reality: we believe (and this belief is widely shared) that for given probabilities of being convicted at trial, the threshold probability that leads an innocent person to switch from going to trial to pleading guilty is larger than the threshold probability of a guilty defendant.

So, on the one hand, the chosen utilities are acceptable in that the impact they have on the equilibrium expresses a fact that is true (at least commonly accepted). On the other hand, given the diversity of humans, it is difficult to impose more restrictions on the utility functions: so we conclude that our utilities fairly translate the context.

And the game in Figure 1.15 becomes the one in Figure 1.18.

CONCLUSION

In the plea bargaining game, we see that what matters is that the description of the game contains the strategic salient characteristics of the utility functions. More generally, the structure of a game can be partly incomplete, providing that the missing part (strategies, utilities) has no impact on the equilibria of the true game. For example, in the envelope game, we did not outline the probability distribution on [0, K]; this is useless as long as we only study Nash equilibria, in that the Nash equilibrium does not depend on it.

According to the used concept of solution, many elements of a structure can be changed or omitted. For example, in the exercises, we propose a first price sealed bid auction (which is a simultaneous game) and a descending Dutch auction which is a long sequential game represented by a long game tree, but both games are identical with regard to the Nash equilibria. So the structure of a game has to be studied together with the game theory tools we use to solve it. It derives that to properly write the structure, it is necessary to know how to solve a game. And this is what we will try to do in the next chapters.

NOTES

1. We will indifferently use the term strategic form game or normal form game. For homogeneity, in this first section, we stick to the term 'strategic form', but in later sections and chapters we use the term 'normal form' more.
2. Vogel, C., April 29, 2005. Rock, paper, payoff: child's play wins auction house an art sale, *The New York Times*.
3. Selten R., 1975. Reexamination of the perfectness concept for equilibrium points in extensive games, *International Journal of Game Theory*, 4, 25–55.
4. See for example Fehr, E. and Schmidt, K., 1999. A theory of fairness, competition, and cooperation, *Quarterly Journal of Economics*, 114 (3), 817–868.
5. Berg, J., Dickhaut, J. and McCabe, K., 1995. Trust, reciprocity and social history, *Games and Economic Behavior*, 10, 122–142.
6. See for example Dufwenberg, M. and Kirchsteiger, G., 2004. A theory of sequential reciprocity, *Games and Economic Behavior*, 47 (2), 268–298.
7. Shubik, M. 1971. The dollar auction game: a paradox in noncooperative behavior and escalation, *The Journal of Conflict Resolution*, 15 (1), 109–111.
8. Rosenthal, R., 1981. Games of perfect information, predatory pricing, and the chain store paradox, *Journal of Economic Theory*, 25, 92–100.
9. Maximiano, S., Sloof, R. and Sonnemans, J., 2013. Gift exchange and the separation of ownership and control, *Games and Economic Behavior*, 27, 41–60.
10. Johnson, E.J., Camerer, C., Sen, S. and Rymon, T., 2002. Detecting failures of backward induction: monitoring information search in sequential bargaining, *Journal of Economic Theory*, 104, 16–47.
11. Kuhn, H.W., 1953. Extensive games and the problem of information, in H.W. Kuhn and A.W. Tucker (eds), *Contribution to the theory of games*, Vol. II, Annals of Mathematics Studies, 28, pp. 193–216, Princeton, Princeton University Press.
12. See Section 3.1 for the meaning of perfect recall.
13. We call it French plea bargaining, because in France, plea bargaining only applies to offences such that the prosecutor can at most require imprisonment for a term not exceeding one year, which explains why his role is rather codified.

Chapter 1

Exercises

♣ EXERCISE 1 EASY STRATEGY SETS

Questions

1. In the next chapters we will study the **traveller's dilemma game** (Basu, 1994),[1] defined as follows:

    Two travellers have identical luggage (they bought it at the same price and they value it in the same way) unfortunately damaged by an airline company. The company agrees to pay for the damage and proposes the following game. First each traveller i=1, 2 asks for a given amount x_i (without knowing the amount proposed by the other traveller), where x_i has to be an integer in [2, 100]. Second, the company observes the two offers and reimburses the travellers in the following way:

    - if $x_1=x_2$, each traveller gets x_1
    - if $x_i>x_j$ i≠j=1, 2, the traveller j gets x_j+P, where P is a fixed amount higher than 1, and the traveller i only gets max (0, x_j–P) (in Basu's original version, he simply gets x_j–P, which can be negative).

    So the company wants to reward the traveller who asks for the lowest amount and to "punish" the one who asks "too much". The aim is to induce the two travellers to not ask for too high amounts. Write the pure strategy sets S_1 and S_2 of traveller 1 and traveller 2.

2. Consider **Bertrand's duopoly game**.

    Two firms set a price for an homogenous good. Firm 1 fixes the price p_1, firm 2 the price p_2. D(p) is the demand for a price p. D(p)=max(0, K–ap), where K and a are two positive constants. c(<K/a) is the marginal (and mean) cost of production (the same for both firms).

    $u_i(p_1, p_2)$ is firm i's profit function (utility), i=1, 2.

    - if $p_1=p_2$, both firms share the demand: $u_1(p_1, p_2)$=max(0, K–ap_1)(p_1–c)/2=$u_2(p_1, p_2)$
    - if $p_1>p_2$, the demand goes to the firm 2, hence: $u_1(p_1, p_2)$=0, $u_2(p_1, p_2)$= max(0, K–ap_2)(p_2–c)
    - if $p_1<p_2$, $u_2(p_1, p_2)$=0, $u_1(p_1, p_2)$=max(0, K–ap_1)(p_1–c)

    Write the pure strategy sets S_1 and S_2 of firm 1 and firm 2, when prices are real numbers.

3. Consider **the guessing game**. It is an N player game.

   Each player has to choose a real number in [0, 100]. The winner is the player closest to p times the average of all chosen numbers, where p belongs to]0, 1[. In case of a tie, all the winners share equally the amount of the win. Write the pure strategy set S_i of each player i, i from 1 to N.

♣ EXERCISE 2 CHILDREN GAME, HOW TO SHARE A PIE EQUALLY

Consider a pie, a mother and her two daughters; each daughter always fears that the other may get a bigger piece of pie.

Questions
Find a game the mother can propose to her daughters so that each child gets exactly half of the pie and is sure to get it.

♣ ♣ EXERCISE 3 CHILDREN GAME, THE BAZOOKA GAME, AN ENDLESS GAME

The bazooka game is a two children game. It's a mime game. At the first stage (stage 0), the children mime how they load their bazooka two times. Then, at stage 1, they can:

- Shoot (action S), in which case they shout "Bazooka!" (or it's not so funny)
- Load the bazooka a third time (action L)
- Protect themselves (action P)

Of course these three actions correspond to different mimes.

   If a child loads when the other shoots, she is dead, so has lost the game. If a child loads whereas the other protects herself, nothing happens but, at the next stage, the child who loaded three times has a 'superbazooka' and kills the other (who has only loaded up two times): so the child who loads wins the game. If a child shoots whereas the other protects herself, nothing happens but, at the next stage, the child who protected herself wins the game, because her bazooka is still loaded and she kills the other (whose bazooka is discharged). If both children load, the game begins again at stage 0. If both shoot, the game begins again at stage 0. If both protect themselves the game begins again at stage 1.

Questions
1. Write the extensive form of the game.
2. Write the normal form of the game, by calling g_1 and g_2 the payoff player 1 and player 2 get in the game.

♣ EXERCISE 4 SYNDICATE GAME: WHO WILL BE THE FREE RIDER?

Consider a syndicate game with three workers. Suppose that at least two workers have to join the syndicate for the syndicate to be created. When the syndicate exists, it is more profitable to benefit from the work done by the members in the syndicate than to join the syndicate (free rider problem). Yet the worst situation is the one where there is no syndicate.

EXERCISES

Questions
1. Give the strategy set and the preference relation of each player.
2. Write the game in normal form.
3. Let the game talk.

♣ EXERCISE 5 ROSENTHAL'S CENTIPEDE (PRE-EMPTION) GAME, REDUCED NORMAL FORM GAME

Consider the pre-emption game depicted in Figure E1.1.

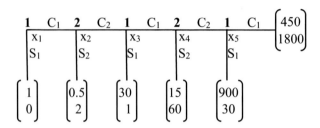

Figure E1.1

Questions
1. Let the game talk. Give Card S_i, i=1, 2. Then write the game in normal form.
2. Find the equivalent strategies and give the reduced normal form game.

♣♣ EXERCISE 6 DUEL GUESSING GAME, A ZERO SUM GAME

Two players play a guessing game. They have to find the name of a hidden item thanks to clues that are progressively revealed to both players. The game goes as follows. In period 1, a first clue is revealed to both players and it is player 1's turn to play. She chooses to answer (i.e. give a name for the hidden item) or to pass. If she gives the good answer (R), she wins the game (payoff 1) and player 2 loses the game (payoff 0). If she gives a wrong answer (W), she loses the game and player 2 wins the game. If she passes, a second clue is revealed to both players and it is player 2's turn to play. Either he answers or he passes. If he gives the right answer he wins the game and player 1 loses the game; if he gives a wrong answer he loses the game and player 1 wins the game. If he passes, a third clue is revealed to both players and it is player 1's turn to play. And so on. At each period, a new clue is revealed to both players. At each odd (even) period it is player 1's (player 2's) turn to play. Each player can answer or pass: if he gives the good answer he wins the game; if not he loses it. If he passes, the game goes on with a new clue. The probability to guess the name of the item grows in the number of clues and a player is supposed to answer when the probability to give the right answer is equal to 1.

Questions
1. We suppose that both players have the same ability to guess and that each new clue increases the probability to give the good answer by 0.1. So we get the data in Table E1.1. Give the extensive form of the game. How many pure strategies does player 1 have? Player 2?

2. What are the equivalent strategies? Give the reduced normal form game.
3. Now suppose that player 1 guesses better than player 2, so that an additional clue increases her probability to give the good answer by 0.15 (Table E1.2). Write the new extensive form game.

Table E1.1

Number of clues	Probability of giving the right answer
1	0.1
2	0.2
3	0.3
4	0.4
5	0.5
6	0.6
7	0.7
8	0.8
9	0.9
10	1

Table E1.2

Number of clues	Probability that player 1 gives the right answer	Probability that player 2 gives the right answer
1	0.15	0.1
2	0.30	0.2
3	0.45	0.3
4	0.60	0.4
5	0.75	0.5
6	0.90	0.6
7	1	0.7
8	1	0.8
9	1	0.9
10	1	1

♣♣ EXERCISE 7 AKERLOF'S LEMON CAR, EXPERIENCE GOOD MODEL, SWITCHING FROM AN EXTENSIVE FORM SIGNALLING GAME TO ITS NORMAL FORM

Consider an experience good market, close to Akerlof's[2] model. A seller wants to sell a car to a buyer. The car can be of two different qualities t_1 and t_2, with $t_1<t_2$. The seller's reservation price for a good of quality t_i is h_i, i=1, 2, with $h_1<h_2$. The seller sets a price for her good. The buyer observes the price and accepts (A) or refuses (R) the transaction. The buyer's reservation price for a good of quality t_i is H_i, i=1, 2 with $H_1<H_2$. The buyer ignores the quality during the transaction, but has a prior probability distribution over the qualities, that is common knowledge of both players; the probability distribution assigns probability ρ_i to the quality t_i, with $0<\rho_i<1$, i=1, 2 and $\rho_1+\rho_2=1$. It is assumed that $H_2>h_2>H_1>h_1$ and $\rho_1H_1+\rho_2H_2<h_2$.

The buyer's payoff, when he buys with probability q a car of quality t_i at price p, is $(H_i-p)q$. The seller's payoff is $qp+(1-q)h_i=q(p-h_i)+h_i$, i=1, 2. Given that the constant h_i appears in all the payoffs, we can suppress it.

We here work with the values $h_1=40$, $H_1=60$, $h_2=100$, $H_2=120$, $\rho_1=0.4$, $\rho_2=0.6$.

Questions
1. Give the extensive form game of this experience good model, when the seller can only set two prices, 55 and 105.
2. Give the pure strategies of each player. Write the game in normal form.

EXERCISES

♣♣♣ EXERCISE 8 BEHAVIOURAL STRATEGIES AND MIXED STRATEGIES, HOW TO SWITCH FROM THE FIRST TO THE SECOND AND VICE VERSA?

Questions
1. Consider the game in Figure E1.2. Give the pure strategies of player 1 and player 2. Consider a profile of mixed strategies. Show that it is possible to find a profile of equivalent behavioural strategies, i.e. a profile of behavioural strategies that leads to the same payoffs than the profile of mixed strategies.

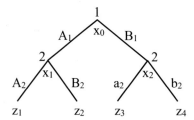

Figure E1.2

2. Consider a profile of behavioural strategies. Show that it is possible to find a profile of equivalent mixed strategies, i.e. a profile of mixed strategies that leads to the same payoffs than the profile of behavioural strategies. Comment.
3. First application: Consider the pre-emption game in Figure E1.1, and the mixed strategies:
$p_1(S_1/x_1, S_1/x_3, S_1/x_5)=p_1(S_1/x_1, S_1/x_3, C_1/x_5)=p_1(S_1/x_1, C_1/x_3, S_1/x_5)=p_1(S_1/x_1, C_1/x_3, C_1/x_5)=p_1(C_1/x_1, S_1/x_3, S_1/x_5)=p_1(C_1/x_1, S_1/x_3, C_1/x_5)=p_1(C_1/x_1, C_1/x_3, S_1/x_5)=1/7$, $p_1(C_1/x_1, C_1/x_3, C_1/x_5)=0$ and $p_2(S_2/x_2, S_2/x_4)=p_2(S_2/x_2, C_2/x_4)=p_2(C_2/x_2, S_2/x_4)=2/7$, $p_2(C_2/x_2, C_2/x_4)=1/7$.

Find the profile of behavioural strategies that leads to the same payoffs.
4. Second application: now consider the reduced normal form of this pre-emption game in Matrix A1.3b. Consider the mixed strategies in this reduced normal form given by:
$p_1(S_1/x_1)=½$, $p_1(C_1/x_1 S_1/x_3)=1/3$, $p_1(C_1/x_1, C_1/x_3, S_1/x_5)=1/6$, $p_1(C_1/x_1, C_1/x_3, C_1/x_5)=0$
$p_2(S_2/x_2)=½$, $p_2(C_2/x_2, S_2/x_4)=1/3$, $p_2(C_2/x_2, C_2/x_4)=1/6$. Show that it is possible to find, in the extensive form game, a profile of behavioural strategies that leads to the same payoffs.

The answers to questions 3 and 4 will be used later, in an exercise of chapter 6.

♣♣♣ EXERCISE 9 DUTCH AUCTION AND FIRST PRICE SEALED BID AUCTION, STRATEGIC EQUIVALENCE OF DIFFERENT GAMES

Two players (J1 and J2) bid for an object whose value, for each player, is distributed on [0, V]. Each player knows the value he assigns to the object but ignores the opponent's value (except its distribution).

In a first price sealed bid auction (FPSBA), each player proposes a price for the good, without knowing the price proposed by the other bidder. The player with the highest bid gets the object and pays the bid he proposed. If both make the same bid, each gets the object and pays his bid with probability ½.

In a Dutch auction the auctioneer proposes a highest possible price, say V. Each player can accept or refuse this price. If one bidder accepts, he pays V, gets the object and the auction stops. If both accept, each gets the object with probability ½, pays V if he gets the object, and the auction stops. If both refuse, the auctioneer proposes the immediate lower price, V–0.01. If one bidder accepts, he pays V–0.01, gets the object and the auction stops. If both accept, each gets the object with probability ½, pays V–0.01 if he gets the object, and the auction stops. If both refuse, the auctioneer proposes the immediate lower price, V–0.02, and so on, till the auctioneer proposes 0, which is necessarily accepted by both bidders.

In this exercise, we simplify the context. We suppose that for each player, the object is of value 0, 1 or 2, each with a positive probability. In the first price sealed bid auction, the bidders can bid 0, 1 or 2 (we suppose that they never bid more than the value they assign to the object). In the Dutch auction, the auctioneer first proposes 2, then 1 if necessary, then 0 if necessary.

Questions

1. Give the extensive form games of the first price sealed bid and the Dutch auctions.
2. Give the size of the pure strategy set of each player in both games and propose a strategy for player 1 in each game. Comment. Why do these games seem different?
3. Count the number of terminal nodes in both games. Why are these numbers different? What information do we lose when switching from the FPSBA to the Dutch auction?
4. We will see in Exercise 28 (chapter 3) that these two games lead to the same Nash equilibria. Comment.

NOTES

1 Basu, K., 1994. The traveler's dilemma: paradoxes of rationality in game theory, *American Economic Review*, 84 (2), 391–395.
2 Akerlof, G.A., 1970. The market for "lemons": quality uncertainty and the market mechanism. *The Quarterly Journal of Economics*, 84, (3), 488–500.

Chapter 2

Dominance

INTRODUCTION

Dominance is an old and rather intuitive concept, and it is surely one of the best accepted concepts in game theory, at least to some extent. The idea is simple: if, for a given player, a strategy α leads to a better payoff than a strategy β, regardless of the way other people play, then α dominates β, and logically, the player should never play β, given that he always gets more with α. Yet things are far from being obvious. Suppose that another strategy δ ensures a lower payoff than a strategy γ, except if strategy β is played: if you agree that β should not be played – i.e. if you eliminate β from the game – the δ becomes dominated by γ: we will say that it is iteratively dominated. So you might eliminate δ too. But is this really a good idea?

In this chapter we try to capture the intuitive side but also the unintuitive side of dominance. So we first discuss games where dominance does a good job, i.e. leads to a result that fits with something which seems natural. Usually, non iterated dominance is easily accepted, even if we have to mention some exceptions. I would say that even some iterated dominance seems natural and leads to a behaviour that partly fits with intuition. But, if one needs many iterations, things become less intuitive, in that the obtained strategies rest too heavily on an assumption of crossed rationality, which is completely strange.

In the first section, we define dominance and apply it to all pay auctions and to the Fort Boyard sticks game. In this section dominance appears as a natural tool most people agree with. In section 2 we turn to iterated dominance and we apply it to the ascending all pay auction. Here results are more mitigated but still plausible. Things change in the third section where we show that iterated dominance may lead to strange phenomena. We mainly refer to the envelope game. We show that the strange result obtained derives from the high degree of crossed rationality required by iterated elimination. Yet despite this flaw, we show in section 4 that dominance is a quite powerful tool that can lead to optimal behaviour in some complicated games. We show that in the prisoners' disks game, iterated strict dominance leads to suppressing branches in the game tree so as to progressively get a structure that naturally leads to the optimal behaviour. We conclude with experimental observations about a game similar to the Fort Boyard sticks game.

DOMINANCE

1 NON ITERATED DOMINANCE

1.1 Definitions, strict and weak dominance

Dominance is defined in both mixed and behavioural strategies and may be strict or weak.

Definition 1. Strict dominance

A mixed strategy p_i, $p_i \in P_i$, i from 1 to N, is strictly dominated by another strategy $p_i' \in P_i$ if:

$$\forall p_{-i} \in P_{-i} \quad u_i(p_i', p_{-i}) > u_i(p_i, p_{-i})$$

A behavioural strategy $\pi_i \in \Pi_i$, i from 1 to N, is strictly dominated by another strategy $\pi_i' \in \Pi_i$ if: $\forall \pi_{-i} \in \Pi_{-i} \quad u_i(\pi_i', \pi_{-i}) > u_i(\pi_i, \pi_{-i})$

Definition 2. Weak dominance

A mixed strategy p_i, $p_i \in P_i$, i from 1 to N, is weakly dominated by another strategy $p_i' \in P_i$ if:

$$\forall p_{-i} \in P_{-i} \quad u_i(p_i', p_{-i}) \geq u_i(p_i, p_{-i})$$

And $\exists\, p_{-i} \in P_{-i} / u_i(p_i', p_{-i}) > u_i(p_i, p_{-i})$

A behavioural strategy $\pi_i \in \Pi_i$, i from 1 to N, is weakly dominated by another strategy $\pi_i' \in \Pi_i$ if:

$$\forall \pi_{-i} \in \Pi_{-i} \quad u_i(\pi_i', \pi_{-i}) \geq u_i(\pi_i, \pi_{-i})$$

And $\exists\, \pi_{-i} \in \Pi_{-i} / u_i(\pi_i', \pi_{-i}) > u_i(\pi_i, \pi_{-i})$

---

p_{-i} and π_{-i} are profiles of strategies of all the players except player i. So p_i is strictly dominated by p_i' if player i gets a lower utility with p_i than with p_i' whatever the other players play. It is weakly dominated by it, if he gets a lower or the same utility with p_i than with p_i' whatever the other players play, and if he prefers playing p_i' than p_i for at least one profile of strategies of the other players (p_i and p_i' can be replaced with π_i and π_i').

1.2 Dominance in normal form games

1.2.1 Existence of dominated strategies

First observe that in many games, *there are no dominated strategies*.

In the chicken game for example (see chapter 1, section 1.1.2), recalled in Matrix 2.1, there is neither a strictly nor a weakly dominated strategy.

		Player 2 A	Player 2 B
Player 1	A	(u, u')	(v, 1)
	B	(1, v')	(0, 0)

Matrix 2.1

DOMINANCE

with 1>u>v>0 and 1>u'>v'>0 (A means "swerve", B means "go straight").

As a matter of fact, for player 1, A is neither strictly nor weakly dominated by B, because A leads to a higher payoff than B when the opponent plays B (v>0) (she is better off swerving than going straight when the opponent goes straight). And B is neither strictly nor weakly dominated by A, because B leads to a higher payoff than A when the opponent plays A (1>u) (she is better off going straight than swerving when the opponent swerves).
And by symmetry the same is true for player 2.

This fact is very common: in many economic, law or political situations, dominance is of no help because there is no dominated strategy: in the French plea bargaining game for example, "plead guilty" is not dominated by "going to trial" because pleading guilty is the defendant's best strategy when she is convicted at trial, and "going to trial" is not dominated by "plead guilty" because going to trial is the defendant's best strategy if she is released at trial.

1.2.2 Normal form games with dominated strategies

Let us illustrate the dominance concepts on the first price sealed bid all pay auction game introduced in chapter 1 (section 1.1.3). The normal form of this game, for V=3, M=5 is recalled in Matrix 2.2a.

		Player 2					
		0	1	2	3	4	5
Player 1	0	(6.5,6.5)	(5,7)	(5,6)	(5,5)	(5,4)	(5,3)
	1	(7,5)	(5.5,5.5)	(4,6)	(4,5)	(4,4)	(4,3)
	2	(6,5)	(6,4)	(4.5,4.5)	(3,5)	(3,4)	(3,3)
	3	(5,5)	(5,4)	(5,3)	(3.5,3.5)	(2,4)	(2,3)
	4	(4,5)	(4,4)	(4,3)	(4,2)	(2.5,2.5)	(1,3)
	5	(3,5)	(3,4)	(3,3)	(3,2)	(3,1)	(1.5,1.5)

Matrix 2.2a

For player 1, the strategies "bid 4" and "bid 5" are strictly dominated by the strategy "bid 0", because, regardless of the strategy played by player 2, bidding 0 leads to the payoff 5 or 6.5, whereas bidding 4 or 5 leads to a payoff lower than 5. Symmetrically, the strategies "bid 4" and "bid 5" are strictly dominated by the strategy "bid 0" for player 2, and *it is not difficult to agree with the fact that there is no reason to play 4 or 5.*

> ### ➤ STUDENTS' INSERT
>
> *To prove that a strategy is strictly dominated by another one, it is enough to show that it leads to a lower payoff regardless of the pure strategies played by the other players (because it then automatically leads to a lower payoff for any profile of mixed (behavioural) strategies of the other players):* $u_1(0,0)=6.5>u_1(4,0)=4$, $u_1(0,1)=5>u_1(4,1)=4$, $u_1(0,2)=5>u_1(4,2)=4$, $u_1(0,3)=5>u_1(4,3)=4$, $u_1(0,4)=5>u_1(4,4)=2.5$ *and* $u_1(0,5)=5>u_1(4,5)=1$. *So, for player 1, bidding 4 is strictly dominated by bidding 0.*

DOMINANCE

Bidding 3 is weakly dominated by bidding 0, in that both strategies lead to a payoff 5 if the opponent bids 1 or 2 (or a mixed strategy with these two bids in the support), but bidding 3 leads to a lower payoff than bidding 0 for any other strategy of the opponent. *Usually, people also agree that there is no reason to bid 3.*

> **STUDENTS' INSERT**
>
> To prove that a strategy is weakly dominated by another one, it is enough to show that it leads to a lower or equal payoff regardless of the **pure strategies** played by the other players (because it then automatically leads to a lower or equal payoff for any profile of mixed (behavioural) strategies), and that there exists at least one profile of **pure strategies** of the other players for which it leads to a lower payoff.
> Hence $u_1(0,0)=6.5>u_1(3,0)=5$, $u_1(0,1)=5=u_1(3,1)$, $u_1(0,2)=5=u_1(3,2)$, $u_1(0,3)=5>u_1(3,3)=3.5$, $u_1(0,4)=5>u_1(3,4)=2$, $u_1(0,5)=5>u_1(3,5)=2$. So, for player 1, bidding 3 is weakly dominated by bidding 0.

Consequently, for most people, it seems reasonable to restrict attention to a smaller game, depicted in Matrix 2.2b:

		Player 2		
		0	1	2
Player 1	0	(6.5,6.5)	(5,7)	(5,6)
	1	(7,5)	(5.5,5.5)	(4,6)
	2	(6,5)	(6,4)	(4.5,4.5)

Matrix 2.2b

Yet eliminating weakly dominated strategies is not always obvious. Consider for example the game in Matrix 2.3.

		Player 2	
		A_2	B_2
Player 1	A_1	(1,1)	(100,100)
	B_1	(2,2)	(100,0)

Matrix 2.3

A_1 is weakly dominated by B_1, in that it leads to the same payoff for player 1 if player 2 plays B_2 (100=100) but it leads to a lower payoff if player 2 plays A_2 (1<2). Yet, if player 2 is sure that player 1 will not play A_1, he has no motivation to play B_2, and probably both players get, at the end, the payoff 2. *But should player 2 really be sure that player 1 eliminates A_1?* Player 1, like player 2, is better off with a payoff 100; and she knows that player 2 is motivated to play B_2, to get 100, if he is sure that player 1 plays A_1. And player 1 is pleased to play A_1 if player 2 plays B_2, because A_1, in this case, leads to her highest payoff (B_1 does no better). Consequently, not only should player 1 not eliminate A_1, but she should be able to convince player 2 that she plays the weakly dominated strategy. This would be optimal for both players.

Some people even criticize the elimination of strictly dominated strategies. This is especially true for the prisoner's dilemma. In this game, reproduced in Matrices 2.4a, 2.4b and 2.4c (three

65

DOMINANCE

variants of the same game), denouncing strictly dominates cooperating because a player always gets more by denouncing than by cooperating, regardless of the behaviour of the opponent. Yet eliminating cooperation leads both players to a social suboptimal solution, and this has often been criticized, namely in experimental game theory. *Well, is this criticism acceptable?*

A first comment is that each player observes that he is better off denouncing the opponent, regardless of the opponent's behaviour. And each player observes that the opponent is also better off denouncing, regardless of his behaviour. And of course, if the opponent denounces, it is all the more better to denounce him too. To my mind, at least in Matrix 2.4a, these facts are enough to incite to denounce, so to reject the criticism.

Player 2

	C_2	D_2
C_1	(2,2)	(0,3)
D_1	(3,0)	(1.9,1.9)

Player 1

Matrix 2.4a

Player 2

	C_2	D_2
C_1	(7,7)	(0,7.1)
D_1	(7.1,0)	(4,4)

Player 1

Matrix 2.4b

Player 2

	C_2	D_2
C_1	(7,7)	(0,7.1)
D_1	(7.1,0)	(0.1,0.1)

Player 1

Matrix 2.4c

A second comment is that, despite the games in Matrices 2.4b and 2.4c being strategically (from a dominance point of view) equivalent to the game in Matrix 2.4a, we would not be surprised that, *even without introducing feelings – remember that they are supposed to be included in the payoffs* – more people cooperate in these games for a reason not taken into account by the dominance concept: *the difference in payoffs*. So in Matrix 2.4a, denouncing, as compared to cooperating, always leads to a payoff that is at least 50% higher than the payoff achieved by cooperating (3>>2, and 1.9>>0). In Matrix 2.4b and 2.4c, this is not true.

Does a player really take into account a difference in payoffs equal to 1 or 2%? If people could talk before playing, I am quite sure that they may agree on (C_1,C_2) in both Matrices 2.4b and 2.4c (and especially Matrix 2.4c), because they may think that the additional payoff obtained by deviating is not enough to destroy their efforts to converge on the high payoffs (7,7).

Yet there is a difference between Matrix 2.4b and Matrix 2.4c. With regard to the game in Matrix 2.4b, *I may perhaps still denounce* for the following reason. The difference between 7 and 7.1 is not large enough for me to destroy the coordination on (C_1,C_2): *7 is enough for me*. But I am not sure that the other player shares the same point of view! He might play D, because D leads to higher payoffs than C, regardless of my behaviour. So, by playing C, I would get 0 instead of the payoff 4 I get by playing D, and 4–0 is an important difference. What is more, *it is enough to fear*

DOMINANCE

that the opponent fears that I play D, to be induced to play D, because if he fears that I play D, he plays D ...

This is not true in Matrix 2.4c. In Matrix 2.4c, even if you fear that the opponent plays D, this does not necessarily motivate you to play D. As a matter of fact, if you cooperate, either the opponent also cooperates and you are happy (because the difference between 7 and 7.1 is not enough to destroy the coordination on (C_1,C_2)), or he denounces in which case you do not lose enough to be deceived (because cooperating leads to the payoff 0 and denouncing leads to 0.1, i.e. almost the same payoff).

So, despite the three situations in Matrices 2.4a, 2.4b and 2.4c being strategically equivalent – they are three prisoners' dilemmas – I am not sure that they will be played in the same way in reality because of a fact which is not often taken into account in game theory: a loose appreciation of small payoff differences.

1.3 Dominance in extensive form games

1.3.1 Strict and weak dominance in the ascending all pay auction/war of attrition game

We now apply dominance to extensive form games.

Consider the ascending all pay auction game, with V=10 and M=7, depicted in Figure 2.1.

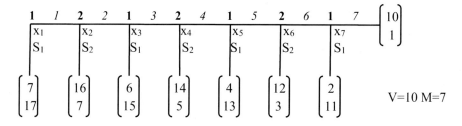

Figure 2.1

In this game, $(1/x_1, 3/x_3, 5/x_5, 7/x_7)$ strictly dominates any strategy $(S_1/x_1, */x_3, */x_5, */x_7)$, where * is any available action, and weakly dominates any strategy $(1/x_1, S_1/x_3, */x_5, */x_7)$, $(1/x_1, 3/x_3, S_1/x_5, */x_7)$ and $(1/x_1, 3/x_3, 5/x_5, S_1/x_7)$.

Let us justify it:

> ### ➤ STUDENTS' INSERT
>
> $(1/x_1, 3/x_3, 5/x_5, 7/x_7)$ at worst leads to the payoff 10 (when player 2 goes on at each decision node), whereas $(S_1/x_1, */x_3, */x_5, */x_7)$ leads to the payoff 7(<10) regardless of player 2's strategy. So $(1/x_1, 3/x_3, 5/x_5, 7/x_7)$ strictly dominates $(S_1/x_1, */x_3, */x_5, */x_7)$.
>
> To prove that $(1/x_1, 3/x_3, 5/x_5, 7/x_7)$ weakly dominates any strategy $(1/x_1, S_1/x_3, */x_5, */x_7)$, first observe that all the strategies lead to the same payoff 16 if player 2 stops at x_2. Then observe that, as soon as player 2 goes on at x_2, the strategies $(1/x_1, S_1/x_3, */x_5, */x_7)$ lead to the payoff 6, whereas $(1/x_1,3/x_3,5/x_5,7/x_7)$ leads at worst to the payoff 10 (>6)(when player 2 goes on at each further decision node). So $(1/x_1,3/x_3,5/x_5,7/x_7)$ weakly dominates the strategies $(1/x_1,S_1/x_3, */x_5, */x_7)$.

DOMINANCE

> In a similar way, to prove that $(1/x_1,3/x_3,5/x_5,7/x_7)$ weakly dominates any strategy $(1/x_1, 3/x_3, S_1/x_5, */x_7)$, first observe that all the strategies lead to the same payoff if player 2 stops at x_2 or at x_4. Then observe that, if player 2 goes on at x_2 and x_4, the strategies $(1/x_1,3/x_3,S_1/x_5,*/x_7)$ lead to the payoff 4, whereas $(1/x_1,3/x_3,5/x_5,7/x_7)$ leads at worst to the payoff 10 (>4). So $(1/x_1,3/x_3, 5/x_5,7/x_7)$ weakly dominates the strategies $(1/x_1,3/x_3,S_1/x_5,*/x_7)$. Finally, to prove that $(1/x_1, 3/x_3,5/x_5,7/x_7)$ weakly dominates $(1/x_1,3/x_3, 5/x_5,S_1/x_7)$, observe that both strategies lead to the same payoff if player 2 stops at x_2, x_4 or x_6. Then observe that, if player 2 goes on at each decision node $(1/x_1,3/x_3,5/x_5,S_1/x_7)$ leads to the payoff 2 whereas $(1/x_1,3/x_3,5/x_5, 7/x_7)$ leads to the payoff 10 (>2). So $(1/x_1,3/x_3,5/x_5,7/x_7)$ weakly dominates $(1/x_1,3/x_3,5/x_5,S_1/x_7)$.

The above reasoning shows that *it is easier in extensive form games to get weakly dominated strategies than strictly dominated ones*. This is due to the fact that many parts of the game tree are not activated. For example, in the previous game, when player 2 stops at x_2, player 1 is no longer called on to play at further nodes. Consequently all the actions she chooses after x_2 have no impact on the payoffs, so strategies that only differ in those actions can at best weakly dominate one another.

The sticks game illustrates this property well.

1.3.2 Weak dominance and the Fort Boyard sticks game

1.3.2.1 The weakly dominant strategy

Let us study the game (see chapter 1, section 2.2.2) with 6 sticks, given in Figure 2.2.

We first show that for the candidate, player C, the strategy:

> *take 1 stick in front of 6 sticks, 3 in front of 4 sticks, 2 in front of 3 sticks, 1 in front of 2 sticks and 1 in front of 1 stick, is weakly dominant, i.e. it weakly dominates any non equivalent strategy.*[1,2]

First, this strategy *leads player C to always win* the game. As a matter of fact, regardless of player M's behaviour, player C's strategy leaves player M in front of 5 sticks and then, regardless of player M's behaviour, player C's strategy leaves player M in front of 1 stick that he is compelled to take.

Second, the other (non equivalent strategies) are weakly dominated by this strategy. For example, if player C takes 3 sticks in front of 6 sticks (and 3, 2, 1, 1 sticks respectively in front of 4, 3, 2, 1 sticks), player 1 may still win the game namely if player M, after player C's first action, takes the three remaining sticks; but she may also lose the game if, after this action, player M only takes 2 sticks. More generally, as soon as the candidate's strategy allows player M to reach a node with 4, 3 or 2 sticks, which happens as soon as she does not play an equivalent strategy to the one given above, player M can take advantage of his position to lead player C to take the last stick (by taking 3, respectively 2 and 1 sticks when he is in front of 4, respectively 3 and 2 sticks.)

In the same way, for player M, the strategy: *take x sticks in front of 5 sticks, 3 in front of 4 sticks, 2 in front of 3 sticks, 1 in front of 2 sticks and 1 in front of 1 stick, weakly dominates any non equivalent strategy (x may be equal to 1, 2 or 3 sticks).*

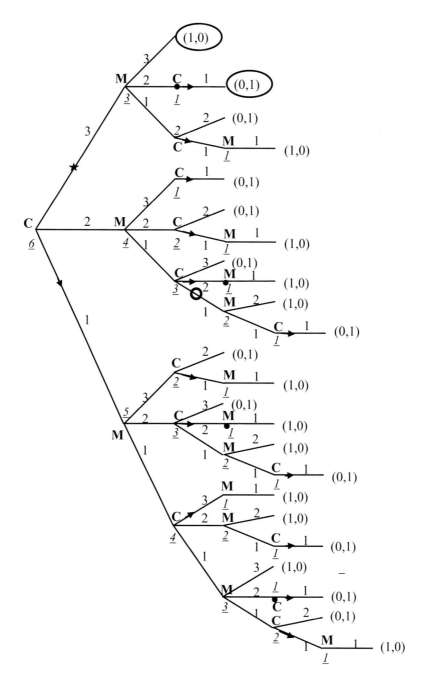

Figure 2.2 C's weakly dominant strategy is the one with the arrows. The (proposed) equivalent strategy is the same as the weakly dominant strategy, except in front of 3 sticks after taking 2 sticks in front of 6 sticks, where the chosen action is the one with a circle. The (proposed) weakly dominated strategy is the same as the weakly dominant strategy, except in front of 6 sticks, where the chosen action is the one with a star. This weakly dominated strategy (possibly) leads to the two circled payoff vectors.

DOMINANCE

Of course, the above result can be generalized when there are more than 6 sticks.

The idea is to observe that being in front of 5 sticks is a bad event because whatever you do, you will lose (either you take 1 stick and the other takes 3 sticks, either you take 2 sticks and the other takes 2 sticks, either you take 3 sticks and the other takes 1 stick: either way, you take the last stick).

So, as a consequence, when you are in front of 6, 7 or 8 sticks, you should take 1, 2 and 3 sticks (respectively) so that the other player is in front of 5 sticks.

Being in front of 9 sticks is again a bad occurrence because whatever you do, the other player can bring you in front of 5 sticks and you will lose.

So, if you are in front of 10, 11 or 12 sticks you should respectively take 1, 2 and 3 sticks so that the other player will be in front of 9 sticks.

This automatically leads to the result:

> **Dominance in the sticks game**
> *k is an integer. The 3 strategies:*
> *Take 1 stick in front of 1 and $4k+2$ sticks, take 2 sticks in front of $4k+3$ sticks, take 3 sticks in front of $4k$ sticks ($k \geq 1$), take x sticks in front of $4k+1$ sticks ($k \geq 1$), where x is equal to 1, 2 or 3, weakly dominates any non equivalent strategy.*

1.3.2.2 Dominance and rules of behaviour

Let us make a practical comment. When you are playing the Fort Boyard sticks game, it is almost impossible for you to count the number of sticks at each round, *because things go too fast (the game lasts less than 1 minute)*. So to win, it is necessary to find an *easy rule of behaviour* that fits with the weakly dominant strategy, but that does not compel you to count the number of sticks at each round. *And such a rule exists*. In fact you only have to count the number of sticks until the opponent is in front of $4k+1$ sticks. For example, if there are 20 sticks (at the beginning of the game), it is enough, if you are the candidate (and so the first player) to know that you are in front of $4k$ sticks; so you take 3 sticks and you know that the opponent is in front of $4(k-1)+1$ sticks. *Then, in the remainder of the game, it is enough, at each of your decision rounds, to complete to 4 the number of sticks taken by the opponent at the previous round.*

> *Hence the rule is: as soon as possible, lead your opponent to a state where he is in front of $4k+1$ sticks. Then, at each of your decision rounds, complete to 4 the number of sticks your opponent took at the previous round.*

This rule is illustrated in Figure 2.3.

Two things are good in this game. First, playing the weakly dominant strategy, provided you have the possibility of being in front of a number of sticks different from $4k+1$, *leads to winning the game easily*. Second, *you win the game regardless of the rationality of the other player* because, regardless of what he does in front of $4k+1$ sticks, you will win the game. This is a very positive fact, not often observed in game theoretical contexts. It follows that the *above weakly dominant way of playing is very robust*.

DOMINANCE

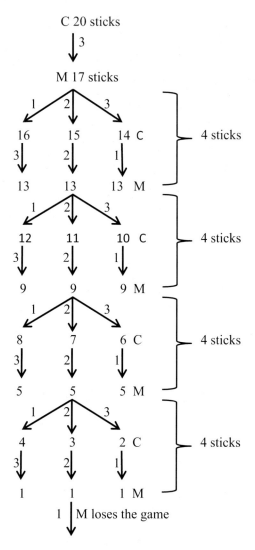

■ **Figure 2.3** C, in front of 20 sticks, takes 3 sticks (first arrow). So M is in front of 17 sticks and takes 1, 2 or 3 sticks (the three arrows). Hence C is in front of 16, 15 or 14 sticks. She systematically takes the number of sticks such that the sum of her taken sticks and M's taken sticks is equal to 4. So she takes 3, respectively 2 and 1 sticks (the three arrows) when M took 1, respectively 2 and 3 sticks. Hence M is systematically in front of 17–4=13 sticks. And so on, until M takes the last stick.

But there is an additional observation: given that usually the candidate starts the game with a number of sticks different from 4k+1, *why does the candidate lose in many cases*? The easiest response is that she does not know the weakly dominant strategy, and has no time to learn it. By the way, it is funny to observe that a way to play the game appeared on the Internet a few years ago, and as a consequence, the game is less often proposed to candidates because some of them may now know the good way to play![3] *Yet even if you know the rule, it is still very difficult to calculate the number of sticks at the beginning of the game because you discover the sticks only at the*

DOMINANCE

moment you have to play, and so you have no time to count them: yet this number is essential to start in the good way. And there is another difficulty: when the professional player, M, takes the sticks, *he plays very fast and even hides (just a little) the sticks he takes with his hand.* So it is not so easy to see how many sticks he takes! It follows that this game still leads to an uncertain outcome, even if everybody knows the rule (see also the conclusion of this chapter).

2 ITERATED DOMINANCE

Theoretically, in the sticks game, you can win the game just by playing the weakly dominant strategy (providing that you have the opportunity to lead the opponent to 4k+1 sticks). So the dominance tool is very powerful and quite convincing.

Yet, in most games, things are less obvious, because *dominance becomes powerful only if it is applied in an iterative way.* And this, as we will see later, is not necessarily natural.

2.1 Iterated dominance, the order matters

> **Property 1:** The order of elimination of strictly dominated strategies has no impact on the remaining strategies. But the remaining strategies, *after elimination of weakly dominated strategies,* may depend on the order of elimination.

We illustrate iterative elimination of dominated strategies and the impact of the order of elimination on the game given in Matrix 2.5.

		Player 2		
		A_2	B_2	C_2
	A_1	(7,8)	(3,3)	(0,8)
Player 1	B_1	(0,0)	(4,4)	(1,4)
	C_1	(7,1)	(3,0)	(1,1)

Matrix 2.5

> ➤ **STUDENTS' INSERT**
>
> First observe that A_2 is weakly dominated by C_2 (8=8, 0<4 and 1=1), B_2 is weakly dominated by C_2 (3<8, 4=4, 0<1) and A_1 is weakly dominated by C_1 (7=7, 3=3, 0<1). By deleting A_2, B_2 and A_1, we get the outcomes **(1,4)** and **(1,1)**.
>
> But we can proceed differently, in different iterative ways.
>
> For example, we can first eliminate A_2 (weakly dominated by C_2) and A_1 (weakly dominated by C_1) and observe that, after these eliminations, C_1 becomes weakly dominated by B_1 (3<4 and 1=1). So we delete C_1 and we get the outcomes **(4,4)** and **(1,4)**.

> But we can also first eliminate A$_2$ (weakly dominated by C$_2$). After this deletion, we observe again that C$_1$ is weakly dominated by B$_1$. We delete C$_1$ and then we observe that B$_2$ is (still) weakly dominated by C$_2$ (3<8, 4=4). We eliminate B$_2$ and we observe that A$_1$ becomes strictly dominated by B$_1$ (0<1); after deleting A$_1$, there remains the outcome **(1,4)**.
>
> Finally we can first eliminate B$_2$ (weakly dominated by C$_2$) and A$_1$ (weakly dominated by C$_1$). After these deletions, we observe that B$_1$ becomes weakly dominated by C$_1$ (0<7, 1=1). After eliminating B$_1$ there remain the outcomes **(7,1)** and **(1,1)**.
>
> So it clearly derives that, according to the chosen order of elimination of weakly dominated strategies, we do not get the same remaining outcomes. Moreover, people are not similar with regard to the preferred iteration, so may select different processes of elimination. For example, in the studied game, player 1 prefers the sequence of eliminations leading to the outcomes (7,1) and (1,1) and player 2 prefers the one leading to the outcomes (4,4) and (1,4).

The fact that the order matters is problematic. But things are even worse in other games, where iterative elimination may lead to a unconvincing result (see section 3.1).

2.2 Iterated dominance and first doubts

We now apply iterated dominance to extensive form games.

Let us first come back to the ascending all pay auction in Figure 2.1. We already know that the strategy (1/x$_1$,3/x$_3$,5/x$_5$,7/x$_7$) weakly or strictly dominates all the other player 1's strategies, so we suppress all these dominated strategies. After their elimination, player 2's family of strategies (S$_2$/x$_2$,*/x$_4$,*/x$_6$) strictly dominates the family of strategies (C$_2$/x$_2$,*/x$_4$,*/x$_6$), given that player 2 gets 7 with the first strategies and at best 5 with the seconds. After elimination of the strategies (C$_2$/x$_2$,*/x$_4$,*/x$_6$), we observe that player 1 bids 1 and that player 2 stops the game after this first bid. So player 1 gets the object at price 1, which is quite convincing: given that player 1 is happy to go on in the whole game, it is natural that player 2 immediately stops the game, as he easily discovers that he can only lose the auction. So the result is easily accepted: observe that it only requires one iteration.

We now turn to the ascending all pay auction game with V=3 and M=5 introduced in chapter 1 (section 1.2.2) and depicted in Figure 2.4. In this game there is no strategy for player 1 that strictly or weakly dominates all the others (by contrast to the game in Figure 2.1).

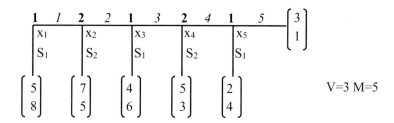

Figure 2.4

DOMINANCE

> **STUDENTS' INSERT**

$(1/x_1, 3/x_3, S_1/x_5)$ is weakly dominated by $(1/x_1, 3/x_3, 5/x_5)$ because both strategies lead to the same payoff if player 2 stops at x_2 or x_4, but the first strategy leads to the payoff 2, whereas the second leads to the payoff 3, if player 2 goes on at each round of play.

It follows that player 1 bids 5 if she reaches node x_5.

Observe that removing this weakly dominated strategy does not amount to cutting the last branch S_1, in that playing $(S_1/x_1, S_1/x_3, S_1/x_5)$ for example, is not yet dominated at this level of reasoning. So we cannot suppress the action S_1 at x_5 at this level. We only eliminate the strategy $(1/x_1, 3/x_3, S_1/x_5)$.

It follows from this elimination that $(2/x_2, 4/x_4)$ is weakly dominated by $(2/x_2, S_2/x_4)$ because both strategies lead to the same payoff if player 1 stops at x_1 or x_3, but the first leads to the payoff 1, whereas the second leads to the payoff 3, if player 1 goes on at nodes x_1 and x_3, given that she bids 5 at x_5.

So if player 2 reaches x_4, he stops the game. It follows that the strategies $(1/x_1, S_1/x_3, */x_5)$ are weakly dominated by $(1/x_1, 3/x_3, 5/x_5)$ because all strategies lead to the same payoff if player 2 stops at x_2, but the first strategies lead to the payoff 4, whereas the second leads to the payoff 5, given that player 2 stops at node x_4 when he reaches this node (* means any available action).

So player 1 goes on if she reaches x_3. It follows that $(2/x_2, S_2/x_4)$ is weakly dominated by $(S_2/x_2, S_2/x_4)$ because both strategies lead to the same payoff if player 1 stops at x_1, but the first leads to the payoff 3, whereas the second leads to the payoff 5, if player 1 goes on at node x_1 given that she also goes on at x_3 and x_5.

So if player 2 reaches x_2, he stops the game. Finally the strategies $(S_1/x_1, */x_3, */x_5)$ are strictly dominated by $(1/x_1, 3/x_3, 5/x_5)$, given that player 2 stops at x_2 (5 < 7).

So the iterative elimination of weakly and strictly dominated strategies allows player 1 to get the object at the very low price 1.

Well, can we be happy with this result? *From a theoretical point of view*, one thing is logical: *the fact that the game stops very quickly.* As a matter of fact, given that one of the players will lose the auction, he has to logically stop the game as soon as possible. *Usually both players know* that one player will lose and that it is best for him to stop as soon as possible. But the problem is that *this does not help them to play the game*, given that usually no player knows *who* will lose the auction (and usually expects that it is the opponent). So no one knows who the loser is and consequently nobody knows if he should stop the game (we come back to this point in chapter 3). *In other words, theory is nice, logical, but not helpful.*

Iterative elimination also claims that player 2 is the loser: is this fact convincing? As it is easy to observe, the last player in this game has an extra power, with regard to iterated dominance. Given that overbidding is better than stopping at the last round, the last player is automatically the one who wins the game. And this fact, even from a theoretical point of view, is rather *artificial:* a slight change in the increment may change the last player and hence the winner of the auction (see also chapter 4, section 3.1 for more comments on the power of the last player).

DOMINANCE

3 CROSSED RATIONALITY AND LIMITS OF ITERATED DOMINANCE

In the sticks game, there is no problem. A simple strategy weakly dominates all the non equivalent ones and leads the player to win the game, regardless of the play of the opponent and without iteration. So the obtained result is robust and we can reasonably expect it to be played by players able to calculate it. In the ascending all pay auction in Figure 2.1, there is no problem either. One convincing iteration leads player 1 to bid 1 and player 2 to stop the game (given that player 1 is always better off going on, which compels player 2 to stop immediately).

By contrast, in the game of Figure 2.4, the result linked to iterative elimination is a bit less convincing, given that it requires four iterations and gives perhaps too much power to the last player in the game. We now show that the intuitive content of the result obtained by iterated dominance diminishes much with the number of iterations.

3.1 Envelope game: K–1 iterations for a strange result

We switch to the envelope game (introduced in chapter 1 section 2.2.2). We recall that each player gets an envelope with a number between 1 and K (included), K being common knowledge. The players only observe their own number and decide simultaneously whether or not they want to exchange the envelope with the opponent. The exchange takes place only if both want it. Then the players get in dollars the number in their envelope.

The puzzling question is when should a player be ready to exchange his envelope? What does iterative elimination of weakly dominated strategies say? Well, something very strange!

We recall that a strategy has to say what you do, exchange or not, for any number between 1 and K. For ease of notations we say "exchange x" for "exchange an envelope with the number x".

It is obvious that a strategy that leads to exchange K is weakly dominated by the strategy that, all else being equal, does not exchange K. As a matter of fact, exchanging K can only lead, either to the same payoff than not exchanging (if the other does not want to exchange or if he also has K in his envelope and wants to exchange it), or to a lower payoff (if the other player wants to exchange but has a lower number in his envelope).

Once all the strategies that lead to exchange K are suppressed, it appears that a strategy that exchanges K–1 is weakly dominated by the strategy which, all else being equal, does not exchange K–1. As a matter of fact, given that nobody exchanges K, if the other player wants to exchange, he necessarily has an amount lower than K. Consequently exchanging K–1 leads at best to the payoff K–1 (if the other player does not want to exchange or has K–1 in his envelope and wants to exchange); in all other cases it leads to losing money.

But if one suppresses the strategies that exchange K and K–1, then a strategy that leads to exchange K–2 is weakly dominated by the strategy that, all else being equal, does not exchange it. And it is easy to see that this elimination process goes on, until it reaches the number 2. It follows that no player exchanges his envelope unless it contains 1, he can keep or exchange: in any case he gets 1 because the other player doesn't exchange a number higher than 1.

So a K–1 steps iterative elimination of weakly dominated strategies leads to never exchanging the envelope, except if the number is 1. Is this result realistic? Does it help us to solve the game? If K=100,000 and if you have 50 in our envelope, do you really not try to exchange? What is reasonable in such a game? Let us see what happens in reality.

The histogram in Figure 2.5 shows the maximal number exchanged by my (146) L3 students for K=100,000(euros).

75

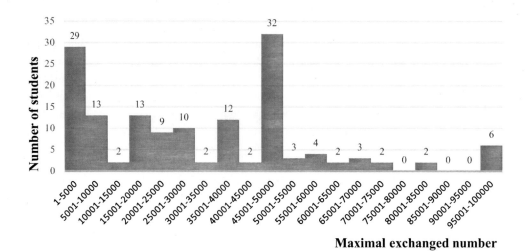

Figure 2.5 Envelope game, maximal number exchanged for K=100,000

First observe that the histogram does not completely contradict the above theoretical result, in that about 1/5th of the (146) L3 students stopped exchanging when the number is higher than 5,000, 13% stopped exchanging when the number is higher than 1,000 and even 1.4% only exchanged 1. But the histogram in Figure 2.5 also reveals that many players exchange up to a number around K/2 (36.3% exchange up to a number in [40,000, 60,000] and 19.2% exchange up to 50,000).

Many of my students confessed that they wrongly expected that the opponent exchanges all his possible envelopes, so that the mean expected number in the opponent's envelope is 50,000, so 19.2% exchanged up to this number. But many of them also realized that exchanging an amount close to K/2 is in fact not very reasonable, given that it amounts to expecting that the other player's exchanged number is between 1 and K with the same probability. This is unreasonable, because you expect that the opponent exchanges all his numbers whereas you only exchange numbers lower than K/2. So some students often made one additional step of reasoning. They said: "let me suppose that the opponent only exchanges numbers lower than K/2 (as I wanted to do). Then the expected value I get if I exchange is about K/4 (if the numbers are uniformly distributed on [0, K/2]) and I should not exchange numbers higher than K/4". So 21.2% of my students exchanged up to a number in [20,000, 30,000]. This is a *level-k reasoning* (see chapter 6 for more details). And often students stopped at this level, but sometimes they made the following observation: "Of course, if others think like me, I should only exchange lower values, but do they?"

These observations share a common point with our above reasoning: *the notion of iterative elimination is rather natural, at least for some players, but only for two or three steps*.

But they also point out many differences. I return to the mean expected approach chosen by my students and by level-*k* reasoning. Why do people reason in this way? The iterative elimination of weakly dominated strategies does not proceed in this way; it never refers to a mean calculus. So, why do people make such calculi? In the proposed game, such a calculus seems baseless because the game does not specify with which probability each number is selected. Students reason as if each number is chosen with the same probability, something which is not specified in the description of the game.

So why do students reason in this way? *Well, the aim of this game is to win money, not to do a sophisticated reasoning! A positive impact of a level-k reasoning is that you need few steps to come to the conclusion to only exchange low numbers:* at the first step, you are ready to exchange up to K/2. Then, if you suppose that the opponent plays in this way, you are ready to exchange up to K/4=K/2². And so on. In five steps, you are ready to exchange up to $K/2^5$=3,125, and you only need 16 steps to only exchange the number 1 (by contrast to the 99,999 steps required by iterative elimination of weakly dominated strategies)! Given that in a real game, you do not have enough time to make a reasoning that needs 99,999 steps, it seems cleverer to adopt a reasoning that comes to the same conclusion in fewer steps!

The reality sees many situations where a simple rule of behaviours which seems theoretically baseless, leads to the optimal situation (see section 4).

Now observe that most of my students, even if adopting a level-*k* reasoning, stopped at a level lower than 5 (given that 4/5th of the students exchanged up to a number higher than 5,000). Why is that? Were they not able to go further? They were surely, but they surely thought that *the opponent was not able to*. If you think that your opponent is not so clever, and that he exchanges all the numbers up to 50,000, it would be a pity to not exchange your envelope up to 5000 in order to try to benefit from his generosity! (See also chapter 3, section 5.1 for more comments.) In other words, an *iterative reasoning rests on a high degree of crossed rationality*, where by crossed rationality, we mean that players are rational but also know that their opponents are rational and that the opponents know that the others know that they are rational, and so on.

3.2 Levels of crossed rationality in theory and reality

We now study crossed rationality in iterated dominance. So let us come back to the iterative elimination of dominated strategies in the envelope game.

> **STUDENTS' INSERT**

To show that iterative elimination of dominated strategies rests on a too high degree of crossed rationality, just consider an envelope game with K=5.

We call E_m any strategy that asks to exchange m, and $\overline{E_m}$ any strategy that asks to not exchange m. It is obvious that both players, if they are rational, eliminate E_5.

It is rational for a player to eliminate E_4 as soon as his opponent does not exchange 5, i.e. as soon as the opponent eliminated E_5. But how can a player know that the opponent eliminated E_5? If and only if he knows that the opponent is rational.

Consequently player 1, to eliminate E_4, needs to be rational and needs to know that player 2 is rational. In the same way, player 2, to eliminate E_4, needs to be rational and needs to know that player 1 is rational.

In the same way, it is rational to eliminate E_3 as soon as the opponent eliminated E_4. That is to say, player 1 eliminates E_3 if she is rational and if she knows that player 2 eliminated E_4, i.e. if she knows that player 2 is rational and knows that player 1 is rational. In the same way, player 2 eliminates E_3 if he is rational and if he knows that player 1 is rational and that player 1 knows that player 2 is rational.

And it is rational to eliminate E_2 as soon as the opponent eliminated E_3. That is to say, player 1 eliminates E_2 if she is rational and if she knows that player 2 eliminated E_3, i.e. if she knows that

DOMINANCE

> *player 2 is rational and that player 2 knows that player 1 is rational and that player 1 knows that player 2 is rational. In the same way, player 2 eliminates E_2 if he is rational and if he* knows *that player 1 is rational and that player 1* knows *that player 2 is rational and that player 2* knows *that player 1 is rational. We say that eliminating E_2 needs* **three levels of crossed rationality**. *As you can see, each step of reasoning needs an additional level of crossed rationality. And of course the more levels of crossed rationality are needed, the less realistic the obtained result seems.*

It follows from above that each iteration needs an additional level of crossed rationality. So, even for K=100 we need 98 levels of crossed rationality to get the conclusion to only exchange 1! The above reasoning clearly explains why in reality people stop iterative reasoning after two or three steps, because it seems difficult to expect more crossed rationality.

This reasoning also leads to the following observation:

Game theory often works with rational players; it even often assumes that this rationality is common knowledge. In that case, *there is no reason not to do a reasoning based on 1,000 – even an infinity – levels of crossed rationality.* If rationality is common knowledge, then we can accept the result – only the number 1 can be exchanged – as soon as we accept elimination of weakly dominated strategies.

But this is of course not convincing: common knowledge of rationality is surely too strong an assumption. In a more funny way, what makes the envelope game interesting is precisely the fact that we do not know how rational the opponent is!

3.3 Crossed rationality in extensive form games, a logical inconsistency

The above observation also explains why, in the ascending all pay auction in Figure 2.4, it is not so sure that player 1 will systematically go on and that player 2 will systematically give up. But, by contrast to the simultaneous game above, in a sequential game it becomes more difficult – *even if we a priori accept that rationality is common knowledge* – to accept the iterative elimination of dominated strategies, because, in an extensive form game, *the way players played in the past may contradict the assumption of common knowledge of rationality.*

So for example, player 1 at x_3 might be better off stopping the game and getting 4, rather than pursuing the game because she cannot be sure that player 2 stops the game at x_4. As a matter of fact, player 1 has to go on at x_3 only if she is sure that player 2 stops at x_4; i.e. only if she knows that player 2 is rational and knows that player 1 is rational and hence will not stop at x_5. Yet, can player 1, when being at x_3, expect that player 2 will have a sufficient level of crossed rationality to stop at x_4, *given that he did not stop at x_2?* If rationality were common knowledge, *player 2 should have stopped at x_2*, but he did not. So *common knowledge of rationality is broken by the fact that player 2 did not stop at x_2, as he should have done.*

In normal form games, you can reasonably fear that common knowledge of rationality is not fulfilled, *but you have no proof of it before playing, because the players play only one time, after the reasoning. But in extensive form games this is no longer true. As soon as somebody behaved in an unexpected way in the past, the players have the proof that common knowledge of rationality is disrupted. And they have to react to this fact, which is very difficult.*

In particular, what do we learn from past actions that do not fit the assumption of common knowledge of rationality? With what level of rationality is player 2's behaviour compatible? Has player 1 to fear that player 2 goes on at x_4 or can he expect that he will stop at this node? Perhaps player 2 is fully rational but thinks that player 1 is not rational. Perhaps he is systematically irrational and simply always goes on – so he goes on at x_4 – perhaps he is unable to understand the game at the beginning but becomes more rational as the game goes on – so he stops at x_4 because he can understand that player 1 goes on at x_5. Once common knowledge of rationality is broken, many things become possible and, unfortunately, game theory poorly helps guide the behaviour (player 1 at x_3 cannot know if it is best to stop or to go on).

As a byproduct, to our mind, it's not so astonishing that most of game theory focuses on normal form games, given that the above difficulty only appears in extensive form games.

4 DOMINANCE AND STRUCTURE OF A GAME, A COME BACK TO THE PRISONER'S DISKS GAME

4.1 Solving the game by iterative elimination of strictly dominated strategies

Be that as it may, dominance is a strong interesting tool we work with even if it has flaws. In fact, in some special games, taking into account the power of dominance can lead to simplifying the structure of the game, in a way that highly eases its resolution.

As we will see later (chapters 3–5), playing in a given way has no impact on the structure of the game in that it only changes the probabilities assigned to some nodes in some information sets. And these probabilities are difficult to establish at many information sets because these sets should not be reached if people behave according to a given profile.

By contrast, if we eliminate the branches that correspond to strictly dominated strategies (for example because, as we will see in chapter 3, this elimination does not eliminate any Nash equilibrium path), then the evolution of time may become equivalent to the evolution of available strategies (because only some actions[4] are not strictly dominated for each player at each period) and lead to the evolution of the structure of the information sets.

For example, in the prisoners' disks game, each player, at the beginning of the play, has only one information set where he knows the colour of his disk, but after the elimination of the strictly dominated actions, the number of these sets grows.

The prisoners' disks game is in fact a very nice (and special) game with regard to the iterative elimination of strictly dominated strategies.

Figure 2.6 focuses on the game with only two prisoners and at least one black disk. The structure of this game, as discussed in chapter 1, is really complicated even if we only represent two periods of the game (which is not in the spirit of the game but we allow it for more clarity): the cardinal of each player's strategy set is $3\times(3\times3) \times 3 \times (3\times3)$.

But let us now eliminate strictly dominated strategies.

■ *We first observe that, in the first period of play, when observing N–1 white disks, the player knows that he has a black disk and so he has to play OB (go out and announce a black disk). Any other strategy that starts differently is strictly dominated, given that it can only yield a lower payoff than OB.[5] What is important, and special to this game, is that this strategy is not weakly but strictly dominated, because the information set where a player observes N–1 white disks is reached with positive probability.*

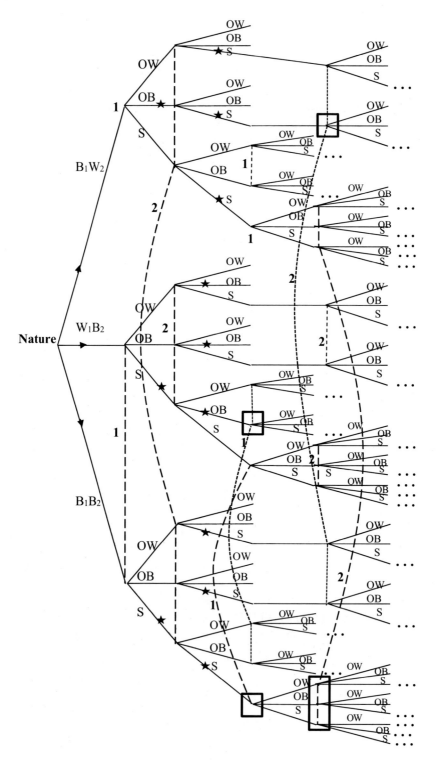

Figure 2.6 In period 1, only the actions with the stars are part of non strictly dominated strategies. It follows that the nodes in the squares are the nodes where the players can be in the second period.

DOMINANCE

The elimination of the strictly dominated strategies has a strong impact with regard to the structure of the game and the associated strategies. It not only divides the strategy set by 3 (because we switch from three choices at a given set to one choice), but it leads to a much stronger reduction.

For example, in the game with only two prisoners in Figure 2.6, in which the size of player 1's strategy set is 3×(3×3) x 3 x (3×3), player 1, in front of a white disk, knows that she has a black disk. So if we eliminate S (stay in) and OW (go out and say white), we not only divide the number of strategies by 3, but we divide it by 3 x (3×3), because the two information sets after the inappropriate action S also disappear.

Let us insist on the fact that this game is very special; it is for example highly different from the ascending all pay auction in Figure 2.4. In the latter, we cannot delete branches because the iterative process starts with actions *at the last nodes, and not with actions at the first nodes.* And the status of these last actions depends on what has been played before: so for example stopping at x_5 belongs to a weakly dominated strategy, *only if* player 1 goes on at x_1 and x_3.

What is more, in the prisoners' disks game, the deletion of the strictly dominated actions also has a strong impact on the information sets of the other players. Given that a prisoner plays in a different way at different information sets, the other players get information by observing his action. And information sets that follow actions that cannot be rational (for example playing OW in period 1) completely disappear. Of course these information sets are not normally reached when people play optimally, but if they are not eliminated, one has to give an action at each of them, which is difficult because somebody played irrationally in the past.

■ *We also see that in the first period of play, if a player does not observe N–1 white disks, then he does not know the colour of his disk and, given the structure of the payoffs,[6] he has to stay in (strictly dominant strategy); so we can delete all the strategies that start with OB and OW in this context.*
■ So, finally, after this first round of eliminations, the number of strategies for each player drastically shrinks.

Observe also that, given that each information set in period 1 only leads to one possible action (S or OB), each information set only contains 1 node (where the player knows the colour of his disk) or 2 nodes (where he doesn't).

Moreover the number of decision nodes for each player is lower in period 2 than in period 1. Given that only one action is not part of a strictly dominated strategy at any information set in period 1, one node in period 1 leads to at most 1 node in period 2, and given that there is one node in period 1 where the player has to go out (he plays OB), the number of decision nodes at the beginning of period 2, diminishes by 1. So, in an N player game, it is $(2^N-1)-1$. *(In passing, note for further discussion that the previous reasoning proves that if there is only one black disk, the person with the black disk goes out at period 1).*

This reduction is not the only change. Some of the two nodes information sets in period 1 now split into singletons: the sets of states that lead a player to observe N–2 white disks now split in two. As a matter of fact such an information set, in the first period, contains the state where the player's disk is black (so the only non strictly dominated action is S for each player) and the state where his disk is white (in this state the prisoner with the black disk went out). So given that the two states do not lead to the same sequence of actions in period 1, they can be distinguished and become two singleton information sets at the beginning of period 2.

DOMINANCE

So in period 2, if N=2, the player knows the colour of his disk and a unique strictly dominant strategy induces him to go out, saying the right colour of his disk. So, after elimination of all strictly dominated strategies for N=2, we get the new game tree in Figure 2.7.

More generally, when a player observes N–2 white disks, then at the beginning of period 2, he goes out announcing white if there is only 1 black disk (because the only black disk went out in period 1), and he goes out announcing black (if nobody went out in period 1). *(So in passing, note for further discussion that, regardless of the value of N, if there is 1 black disk, the black disk goes out in period 1 and the white disks go out in period 2, and if there are 2 black disks, the two black disks go out in period 2).*

If, for N>2, a player sees less than N–2 white disks in period 1, then everybody plays S in period 1, which leaves a player in period 2 with the same information sets than in period 1 (just a step further): any non strictly dominated strategy asks him to play S in period 2.

So at the beginning of period 3, a player is still in the game if and only if he did not go out in period 1 (1 decision node) or in period 2 (he goes out at 2(N–1) decision nodes, given that he observes N–2 white disks at 2(N–1) nodes (because the black disk may belong to any of the N–1 other players, and his disk is either black or white). So his number of decision nodes at the beginning of period 3 is: $(2^N-1)-1-2(N-1)=2^N - 2C_{N-1}^0 - 2C_{N-1}^1$.

This context is illustrated for N=3 in Figure 2.8.

As a consequence, given that the number of decision nodes shrinks from one period to the other (even if the number of information sets grows because of more information), ***the game inevitably stops after a finite number of periods.***

Let us be more precise. If N≥3, if a player observes N–3 white disks (hence 2 black disks), and if he is at the beginning of period 3, his decision nodes belong to singleton information sets because if his disk is white, two prisoners (the two with the black disks) went out in the preceding period, and if it is black, nobody went out; so the player knows the colour of his disk and has to go out. So, given that the 2 black disks correspond to C_{N-1}^2 configurations, our player goes out in

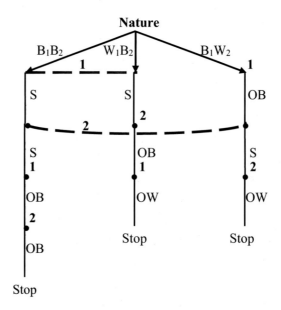

■ Figure 2.7

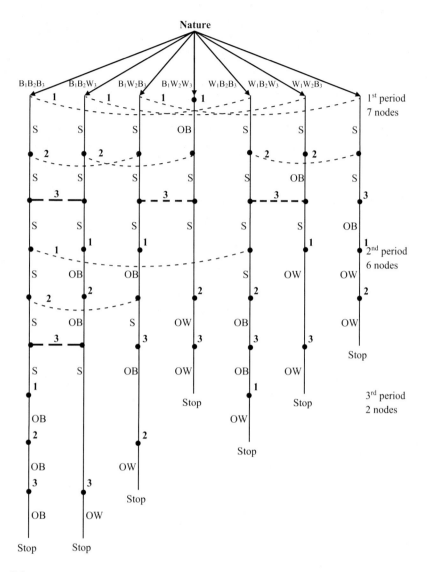

Figure 2.8

period 3 at $2C_{N-1}^2$ nodes. *(In passing, note that regardless of N, if there are 2 black disks, they go out in period 2 and the white disks go out in period 3, and when there are 3 black disks, the 3 black disks go out in period 3.)*

It follows that, at the beginning of period 4, there remain $2^N - 2C_{N-1}^0 - 2C_{N-1}^1 - 2C_{N-1}^2$ decision nodes. And so on…. At the beginning of step N, there remain: $2^N - 2C_{N-1}^0 - 2C_{N-1}^1 - 2C_{N-1}^2 \cdots 2C_{N-1}^{N-2} = 2^N - 2(2^{N-1} - 1) = 2$ states. In these two states, the player observes N–1 black disks, but he can distinguish the two states, given that the other players either went out in the previous period (because his disk is white) or stayed in, because his disk is black. So he knows the colour of his disk and goes out. *(In passing, note for further discussion that regardless of N, if there are N–1 black disks, the N–1 black disks go out in period N–1, and the white disk goes out in period N, and if there are N black disks, all the black disks go out in period N.)*

DOMINANCE

To summarize, eliminating strictly dominated actions allows the reduction of the number of decision nodes at each period. This reduction is parallel with a growing number of going out actions and a growing information in the game. So iterative elimination of strictly dominated strategies is very powerful and appreciated by all the players, given that it corresponds to the optimal way to play the game: it ensures that each player goes out as fast as possible.

But this also shows how much this game is a specific one. At each step, strict dominance does not depend on what happens in the future, and this makes the game easier to solve. The behaviour in the past, considered as rational (because one eliminates the strictly dominated actions in the past) diffuses enough information on the right way to play in the future. In other words, what is nice about this game is that we solve it by following the arrow of time. We can ignore the way people play in the future while studying the behaviour in a given period. The past (rational) actions are the only necessary information for establishing the rational future behaviour, so we can say that the prisoners' disks game is an illustration of forward induction (see chapter 6).

4.2 Dominance and easy rule of behaviour

We observe that iterative elimination of strictly dominated actions leads to an interesting result which is:

Regardless of the value of N:

When there is 1 black disk, he goes out in period 1 and the white disks go out in period 2.
When there are 2 black disks, they go out in period 2 and the white disks go out in period 3.
When there are 3 black disks, they go out in period 3 and the white disks go out in period 4.
...
When there are N–1 black disks, they go out in period N–1 and the white disk goes out in period N.
When there are N black disks, they go out in period N.

So the rational behaviour, which fits with the successive elimination of strictly dominated strategies, can be summarized in Table 2.1.[7]

It is easy to check these optimal going out periods for N=3 prisoners in the game in Figure 2.8.

Table 2.1

Number of prisoners	Number of black disks		1	2	3	4	5
2	Optimal going out period for the black disks=t_B		1	2			
	Optimal going out period for the white disks=t_W		2				
3	t_B		1	2	3		
	t_W		2	3			
4	t_B		1	2	3	4	
	t_W		2	3	4		
5	t_B		1	2	3	4	5
	t_W		2	3	4	5	

DOMINANCE

Yet people are generally not able to do all the iterative eliminations of strictly dominated strategies and are far from knowing if the other players are able to do them. In other words, given that the number of steps of reasoning grows with the number of black disks, we cannot be sure that people can calculate the optimal going out period and announce the good colour, because of a lack of crossed rationality.

And even if you distribute Table 2.1 to the prisoners and say that it gives the best period for going out, it will be of no help to them. As a matter of fact, when, for example, there are 3 black disks and 5 prisoners, a prisoner with a black disk does not know if there are 3 or 2 black disks, and so he does not know when he has to go out and what colour he has to announce.

But fortunately, we can propose[8] an algorithm of behaviour that fits perfectly with the above table and automatically leads the players to go out at the optimal period.

Optimal algorithm of behaviour:

t:=1 If a prisoner observes t−1 black disks, he goes out at period t announcing a black disk. If not, he stays in the game (in the room).

t:=t+1

If there is no prisoner left, stop.

If not:

If the prisoner sees, at the beginning of the game, t−1 black disks, then:

- If these prisoners with the black disks went out at period t−1, he goes out at period t announcing a white disk;
- If nobody went out at period t−1, he goes out at period t announcing a black disk.

If not, he stays in the game (in the room).

This algorithm works exactly like the optimal reasoning without requiring any reasoning from the players. To illustrate it, let us again take 5 prisoners and 3 black disks.

At period 1 (t=1), no prisoner observes t−1=0 black disk, so everybody stays in the game (in the room).

And they turn to period (1+1)=2. At period 2, all the prisoners are still present and no prisoner observes exactly 1 (t−1=1) black disk. So everybody stays in the room.

And they turn to period 2+1=3. At period 3, all the prisoners are still present.

A prisoner with a black disk sees exactly t−1=2 black disks. So, given that nobody went out in period t−1=2, he goes out and announces a black disk.

A prisoner with a white disk does not observe (t−1)=2 black disks. So he stays in the game.

And they turn to period 3+1=4. At period 4, there are still prisoners in the room (the ones with the white disks). Each of them, at the beginning of the game, observed exactly 3 black disks. Given that the prisoners with the black disks went out at period (t−1=3), they go out at period 4 and announce a white disk.

Perhaps you find the algorithm a little difficult for people that are not used to working with algorithms. Well, we can summarize it in an even easier rule of behaviour:

> *Easy and optimal rule of behaviour:*
> When you observe k black disks at the beginning of the game, observe what happens at period k and be ready to jump at period k+1. More precisely, if you observe, at period k, that nobody goes out, go out at period k+1 and announce black. If you observe, at period k, that the k black disks go out, go out at period k+1 and announce white.

That's not too demanding is it?

So, like in the sticks game, we have again found an easy rule of behaviour that fits with a rational reasoning, which can be easily adopted by players who are not able to do the rational reasoning. I think that trying to translate complicated reasoning in easy rules of behaviour is a very interesting job in game theory! I even suspect that many easy rules of behaviour regularly adopted in management, political or economical contexts, in fact correspond to Nash equilibria, that are sometimes difficult to establish (for example the way people behave in negotiations, the frequency of the negotiations and meetings).

Let us conclude with the following remark:

The algorithm (and the associated easy rule of behaviour) has a flaw: it doesn't specify what we have to do at period t, when we observed t–1 black disks at the beginning of the game, but we are neither in the case where all the black disks went out in period t–1, nor in the case where nobody went out in period t–1.

In other words, the algorithm does not specify what to do if somebody made an error in the past. It implicitly assumes that everybody is able to follow the algorithm (or the associated easy rule of behaviour).

How can we reply to this criticism?

I would first say that without the algorithm (or the easy rule of behaviour) people have to be very rational and have to believe in many levels of crossed rationality to go out at the optimal date announcing the right colour: so they have almost no positive probability to do so.

With the rule, almost everybody – each person able to count and to observe the other players' behaviour – is able to go out announcing the right colour, so the rule is a very good one, with regard to its effectiveness; an optimal behaviour is obtained with even a low rationality.

What is more, let us also observe that, if N is high and the number of black disks is high enough (higher than 4 for example), then, even if one or a few persons behave in a wrong way, this should not disturb the other players too much.

For example, if N=10 and K, the number of black disks, is equal to 6, then a player with a white disk, who observes that only 5 black disks go out in period 6 instead of 6, may reasonably deduce that the person with the black disk who didn't go out didn't understand the rule, and this should not prevent him from going out at period 7 announcing white. Well, in other words, if the observed actions are still close to the expected ones, one may still adopt the behaviour proposed in the rule, by not taking into account the actions that seem obviously wrong. Usually players will surely come to this conclusion and adapt their behaviour.

CONCLUSION

I recently learned, thanks to the papers of Dufwenberg et al. (2010)[9] and Gneezy et al. (2010)[10] that the sticks game, played in the French TV show "Fort Boyard", has a companion game, called Race Game or Game of 21, played in an American TV show *The Survivor*! In this game, the winner is the player who first reaches a given number, for example 21, by proceeding as follows: player 1 announces 1 or 2, player 2 adds 1 or 2 to this number, player 1 adds 1 or 2 to the previous amount, player 2 adds 1 or 2 to the previous amount and so on, till one reaches 21. The player who reaches 21 wins the game.

More generally, a race game is defined by two numbers: the amount to reach and the maximum number you can add each time it's your turn to play. So the game of 21 is defined by $G(21,2)$; Dufwenberg et al. also studied the game $G(6,2)$ and Gneezy et al. focused on $G(15,3)$ and $G(17,4)$. The weakly dominant strategy in a game $G(M,n)$ consists to reach as soon as possible a number $M-(n+1)k$ (k is an integer), and then complete to $n+1$ the number added by the opponent (if he adds x, add $n+1-x$). The reason for this dominance goes as follows: if your opponent is in front of the amount $M-(n+1)$ he can only lose the game, because whatever he chooses to add (from 1 to n), you can complete to $n+1$ and win the game. If your opponent is in front of the amount $M-2(n+1)$, he can only lose the game, because whatever he chooses to add (from 1 to n), you can complete to $n+1$ so that your opponent is in front of $M-(n+1)$, a state where he loses the game. And so on. The similarity with our sticks game is obvious.

In the race game and the Fort Boyard sticks game, the dominant strategy seems rather easy to find: but are people really able to find it? How do people learn this strategy? Dufwenberg et al. and Gneezy et al. studied this way of learning in experimental sessions. Gneezy et al. first asked the players to play the race game $G(15,3)$, where a weakly dominant strategy consists of bringing the opponent in front of the amounts 3, 7 and 11. The players played the same game 20 times. Gneezy et al. observed that, in early rounds, players were not able to understand the strategic value of 3 and 7. So at their first decision round, they played randomly, just in order to see what would happen next. The first thing they learned is the strategic value of 11, namely because they observed that they lost the game when faced with this amount (two rounds were enough to understand this fact). They inferred later, as a consequence, that 7 is also a bad state (five rounds were necessary). And they inferred later (eight rounds were necessary) that 3 is also a bad state. After 10 rounds almost everybody played optimally. So the players systematically first solved easy games (how to play after reaching an amount close to 11) before being able to best react in front of lower amounts (which is more difficult). Dufwenberg et al. proved a similar fact. To do so, they first asked some players to play five times the easy game $G(6,2)$ before turning five times to the more complicated game $G(21,2)$; and they asked other players to first play $G(21,2)$ five times before turning to $G(6,2)$ five times. The players who first played the easy game $G(6,2)$ played the game $G(21,2)$ better than the players who started with this latter game. A similar fact was observed by Gneezy et al. After having played the game $G(15,3)$ 20 times, the players were asked to play the game $G(17,4)$: Gneezy et al. observed that they learned the weakly dominant strategy faster than they learned it in the game $G(15,3)$. What is more, whereas in the game $G(15,3)$, people first learned that 11 is a bad state, and after discovered that 7 and 3 are also bad states, in the game $G(17,4)$ they almost learned at the same time that 12, 12–5=7 and 12–5–5=2 are bad states. So, by contrast to what happened in the first game where they progressively discovered the bad states, it happens as if *they know that there is a strategy to find* and they discover it quickly after having understood that 12 is a bad state.

DOMINANCE

So both studies highlight the fact that people first try to solve easy games (or easy parts of the game, at the end of the game) before trying to understand more difficult ones; so they progressively learn a dominant strategy. And both studies also show that players are able to export their newly acquired knowledge to similar new games (so learning faster dominant strategies in similar games). See also the conclusion of chapter 4.

NOTES

1. An equivalent strategy can be: player C plays as above, except that she takes 1 stick in front of 3 sticks, but only after having taken 2 sticks in front of 6 sticks (as illustrated in the game in Figure 2.2). This change has of course no impact, given that player C takes 1 stick in front of 6 sticks, so she will never be at the node where she plays differently.
2. A strategy is strictly dominant if it strictly dominates any non equivalent strategy.
3. In fact the Fort Boyard game is a variant of a Nim game, and Nim games are very old games that have been solved a long time ago. Yet people usually do not know this fact: Fort Boyard candidates surely do not know the solution of Nim games when playing the sticks game.
4. In the following paragraphs, we often write, for ease of notation, "an action is dominated", instead of "an action is part of a dominated strategy".
5. We can introduce discount factors to avoid the case in which a prisoner who knows his colour waits an additional period to go out.
6. If you go out announcing a colour at random, this colour is wrong with a positive probability, so you stay in jail forever with positive probability, an event which is much more dramatic than staying in jail for one year (we of course suppose that the prisoners are not very old persons).
7. This table is from Umbhauer, G., 2004. *Théorie des jeux*, Editions Vuibert, Paris.
8. This is from Umbhauer, G., 2004. *Théorie des jeux*, Editions Vuibert, Paris.
9. Dufwenberg, M., Sundaram, R., Butler, D.J., 2010. Epiphany in the game of 21, *Journal of Economic Behaviour and Organization*, 75, 132–143.
10. Gneezy, U., Rustichini, A., Vostroknutov, A., 2010. Experience and insight in the race game, *Journal of Economic Behaviour and Organization*, 75, 144–155.

Chapter 2

Exercises

♣ EXERCISE 1 DOMINANCE IN A GAME IN NORMAL FORM

Consider the normal form game in Matrix E2.1

		Player 2			
		A_2	B_2	C_2	D_2
Player 1	A_1	(5,3)	(1,1)	(1,2)	(0,2)
	B_1	(3,0)	(1,2)	(2,5)	(1,3)
	C_1	(2,1)	(4,0)	(1,0)	(0,4)
	D_1	(0,0)	(1,0)	(0,1)	(4,1)

Matrix E2.1

Question

Eliminate iteratively the strictly and weakly dominated strategies.

♣ EXERCISE 2 DOMINANCE BY A MIXED STRATEGY

Consider the normal form game in Matrix E2.2

		Player 2			
		A_2	B_2	C_2	D_2
Player 1	A_1	(5,0)	(1,1)	(0,2)	(0,2)
	B_1	(1,6)	(1,4)	(2,5)	(1,3)
	C_1	(2,0)	(4,0)	(1,0)	(0,4)
	D_1	(0,1)	(1,0)	(0,1)	(4,0)

Matrix E2.2

Questions
1. Show that B_2 is strictly dominated by a mixed strategy. After eliminating B_2, show that C_1 is strictly dominated by a mixed strategy.
2. Eliminate iteratively the other strictly and weakly dominated strategies.

EXERCISES

♣ EXERCISE 3 ITERATED DOMINANCE, THE ORDER MATTERS

Consider the normal form game in Matrix E2.3

		Player 2		
	A_2	B_2	C_2	D_2
A_1	(1,1)	(3,3)	(2,2)	(0,2)
B_1	(1,2)	(3,0)	(2,5)	(1,5)
C_1	(1,0)	(0,0)	(2,2)	(4,1)
D_1	(4,0)	(2,1)	(1,0)	(0,4)

Player 1 (rows)

Matrix E2.3

Question

Eliminate iteratively the strictly and weakly dominated strategies. Show that the order of elimination matters. Comment.

♣♣ EXERCISE 4 ITERATED DOMINANCE IN ASYMMETRIC ALL PAY AUCTIONS

Consider an asymmetric first price all pay auction game: each player has an initial amount of money (M=5) and can bid a certain amount x/2, where x is an integer from 0 to 10. Only the player who makes the highest bid gets the object, but each player pays his bid. Player 1 values the object 4 (V_1=4) and player 2 values the object 3 (V_2=3).

Questions

1. Show that $x_2 > 6$ is strictly dominated by $x_2 = 6$, and that $x_1 > 8$ is strictly dominated by $x_1 = 8$. After eliminating these strategies, show that $x_1 = 8$ is strictly dominated by $x_1 = 7$. You get the normal form game in Matrix E2.4a:

	Player 2						
	0	0.5	1	1.5	2	2.5	3
0	(7,6.5)	(5,7.5)	(5,7)	(5,6.5)	(5,6)	(5,5.5)	(5,5)
0.5	(8.5,5)	(6.5,6)	(4.5,7)	(4.5,6.5)	(4.5,6)	(4.5,5.5)	(4.5,5)
1	(8,5)	(8,4.5)	(6,5.5)	(4,6.5)	(4,6)	(4,5.5)	(4,5)
1.5	(7.5,5)	(7.5,4.5)	(7.5,4)	(5.5,5)	(3.5,6)	(3.5,5.5)	(3.5,5)
2	(7,5)	(7,4.5)	(7,4)	(7,3.5)	(5,4.5)	(3,5.5)	(3,5)
2.5	(6.5,5)	(6.5,4.5)	(6.5,4)	(6.5,3.5)	(6.5,3)	(4.5,4)	(2.5,5)
3	(6,5)	(6,4.5)	(6,4)	(6,3.5)	(6,3)	(6,2.5)	(4,3.5)
3.5	(5.5,5)	(5.5,4.5)	(5.5,4)	(5.5,3.5)	(5.5,3)	(5.5,2.5)	(5.5,2)

Player 1 (rows)

Matrix E2.4a

2. Eliminate iteratively the weakly and strictly dominated strategies, and show that the matrix of remaining strategies is Matrix E2.4b:

		Player 2		
		0	1	2
Player 1	0.5	(8.5,5)	(4.5,7)	(4.5,6)
	1.5	(7.5,5)	(7.5,4)	(3.5,6)
	2.5	(6.5,5)	(6.5,4)	(6.5,3)

Matrix E2.4b

3. Comment on the elimination process.

♣ EXERCISE 5 DOMINANCE AND VALUE OF INFORMATION

Consider the two extensive form games in Figures E2.1a and E2.1b.

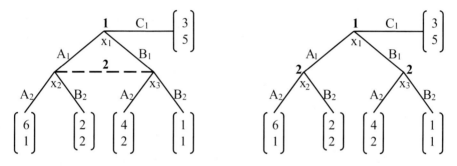

Figure E2.1a   **Figure E2.1b**

Questions
1. Explain why B_1 is strictly dominated in the game in Figure E2.1a but not in the game in Figure E2.1b.
2. Eliminate iteratively the dominated strategies in Figures E2.1a and E2.1b.
3. Comment on the value of information.

EXERCISES

♣ EXERCISE 6 STACKELBERG FIRST PRICE ALL PAY AUCTION

Consider the Stackelberg first price all pay auction in Figure E2.2.

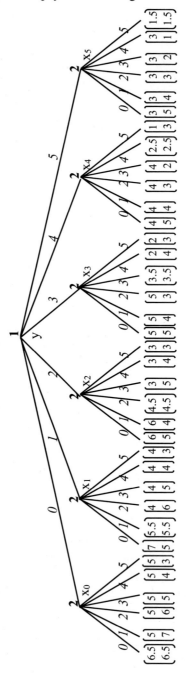

■ Figure E2.2

Question

Eliminate iteratively the strictly and weakly dominated strategies. Show that the order of eliminations matters.

♣♣ EXERCISE 7 BURNING MONEY

The *burning money game* has been proposed by Van Damme (1989)[1] and Ben-Porath and Dekel (1992).[2] Two players play a battle of the sexes game (where the payoffs are in dollars), described in Matrix E2.5a, but player 1 has the possibility of burning (or not), b (or $=\bar{b}$), 2 dollars before starting the game. If she burns 2 dollars, the payoffs, at the end of the game, are the ones in Matrix E2.5b. The extensive form of the game is given in Figure E2.3.

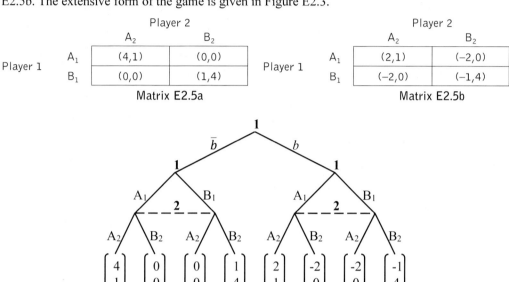

Figure E2.3

Questions

Eliminate iteratively all the dominated strategies. Comment. What happens if people do not make many iterations?

♣♣ EXERCISE 8 PRE-EMPTION GAME IN EXTENSIVE FORM AND NORMAL FORM, AND CROSSED RATIONALITY

Consider the pre-emption game studied in Exercise 5 in chapter 1, and reproduced in Figure E2.4. The normal form of the game is recalled in Matrix E2.6.

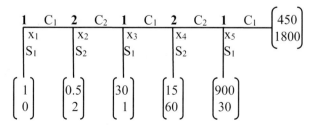

Figure E2.4

EXERCISES

		Player 2			
		$(S_2/x_2 S_2/x_4)$	$(S_2/x_2 C_2/x_4)$	$(C_2/x_2 S_2/x_4)$	$(C_2/x_2 C_2/x_4)$
	$(S_1/x_1 S_1/x_3 S_1/x_5)$	(1,0)	(1,0)	(1,0)	(1,0)
	$(S_1/x_1 S_1/x_3 C_1/x_5)$	(1,0)	(1,0)	(1,0)	(1,0)
	$(S_1/x_1 C_1/x_3 S_1/x_5)$	(1,0)	(1,0)	(1,0)	(1,0)
Player 1	$(S_1/x_1 C_1/x_3 C_1/x_5)$	(1,0)	(1,0)	(1,0)	(1,0)
	$(C_1/x_1 S_1/x_3 S_1/x_5)$	(0.5,2)	(0.5,2)	(30,1)	(30,1)
	$(C_1/x_1 S_1/x_3 C_1/x_5)$	(0.5,2)	(0.5,2)	(30,1)	(30,1)
	$(C_1/x_1 C_1/x_3 S_1/x_5)$	(0.5,2)	(0.5,2)	(15,60)	(900,30)
	$(C_1/x_1 C_1/x_3 C_1/x_5)$	(0.5,2)	(0.5,2)	(15,60)	(450,1800)

Matrix E2.6

Questions
1. Eliminate iteratively the weakly and strictly dominated strategies in the normal form game.
2. Eliminate iteratively the weakly and strictly dominated strategies in the extensive form game. Do the iterative processes follow the same steps in both forms?
3. Why is there a logical inconsistency in the extensive form approach? What about crossed rationality?

♣ EXERCISE 9 GUESSING GAME AND CROSSED RATIONALITY

Remember the N-player guessing game presented in Exercise 1 of chapter 1: each player chooses a real number in [0,100]. N is unknown but is usually supposed to be large, so that the mean of the chosen numbers does not depend a lot on the value chosen by an individual. The winner is the player closest to p times the average of all chosen numbers, where p belongs to]0,1[. In case of a tie, all the winners share equally the amount of the win.

Questions
1. Eliminate iteratively the strictly and weakly dominated strategies for p=2/3.
2. What about the levels of crossed rationality?

♣ ♣ EXERCISE 10 BERTRAND DUOPOLY

See the Bertrand duopoly presented in Exercise 1 in chapter 1. $u_i(p_1,p_2)$ is firm i's profit function, p_i is the price set by firm i, i=1,2 and D(p)=max (0, K−ap) is the demand for price p. We get:

- if $p_1=p_2$, $u_1(p_1,p_2)=\max(0,K-ap_1)(p_1-c)/2=u_2(p_1,p_2)$
- if $p_1>p_2$, $u_1(p_1,p_2)=0$, $u_2(p_1,p_2)=\max(0,K-ap_2)(p_2-c)$
- if $p_1<p_2$, $u_2(p_1,p_2)=0$, $u_1(p_1,p_2)=\max(0,K-ap_1)(p_1-c)$

Suppose that the prices are in dollars. So each player may a priori play any price 0, 0.01, 0.02,....∞

Questions
Eliminate iteratively the strictly and weakly dominated strategies. Comment on the result.

EXERCISES

♣ EXERCISE 11 TRAVELLER'S DILEMMA (BASU'S VERSION)

See the traveller's dilemma exposed in Exercise 1 of chapter 1. Recall that each traveller $i=1,2$ asks for a given integer x_i in [2,100] (for his damaged luggage), and that the company reimburses the travellers in the following way:

- If $x_1=x_2$, each traveller gets x_1
- If $x_i>x_j$ $i\neq j=1,2$, traveller j gets x_j+P, where P is a fixed amount higher than 1, and traveller i only gets max $(0, x_j-P)$; P is common knowledge.

Questions:
1. What about the company if a player never plays an (iterative) dominated strategy?
2. Does the same result hold if we replace max $(0, x_j-P)$ by x_j-P (Basu's rule)?
3. What about the levels of crossed rationality?

♣ EXERCISE 12 TRAVELLER'S DILEMMA, THE STUDENTS' VERSION

The first time my students played the traveller's dilemma, some of them *wrongly understood the reimbursement rule*. They worked with the following, now called students' reimbursement rule:

- If $x_1=x_2$, each traveller gets x_1.
- If $x_i>x_j$ $i\neq j=1,2$ traveller j gets x_j+P (with P>1), but traveller i gets max $(0, \boxed{x_i}-P)$. So the traveller who asks for more is still punished, but the penalty P is deduced from the amount he asks for.

That is why I asked other students to play the above game, with one modification: if $x_i>x_j$ $i\neq j=1,2$ traveller j gets x_j+P and traveller i gets $\boxed{x_i}-P$. We now study this game, I will still call it the students' variant, despite my modification (this modification leads to a more easy study of equilibria).

Questions
1. Suppose P<49. If a player never plays a dominated strategy, does the company achieve its aim?
2. What about the levels of crossed rationality?

♣ ♣ ♣ EXERCISE 13 DUEL GUESSING GAME, THE COWBOY STORY

The dual guessing game is presented in Exercise 6 in chapter 1 and its extensive form is recalled in Figure E2.5a.

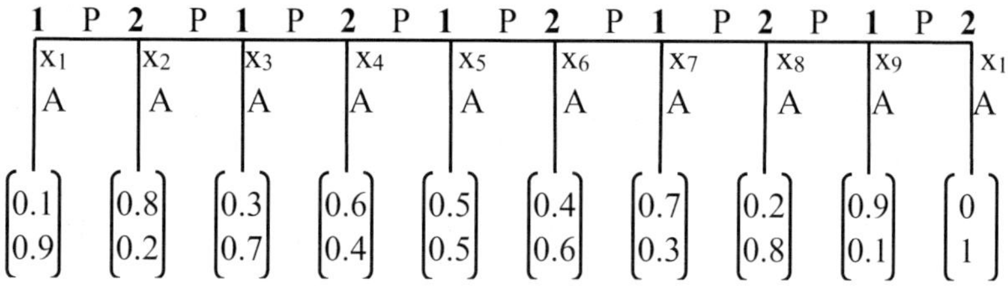

```
   1   P   2   P   1   P   2   P   1   P   2   P   1   P   2   P   1   P   2
  |x1     |x2     |x3     |x4     |x5     |x6     |x7     |x8     |x9     |x10
  |A      |A      |A      |A      |A      |A      |A      |A      |A      |A

  [0.1]   [0.8]   [0.3]   [0.6]   [0.5]   [0.4]   [0.7]   [0.2]   [0.9]   [0 ]
  [0.9]   [0.2]   [0.7]   [0.4]   [0.5]   [0.6]   [0.3]   [0.8]   [0.1]   [1 ]
```

Figure E2.5a

EXERCISES

Questions
1. Eliminate iteratively the strictly and weakly dominated strategies of this game. Show that there are different ways of proceeding.
2. Switch to the reduced normal form of this game. Eliminate iteratively the strictly and weakly dominated strategies of this game. Find again the different ways mentioned in question 1.
3. Now suppose that player 1 guesses better than player 2, so that an additional clue raises her probability to give the good answer by 0.15 (see Exercise 6 in chapter 1). The new extensive form game is recalled in Figure E2.5b. Eliminate iteratively the dominated strategies.

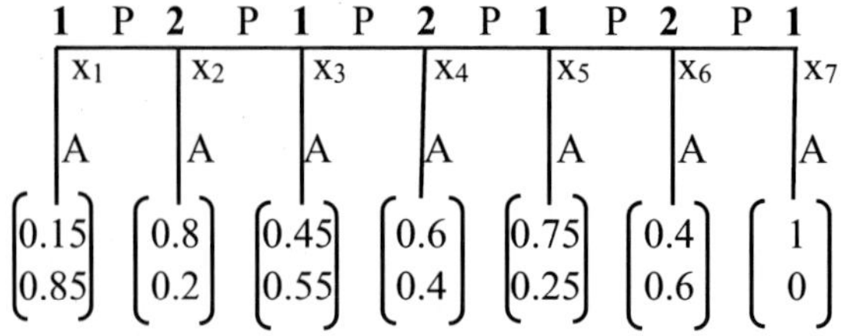

$$\begin{array}{ccccccccccccc}
1 & P & 2 & P & 1 & P & 2 & P & 1 & P & 2 & P & 1 \\
\mid x_1 & & \mid x_2 & & \mid x_3 & & \mid x_4 & & \mid x_5 & & \mid x_6 & & \mid x_7 \\
\mid A & & \mid A & & \mid A & & \mid A & & \mid A & & \mid A & & \mid A \\
\begin{pmatrix}0.15\\0.85\end{pmatrix} & & \begin{pmatrix}0.8\\0.2\end{pmatrix} & & \begin{pmatrix}0.45\\0.55\end{pmatrix} & & \begin{pmatrix}0.6\\0.4\end{pmatrix} & & \begin{pmatrix}0.75\\0.25\end{pmatrix} & & \begin{pmatrix}0.4\\0.6\end{pmatrix} & & \begin{pmatrix}1\\0\end{pmatrix}
\end{array}$$

■ **Figure E2.5b**

4. Comment: why is this game a duel game? Do you know the cowboy story? Generalize the above studies to give the characteristics of the remaining strategies after the iterative elimination of weakly and strictly dominated strategies.

♣ ♣ ♣ EXERCISE 14 SECOND PRICE SEALED BID AUCTION

The game is an N–player second price sealed bid auction (SPSBA). Each player i values an object x_i, with x_i distributed on $[0,V]$. The distributions are independent among the players and common knowledge but each player only knows his own value. The players simultaneously bid for the object. The highest bidder gets the object but he only pays the second highest price. The other players do not get the object and pay nothing. If K players propose the highest bid each of them gets the object and pays the second highest price with probability $1/K$. If all the players play the same price, they all get the object and pay the common proposed price with probability $1/N$.

Questions
1. Show that the bidding function, $b(x_i)=x_i$, weakly dominates each other strategy, when N=2, regardless of the value distribution functions.
2. Turn to N>2. Observe that the payoff rule "If K players propose the highest price, each of them gets the object and pays the second highest price with probability $1/K$" is somewhat ambiguous. If we rank the bids from the highest to the lowest, the first K prices are p, and the K+1th price is p'<p. So the second highest price is either p' (first payoff rule), or it is p (second payoff rule), *because it is the second price in the list*. Does the bidding function $b(x_i)=x_i$, i from 1 to N, still dominate all the others, when you opt for the first payoff rule? What problem do you encounter? Comment.

EXERCISES

♣ ♣ ♣ EXERCISE 15 ALMOST COMMON VALUE AUCTION, BIKHCHANDANI AND KLEMPERER'S[3] RESULT

The game is a two player **second price almost common value auction**. Two players participate in the auction of an object of common value V, which means that both players value the object V. Player i, i=1,2, privately observes a signal x_i distributed on [0, 1], and the common value of the object is $V=V(x_1,x_2)$. We focus on the wallet game, where $V(x_1,x_2)=x_1+x_2$ (all happens as if the wallet is compounded of two parts, and each player only observes one of the parts). *(See Exercises 20 to 24 in chapter 3 for more information on the wallet game.)*

In the almost common value auction, one player has a valuation advantage over the other. Player 2, the regular player, values the wallet at x_1+x_2, whereas player 1, the advantaged player, has an additional private value K for the wallet, so that she values it x_1+x_2+K.

The game goes as follows: given his private signal x_i, each player i, i=1,2, makes an offer for the wallet. The winner is the player who makes the highest offer: she gets the wallet and pays the bid proposed by the loser (second price). The loser doesn't get the object and pays nothing.

Question

Bikhchandani (1988)[4] showed that in this auction the advantaged player always wins the auction at a very low price, even for K close to 0. Prove this result by eliminating iteratively the weakly dominated strategies.

NOTES

1. Van Damme, E., 1989. Stable equilibria and forward induction. *Journal of Economic Theory*, 48, 476–496.
2. Ben-Porath, E. and Dekel, E., 1992. Signaling future actions and the potential for sacrifice. *Journal of Economic Theory*, 57, 36–51.
3. Klemperer, P.D., 1998. Auctions with almost common values: the wallet game and its applications, *European Economic Review*, May 42 (3–5), 757–769.
4. Bikhchandani, S., 1988. Reputation in repeated second-price auctions. *Journal of Economic Theory*, 46, 97–119.

Chapter 3

Nash Equilibrium

INTRODUCTION

Nash equilibrium is a very simple concept, so simple that it is now applied in almost every field of economics: a strategy profile is a Nash equilibrium if each player is happy to play his strategy, given the strategies played by the others. In other words, a Nash equilibrium requires minimal stability: if a strategy profile isn't a Nash equilibrium, then at least one player is better off switching to another strategy without any help or co-deviation from somebody else, so it has indeed very little stability.

So, how is it possible that not everybody agrees with this concept? Well, the idea behind the concept is fine, strong, simple and often very efficient. But it has also numerous limits. One flaw is that the Nash concept only ensures stability against unilateral deviations. Another flaw is that it only tests if an action does worse or better than another; it does not take into account other possible comparisons in the game. Nash equilibria may also not help us to play: for example, when there are many Nash equilibria, how can a player guess that the other players coordinate on a given Nash equilibrium profile? If you can't be sure that the others conform to a given Nash equilibrium profile, you generally have no reason to conform to it.

Be that as it may, Nash equilibrium is a very stimulating concept. We will first focus on the positive sides of the concept before highlighting some of its limits.

In section 1 we define the concept and apply it to normal form games, namely to the first price sealed bid all pay auction. We then apply it to extensive form games in section 2, namely to the ascending all pay auction. In the proposed games Nash equilibria at least partially fit with intuition. In section 3, we show that Nash equilibria often give good hints on the way to play economic games, namely auctions, where intuition is of little help. We also show another nice asset: in some games, a complicated Nash equilibrium may correspond to an easy to learn rule of behaviour. We illustrate this property with the prisoners' disks game.

In section 4 however, we start moderating our enthusiasm. First we show that the multiplicity of Nash equilibria may prevent the possibility of spontaneously coordinating on a given Nash equilibrium, even if players are allowed to talk before playing. Our remarks open on focal points and forward induction. We also show that Nash's approach of out of equilibrium actions can lead to incredible, unintuitive out of equilibrium actions. In section 5, we highlight another drawback. Even if the game has a unique easy Nash equilibrium, it may be of no use, because people obviously will not play it. This will be true in the envelope game, but also in asymmetric auction games. In the envelope game, the difference between the Nash equilibrium and the real way to play stems from the possibility, in real life only, of sharing inconsistent beliefs. In the studied asymmetric auction, Nash equilibria either may be strange, or may be risky. This

NASH EQUILIBRIUM

leads us to discuss the consistency of beliefs and alternative approaches of games, like cautious behaviour and risk dominance. We conclude with the meaning of the probabilities in a mixed Nash equilibrium.

1 NASH EQUILIBRIUM, A FIRST APPROACH

1.1 Definition and existence of Nash equilibrium

Nash equilibria[1] are usually defined in mixed or behavioural strategies, but you can also focus on pure strategy equilibria. Either way the definition is the same:

> **Definition 1. Nash equilibrium, Nash equilibrium path, Nash equilibrium outcome**
> A pure strategy profile s* is a *Nash equilibrium* if and only if:
> $$\forall i \in \mathcal{N}, \forall s_i \in S_i \ u_i(s_i^*, s_{-i}^*) \geq u_i(s_i, s_{-i}^*)$$
> A mixed strategy profile p* is a Nash equilibrium if and only if:
> $$\forall i \in \mathcal{N}, \forall p_i \in P_i \ u_i(p_i^*, p_{-i}^*) \geq u_i(p_i, p_{-i}^*)$$
> A behavioural strategy profile π^* is a Nash equilibrium if and only if:
> $$\forall i \in \mathcal{N}, \forall \pi_i \in \Pi_i \ u_i(\pi_i^*, \pi_{-i}^*) \geq u_i(\pi_i, \pi_{-i}^*)$$
> We say that an action belongs to the *Nash equilibrium path*, if it is played with positive probability when everybody plays in accordance with the Nash equilibrium. The *support of player i's Nash equilibrium strategy* is the set of pure strategies he plays with positive probability at equilibrium. The *Nash equilibrium outcome* is the distribution of probabilities on the payoff vectors induced by the Nash equilibrium.

As explained in the introduction, the idea behind Nash equilibrium is easy and intuitive: in a Nash equilibrium each player plays optimally given the behaviour of the others. No player is tempted to deviate unilaterally; everybody is happy to play his equilibrium strategy, given that the others also play theirs. That is why the Nash equilibrium is sometimes called a *rest point*. Nash's stability is often viewed as a minimal one: if a profile is not a Nash equilibrium, then at least one player is better off deviating from it.

The Nash equilibrium has a nice technical property:

> **Property 1. Existence**
> Any finite game has at least one Nash equilibrium.

The proof rests on the fact that a Nash equilibrium in mixed strategies is a fixed point of a correspondence defined on a convex set.

NASH EQUILIBRIUM

1.2 Pure strategy Nash equilibria in normal form games and dominated strategies

We first apply this concept to easy normal form games, the coordination games in Matrices 3.1 and 3.2:

Matrix 3.1
Player 2
	A_2	B_2
A_1	($\underline{1}$,*1*)	(0,0)
B_1	(0,0)	(1,1)

Matrix 3.2
Player 2
	A_2	B_2
A_1	(1,3)	(0,0)
B_1	(0,0)	(3,1)

In the pure coordination problem (Matrix 3.1), and in the battle of the sexes (Matrix 3.2), there are two obvious pure strategy Nash equilibria: (A_1,A_2) and (B_1,B_2).

> **STUDENTS' INSERT**

(A_1,A_2) is a Nash equilibrium of the game of Matrix 3.1 because:

- A_1 is a best response to player 2's strategy A_2 because A_1 yields the payoff 1 whereas B_1 only yields the payoff 0 when player 2 plays A_2: the compared payoffs are underlined in Matrix 3.1;
- and A_2 is a best response to player 1's strategy A_1 because A_2 yields the payoff 1 whereas B_2 only yields the payoff 0 when player 1 plays A_1: the compared payoffs are in italics in Matrix 3.1.

A similar proof holds for the equilibrium (B_1,B_2) and for the game in Matrix 3.2.

These Nash equilibrium profiles fit completely with intuition, so on the one hand we can agree with Nash's choices, but on the other hand, we can say that we do not need the Nash concept to come to these easy conclusions! In these games, the Nash equilibria are the *Pareto optimal profiles* (i.e. profiles such that no player can do better without harming another one).

But this is not always the case, as can be observed in the game in Matrix 3.3, introduced in chapter 2:

Matrix 3.3
Player 2
	A_2	B_2	C_2
A_1	(7,8)	(3,3)	(0,8)
B_1	(0,0)	(4,4)	($\underline{1}$,4)
C_1	(7,1)	(3,0)	($\underline{1}$,$\underline{1}$)

In this game, there are five pure strategy Nash equilibria: (A_1,A_2), (B_1,B_2), (B_1,C_2), (C_1,A_2) and (C_1,C_2). Only the first is Pareto optimal: (C_1,C_2) for example is very far from being Pareto optimal (both players can do better by switching to (B_1,B_2) or to (A_1,A_2) without harming anybody).

(C_1,C_2) is a Nash equilibrium because player 1 is best off choosing C_1 when player 2 plays C_2 (she gets 1 with C_1, 1 with B_1 and 0 with A_1 and $1 \geq 1 > 0$ (underlined payoffs)), and player 2 is best off choosing C_2 when player 1 plays C_1 (he gets 1 with C_2, 0 with B_2 and 1 with A_2 (payoffs in

NASH EQUILIBRIUM

italic)). But is this equilibrium as intuitive as (A_1,A_2)? *More generally, aren't there some equilibria which have a higher probability of being played?*

A first reaction is to say that players will not coordinate on (C_1,C_2) because this equilibrium is not interesting from a payoff point of view. But this reaction is perhaps a bad one. As a matter of fact (C_1,C_2) has very nice properties not shared by other equilibria. Especially, C_2 weakly dominates both A_2 and B_2, so player 2 is rather pleased to play C_2, in that this action does better or at least as well as A_2 and B_2. It derives that player 1 can reasonably fear that player 2 plays C_2, which justifies her action C_1 (which, moreover, weakly dominates A_1).

By contrast, the Pareto optimal Nash equilibrium (A_1,A_2) we may select as a first reaction seems more fragile. A_1 and A_2 are weakly dominated by C_1 and C_2, even without any iteration. So the players, individually, can only be better off (or get the same payoff) switching from A_1 to C_1 and from A_2 to C_2, even if collectively such a switch is bad for both players.

Let us make an additional observation: Nash equilibria seem to be linked to dominance. As a matter of fact (B_1,B_2), (B_1,C_2), (C_1,A_2) and (C_1,C_2) are all profiles that resist at least one iterative elimination of weakly dominated strategies (see chapter 2). Only the Pareto optimal Nash equilibrium (A_1,A_2) never resists an iterative elimination of weakly dominated strategies. This leads us to outline the links between Nash equilibria and dominated strategies:

Property 2. Links between Nash equilibria and dominated strategies

The iterative elimination of strictly dominated strategies, regardless of the order of the iterations, never eliminates a Nash equilibrium of the game.

The iterative elimination of weakly and strictly dominated strategies may eliminate a Nash equilibrium of the game. But, if the game is finite, then, regardless of the order of elimination, the remaining game obtained after the eliminations always has at least one Nash equilibrium of the studied game. Moreover all the Nash equilibria of the remaining game are Nash equilibria of the studied game.

> **STUDENTS' INSERT**
>
> Let us check these properties on the game in Matrix 3.3.
> By first eliminating A_1 (weakly dominated by C_1) and A_2 (weakly dominated by C_2), and second eliminating C_1 (weakly dominated by B_1), we get the couples (B_1,B_2) and (B_1,C_2), which are both Nash equilibria of the game in Matrix 3.3, but we eliminate all the other Nash equilibria of the game.
> Similar observations follow the other orders of eliminations.

To illustrate the first point of property 2, we return to the prisoner's dilemma. Consider any version of this game, for example the one given in Matrix 3.4:

		Player 2 A_2	Player 2 B_2
Player 1	A_1	(7,7)	(0,8)
	B_1	(8,0)	(1,1)

Matrix 3.4

NASH EQUILIBRIUM

The only pure strategy Nash equilibrium (more generally the only Nash equilibrium) of this game is (B_1,B_2) (B_1 is the best answer to B_2 and B_2 is the best answer to B_1). Especially (A_1,A_2) is not a Nash equilibrium because B_1 (and not A_1) is the best response to A_2 (and the best response to A_1 is B_2 and not A_2).

We already know that A_1 and A_2 are strictly dominated by B_1 and B_2. So we observe that eliminating strictly dominated actions (even iteratively and regardless of the order of the iterations) does not eliminate any Nash equilibrium of the game.

The fact that the Nash equilibrium is the only non Pareto optimal couple, and that it is *Pareto dominated* by the non equilibrium couple (A_1,A_2), is the *dilemma* of the prisoners. We have already commented on this dilemma in chapter 1 and chapter 2 (see section 1.2.2).

Up to now we have only worked with pure strategies. You may wonder if a pure strategy Nash equilibrium resists the introduction of mixed strategies? The answer is of course yes, because otherwise this equilibrium concept would make no sense. As a matter of fact, suppose that a player has three actions, A, B and C, and that he plays A at the pure strategy equilibrium. Consequently A does better than B and C given the strategies of the other players. It follows that A also does better than any mixture of A, B and C, which ensures that A is still the best answer, even if you allow players to play in mixed strategies. So studying pure strategy Nash equilibria does in no way mean that people are only allowed to play pure strategies. We could redefine a pure strategy Nash equilibrium in this way:

> A pure strategy profile s* is a *Nash equilibrium* if and only if:
> $\forall i \in \mathcal{N}, \forall p_i \in P_i \; u_i(s_i^*, s_{-i}^*) \geq u_i(p_i, s_{-i}^*)$

1.3 Mixed strategy Nash equilibria in normal form games

Pure strategy Nash equilibria do not always exist. The first price sealed-bid all pay auction with M=5, V=3, studied in chapters 1 and 2, is a game with no pure strategy Nash equilibrium. It is recalled in Matrix 3.5a.

		Player 2					
		0	1	2	3	4	5
Player 1	0	(6.5,6.5)	(5,7)	(5,6)	(5,5)	(5,4)	(5,3)
	1	(7,5)	(5.5,5.5)	(4,6)	(4,5)	(4,4)	(4,3)
	2	(6,5)	(6,4)	(4.5,4.5)	(3,5)	(3,4)	(3,3)
	3	(5,5)	(5,4)	(5,3)	(3.5,3.5)	(2,4)	(2,3)
	4	(4,5)	(4,4)	(4,3)	(4,2)	(2.5,2.5)	(1,3)
	5	(3,5)	(3,4)	(3,3)	(3,2)	(3,1)	(1.5,1.5)

Matrix 3.5a

We know that we can eliminate the strictly dominated strategies without eliminating any Nash equilibrium, so we can focus on the game in Matrix 3.5b:

NASH EQUILIBRIUM

		Player 2		
	0	1	2	3
0	(6.5,6.5)	(5,7)	(5,6)	(5,5)
1	(7,5)	(5.5,5.5)	(4,6)	(4,5)
2	(6,5)	(6,4)	(4.5,4.5)	(3,5)
3	(5,5)	(5,4)	(5,3)	(3.5,3.5)

Player 1 labels rows.

Matrix 3.5b

We already observed in chapter 1 that no couple of (pure) actions is (unilaterally) stable. For example, if player 1 bids 0, player 2 is best off bidding 1, but if player 2 bids 1, player 1 is best off bidding 2. But, if player 1 bids 2, then player 2 is for example best off bidding 0, in which case it is better for player 1 to bid 1 and so on. This simply means that there is no pure strategy Nash equilibrium. Let us prove it:

> **STUDENTS' INSERT**

Suppose that (a,b) is a Nash equilibrium.
- If a=b, then, if a=0 or 1, it is better for player 2 to bid a+1 (because 7>6.5 and 6>5.5), so b=a is not his best response, and if a=2 or 3, it is better for player 2 to bid 0 (b=a is not his best response).
- If a<b, then, if b=1, it is better for player 1 to bid b+1 (hence a<b is not her best response). If b>1, if a is different from 0, then it is better for player 1 to bid 0, hence a is not her best response; and if a=0, then it is better for player 2 to bid 1; hence b is not his best response.

It follows that people usually quickly agree that playing in a mixed way is surely the best way to play. But how? What should we do to fix the probabilities? This is a very difficult question and *we are not sure that people necessarily agree with the probabilities as they are defined in a Nash equilibrium* (see the conclusion and chapter 6).

But for the moment, let us stick to Nash's logic. To establish mixed strategy Nash equilibria, we simply have to take into account a logical property:

Property 3

If two actions A and B are played with positive probability at equilibrium, then they necessarily lead to the same expected payoff, and this payoff is necessarily higher or equal to the expected payoff the player can obtain with any other pure strategy. This immediately derives from the definition of the Nash equilibrium: if A leads to a lower payoff than B, A cannot be played at equilibrium, because adding to B's probability the probability assigned to A (and not playing A) raises the player's payoff.

NASH EQUILIBRIUM

> **STUDENTS' INSERT**
>
> We construct a Nash equilibrium in which both players assign a positive probability to the three bids 0, 1 and 2. Suppose that player 1, respectively player 2, bids 0, 1 and 2, with probabilities p_0, p_1 and p_2, respectively with probabilities q_0, q_1 and q_2. To bid 0, 1 and 2 with positive probability, player 1 has to get the same payoff with the three bids, and this payoff has to be higher than or equal to the one obtained with bid 3, i.e. $6.5q_0+5q_1+5q_2=7q_0+5.5q_1+4q_2=6q_0+6q_1+4.5q_2 \geq 5q_0+5q_1+5q_2$ and $q_0+q_1+q_2=1$. This system of equations has a unique solution $q_0=q_1=q_2=1/3$.
>
> Consequently, for player 1 to optimally play her three bids 0, 1 and 2, it is necessary that player 2 bids 0, 1 and 2 with the same probability 1/3. But, for player 2 to be able to bid 0, 1 and 2 with positive probability, it is necessary that the three bids lead to the same expected payoff, and that this payoff is higher than or equal to the one obtained with bid 3. i.e.: $6.5p_0+5p_1+5p_2=7p_0+5.5p_1+4p_2=6p_0+6p_1+4.5p_2 \geq 5p_0+5p_1+5p_2$ and $p_0+p_1+p_2=1$. This system has the unique solution $p_0=p_1=p_2=1/3$. And player 1 can play these probabilities, because she is indifferent between the bids 0, 1 and 2 (and better off than with the bid 3) as soon as player 2 bids 0, 1 and 2, each with probability 1/3.
>
> So we get a mixed strategy Nash equilibrium, in which each player bids 0, 1 and 2 with the same probability 1/3. We will write it: $((p_0=p_1=p_2=1/3, p_3=0), (q_0=q_1=q_2=1/3, q_3=0))$, where p_3 and q_3 are the probabilities assigned to bid 3 respectively by player 1 and 2.
>
> Player 1's and player 2's expected payoff is $(6.5+5+5)/3=5.5$. The payoff obtained with bid 3 is only $5q_0+5q_1+5q_2=5p_0+5p_1+5p_2=5$.

The way we obtained the equilibrium highlights the following rather strange property:

> **Property 4**
>
> In a mixed Nash equilibrium, the values of player i's equilibrium probabilities have no impact on player i's payoff: they are linked to the other players' payoffs.

For example, in the previous equilibrium, it is because player 2 assigns probability 1/3 to each of his bids 0, 1 and 2 that player 1 gets the same payoff with the bids 0, 1 and 2. These probabilities 1/3 are of no importance for player 2. As soon as player 1 behaves in the equilibrium way, player 2 *is completely indifferent between the three bids* 0, 1, and 2: hence, as regards his own payoff, he could bid 0 with probability 0.5, 1 with probability 0.4 and 2 with probability 0.1, for example. If he chooses to play the three bids with probability 1/3, it is not because it is best for him to play in this way, but because *player 1 needs the three probabilities 1/3* to be indifferent between her own bids 0, 1 and 2. And the same observation is true for player 1.

So player 2's probabilities are only necessary to justify player 1's behaviour, and player 1's probabilities are only necessary to justify player 2's behaviour. More generally, in an N player game mixed Nash equilibrium, the only *role of a player's probabilities is to stabilize (justify) the strategies of the N–1 other players*.

This is a logical property from a Nash point of view: if a player plays several actions at equilibrium, they necessarily provide the same payoff (unless he would only play the action that

NASH EQUILIBRIUM

leads to the highest payoff): so the player is necessarily indifferent with regard to the probabilities he assigns to these actions.

But what is logical from the Nash point of view is perhaps less logical in reality, which may explain why many students, when asked to comment on the values of given probabilities in a Nash equilibrium, often make the *following wrong comment: if a player plays A with probability 4/5 and B with probability 1/5, he appreciates A more than B. This is a lasting error, and I think that lasting errors should command attention: I will come back to this point in chapter 6.*

For the moment, let us practise mixed strategy Nash equilibria.

> **STUDENTS' INSERT**
>
> What is funny to do and not often done, is to show that there exists no other mixed strategy Nash equilibrium in the above all pay auction.
>
> So, let us first observe that player 1 cannot bid 3 at equilibrium. Why is that?
>
> Bidding 3 does worse than bidding 0, except if player 2 only bids 1 and 2. So let us suppose that player 2 only bids 1 and 2 with positive probability. Then player 1, at equilibrium, will not bid 1, because bidding 2 does always better (6>5.5 and 4.5>4). So player 1 bids at most 0, 2 and 3 with positive probability. But then, at equilibrium, player 2 never bids 2, because bidding 0 does always better (6.5>6, 5>4.5 and 5>3). So player 2 can only bid 1 at equilibrium, which induces player 1 to only play 2 and not 3. For reasons of symmetry, we also deduce that player 2 cannot bid 3 at equilibrium.
>
> Now let us show that player 1 necessarily bids 0 with positive probability at equilibrium. We proceed by contradiction and suppose that she doesn't bid 0 at equilibrium. If so, given that she does not bid 3, player 2 will never bid 1, because bidding 2 always does better (6>5.5 and 4.5>4). But in that case player 1 cannot bid 2, because, if player 2 only plays 0 and 2, player 1 is better off bidding 0 than bidding 2 (6.5>6, 5>4.5). So, at the end, player 1 only plays 1, which leads player 2 to bid 2, which leads player 1 back to 0 or 3, a contradiction. So, player 1 bids 0 with positive probability at equilibrium, and player 2 also, by symmetry.
>
> It follows that player 2 bids 2 with positive probability, because if not, player 1 would not bid 0 at equilibrium, because bidding 1 would lead to a higher payoff (7>6.5 and 5.5>5). By symmetry, player 1 also bids 2 with positive probability. It follows that player 2 also necessarily bids 1 with positive probability, unless player 1 would not bid 2 (she would get more with bid 0 than with bid 2 (6.5>6 and 5>4.5). By symmetry, player 1 also bids 1 with positive probability.
>
> So the conclusion is that, at the equilibria of this game, both players bid with positive probability each bid that is not strictly or weakly dominated. And, given our previous proof, we know that there is a unique equilibrium of this kind, in which the players play each bid with the same probability.

This result *is robust in that the structure of the equilibrium (bid all non dominated bids with a same probability) does not vary with the capacity to pay, M, the value of the object, V, or the size of the increment.* For example, if we switch from the bid increment 1 to the increment 0.5, the only Nash equilibrium consists in bidding all prices from 0 to 2.5 with probability 1/6. More generally, when the bid increment is Δ, the only Nash equilibrium consists of playing all the prices from 0 to 3–Δ with the same probability $\Delta/3$. So at equilibrium each player has the expected payoff: $[6.5 + 5(\frac{3}{\Delta} - 1)]/(\frac{3}{\Delta}) = 0.5\Delta + 5$.

So we may conjecture that if the increment goes to 0; i.e. if we *switch to a continuous game*, each player should play any price in the interval [0,3] with the same probability and get a payoff that

NASH EQUILIBRIUM

goes to 5. That is to say, it seems that the equilibrium strategy, for V=3, consists in playing the uniform distribution on [0,3].

Let us prove that this conjecture is right.

> **➤ STUDENTS' INSERT**
>
> This proof shows how to approach continuous games.
>
> If player 2 plays the uniform distribution on [0,3], player 1, by bidding b (necessarily ≤3), wins the auction if and only if player 2 plays less than b, so she gets the expected payoff: (5–b)+3F(b), where F(x) is player 2's cumulative distribution function. F(b)=b/3, given player 2's density function. So player 1 gets 5–b+3b/3=5 regardless of b, and she can play each bid with any probability; it follows that playing the uniform distribution on [0,3] is a best answer.
>
> More generally, we show that, if the value of the object is V and the capacity to pay M, playing the uniform distribution on [0,V] is the unique mixed Nash equilibrium of the game that assigns a positive weight to each bid b in [0,V].
>
> Given that 0 and any bid b in [0,V] are played with positive probability, player 1 gets the same payoff with 0 and b. The payoff obtained with 0 is M, the payoff obtained with b is $M - b + V \int_0^b f(x)dx$
>
> =(M–b)+VF(b), where f(x) and F(x) are player 2's equilibrium density function and cumulative distribution function; we need M=M–b+VF(b). It follows F(b)=b/V, so player 2 plays the uniform distribution on [0, V]. By symmetry, player 1 plays the same distribution. Each bid b is played with probability dv/V (observe that dv is the increment, which in our discrete version is 1, leading to the probabilities 1/V=1/3).

What about the intuition behind the result? We find the nature of the result (the play of each non dominated bid with the same probability), its robustness with regard to the choice of the increment, its continuity in the increment going to 0, quite interesting from a pragmatic point of view and rather intuitive, so we think that it can help the players to bid in such an auction game.

2 NASH EQUILIBRIA IN EXTENSIVE FORM GAMES

We now discuss the Nash equilibrium concept in extensive form games. *Here, I have to mention that my students often skip this section* because of the nice property below:

> **Property 5. Nash equilibrium, normal form and extensive form games**
> Nash equilibrium outcomes do not depend on the chosen representation of the game, i.e. they are the same in the normal and in the extensive form of a game.

As a consequence, my students often promptly write the normal form matrix that corresponds to the extensive form game under study, and study the Nash equilibria in the matrix game, because they usually find this way much easier!

Yet there are at least two good reasons to generally not do so. First, as soon as a player plays several times in a game or if there are more than three players, the normal form matrix becomes a nightmare. Second, the extensive form, better than the strategic form, highlights the way the Nash

NASH EQUILIBRIUM

equilibrium concept deals with some parts of the game (namely those that should not be reached when players stick to the Nash equilibrium path): so an extensive form game often *talks better* about behaviour than a matrix.

2.1 Nash equilibria in the ascending all pay auction/war of attrition game

The ascending all pay auction, studied in chapter 1 and 2 with V=3 and M=5, is reproduced in Figure 3.1.

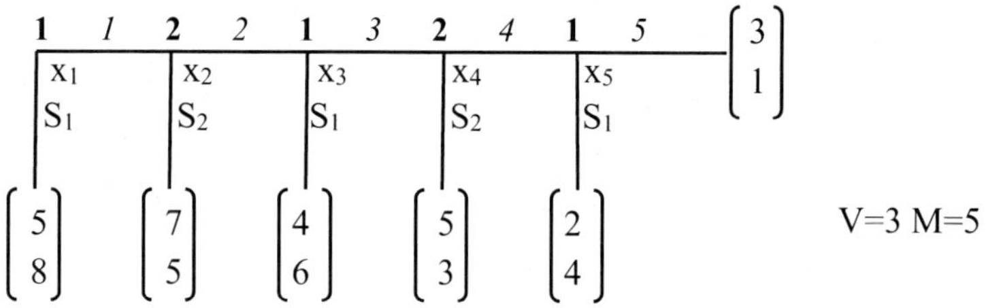

$$
\begin{array}{c}
\underset{S_1}{\overset{1}{|x_1}} \xrightarrow{1} \underset{S_2}{\overset{2}{|x_2}} \xrightarrow{2} \underset{S_1}{\overset{1}{|x_3}} \xrightarrow{3} \underset{S_2}{\overset{2}{|x_4}} \xrightarrow{4} \underset{S_1}{\overset{1}{|x_5}} \xrightarrow{5} \begin{pmatrix}3\\1\end{pmatrix}\\
\begin{pmatrix}5\\8\end{pmatrix} \quad \begin{pmatrix}7\\5\end{pmatrix} \quad \begin{pmatrix}4\\6\end{pmatrix} \quad \begin{pmatrix}5\\3\end{pmatrix} \quad \begin{pmatrix}2\\4\end{pmatrix}
\end{array}
\qquad V=3\ M=5
$$

Figure 3.1

> **STUDENTS' INSERT**

We show that $E_1=((S_1/x_1,S_1/x_3,S_1/x_5), (2/x_2,4/x_4))$ and $E_2=((1/x_1,3/x_3, 5/x_5), (S_2/x_2,S_2/x_4))$ are two Nash equilibria of the game.

Let us start with E_1.

Player 1 knows that player 2 always overbids. Hence if she bids 1, she can at best get 4 by stopping at x_3, because player 2 bids 2 at x_2 and 4 at x_4. So player 1 is better off stopping immediately at x_1 in order to get 5. Given that she necessarily stops at x_1, her actions at x_3 and x_5 have no impact on her payoff. Hence $(S_1/x_1,S_1/x_3,S_1/x_5)$ is a best response of player 1.

Given that player 1 stops at x_1, player 2's behaviour has no impact on his payoff given that he gets 8 whatever he does. It follows that overbidding at each decision node is one of the best responses of player 2.

We now justify E_2.

Player 2 knows that player 1 always overbids. Given that player 1 bids 1 at x_1, player 2 is called on to play at x_2. He knows that if he bids 2, player 1 overbids 3 at x_3 and 5 at x_5 so he gets at best the payoff 3, which is lower than the payoff 5 he gets by stopping at x_2. So he necessarily stops at x_2. It derives that what he does at x_4 is of no importance, hence $(S_2/x_2,S_2/x_4)$ is a player 2's best response. Now, given that player 2 stops at x_2, it is best for player 1 to bid 1 at x_1 because she gets her maximal payoff 7. What she does at x_3 and x_5 is of no importance, because these nodes are not reached given that player 2 stops at x_2. So $(1/x_1,3/x_3,5/x_5)$ is a player 1's best response.

Let's discuss both equilibria.

We start with a comment on best responses. Look, for example, at the strategy $(1/x_1,3/x_3,5/x_5)$ in E_2. We said that player 1's behaviour at x_3 and x_5 has no impact on her payoffs because, given player 2's behaviour, x_3 and x_5 are never reached. So why do we choose the strategy $(1/x_1,3/x_3, 5/x_5)$? Why don't we choose $(1/x_1,S_1/x_3,S_1/x_5)$ for example? Well $(1/x_1,S_1/x_3,S_1/x_5)$ is also a best response for player 1, but *if player 1 plays in this way, then $(S_2/x_2,S_2/x_4)$ is no longer a best response*

NASH EQUILIBRIUM

for player 2. As a matter of fact, if player 1 goes on at x_1 but stops at x_3, then player 2 reaches x_2 and it is better for him to overbid, because he will get 6 instead of 5 (the payoff obtained by stopping). *So the local strategies of a player in an equilibrium do not all have the same role.* With regard to $(1/x_1, 3/x_3, 5/x_5)$ in E_2, $1/x_1$ *has a direct optimal impact on player 1's payoff*, whereas $3/x_3$ and $5/x_5$ just justify player 2's strategy, so have *no direct impact on player 1's payoff* (they have an indirect impact in that player 2's induced stopping strategy benefits player 1).

> So, and this is a very general observation, many local strategies of player i's Nash behavioural strategy are not necessary for the strategy to be a best response, but they are necessary for the strategies of other players to be best responses.

We now focus on a common point of both equilibria: the game either stops immediately or at player 2's first decision node. As already observed in chapter 2, this fact is logical from a theoretical (rational) point of view. In an all pay auction, if you don't win the object, you only lose money by bidding: so the loser should stop immediately, and this happens in all Nash equilibria. As a consequence of this logical stop, the winner of the auction gets a very high payoff, in that he has to make a very low bid to get the object (player 1 pays 1 in E_2 and player 2 pays nothing in E_1).

E_1 and E_2 differ with regard to the player who wins the auction: player 2 wins in E_1, player 1 wins in E_2. This result differs from the one obtained with iterative elimination of (mostly weakly) dominated strategies, which only led to E_2. So we have again an illustration of the property that the elimination of weakly dominated strategies may suppress Nash equilibria.

But how can we get two Nash equilibria that are diametrically opposed in terms of winners and losers? This is precisely *due to the high degree of liberty offered by unilateral stability*: when you test the optimality of the strategy of a player, you take the strategies of the other players *as fixed*. It is as if *you are sure* that *the others are sticking to their equilibrium strategies*. So, in E_1, player 1 takes for granted that player 2 will always overbid, so it is best for her to stop immediately, which, in turn, justifies player 2's overbidding behaviour, which doesn't cost him anything because player 1 stops before he can bid. In a similar way, in E_2, player 2 takes for granted that player 1 always overbids, so it is best for him to stop immediately, which, in turn, justifies player 1's overbidding behaviour, given that this behaviour leads to a very low price (1) for an object of value 3.

Well, *an equilibrium looks for behaviour that equilibrates*: one player has to anticipate (to hope) that the other always stops in order to justify his overbidding, and the other has to anticipate (to fear) that the other always overbids in order to justify his always stopping. It is as if one player is anxious (he fears that the opponent overbids and so prefers stopping) and the other plays with this anxiety (he threatens to overbid, hoping that the other stops). In a Nash equilibrium, *anticipations are right* and the associated strategies are therefore optimal.

It also follows from the above analysis that the Nash equilibrium path may contain very few actions. So in a Nash equilibrium, a large part of the game tree is not reached. We come back to this point later.

We also observe that the Nash equilibrium reasoning, by contrast to the iterated dominance process, *does not start with actions at the last nodes*. As a matter of fact we justified both equilibria almost only by focusing on the nodes at the beginning of the game. So it is obvious that we do not necessarily get the same results.

Are there other Nash equilibria? Yes. In general, looking for all the Nash equilibria in a large extensive form game is not obvious (turning to subgame perfection equilibrium criteria will be easier). In our game there are many equilibria but only two equilibrium paths (S_1) and $(1,S_2)$, because only these two paths are reasonable from a rational point of view (the loser stops as soon as possible).

Let us propose the profile E_3: $((S_1/x_1, 3/x_3, 5/x_5), (2/x_2, S_2/x_4))$.

E_3 is a Nash equilibrium: as soon as player 1 stops at x_1, player 2's behaviour has no impact on his payoffs, hence all player 2's strategies are automatically best responses. And, as soon as player 2 overbids at x_2 even if he stops at x_4, stopping at x_1 is optimal for player 1 (because she gets 5 by stopping and at best 5 by bidding 1). So $(S_1/x_1, 3/x_3, 5/x_5)$ is player 1's best response.

Are all the Nash equilibria intuitive? Let us try to answer by analysing E_3. To justify this equilibrium, we could say: player 1 cannot be absolutely sure that player 2 will stop the game at x_2. She may believe that player 2 is rational enough to discover, once at x_4, that player 1 necessarily overbids at x_5, so that he will stop at x_4. But she may not be sure that player 2 is clever enough to deduce that this induces player 1 to go on at x_3: in other terms, player 1 is not sure that player 2 already stops at x_2. She may fear that player 2 overbids at x_2, which can lead her to stop at x_1, because she cannot get more than 5 in that case. But, if she should be at x_3 and x_5 (which is normally not possible given her action at x_1), she overbids because she expects player 2 to stop at x_4. So it is possible to propose a rational explanation, even for actions out of the equilibrium path.

E_1 and E_2 are even easier to justify. If you fear that the opponent always overbids, it is best to stop immediately. And if you hope that the opponent is anxious so that he stops each time he is called on to play, it is rational to always overbid.

Well, what should we play in this game if we were player 1? I propose to bid 1. Why? As stated above, we can't be sure that player 2 will stop at x_2, but we can reasonably hope that he will stop at x_4, in which case we get 5, that is to say as much as if we stop at x_1. But perhaps we are lucky and player 2 already stops at x_2, because he is cleverer or more cautious than we thought, in which case we get more than 5. But of course, we face a risk: if player 2 is stupid or angry (because we did not stop at x_1 and x_3 as he hoped), and chooses to always overbid, we only get 3.

What can we deduce from these stories? Well, to be honest, not a lot. You must be very cautious: stories are often *a nice packaging for Nash equilibria, but not much more*. First, a plausible story for each Nash equilibrium doesn't exist. In our game for example, $((S_1/x_1, 3/x_3, S_1/x_5), (2/x_2, S_2/x_4))$ is a Nash equilibrium, but it is difficult to find a nice story that justifies all the out of equilibrium actions (why does player 2 stop at x_4 knowing that player 1 stops at x_5)? Second, many nice stories exist that are not compatible with a Nash equilibrium. For example, imagine a much longer ascending all pay auction game, for example M=100 and V=60, as in the game proposed to my students (see chapter 1 section 1.2.2 and chapter 4 section 3.1). In that case, many stories may lead both players to overbid at early decision nodes, because both players hope that the opponent will stop first: so 44.4% of my students advise both players to bid till 30 or more, which is incompatible with a Nash equilibrium. Third, the Nash equilibrium concept doesn't require that you justify Nash equilibrium actions that are out of the equilibrium path (because they have no impact on the payoffs and are therefore *automatically* optimal).

We observe here a strong difference between game theory and game reality: theory rightly argues that the loser should not invest money. *But reality says that players have no idea about who will be the loser in the game* when they start playing, so both players may overbid at the first decision nodes because *they both hope to be the winner*. Of course, these anticipations are

NASH EQUILIBRIUM

inconsistent, because there is a winner and a loser, an inconsistency which is not possible at equilibrium. We come back to this point in section 4.

2.2 Same Nash equilibria in normal form games and extensive form games

We now illustrate property 4. According to this property, the Nash equilibria of the game in extensive form are the same as the ones in normal form. The normal form of the game in Figure 3.1 is given in Matrix 3.6.

		Player 2			
		$S_2/x_2 S_2/x_4$	$S_2/x_2 4/x_4$	$2/x_2 S_2/x_4$	$2/x_2 4/x_4$
Player 1	$(S_1/x_1 S_1/x_3 S_1/x_5)$	(5,8)	(5,8)	(5,8)	(5,8)
	$(S_1/x_1 S_1/x_3 5/x_5)$	(5,8)	(5,8)	(5,8)	(5,8)
	$(S_1/x_1 3/x_3 S_1/x_5)$	(5,8)	(5,8)	(5,8)	(5,8)
	$(S_1/x_1 3/x_3 5/x_5)$	(5,8)	(5,8)	(5,8)	(5,8)
	$(1/x_1 S_1/x_3 S_1/x_5)$	(7,5)	(7,5)	(4,6)	(4,6)
	$(1/x_1 S_1/x_3 5/x_5)$	(7,5)	(7,5)	(4,6)	(4,6)
	$(1/x_1 3/x_3 S_1/x_5)$	(7,5)	(7,5)	(5,3)	(2,4)
	$(1/x_1 3/x_3 5/x_5)$	(7,5)	(7,5)	(5,3)	(3,1)

Matrix 3.6

All the strategy profiles which correspond to the encircled couples of outcomes are Nash equilibria of the game. So the game, both in normal and extensive form, has 12 pure strategy Nash equilibria (and two equilibrium paths). The equilibria E_1, E_2 and E_3 are the strategy profiles that correspond to the bold payoffs in italics.

We stress that this equivalence of results between the normal form and the extensive form – which seems a natural property – is not frequent in game theory. Most equilibrium concepts are either developed for the normal or the extensive form, but not for both. So it is a strong property.

> This equivalence goes even further. Reduced normal form games have the same Nash equilibrium outcomes as the associated normal form games.

Let us develop the equivalence of Nash equilibria in normal form and extensive form on the easier game given in Figure 3.2, which is a small two-step sequential all pay auction, in which player 1 can only bid 2 prices, 1 and 3, and player 2, after each price, has only two possible actions, leave the game or overbid by 1 unity.

NASH EQUILIBRIUM

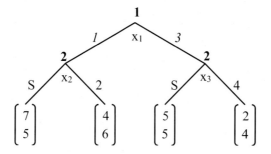

Figure 3.2

First we study this game in extensive form.
Let us look for all the equilibria in behavioural strategies.

> **STUDENTS' INSERT**
>
> **Case 1: player 1 only bids 1**
> In this case player 2 reaches only x_2 and $\pi_{2x2}(S)=0$, because player 2 prefers bidding 2 to leaving the game (6>5). Given that x_3 is not reached, player 2 may play any action at this node, so $\pi_{2x3}(S)$ can be chosen between 0 and 1. Yet, for player 1's strategy to be a best response to player 2's strategy, it is necessary that 4 (the payoff she gets with bid 1) is higher than or equal to $5\pi_{2x3}(S)+2(1-\pi_{2x3}(S))$ (the payoff she gets with bid 3); we get $\pi_{2x3}(S) \leq 2/3$. It follows a first family of behavioural Nash equilibria: $\{(\pi_1(1)=1, (\pi_{2x2}(2)=1, 0 \leq \pi_{2x3}(S) \leq 2/3))\}$. This first family includes the pure strategy Nash equilibrium $(1, (2/x_2, 4/x_3))$.
>
> **Case 2: player 1 only bids 3**
> In this case player 2 reaches x_3 and $\pi_{2x3}(S)=1$ because 5>4. Given that x_2 is not reached, player 2 may play what he wants at this node, so $0 \leq \pi_{2x2}(S) \leq 1$. Yet, for player 1's strategy to be a best response to player 2's strategy, it is necessary that the payoff she gets with bid 3, i.e. 5, is higher than or equal to the payoff obtained with bid 1, i.e. $7\pi_{2x2}(S)+4(1-\pi_{2x2}(S))$. We get $\pi_{2x2}(S) \leq 1/3$. It follows a second family of behavioural Nash equilibria:
> $\{(\pi_1(3)=1, (0 \leq \pi_{2x2}(S) \leq 1/3, \pi_{2x3}(S)=1))\}$. This second family includes the pure strategy Nash equilibrium $(3, (2/x_2, S/x_3))$
>
> **Case 3: player 1 bids 1 and 3 with positive probability**
> In this case player 2 reaches x_2 and $\pi_{2x2}(2)=1$ because 6>5. He also reaches x_3 and $\pi_{2x3}(S)=1$ because 5>4. It follows that player 1 only bids 3 because 5>4, so there is no Nash equilibrium in which player 1 both bids 1 and 3 with positive probability.

The same game in normal form is given in Matrix 3.7:

		Player 2			
		$(S/x_2, S/x_3)$	$(S/x_2, 4/x_3)$	$(2/x_2, S/x_3)$	$(2/x_2, 4/x_3)$
Player 1	1	(7,5)	(7,5)	(4,6)	(4,6)
	3	(5,5)	(2,4)	(5,5)	(2,4)

Matrix 3.7

NASH EQUILIBRIUM

Let us look for all the Nash equilibria. We call p the probability player 1 assigns to bid 1 and (1–p) the probability she assigns to bid 3. We call q_1, respectively q_2, q_3 and q_4 the probabilities assigned to (S/x_2,S/x_3), respectively (S/x_2,4/x_3),(2/x_2,S/x_3) and (2/x_2,4/x_3).

Case 1: player 1 only bids 1 (p=1)
In this case (2/x_2,S/x_3) and (2/x_2,4/x_3) are player 2's best responses, so $q_3+q_4=1$. But for player 1's strategy to be a best response to player 2's strategy, it is necessary that player 1 gets more (or the same amount) with bid 1 than with bid 3, hence $4 \geq 5q_3+2(1-q_3)$, i.e. $q_3 \leq 2/3$.
   We get a first family of Nash equilibria: {(1, (0,0,q_3, 1–q_3)), with $0 \leq q_3 \leq 2/3$}.
   This first family includes the pure strategy Nash equilibrium (1, (2/x_2, 4/x_3)).

Case 2: player 1 only bids 3 (p=0)
In this case (S/x_2,S/x_3) and (2/x_2,S/x_3) are player 2's best responses and $q_1+q_3=1$. But for player 1's strategy to be a best response to player 2's strategy, it is necessary that player 1 gets more (or the same payoff) with bid 3 than with bid 1, hence $5 \geq 7q_1+4(1-q_1)$, i.e. $q_1 \leq 1/3$.
   So we get a second family of Nash equilibria: {(3, (q_1,0,1–q_1, 0)), with $0 \leq q_1 \leq 1/3$}.
   This second family includes the pure strategy Nash equilibrium (3, (2/x_2, S/x_3)).

Case 3: player 1 bids 1 and 3 with positive probability (0<p<1)
It follows that (2/x_2,S/x_3) is player 2's best response, so $q_3=1$. But player 1's strategy is not a best response to player 2's strategy, given that she gets a higher payoff with bid 3 than with bid 1. So she cannot play both bids with positive probability at equilibrium.

We now observe that the behavioural strategy families and the mixed strategy families design the same behaviour.
   Consider the first family of behavioural Nash equilibria:

$$\{(\pi_1(1)=1, (\pi_{2x2}(2)=1, 0 \leq \pi_{2x3}(S) \leq 2/3))\}$$

We have:

$\pi_1(1)=1 \Leftrightarrow p=1$

$\pi_{2x2}(2)=1 \Leftrightarrow q(2/x_2,*/x_3)=1$ (where * means any available action) $\Leftrightarrow q_3+q_4=1$

$0 \leq \pi_{2x3}(S) \leq 2/3 \Leftrightarrow 0 \leq q(*/x_2,S/x_3) \leq 2/3 \Leftrightarrow 0 \leq q_1+q_3 \leq 2/3$.

These equivalences lead to $p_1=1$, $q_1=q_2=0$ and $0 \leq q_3 \leq 2/3$ i.e. to the first family of mixed Nash equilibria: {(1, (0,0,q_3, 1–q_3)), with $0 \leq q_3 \leq 2/3$}.
   Similar observations hold for the second family of behavioural strategies:

$$\{(\pi_1(3)=1, (0 \leq \pi_{2x2}(S) \leq 1/3, \pi_{2x3}(S)=1))\}$$

$\pi_1(3)=1 \Leftrightarrow p=0$, $0 \leq \pi_{2x2}(S) \leq 1/3 \Leftrightarrow 0 \leq q(S/x_2,*/x_3) \leq 1/3 \Leftrightarrow 0 \leq q_1+q_2 \leq 1/3$, $\pi_{2x3}(S)=1 \Leftrightarrow q(*/x_2,S/x_3)=1$ $\Leftrightarrow q_1+q_3=1$. The equivalences lead to $q_2=q_4=0$ and $0 \leq q_1 \leq 1/3$. So we get the second family of Nash equilibria {(3, (q_1, 0, 1–q_1, 0)), with $0 \leq q_1 \leq 1/3$)}.

NASH EQUILIBRIUM

3 NASH EQUILIBRIA DO A GOOD JOB

We showed in section 1.3 that in the first price sealed bid all pay auction, with known values of V and M, the strategy profile that consists of playing all the bids in [0,V] with the same probability is the unique Nash equilibrium, in both the discrete and continuous version of the game. This equilibrium is rather intuitive, robust (it is independent of M and logically linked to V and to the increment), and easy to conceive, so gives a good hint on how to play the game.

This is not an exception. Nash equilibria are often criticized, but in many cases they lead to strategies with nice properties. What is more, sometimes they can be translated into an easy rule of behaviour, easier to play for usual players.

3.1 Nash equilibrium, a good concept in many games

We propose three applications: all pay auctions with incomplete information, first price sealed bid auctions and second price sealed bid auctions.

3.1.1 All pay auctions with incomplete information

Up until now we talked about all pay auctions where both players grant the same value to an object, this value being common knowledge to both players. Yet the Nash equilibrium concept also helps in the incomplete information context defined by: player i, i from 1 to N assigns a value x_i to the object, x_i being uniformly distributed on [0,V]. Player i knows x_i but not x_j, with $j \neq i$, all the distributions are independent, and these facts are common knowledge. x_i is called player i's private information.

We first study the two player game.

There is one pure strategy Nash equilibrium obtained as follows:

We look for player 1's equilibrium bid function $b_1(x_1)$. When she bids b, she gets the object each time player 2 bids less than b. Suppose that player 2's strategy is the bid function $b_2(x_2)$. By playing b, player 1 wins the auction each time $b_2(x_2) \leq b$, i.e. $x_2 \leq b_2^{-1}(b)$. So player 1 maximizes:

$$\max_b x_1 \int_0^{b_2^{-1}(b)} f(x_2) dx_2 - b$$

(where f(.) is the uniform distribution on [0,V])

It follows: $-1 + x_1 \dfrac{1}{b_2'(b_2^{-1}(b))} f(b_2^{-1}(b)) = 0$ at equilibrium, i.e. for $b = b_1(x_1)$.

We look for a symmetric equilibrium, so we have $b_2(x) = b_1(x) = b(x)$ and the previous equation becomes:

$-1 + x_1 \dfrac{1}{b'(x_1)} \dfrac{1}{V} = 0$ Hence b'(x_1)=x_1/V and $\boxed{b_1(x) = b_2(x) = x^2/(2V)}$ (there is no constant because b(0) is necessarily equal to 0).

This equilibrium bid function highlights the impact of incomplete information. It is strictly convex, so, for low values of x, a player bids a very low amount (that grows up to V/2 for the

113

NASH EQUILIBRIUM

highest value of x). Is such a result natural? I would say yes. It is natural for you to bid almost nothing when x is low, both because the object is of low value for you and because there is a strong probability of meeting an opponent who grants more value to the object and hence bids more, so that you lose your bid with a high probability. Observe also that, in contrast to the complete information context, even if the object is worth V for you, you do not bid more than V/2; this derives from the fact that your opponent's value is uniformly distributed on [0,V], so that he may also bid low amounts. Finally observe that incomplete information allows players to switch from a unique mixed strategy equilibrium to a pure strategy equilibrium.

The above result easily generalizes to N players. As you may expect, the convexity of the bid function will become stronger: when N grows, for low but also middle values of x_i, player i may reasonably fear that one of the other players values the object more. So player i proposes low bids, even for middle values of x_i.

To put it more precisely, we look for a pure strategy that assigns b to x_i. Player i gets the object only if all the N–1 other players bid less than b. So, if each of the other players has the strategic bid function b(x) (we only look for a symmetric equilibrium), he gets the object only if each x of each of the N–1 players is lower than $b^{-1}(b)$. Consequently player i maximizes:

$$\max_{b} x_i \left(\int_0^{b^{-1}(b)} f(x)dx \right)^{N-1} - b$$

It follows: $-1 + (N-1)x_i \left(\dfrac{1}{b'(b^{-1}(b))} \right) f(b^{-1}(b)) \left(\int_0^{b^{-1}(b)} f(x)dx \right)^{N-2} = 0$

at equilibrium, i.e. for $b=b_i(x_i)$, equal to $b(x_i)$ in that we only look for a symmetric equilibrium.

It follows:

$$-1 + (N-1)x_i \left(\dfrac{1}{b'(x_i)} \right) \dfrac{1}{V} \left(\dfrac{x_i}{V} \right)^{N-2} = 0.$$

Hence $b'(x_i)=(N-1)x_i^{N-1}/V^{N-1}$ and $b(x)=(N-1)x^N/(NV^{N-1})$. When N grows, the curve of the bid function becomes flatter for low values of x_i but steeper for high values of x_i. This change is intuitive. When your x_i is small, you are more and more convinced that you will lose the auction because the probability of somebody bidding more than you grows with N; so you bid less and less. When x_i goes to V, the probability of somebody valuing the object more than you becomes smaller, but, in order to be sure to win the auction you are compelled to make a high bid. So there is a kind of threshold point: if your value is lower than this threshold, you almost bid nothing (because you are almost sure to lose), and if it is higher than this threshold, then you bid a high amount, high enough to have a high probability of winning the auction.

Example:
For V=3 and N=2, we get b(1)=1/6=0.167, b(2)=0.667, b(2.7)=1.215 and b(3)=1.5.
But for V=3 and N=5, we get b(1)=4/(81x5)=0.01<<0.167, b(2)=128/405=0.316<0.667, b(2.7)=1.417>1.215 and b(3)=2.4>>1.5.

So the Nash bidding equilibrium function: b(x)=(N–1)x^N/(NV^{N-1}) nicely deals with a quite natural dichotomous behaviour, which strengthens when N grows: bid very little as long as your value is below a given threshold, and bid a high amount when your value is above this threshold.

3.1.2 First price sealed bid auctions

The difference between the first price sealed bid auction we study now, and the all pay auction with incomplete information we studied above is that, in the first price sealed bid auction, the players who do not get the object *do not pay anything*.

Intuitively, how should the bids evolve as compared to the all pay auctions bids? It seems intuitive to expect that a player is encouraged to make higher bids given that he doesn't pay them if he doesn't win the auction. But how much should he bid?

We first study the case N=2:

If player 1, who values the object x_1, bids b, she wins as long as player 2 proposes a lower bid; i.e. each time $x_2 \leq b_2^{-1}(b)$, $b_2(x_2)$ is his bid function. So player 1 maximizes:

$$\max_b (x_1 - b) \int_0^{b_2^{-1}(b)} f(x_2) dx_2$$

We get:

$$-b_2^{-1}(b)/V + (x_1 - b)\left(\frac{1}{b_2'(b_2^{-1}(b))} f(b_2^{-1}(b))\right) = 0 \text{ at equilibrium, i.e. for } b=b_1(x_1).$$

In a symmetric equilibrium $b_2(x)=b_1(x)=b(x)$, and the previous equation becomes:

$-x_1/V+(x_1-b(x_1))/(Vb'(x_1))=0$ whose solution is $b(x_1)=x_1/2$, so $\boxed{b_1(x)=b_2(x)=x/2}$.

At equilibrium, each player bids x/2, i.e. half of his object's value, x, regardless of V! This is a strong result, easy to apply and easy to explain to a fellow as follows: If you win, you surely value the object more than your opponent. So, the opponent's value is in [0,x] and its mean value is x/2. Given that your opponent is rational, he will not bid more than his object's value, so his maximal mean bid is x/2. So you don't need to bid more. This explanation is not completely right but it may be accepted.

The bid function checks our conjecture: players bid more in this auction than in the all pay auction: x/2>x^2/2V, each time x<V; both bids are equal only for x=V.

Observe that the first price sealed bid auction bid function is linear: there is no longer a strong dichotomy between low values and high values, and this is logical because the high probability to lose money for low values has disappeared.

What is nice with the Nash concept is that it translates intuitions *and gives precision on these intuitions:* for example, thanks to the Nash concept, we know that a player switches from a strictly convex function, whose degree of convexity depends on the number of players, to a linear function.

We now switch to N players. Player i's maximization programme becomes:

115

NASH EQUILIBRIUM

$$\max_{b}(x_i - b)\left(\int_0^{b^{-1}(b)} f(x)dx\right)^{N-1}$$

where b(x) is each other player's equilibrium bid function given that we look for a symmetric equilibrium. We get:

$$-\left(b^{-1}(b)\right)^{N-1}/V^{N-1} + (x_i - b)\left[\frac{1}{b'(b^{-1}(b))}\right]f(b^{-1}(b))(N-1)(b^{-1}(b))^{N-2}/V^{N-2} = 0$$

at equilibrium, i.e. for b=b$_i$(x$_i$). At the symmetric equilibrium, b=b$_i$(x$_i$)=b(x$_i$), and so we get:
$-x_i^{N-1}+(N-1)(x_i-b(x_i))x_i^{N-2}/b'(x_i)=0$

So $-x_i b'(x_i)+(N-1)(x_i-b(x_i))=0$ and the solution is $\boxed{b(x)=(N-1)x/N}$. By contrast to the N player all pay auction equilibrium, where low valued players were induced to bid less and high valued players to bid more when N grows, here all players, regardless of the value, bid more. The growing coefficient (N–1)/N simply expresses that the competition becomes stronger when N grows. The larger N is, the more you can be sure that there is an opponent that has a value near to yours, which constrains you to propose a bid near to your value in order to not see an opponent get the object.

3.1.3 Second price sealed bid auctions, Nash equilibria and the marginal approach

The difference between the first price sealed bid auction and the second price sealed bid auction is that, in the second price sealed bid auction, the *winner does not pay the price he proposes but the second price*, i.e. the proposed bid that is just below his own bid.

So, if N=2, and player 1 wins the auction by bidding b, she just pays player 2's bid, b$_2$(x$_2$). So player 1 maximizes:

$$\max_{b}\int_0^{b_2^{-1}(b)}(x_1-b_2(x_2))f_2(x_2)dx_2$$

We get: $[x_1-b_2(b_2^{-1}(b)]f_2(b_2^{-1}(b))/b_2'(b_2^{-1}(b))=0$ at equilibrium, i.e. for b=b$_1$(x$_1$).
It derives that $b_1(x_1)=x_1$, *regardless of the density function f$_2$(.) on [0,V] and without assuming that we only look for a symmetric equilibrium.*
In the same way, we get b$_2$(x$_2$)=x$_2$, *without any assumption on the density function f$_1$(.) on [0,V], and without assuming that we look for a symmetric equilibrium.*
This equilibrium is interesting because *first it is very robust* in that it is *the unique equilibrium regardless of the distribution of the values on [0,V] of the two players*, and second it leads the players to *perfectly reveal their information*. This is a nice property. Very often, in contexts where people have private information that may be important for other players or an outsider, we look for a game that spontaneously leads the players to reveal their private information (see the literature on

NASH EQUILIBRIUM

the revelation principle). Here we have an auction that leads each player to reveal his private information, for any distribution of the values. And we don't need to assume that the players know the distribution of the opponent, except that it is on [0, V].

This result is, in fact, not astonishing given that bidding the value is also a weakly dominant strategy (see Exercise 14 in chapter 2). So the Nash Equilibrium agrees with the (non iterative) elimination of weakly dominated strategies, and proposes a very strong result.

> To play best (best responses), just bid the value you grant to the object.

Given that this result neither depends on the density function (on [0,V]) of the other player, nor on the information a player has on this distribution, *we get an equilibrium that really does not require much information to be played.*

And what is more, the result holds regardless of N: *so you do not even need to know the number of your opponents!* As a matter of fact, when there are N players, player i gets the object if all the opponents bid less, and he pays the second highest price. Call j the player who bids the second highest price. Hence we have $b_k(x_k)<b_j(x_j)<b$ with k from 1 to N, k≠i, k≠j. Given that player j may be any opponent, player i's maximization programme becomes:

$$\max_b \sum_{j\neq i, j=1}^{N} \int_0^{b_j^{-1}(b)} \left[\prod_{k\neq i, jk=1}^{N} \int_0^{b_k^{-1}(b_j(x_j))} f(x_k) dx_k \right] (x_i - b_j(x_j)) f_j(x_j) dx_j$$

We get:

$$\sum_{j\neq i, j=1}^{N} [\prod_{k\neq i, jk=1}^{N} \int_0^{b_k^{-1}(b))} f(x_k) dx_k] (x_i - b) f_j(b_j^{-1}(b)) / b_j'(b_j^{-1}(b)) = 0$$

at equilibrium, i.e. for $b=b_i(x_i)$.

$b_i(x_i)=x_i$ is the solution of this equation (see Exercise 14 in chapter 2 for a link with dominance).

So there is a strong difference between second price and first price sealed bid auctions: in the first price sealed bid auction, the competition among players increases in N, whereas N is without impact in the second price sealed-bid auction.

To summarize, we have here one of the nicest games in auction theory: without any information on the number of opponents and their value density functions on [0,V], each player is led to bid his value, hence to reveal his private information. *And this behaviour is easy to justify to someone as follows: if you play your value, you never pay more than what you are ready to pay. And if you win the auction, you will pay less than the price you are ready to pay if the second price is lower than yours, which is generally the case. And it would be a pity to bid less than your value, lose the auction, and see the winner paying less than your value!*

In addition, second price sealed bid auctions have another very special property: you can get the Nash equilibrium with *a marginal approach*. So consider a second price sealed bid auction with N

NASH EQUILIBRIUM

players and look for player i's best response bid b. Observe that when player i, who values the object x_i, switches from b to b+ε, with ε positive and going to 0, *nothing changes* regarding his payoff in front of opponents playing less than b (he wins against them and pays the second price) and *nothing changes* regarding his payoff in front of an opponent bidding more than b (he loses against this opponent). The only change is that he (additionally) wins each time an opponent plays b, i.e. values the object x_i, so he gets an additional payoff equal to (x_i-b) multiplied by the probability of this event. So, for b to be optimal, the additional payoff has to be null, and we get $b=x_i$, i.e. $b_i(x_i)=x_i$. That's a new way to automatically get the equilibrium which can *only be used for second price auctions*, because, *only* for these auctions, a switch to a higher bid does not change the payoff in front of players bidding less (you still pay the second highest price, which is not influenced by your switch to a higher bid). *(See Exercise 24 on the wallet game for another application of the marginal approach.)*

3.2 Nash equilibrium and simple rules of behaviour

Up to now we have shown that the Nash equilibrium concept may be helpful in difficult economic games. We also showed that *a Nash equilibrium may fit with a simple rule of behaviour* (like bidding half of the value in the first price sealed bid auction (for N=2) or bidding the value in the second price sealed bid auction). In the exercises on the wallet game (see the exercises 20 to 24), Nash equilibria fit with other simple rules of behaviour. We now show this property in the prisoners' disks game.

So let us come back to the easy rule of behaviour given in chapter 2:

> *Easy and optimal rule of behaviour:*
> When you observe k black disks at the beginning of the game, stay in the game up to period k+1, but be very careful at period k and ready to jump at period k+1. More precisely, if nobody went out at period k, go out at period k+1 and announce black. If the k black disks went out at period k, go out at period k+1 and announce white.

We showed in chapter 2 that this rule is in accordance with the only strategies that survive iterative elimination of strictly dominated strategies. This iterative elimination led to a unique path of behaviour: it led a black disk to go out at period k (and announce the right colour) and a white disk to go out at period k+1 (and announce the right colour), when there are k black disks.

Given that iterative elimination of strictly dominated strategies does not eliminate a Nash equilibrium, the unique Nash equilibrium path has also to lead to this behaviour.

But, just for fun – and for training – we'll prove directly that the above rule is a Nash equilibrium path, without using the links between strict dominance and Nash equilibria.

To do so, we show that the above rule is a Nash equilibrium path; i.e. that no player is better off unilaterally deviating from it.[2]

What are the deviations to study? Given that we only have to focus on unilateral deviations, we study player i's behaviour, by supposing that the other N–1 players follow the rule.

Suppose that player i observes k black disks and consider another player.

There are four possibilities: either his disk is white, and he observes k black disks (if player i's disk is white), or k+1 black disks (if player i's disk is black). Either his disk is black and he observes k–1 black disks (if player i's disk is white), or k black disks (if player i's disk is black).

Given our rule, player i should stay in up to period k+1 and go out at period k+1 announcing black if the other black disks stayed in in period k, white in the opposite case. We will prove later that by doing so, she says the right colour and leaves the jail. Deviating means that she goes out earlier, or goes out at period k+1 announcing the wrong colour, or goes out later announcing the good or the wrong colour. Clearly, going out later can only lead to a lower payoff, given that a prisoner wants to go out as soon as possible.[3] Going out at period k+1 and announcing the wrong colour is of course silly.

So the main job we have to do is to check that player i cannot be better off going out earlier, before period k+1 and that she leaves jail at period k+1.

To do so, we split the periods up to k+1 into three sets of periods: the periods before k, the period k and the period k+1.

Without loss of generality, we say that player i is the last player to decide at each period (this is of no importance given that the game is simultaneous at each period).

Let us consider the periods before k. Another player observes at least k–1 black disks and so he does not go out before period k. So the information set at which player i evolves is such that only two nodes are reached with positive probability: the node x that follows an uninterrupted sequence of stay in actions and Nature's choice – k black disks, N–k–1 white disks and player i's black disk – and the node y that follows an uninterrupted sequence of stay in actions and Nature's choice – the same k black disks, the same N–k–1 white disks and player i's white disk – and these two nodes have a positive probability. We call a the utility obtained when staying in jail forever, 1 the utility of being released, with a much lower than 0. So, if player i goes out announcing black she gets $p(x)1+p(y)a$, if she goes out announcing white she gets $p(x)a+p(y)1$, where $p(x)$ and $p(y)$ are the probabilities to reach x and y.[4] $p(x)1+p(y)a<0$ and $p(x)a+p(y)1<0$ because $a<<0$, and 0 is lower than the payoff obtained by going out at period k+1 announcing the right colour. So staying in is player i's best answer and she will not go out before period k. This decision context is illustrated in Figure 3.3a.

We now consider period k. The white disks stay in, given that they observe at least k black disks. The prisoners with black disks may observe k–1 black disks when player i's disk is white, so they go out and announce black. If they observe k black disks, which happens if player i's disk is black, they stay in. Yet, given that the decisions are simultaneous, player i does not observe these two types of decisions, so her reached information set again contains two nodes reached with positive probability.

The first node, x', follows Nature's choice (k black disks, N–k–1 white disks, and player i's black disk) and a sequence of uninterrupted stay in decisions up to period k–1 (included), and N–1 stay in decisions in period k. The second node, y', follows Nature's choice (the same k black disks, the same N–k–1 white disks, and player i's white disk) and a sequence of uninterrupted stay in decisions up to period k–1 (included), k black disks' go out decisions in period k, and N–k–1 white disks' stay in decisions in period k. And $p(x')=p(x)$ and $p(y')=p(y)$, given that player i reaches x' if she reaches x, and reaches y' if she reaches y. So, if player i goes out announcing black she gets $p(x')1+p(y')a=p(x)1+p(y)a<0$, if she goes out announcing white she gets $p(x')a+p(y')1= p(x)a+p(y)1<0$, and if she stays in she gets 1 at period k+1. So staying in is player i's best response and she will not go out at period k. This decision context is illustrated in Figure 3.3b.

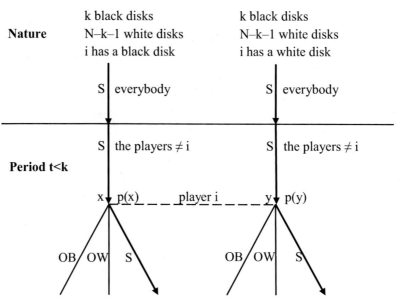

Figure 3.3a

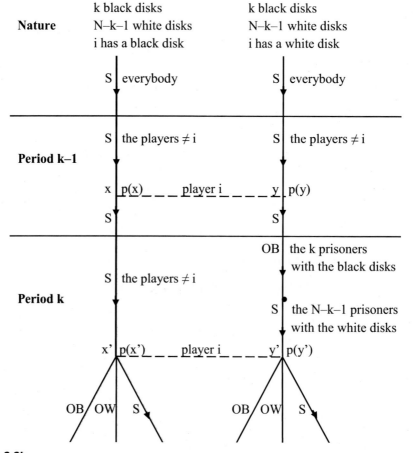

Figure 3.3b

We now consider period k+1. When player i's disk is white, the prisoners with the black disks (k players) went out at period k. When player i's disk is black, nobody went out at period k. So, at period k+1, player i knows the colour of her disk. In other words, the two possibly reached nodes at period k+1 now are in two different information sets. The first node, x", follows Nature's choice (k black disks, N–k–1 white disks, and player i's black disk), a sequence of uninterrupted stay in decisions up to period k (included), k decisions to go out at period k+1 (from the other black disks) and N–k–1 stay in decisions in period k+1 (from the white disks). The second node y" follows Nature's choice (the same k black disks, the same N–k–1 given white disks, and player i' white disk), a sequence of uninterrupted stay in decisions up to period k–1 (included), k black disks' decisions to go out in period k, N–k–1 white disks' stay in decisions in period k, i's stay in decision in period k, and N–k–1 decisions to go out (from the N–K–1 other white disks) at period k+1. And, of course, player i's best decision at node x" is to go out announcing black, and her best decision at node y" is to go out announcing white, because she will leave jail by doing so. This decision context is illustrated in Figure 3.3c.

So we have proven that the easy rule of behaviour is a Nash equilibrium path.

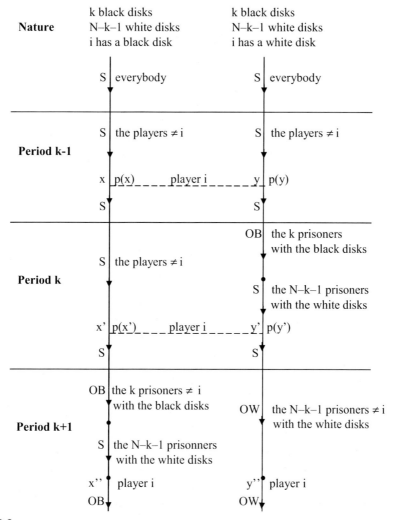

Figure 3.3c

4 MULTIPLICITY OF NASH EQUILIBRIA

Up to now, we have not criticized the Nash equilibrium concept. But unfortunately this concept has some flaws we will now talk about. We will see that the Nash equilibrium concept does not always help us solve an interactive situation.

4.1 Multiplicity in normal form games, focal point and talking

The multiplicity of Nash equilibria may raise a problem: even if people agree to play a Nash equilibrium rather than another strategy profile, how can they converge on the same equilibrium if they play the game only once?

So consider first the coordination game in Matrix 3.1 and its two pure strategy Nash equilibria. Remember that this pure coordination game can be a meeting game: two players want to meet in a town but they have forgotten to fix a meeting point. If the players can't phone or e-mail, they can't know where the other goes and they may never meet.

Focal point theory, as introduced by Shelling[5] in the 1960s may help. Shelling says that players may share a common history that can help them to coordinate; for example, if the players have already met before, they may simply go back to the old meeting place. If the town has an outstanding building (the Eiffel Tower in Paris or the Louvre Museum), they may meet there, because both may think of this building. So the outstanding building, or the old meeting place, becomes a focal point, a point the players, thanks to common culture and common history, focus on *(see Exercise 18 for a thought-provoking coordination role of focal point theory)*.

If there is no focal point, *we can introduce a free talk before the game.* By free talk, we mean that the players can talk together before the game and agree or not on a given way to play, but they are completely free to play what they want when the game begins: in other words they are in no way constrained to play what they proposed to play during the talk. So *free talk* is sometimes called *cheap talk*.

In the pure coordination game in Matrix 3.1, free talk is a very powerful tool: if players are allowed to talk before the game, they will of course have no difficulty with either coordinating on (A_1,A_2) or on (B_1,B_2).

In the game in Matrix 3.2, there is an additional difficulty: how can players converge on an equilibrium when they have different preferences with regard to the equilibria? If, in their common history, players share the knowledge that player 1 is selfish whereas player 2 is altruist, then (B_1,B_2) may become a focal point. Talking before the game may also be helpful. They can for example agree to play according to the following rule: both put a penny on the table, either on heads or on tails. If both pennies are on heads or if both are on tails, they decide to play (A_1,A_2); if not, they decide to play (B_1,B_2). If both agree on the rule, they will coordinate on the outcoming equilibrium. As a matter of fact, if it is for example (A_1,A_2), they will play it because player 2 is most happy to play A_2, and therefore player 1 has no reason to deviate from A_1.

We now look at the more difficult game in Matrix 3.3.

How will players deal with this game? Should we refer to Schelling's focal point theory and say that the players spontaneously focus on (A_1,A_2) because it is the only Pareto optimal profile, very profitable to both players and stable to any unilateral deviation? The answer is not necessarily yes, because there is another focal point: both players know that player 2 is safer by playing C_2, given that C_2 weakly dominates both A_2 and B_2. And this knowledge may encourage player 1 to play C_1

(which also weakly dominates A_1). What is more, even if the opponent plays A because he focuses on (A_1,A_2), C is a best answer. So focal point theory cannot warrant a convergence on (A_1,A_2).

What about talking before the game? The answer is not obvious, *but it is different*. If both agree on (A_1,A_2), they may still prefer playing the safe strategy C. *Yet talking is not neutral, because both players, while choosing their action, now start their reasoning by knowing that they agreed on (A_1,A_2). In other words, (A_1,A_2), as the result of the talk, becomes a focal point and the starting point* of the players' reasoning. So *player 2 has now no reason to fear that player 1 plays B_1*, given that this action is a very bad response to A_2. It follows that he is no more induced to play C_2 because C_2 does better than A_2 only if player 1 plays B_1 (this contrasts with what happens without talking because, without talking, player 2 cannot be sure that player 1 focuses on (A_1,A_2), so he cannot be sure that player 1 does not play B_1). If player 2 is not induced to play C_2, then player 1 is not induced to play C_1, because C_1 does better than A_1 only if player 2 plays C_2. So agreeing on (A_1,A_2) during the talk may perhaps lead to its play, despite the fact that A_1 and A_2 are weakly dominated.

4.2 Talking in extensive form games

In a normal form game, the talk before the game leads to a behaviour that becomes a focal point, a possible starting point of a new reasoning. *In an extensive form game, talking before the game becomes even more interesting* because a behaviour which is not in accordance with the agreed on behaviour, is observed by the opponents *before the end of the game*. So they have to interpret the deviation and play the best reaction, knowing that a *past unexpected action of a player may be a signal on his future actions*. This is what characterizes *forward induction* (see chapter 6).

To see how it works, look again at the ascending all pay auction in Figure 3.1.

Imagine that the two players talk before the game and agree on the fact that player 1 overbids at each decision node and player 2 stops at each decision node. Suppose that, during the play, player 1 bids 1 at x_1 as expected, but player 2 overbids at x_2, an unexpected action. How should player 1 interpret this deviation? If player 2 is rational, he hopes to get more than 5 by deviating, and the only way for him to get more is if player 1 stops at x_3. But, why should she do this? She can reasonably expect that player 2 understands at x_4 that it is best for him to stop, so she overbids at x_3 in order to get 5. So, if players are rational, player 2 seems not able to destabilize the equilibrium they agreed on.

But now imagine that the two players decided, before the game, that player 1 always stops and that player 2 always overbids. What will happen if, during the play, player 1 bids 1 at x_1? How should player 2 interpret this unexpected deviation? If player 1 is rational, then if she deviates, she surely hopes to get more than 5 or at least as much. Hence, if player 2 understands this fact, he deduces that bidding 2 leads player 1 to overbid at x_3 because, on the one hand, it is the only way for her to get at least 5 (the amount she lost by deviating), and on the other hand she is sure to get 5 because player 2 will stop at x_4 (it would be stupid for him to overbid given that player 1 will not stop at x_5). Hence player 2, as a rational player, may prefer stopping at x_2, which encourages player 1 to deviate at x_1 and destroys the equilibrium they agreed on.

It follows that some equilibria (players agreed on before the game) have a good probability to be played, others not, depending on the strategic forward induction reasoning a deviation induces.

In some ways, in our ascending all pay auction game, talking either leads directly to the Nash equilibrium path where player 1 bids 1 and player 2 stops, or it leads to it indirectly, thanks to the (forward induction) interpretation of player 1's deviation at x_1. Hence, in this game, talking is profitable to player 1.

NASH EQUILIBRIUM

4.3 Strange out of equilibrium behaviour, a game with a reduced field of vision

Nash equilibrium (only) provides a profile of actions such that nobody is best off deviating unilaterally from it. One consequence may be the presence, out of the equilibrium path, of very strange actions. This is surely one of the most common criticisms against the Nash equilibrium concept. We briefly expose it below (we come back to it in chapter 4).

Let us consider the Stackelberg first price all pay auction game in Figure 3.4. In this game the strategy profile E* (player 1 bids 0, player 2 bids one unity more in front of any bid lower than 5, and bids 5 if player 1 bids 5) is a Nash equilibrium, which benefits player 2.

E* is a Nash equilibrium because player 1's best bid is 0 given that player 2 systematically overbids and bids 5 if she bids 5 (5>4>3>2>1.5>1), and player 2's behaviour is optimal given player 1's bid: player 2 just has to optimally respond to player 1's bid 0, and that is what he does (7>6.5>6>5>4>3); all other responses – those after player 1's bids 1, 2, 3, 4 and 5 – are of no importance because these bids are not played by player 1.

So player 2 can get his highest possible payoff by threatening to always overbid, in that only one of these threats has to be played, the equilibrium one.

But this equilibrium doesn't seem natural: if player 2 would face 3 or 4 (something that doesn't happen at equilibrium), he would surely not overbid. So intuitively we prefer the three following Nash equilibria:

- Player 1 plays 2, player 2 overbids by one in front of 0 and 1 and bids 0 in each other case.
- Player 1 plays 3, player 2 overbids by one when faced to 0, 1 and 2, and bids 0 in each other case.
- Player 1 plays 0, player 2 overbids by one when faced to 0, 1 and 2, and bids 0 in each other case.

Clearly, we face a limit of the Nash equilibrium concept: *some equilibrium strategies are only here to sustain the equilibrium path, but would never be played if the player who proposes them would really have to take the decision to play them or not.*

As a matter of fact, if player 1 bid 3 instead of 0, it is quite certain that player 2 would bid 0 and not 4 as proposed in E*. *Player 2 can threaten to bid 4 after bid 3 in E* only because this threat leads player 1 to not bid 3.*

> **STUDENTS' INSERT**
>
> Strange behaviour at non-reached nodes is a well-known flaw of Nash's equilibrium. Let us come back to our reduced sequential auction game in Figure 3.2 and especially to the Nash equilibrium in which player 1 bids 1 because player 2 overbids at each node. This Nash equilibrium is not intuitive. **Clearly player 2 can threaten to bid 4 at x_3, i.e. if player 1 bids 3, only because player 1 will not bid 3 at equilibrium.** If player 1 deviated from her equilibrium action and bid 3, player 2 would not execute his threat because he would be better off stopping the game. **In other terms, player 1's deviation would induce player 2's deviation, so we get multilateral deviations.** And it is intuitive to expect that player 1 makes this reasoning even if she is not especially clever. So, clearly, the Nash equilibrium (player 1 bids 1, player 2 always overbids) seems not very robust with regard to real behaviour. We will say later that player 2's decision to always overbid is an **incredible threat**. And we will see that we can avoid Nash equilibria with incredible threats by turning to the concepts of Subgame Perfect Nash Equilibrium (chapter 4) and Perfect Equilibrium (chapter 5).

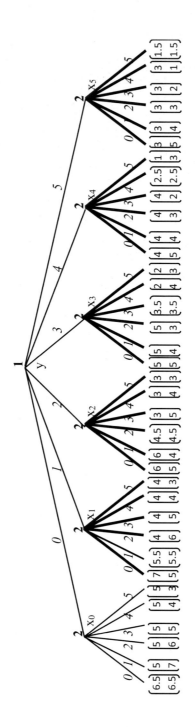

Figure 3.4

NASH EQUILIBRIUM

Clearly, the presence of incredible threats derives from the fact that the Nash concept focuses on unilateral deviations. Due to this way of reasoning, whole parts of a game have no impact; in the game of Figure 3.4, all the actions chosen in the part of the tree in bold lines can be silly, yet they are automatically optimal, because player 1's behaviour does not allow player 2 to reach them, in other words to activate them.

More generally, the stability to unilateral deviations may completely restrict *the field of vision of a game*. To see why, look at the game in Figure 3.5. A_1 is a Nash equilibrium path (player 2 plays A_2 at equilibrium); as a matter of fact, on the one hand, player 1 considers player 2's action as fixed (he plays A_2) and so she is better off playing A_1 (1>0), on the other hand, all the other players, including player 2, automatically play in an optimal way because, considering player 1's action A_1 as granted, they are not called on to play (so all their actions are optimal). So, to establish the Nash equilibrium path, only two branches of the game tree (the two branches in bold lines), and two payoffs, 1 and 0, are taken into consideration. The remaining game tree (which may be as large as you want) is completely out of the field of vision. To get the above equilibrium, we neither need to know the number of players in this part of the game, their strategy sets, nor their payoffs. This is a little strange, isn't it?

But let us make a comment: perturbations will help the Nash equilibrium concept. In chapter 5, we will introduce little perturbations in the strategies, i.e. we will suppose that each player, at any information set, plays each available action with at least a very small probability (due to the perturbation); so automatically all the parts of the game tree will be reached, and so the above strange configuration will automatically disappear, without changing anything in the Nash's unilateral deviation philosophy.

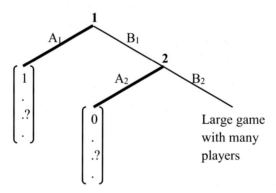

■ **Figure 3.5** The ? in the payoff vectors means that we ignore the number of players and their payoffs (except the payoff of player 1).

NASH EQUILIBRIUM

5 TO PLAY OR NOT TO PLAY A NASH EQUILIBRIUM, CAUTIOUS BEHAVIOUR AND RISK DOMINANCE

In this section we first present a situation where the Nash equilibrium is logical but does not help play the game. Then we turn to another flaw: the Nash concept does not necessarily make the good comparisons in a game; this will lead us to cautious behaviour and risk dominance.

5.1 A logical concept, but not always helpful

We return to the envelope game. We already know that an iterative elimination of weakly dominated strategies leads to never exchanging an envelope with a number higher than 1, but we also know that this iterative process requires too much rationality to be played by the players.

Strangely enough, the Nash equilibrium concept also leads to the unique same outcome: the players get the payoff equal to the number in their first envelope because they do not exchange their envelope if it contains a number higher than 1.

Let us prove this result:

> **STUDENTS' INSERT**

The Nash equilibrium of the envelope game can be established in an original way. We show it for K=5. * in player i's strategy, i=1,2 means any available action, E_i (exchange), $\overline{E}_i$ (not exchange). We'll focus on player 1's strategy.

- We first observe that a strategy (*/1,*/2, */3,*/4, E_1/5) can be a best response to player 2's strategy only if player 2 plays a strategy ($\overline{E}_2$/1, $\overline{E}_2$/2, $\overline{E}_2$/3, $\overline{E}_2$/4, */5) (otherwise, there is for player 1 a positive probability to exchange 5 against a lower number, and she is better off not exchanging 5).

  But a strategy ($\overline{E}_2$/1, $\overline{E}_2$/2, $\overline{E}_2$/3, $\overline{E}_2$/4, */5) is not a best response against player 1's strategy (*/1, */2, */3,*/4, E_1/5) because it is better for player 2 to exchange his envelope when the number is 1, given that he gets more by doing so because player 1 exchanges her 5.

  It follows that, at equilibrium, player 1 necessarily plays a strategy (*/1,*/2, */3,*/4, $\overline{E}_1$/5). For the same reasons, at equilibrium, player 2 necessarily plays a strategy (*/1,*/2, */3,*/4, $\overline{E}_2$/5).

- We now observe that a strategy (*/1, */2, */3, E_1/4, $\overline{E}_1$/5) can be a best response to player 2's strategy only if player 2 plays a strategy ($\overline{E}_2$/1, $\overline{E}_2$/2, $\overline{E}_2$/3, */4, $\overline{E}_2$/5) (otherwise there is for player 1 a positive probability to exchange 4 against a lower number, without any chance to exchange it against a higher number, because we established $\overline{E}_2$/5 in the previous step).

  But a strategy ($\overline{E}_2$/1, $\overline{E}_2$/2, $\overline{E}_2$/3,*/4, $\overline{E}_2$/5) is not a best response against player 1's strategy (*/1, */2, */3, E_1/4, $\overline{E}_1$/5) because, again, it is better for player 2 to exchange his envelope when the number is 1, given that he gets more by doing so because player 1 exchanges her 4.

  It follows that at equilibrium player 1 necessarily plays a strategy (*/1, */2, */3, $\overline{E}_1$/4, $\overline{E}_1$/5). For the same reasons, at equilibrium, player 2 necessarily plays a strategy (*/1, */2, */3, $\overline{E}_2$/4, $\overline{E}_2$/5).

- And so on. We successively show that both players play a strategy (*/1,*/2, $\overline{E}$/3, $\overline{E}$/4, $\overline{E}$/5) and then a strategy (*/1, $\overline{E}$/2, $\overline{E}$/3, $\overline{E}$/4, $\overline{E}$/5). Hence there are 4 Nash equilibria with the same outcome: both players never exchange a number different from 1, and can either exchange or not exchange their number 1.

NASH EQUILIBRIUM

This Nash equilibrium result, "everybody gets the amount in the first envelope", is interesting because it highlights positive points of the Nash concept but also a major difference between theory and reality.

First, this equilibrium result is *robust:* it is easy to see that it holds for any K>1 (the above reasoning easily generalizes). It doesn't depend on the probabilities Nature assigns to the numbers in the envelopes (they are not used to establish the equilibrium). It is only required that each probability is positive. The *distribution of the probabilities may be completely unknown to both players.*

Second, we observe that the proof follows the same steps as the iterative elimination of weakly dominated strategies (we first establish that 5 (more generally K) then 4 (more generally K–1), then...cannot be exchanged at equilibrium), but each step is coupled with the fact that it is better to exchange 1, as long as the other player exchanges a higher number with a positive probability.

So you may get the impression that the Nash concept suffers from the same flaw as iterative elimination of weakly dominated strategies, but this is wrong. As a matter of fact, when we propose a Nash equilibrium, *we do not say that the players make the reasoning that leads to it.* Players just have to be able to detect if it is or isn't interesting for them to deviate unilaterally from a given strategy profile: *that's all.*

So, for example, when faced with the non Nash equilibrium profile ((E_1/1, E_1/2, $\bar{E}_1$/3 $\bar{E}_1$/4, $\overline{E}_1$/5), (E_2/1, E_2/2, E_2/3, $\bar{E}_2$/4, $\bar{E}_2$/5)), player 1 automatically realizes that she is better off deviating by not exchanging her number 3.

By contrast, if both players face a Nash equilibrium profile ((E_1/1, $\bar{E}_1$/2, $\bar{E}_1$/3, $\bar{E}_1$/4, $\bar{E}_1$/5), (E_2/1, $\bar{E}_2$/2, $\bar{E}_2$/3, $\bar{E}_2$/4, $\bar{E}_2$/5)), they realize that they cannot be better off by unilaterally deviating.

So a player has just to be able to establish a best response to a given behaviour of the other. And this is not too demanding.

This is a very positive point for the Nash equilibrium, but it automatically raises a new question. Who or what leads the players to the Nash profile? In this game for example, people will surely not spontaneously come to the Nash equilibrium.

Further, and this is the problem, *even if players know the existence of this unique Nash equilibrium outcome, will they really play it?* It is one thing to know that the only unilaterally stable profile consists in never exchanging a number higher than 1, but it is another thing to conform to this play. Clearly, if K=100,000, even if both players perfectly understand why not exchanging more than 1 is the only Nash equilibrium behaviour, they will not play in this way (remember that only 1.4% of my (146) L3 students refused to exchange any number higher than 1).

Let us insist on the fact that this is not a (theoretical) flaw of the Nash equilibrium. The Nash equilibrium outcome in this game is neither stupid, nor relies on incredible threats. It is perfectly logical: it namely highlights that an exchange (of a number higher than 1) cannot be beneficial to both players.

What we want to stress is that, *despite its logical content, the Nash equilibrium concept is simply not helpful.* Clearly, if K=100,000 and if player 1 has 50 in her envelope, she surely risks the exchange, because she hopes to get more. And this may also be the case for player 2. *In fact, players know that the exchange cannot be beneficial to both players, but each of them hopes to be the player who benefits from the exchange! Reality authorizes each player to think that he is the winner (so one player has inconsistent beliefs), whereas game theory highlights that there is generally a winner and a loser (and recommends to the loser not to exchange).* The real interest of this game is to play *with inconsistent beliefs*: you hope that your opponent exchanges high numbers (perhaps he wrongly thinks that you are a risk taker), so that you can afford to exchange low ones.

For information, the histograms in Figures 3.6a and 3.6b, indicate the highest number my students accept exchanging, for K=100 and K=100,000. Clearly, my students are ready to exchange their envelopes! For K=100, about 1/3 of them are ready to exchange a sum up to 50, and 28.7% are even ready to exchange higher amounts (the mean highest exchanged number is 48, almost K/2). For K=100,000, the students are more cautious, in that the mean highest exchanged number is 33,256, much lower than K/2, but they are still ready to exchange high numbers (19.2% exchange up to 50,000, 15.1% are ready to exchange even higher numbers, and only 13% do not agree to exchange a number higher than 1,000).

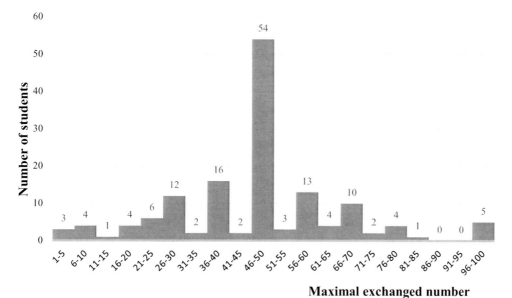

Figure 3.6A Envelope game, maximal number exchanged for K=100

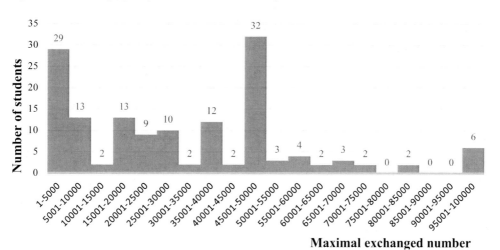

Figure 3.6B Envelope game, maximal number exchanged for K=100,000

NASH EQUILIBRIUM

5.2 Cautious behaviour and risk dominance

5.2.1 Cautious behaviour

Let us make an additional remark. Consider the asymmetric sealed bid first price all pay auction game with M=5, V_1=4 and V_2=3 (*shown in Exercise 4 in chapter 2*) and focus on matrix 3.8 obtained after the iterative elimination of weakly dominated strategies, with increment 0.5.

		Player 2		
		0	1	2
	0.5	(8.5,5)	(4.5,7)	(4.5,6)
Player 1	1.5	(7.5,5)	(7.5,4)	(3.5,6)
	2.5	(6.5,5)	(6.5,4)	(6.5,3)

Matrix 3.8

We call $p_{0.5}$, $p_{1.5}$, $p_{2.5}$, q_0, q_1 and q_2 the probabilities assigned to the bids 0.5, 1.5, 2.5 (by player 1) and to the bids 0, 1 and 2 (by player 2). This game has a unique Nash equilibrium: $p_{0.5}=p_{1.5}=p_{2.5}=1/3$ and $q_0=½$, $q_1=0.25$ and $q_2=0.25$. Player 1's equilibrium payoff is 6.5, player 2's payoff is 5.

When the increment goes to 0, this equilibrium becomes: player 1's strategy is the uniform distribution on $[0,V_2]$, player 2 plays each strategy in $]0,V_2]$ with probability dv/V_1 and plays 0 with probability $1-V_2/V_1$ (*see Exercise 19 in this chapter*).

That is a strange equilibrium. For small increments, if I were player 1, I would bid a little more than 3, because a rational player 2 cannot bid more than 3. So I would be sure to win the auction and to get a payoff close to 6 for sure, i.e. as much as the expected equilibrium payoff for small increments, i.e. $M-V_2+V_1$. *And if I were player 2, I would simply bid 0, given that I would expect player 1 to win the auction. By doing so I would get 5, i.e. M, for sure, as much as the expected payoff in the Nash equilibrium.* Well, something in the Nash equilibrium looks like my behaviour, namely the fact that player 2 assigns a higher probability to the bid 0 than to the other bids. And this probability, $(V_1-V_2)/V_1$, grows with the difference between V_2 and V_1. But it does not go to 1. Nor does player 1, at equilibrium, value more the strategy close to 3 (V_2) than the other ones. But isn't it more interesting to get $M-V_2+V_1$ *for sure* than to get the *expected* payoff $M-V_2+V_1$? And isn't it safer to get M *for sure* than to get the expected payoff M?

Well, in the studied game, my proposed behaviour is close to cautious behaviour, and it may seem more natural than the mixed Nash equilibrium.

Definition 2. Cautious behaviour

A mixed strategy profile p* is a cautious equilibrium if and only if:

$$\forall i \in \mathcal{N}, \forall p_i \in P_i \min_{p_{-i} \in P_{-i}} u_i(p_i^*, p_{-i}) \geq \min_{p_{-i} \in P_{-i}} u_i(p_i, p_{-i})$$

What does cautious behaviour mean? *Simply that you terribly fear your opponents.* In fact you are convinced that their only aim is to minimize your payoff, and so you play *knowing that they react in order to minimize your payoff*; i.e. you choose the strategy that maximizes the minimum the others can impose on you.

NASH EQUILIBRIUM

Well, shouldn't we be paranoid to play in this way? *It depends on the structure of the game.* If you play a zero sum game, i.e. a pure conflict game, then it is rational to be paranoid because other players, to maximize their payoffs, have to minimize yours. But many other games can also reasonably lead to cautious behaviour.

Consider the game in Matrix 3.9 for example:

	Player 2	
	A_2	B_2
A_1	(9,9)	(9,0)
B_1	(0,9)	(10,10)

Player 1

Matrix 3.9

> ### ➤ STUDENTS' INSERT
>
> We establish player 1's cautious strategy. We call p the probability player 1 assigns to A_1 and 1–p the probability she assigns to B_1. If player 2 plays A_2, player 1 gets 9p. If player 2 plays B_2, player 1 gets 9p+10(1–p). If player 2 plays a mixed strategy, she gets something between 9p and 9p+10(1–p), a situation we represent in Figure 3.7.
>
>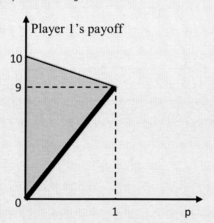
>
> **Figure 3.7**
>
> *The grey area represents all the possible payoffs player 1 can get according to player 2's strategy. Yet, given that player 2 is supposed to choose the strategy that worsens player 1's payoff, player 1 only focuses on player 2's strategy that minimizes her payoff, i.e. the strategy that yields to the bold line. And the maximum payoff she can get if player 2 behaves in this way is 9, a payoff obtained by only playing A_1.*
>
> *Given that the game is symmetric, player 2's cautious strategy is A_2.*

These strategies are easy to understand. By playing A, you are sure to get 9, regardless of the opponent's behaviour. Consequently, even if he is very ill intentioned you get 9, whereas, if you play B, you get 0 (if he is ill intentioned).

Cautious behaviour is not a rarity in reality. Remember (see chapter 1, section 3.3) that some of my students, asked to play the game above, said that they play A, not just because it is a sure way to get 9, but because they feared that the other guy was happy to see them lose a lot of money, so

131

NASH EQUILIBRIUM

hopes that they play B, which leads him to play A. This paranoid view of the game immediately leads to cautious behaviour.

In all pay auctions, cautious behaviour is often not silly. Let us illustrate this fact in the symmetric first price sealed bid all pay auction ($V_1=V_2$) and in the asymmetric one ($V_1 \neq V_2$).

Consider first the symmetric game with any M and V<M.

If player 1 bids b, a player 2's strategy that minimizes player 1's payoff consists in bidding b+ the increment if b<M, and M if b=M, so that the player loses the amount of her bid (when she plays b<M) or gets (V/2<<M) when she plays b=M. Hence, in this game, the cautious strategy consists in bidding 0 in order not to lose money. *Bidding 0 is often compared as a way of not entering the game, a desire some of the students expressed.* And bidding 0 is not silly with regard to the payoffs. As a matter of fact, when the increment goes to 0, the Nash equilibrium payoff goes to M, which is the minimal payoff you get without any risk by bidding 0.

We now turn to the asymmetric all pay auction (we suppose $V_1>V_2$ w.l.o.g). What we said for the symmetric case is still true, despite the asymmetry (just replace V by V_i, i=1,2). So a cautious behaviour consists again in bidding 0. Yet this seems less intuitive for player 1.

In this game we see both a limit of cautious behaviour but also a way to keep cautious behaviour in a more restrictive mode. The limit of cautious behaviour is that you systematically suppose that the only intention of your opponent is to minimize your payoff. And this may be inappropriate, namely if this minimization is harmful for him. For example, in the asymmetric all pay auction with M=5, $V_1=4$, $V_2=3$, if player 1 plays 3, why should player 2 play 3+ the increment, a strategy that is strictly dominated by the bid 0? Such a behaviour would be silly, hence it looks like an incredible threat. In the studied game, a more reasonable way to keep cautious behaviour would be to turn to it only after having eliminated strictly dominated strategies.

For M=5, $V_1=4$, $V_2=3$, the idea is to focus on the game in Matrix 3.10:

		Player 2						
		0	0.5	1	1.5	2	2.5	3
Player 1	0	(7,6.5)	(5,7.5)	(5,7)	(5,6.5)	(5,6)	(5,5.5)	(5,5)
	0.5	(8.5,5)	(6.5,6)	(4.5,7)	(4.5,6.5)	(4.5,6)	(4.5,5.5)	(4.5,5)
	1	(8,5)	(8,4.5)	(6,5.5)	(4,6.5)	(4,6)	(4,5.5)	(4,5)
	1.5	(7.5,5)	(7.5,4.5)	(7.5,4)	(5.5,5)	(3.5,6)	(3.5,5.5)	(3.5,5)
	2	(7,5)	(7,4.5)	(7,4)	(7,3.5)	(5,4.5)	(3,5.5)	(3,5)
	2.5	(6.5,5)	(6.5,4.5)	(6.5,4)	(6.5,3.5)	(6.5,3)	(4.5,4)	(2.5,5)
	3	(6,5)	(6,4.5)	(6,4)	(6,3.5)	(6,3)	(6,2.5)	(4,3.5)
	3.5	(5.5,5)	(5.5,4.5)	(5.5,4)	(5.5,3.5)	(5.5,3)	(5.5,2.5)	(5.5,2)

Matrix 3.10

For each bid of player 1 we have represented in bold circles player 1's payoff when player 2 minimizes it, and for each bid of player 2 we have represented in dotted circles player 2's payoff when player 1 minimizes it. It follows that player 2's cautious strategy consists in bidding 0, because the lowest payoff player 1 can impose on him is 5, and player 1's cautious strategy consists in bidding 3.5, because she gets 5.5 for sure. This behaviour is acceptable, given that it leads to the Nash payoffs when the increment goes to 0, without any risk. And it is in accordance with my above proposed behaviour.

NASH EQUILIBRIUM

Should we conclude that cautious behaviour is a good way to play provided one has dismissed strictly dominated strategies? *Surely not.*

First, let us observe that in some games, cautious behaviour favours inertia. For example, in the symmetric all pay auction, cautious behaviour consists in bidding 0, i.e. in not entering the game: this is surely not something we want to favour in social sciences, notably in economy or management sciences.

Second, let us observe that *the concept is not defined for extensive form games.* So look at the Stackelberg all pay auction in Figure 3.4.

It is easy to establish player 1's cautious behaviour. First, if we do not suppress dominated strategies, after each bid b, player 1 has to anticipate that player 2 bids at least b+1, except for b=5, where she has to expect that player 2 bids 5 too. Hence, player 1 bids 0, and we get the first Nash equilibrium, which is very profitable to player 2; yet remember that we found this equilibrium quite unintuitive because player 2 plays in an incredible way after bids 3, 4 and 5. So let us now restrict player 2's answers to non dominated strategies. In that case, player 2 bids 0 after 3, 4 and 5, but, according to the cautious philosophy, he bids 3 after 2. So, in this new context, player 1 is indifferent between bidding 0 and 3, and it is now possible to reach the previous path (bid 0 followed by bid 1) but also the new path where both players get 5 (bid 3 followed by bid 0). Anyhow, player 1's cautious behaviour eliminates her preferred Nash equilibrium, where she bids 2 and player 2 bids 0. This is not surprising: this equilibrium is dangerous for player 1, given that player 2 is indifferent between bidding 0 and bidding 3 after bid 2. So it is indeed dangerous to bid 2.

But what about player 2? Can we also define a cautious behaviour for him? *Given that he is the last player to play,* he observes player 1's action and *has nothing to fear.* So *he just has to make his best response. Cautious behaviour here makes no sense.*

So clearly, if we want to keep cautious arguments in the extensive form game, we have to adapt the definition of cautiousness, usually only defined for games in normal form.

Last but not least, cautious behaviour, even after elimination of dominated strategies, may be silly as will be shown in the game in Matrix 3.11a.

5.2.2 Ordinal differences, risk dominance

We examine the three games in Matrices 3.11a, 3.11b and 3.11c.

		Player 2 A_2	Player 2 B_2
Player 1	A_1	(1,1)	(1,0.9)
	B_1	(0.9,1)	(100,100)

Matrix 3.11a

		Player 2 A_2	Player 2 B_2
Player 1	A_1	(2,2)	(99,0)
	B_1	(0,99)	(100,100)

Matrix 3.11b

133

NASH EQUILIBRIUM

	Player 2	
	A_2	B_2
A_1	(a,a)	(b,c)
B_1	(c,b)	(d,d)

Player 1

Matrix 3.11c

with $a>c$, $d>b$, $a\geq b$, $d>c$ and $(a-b)<(d-c)$.

Matrix 3.11a is Matrix 3.11c with $a=b=1$, $c=0.9$ and $d=100$.

There are no dominated actions in these games.

In the first two matrices, cautious behaviour leads to play A, in the third matrix the probability to play A grows in the difference $(d-c) - (a-b)$.

To see why, suppose that player 1 assigns probability p to A_1 (and 1–p to B_1) and that player 2 assigns probability q to A_2 (and 1–q to B_2). If player 2 plays A_2, player 1 gets 0.9+0.1p, 2p and $ap+c(1-p)=c+(a-c)p$ respectively in Matrix 3.11a, 3.11b and 3.11c. If player 2 plays B_2, player 1 gets 100–99p, 100–p and $d+(b-d)p$ respectively in Matrix 3.11a, 3.11b and 3.11c. So player 1's range of payoffs in the game of Matrix 3.11a, 3.11b and 3.11c is the grey area respectively represented in Figures 3.8a, 3.8b and 3.8c:

It immediately follows that p=1 and (by symmetry) q=1 are the cautious strategies in Matrices 3.11a and 3.11b and that $p=(d-c)/(a-b+d-c)$, which grows in the difference $(d-c)-(a-b)$, are the cautious strategies in Matrix 3.11c.

In fact, *cautious behaviour is driven by the others' deviations*:

In Matrix 3.11c, when player 1 focuses on the Nash equilibrium (A_1,A_2) she takes into account that player 2 may deviate to B_2 which leads to the loss a–b, and when she focuses on the Nash equilibrium (B_1,B_2), she fears player 2's deviation from B_2 to A_2, and so fears the loss d–c. Given that $p=(d-c)/(a-b+d-c)$ she plays A_1 with a higher probability if the feared loss d–c is higher than the feared loss a–b.

In Matrix 3.11a, player 1, when focusing on (A_1,A_2) doesn't fear player 2's deviation (because she gets 1 with and without deviation), whereas she fears the loss (100–0.9) when focusing on (B_1,B_2). This explains why she only plays A_1.

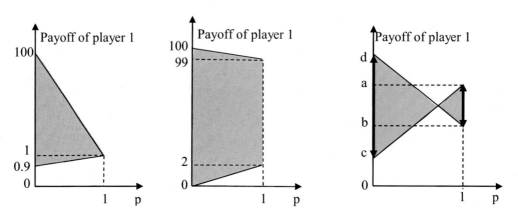

■ **Figures 3.8a, 3.8b, 3.8c** Figure 3.8c can be different (we can have d<a or c>b) but both lines always cross when a>b and intersect at p=1 when a=b.

■ 134

NASH EQUILIBRIUM

In Matrix 3.11b, player 1, when focusing on (A_1,A_2) is happy to see player 2 deviating, whereas she fears the loss (100–0) when focusing on (B_1,B_2). This explains why she only plays A_1.

Yet, at least in the game in Matrix 3.11a, it seems quite strange to play A. As a matter of fact, a player only gets the payoff 1 by playing A, whereas she can get a payoff between 0.9 and 100 by playing B. Moreover, the opponent is exactly in the same situation. And both players are best off when they both play B. Cautious behaviour here is clearly inappropriate in that it only partially analyses the game: as observed above, in Matrix 3.11a, if a player focuses on (A_1,A_2), the opponent's deviation from A to B does not frighten him whereas he fears the opponent's deviation from B to A when he focuses on (B_1, B_2), and this leads him to play A. But the above deviations should not drive the player's decision in the game under study: taking the notations of Matrix 3.11c, a player should be, in Matrix 3.11a, much more interested in the differences a–c (1–0.9) and d–b (100–1). As a matter of fact, by playing A instead of B in Matrix 3.11a, the additional payoff is at most 0.1=1–0.9 (when the opponent plays A too), whereas, by playing B instead of A, the additional payoff can be 99=100–1 (when the opponent plays B too)! *A never leads to a much higher payoff than B, but B can be much more fruitful than A.* Cautious behaviour completely ignores these interesting differences, which are (a–c) and (d–b).

There exist in game theory concepts that especially take these last differences into account, notably the notion of ***risk dominance.***

> **Definition 3. Risk dominance**
> **Risk dominance** compares Nash equilibria by looking for the deviations necessary to destabilize them, i.e. the probability of deviations necessary to switch (directly) from one equilibrium to another. A Nash equilibrium E risk dominates a Nash equilibrium E' if it is more easy to switch from E' to E than from E to E'.

> **➤ STUDENTS' INSERT**
>
> In the game in Matrix 3.11c, (B_1,B_2) risk dominates (A_1,A_2) if and only if d–b>a–c.
>
> To see why, let us start in (A_1,A_2) and suppose that q is the probability that player 2 deviates to B_2. (A_1,A_2) is destabilized as soon as player 1's best response becomes B_1, i.e. as soon as
>
> a(1–q)+bq<c(1–q)+dq, i.e. q>(a–c)/(d–c+a–b).
>
> Similarly, when the players start in (B_1,B_2), the probability q that destabilizes (B_1,B_2) is (d–b)/(d–c+a–b). Hence (B_1,B_2) risk dominates (A_1,A_2) if a–c<d–b.

So risk dominance and cautious behaviour are in conflict in the game in Matrix 3.11a; cautious behaviour favours A, risk dominance favours B, but we guess that experimental games should confirm that players would rather play B in accordance with risk dominance.

In the game in Matrix 3.11b, 2–0>100–99. So *risk dominance and cautious behaviour both favour the equilibrium (A_1,A_2).* Well, from a theoretical point of view, this is not astonishing given that the context in Matrix 3.11b is close to a dominance one: if 99 is replaced by 100, weak dominance leads to (A_1,A_2). But nevertheless, despite the convergence of the two criteria, will real

NASH EQUILIBRIUM

players really play (A_1, A_2)? Are the inequations 2–0>100–99 and 2–99<<100–0 really sufficient to choose A? It seems to me that, at least under some circumstances, players take into account a third difference, namely the difference of payoffs in the equilibria: in Matrix 3.11b, for both players, (B_1, B_2) is much more interesting than (A_1, A_2), in that **d–a=100–2>>0.**

> This last difference is taken into account in some *evolutionary criteria based on imitations*. These criteria work with populations of players that form random couples who play the game. Some imitation evolutionary concepts count the number of mutations necessary to switch from one equilibrium to another when players just compare the obtained payoffs (and switch to the behaviour that leads to the best payoffs). Especially, only two mutations are necessary to switch from a situation where all the couples play (A,A) to the situation where they all play (B,B) if d>a. As a matter of fact, if just two players mutate from A to B, and if they fortunately meet together, they both get d (here 100), whereas all the other players only get a (here 2), which induces everybody to play B!

So, to summarize, the Nash equilibrium concept doesn't take into account all the meaningful comparisons a player can do in a game: if we consider the game in Matrix 3.11c, Nash equilibria only compare a–c to 0 and d–b to 0, whereas cautious behaviour compares a–b and d–c, risk dominance compares a–c and d–b, and evolutionary imitation compares a and d.

CONCLUSION

The Nash equilibrium concept is quite powerful. In many economic, management and political topics it gives nice hints which help in making a decision. Especially in auction theory, more generally in industrial and financial economics, the Nash equilibrium cannot be bypassed. In the exercises of this chapter, you can see the power of the Nash concept in common value auctions (wallet games), its ability to highlight the winner's curse, how it allows to counter Akerlof's reasoning in the lemon car model, and so on.

Yet the Nash equilibrium has limits. We showed in the previous section that it does not take into account ordinal differences which are taken into account by other concepts such as cautious behaviour or risk dominance.

Given that it focuses on unilateral deviations, many actions out of a Nash equilibrium path are also quite strange in that the players would never play them when called on to do so.

Moreover, people may have some problems accepting the meaning of mixed equilibrium strategies. In a Nash equilibrium, when a player plays two actions A and B with positive probabilities, then A and B provide him with the same payoff. So he is completely indifferent with regard to the probabilities he assigns to each action. These probabilities have no meaning for himself; they are chosen to justify the actions of the other players. But this is quite difficult to accept for most people. Usually, when we play two actions A and B with different probabilities, say 9/10 and 1/10, this means that action A seems more appropriate than action B. So it seems that common sense doesn't agree with the logic of mixed Nash equilibria. We will see in chapter 6 that a game theory criterion exists, best reply matching, that fits better with the common sense of a mixed behaviour.

And last but not least, Nash equilibria do not always help us to play. In the ascending all pay auction game and in the envelope game for example, we do not learn how to play. In the envelope game we get a logical equilibrium behaviour that leads to at most exchanging the number 1, in the ascending all pay auction game, we get several opposite equilibria, all characterized by the fact that one player stops immediately. Yet, especially if M and V are large, with M>V, real players usually both overbid for a while, and in the envelope game, people usually are ready to exchange their envelope for amounts much higher than 1. The difference between reality and the Nash concept is that the Nash concept only accepts actions and beliefs that are consistent. Given that the exchange of the envelope may yield a loser and a winner, the loser is not allowed to exchange. Given that, in an all pay auction, the player who doesn't get the object can only lose the money he bids, he is not allowed to bid; as a result, the game stops as soon as the loser takes his first decision. In reality, people know that there is a loser and a winner when they exchange the envelopes, and they know that the loser of the all pay auction loses all the invested money. But nobody knows who the loser will be and everybody expects that it is the opponent. Of course these beliefs are inconsistent, but reality, in contrast to theory, accepts inconsistency. In the envelope game and in the ascending all pay auction, this inconsistency will be harmful for one of the players. But it is worth noting that in other contexts, *inconsistency may be profitable to all the players.* In the centipede game for example (see Exercise 12), the Nash behaviour is to stop immediately. Yet, if players have inconsistent beliefs and go on (because each player supposes, when called on to play, that the opponent goes on one more step), even if one of the players stops at a given level (so the opponent's beliefs were inconsistent at the previous step), both get a high payoff, much higher than the one obtained by stopping immediately. Similar observations hold for the traveller's dilemma (see Exercise 15). *Inconsistency is not incompatible with high payoffs, something we have to keep in mind when looking for strategies that optimize payoffs.*

NOTES

1 Nash, J., 1951. Non cooperative games, *Annals of Mathematics*, 54 (2), September, 286–295.
2 This proof is also in Umbhauer. G., 2004. *Théorie des Jeux*, Editions Vuibert, Paris.
3 We can introduce a discount factor to induce a player to go out as soon as he knows the colour.
4 p(x) and p(y) are the prior (Nature's) probabilities. We can't give their values given that the prisoners' disks game does not specify them: the only thing we need is that they are sufficiently positive, so that p(.)a<<0.
5 Schelling, T.C. 1960. *The strategy of conflict*, Cambridge, Mass. Harvard University Press.

Chapter 3

Exercises

♣ EXERCISE 1 NASH EQUILIBRIA IN A NORMAL FORM GAME

Consider the normal form game in Matrix E3.1

		Player 2		
		A_2	B_2	C_2
	A_1	(1,4)	(2,5)	(5,3)
Player 1	B_1	(7,2)	(1,0)	(4,0)
	C_1	(5,10)	(0,2)	(5,3)

Matrix E3.1

Question
Eliminate iteratively the strictly dominated strategies. Recall the link between Nash equilibria and strictly dominated strategies, find all the Nash equilibria (in pure and mixed strategies) and give the mixed Nash equilibrium expected payoffs.

♣ EXERCISE 2 STORY NORMAL FORM GAMES

Consider the chicken game, the syndicate game with three players, and the matching pennies game, three story games of chapter 1.

Questions
1. Find the pure strategy Nash equilibria of the syndicate game with three players.
2. Find the pure and mixed strategy Nash equilibria of the chicken game and the matching pennies game.

♣ EXERCISE 3 BURNING MONEY

Question
Write the burning money game *(see Exercise 7 in chapter 2)* in normal form and find the pure strategy Nash equilibria. What about the link with the iterative elimination of weakly and strictly dominated strategies?

EXERCISES

♣ EXERCISE 4 MIXED NASH EQUILIBRIA AND WEAK DOMINANCE

Consider the normal form game in Matrix E3.2.

		Player 2			
		A_2	B_2	C_2	D_2
	A_1	(3,7)	(1,2)	(0,3)	(6,1)
Player 1	B_1	(1,2)	(5,2)	(3,0)	(2,0)
	C_1	(4,3)	(2,4)	(4,6)	(5,3)

Matrix E3.2

Question

Find all the Nash equilibria.

♣ EXERCISE 5 A UNIQUE MIXED NASH EQUILIBRIUM

Consider the normal form game in Matrix E3.3.

		Player 2		
		A_2	B_2	C_2
	A_1	(3,3)	(0,4)	(0,2)
Player 1	B_1	(4,0)	(1,1)	(−2,2)
	C_1	(2,0)	(2,−2)	(−1,−1)

Matrix E3.3

Questions

This game has no pure strategy Nash equilibrium. Show that there exists a unique mixed Nash equilibrium. To do so, show successively:

1. player 2 can't play A_2 and B_2 with positive probabilities and C_2 with probability 0.
2. player 2 can't play A_2 and C_2 with positive probabilities and B_2 with probability 0.
3. player 2 can't play B_2 and C_2 with positive probabilities and A_2 with probability 0.
4. Conclude.

♣ EXERCISE 6 FRENCH VARIANT OF THE ROCK PAPER SCISSORS GAME

The unique Nash equilibrium of the rock paper scissors game (see chapter 1 and Matrix E3.4a) consists in playing each action with probability 1/3 (to get this equilibrium, proceed as in Exercise 5).

		Player 2		
		rock	scissors	paper
	rock	(0,0)	(1,−1)	(−1,1)
Player 1	scissors	(−1,1)	(0,0)	(1,−1)
	paper	(1,−1)	(−1,1)	(0,0)

Matrix E3.4a

EXERCISES

In the French version of this children's game, there is a fourth possible action, *the well*: the rock and the scissors fall into the well, but the paper covers the well. So we get the game in Matrix E3.4b.

		Player 2			
		rock	scissors	paper	well
Player 1	rock	(0,0)	(1,−1)	(−1,1)	(−1,1)
	scissors	(−1,1)	(0,0)	(1,−1)	(−1,1)
	paper	(1,−1)	(−1,1)	(0,0)	(1,−1)
	well	(1,−1)	(1,−1)	(−1,1)	(0,0)

Matrix E3.4b

Questions

1. First show that rock is a weakly dominated strategy. Does it mean that it can't be played at a Nash equilibrium?
2. The aim is to show that rock will not be played at equilibrium. Proceed as follows. Call p_1, p_2, p_3 and p_4, respectively q_1, q_2, q_3 and q_4 the probabilities assigned to rock, scissors, paper and well by player 1, respectively by player 2.
    a) First show that player 1 can play rock at equilibrium only if player 2 exclusively plays scissors and/or paper ($q_1=q_4=0$). Deduce that player 1 plays paper with probability 0 ($p_3=0$).
    b) Deduce that player 2 will not play scissors at equilibrium ($q_2=0$). Deduce that player 1 can't play rock.
3. Given that rock can't be played at equilibrium, what can you deduce about the equilibria of the game? What is the strategic meaning of the French variant?

♣♣ EXERCISE 7 BAZOOKA GAME

Remember the children's Bazooka game *(see Exercise 3 in chapter 1)*.

It is an endless game, i.e. a game that possibly lasts an infinite number of periods; its normal form is given in Matrix E3.5.

		Player 2		
		L_2	P_2	S_2
Player 1	L_1	(g_1,g_2)	(1,0)	(0,1)
	P_1	(0,1)	(g_1,g_2)	(1,0)
	S_1	(1,0)	(0,1)	(g_1,g_2)

Matrix E3.5

The special fact is that g_1 and g_2 are the payoffs of the game, because, after (L_1,L_2), (P_1,P_2) and (S_1,S_2), the players play again the same game.

Questions

1. Find the Nash equilibrium of the game.
2. Introduce discount factors, so as to take into account that a child may be patient or impatient. Find again the Nash equilibrium. Do patient children benefit from their patience?

EXERCISES

♣ EXERCISE 8 PURE STRATEGY NASH EQUILIBRIA IN AN EXTENSIVE FORM GAME

Consider the extensive form game in Figure E3.1.

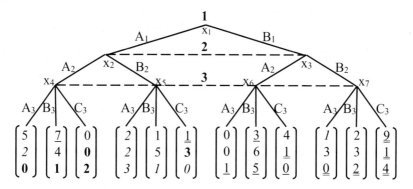

Figure E3.1

Questions
1. Show that there does not exist a Nash equilibrium in which player 1 plays A_1 and player 2 plays A_2.
2. Find all the Nash equilibria in pure strategies.

♣ EXERCISE 9 GIFT EXCHANGE GAME

Consider the gift exchange game *(chapter 1, section 1.2.2)*. Its extensive form is reproduced in Figure E3.2:

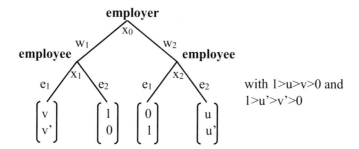

Figure E3.2 The 1st, respectively the 2nd coordinate of each payoff vector is the employer's payoff, respectively the employee's payoff.

Question
Find all the Nash equilibria. How does the obtained result generalize?

EXERCISES

♣♣ EXERCISE 10 BEHAVIOURAL NASH EQUILIBRIA IN AN EXTENSIVE FORM GAME

Consider the extensive form game in Figure E3.3.

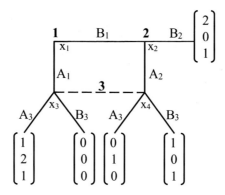

■ Figure E3.3

Question
Find all the Nash equilibria.

♣♣ EXERCISE 11 DUEL GUESSING GAME

Consider the duel guessing game (see *exercises 6 and 13 in chapters 1 and 2*).

Question
Find the pure strategy Nash equilibrium in the reduced normal form game. Compare with the iterative elimination of dominated strategies. Generalize.

♣♣ EXERCISE 12 PRE-EMPTION GAME (IN EXTENSIVE AND NORMAL FORMS)

We come back to the pre-emption game in Figure E3.4 *(see exercises 5 and 8 in chapters 1 and 2)*. The only strategy profile surviving iterative elimination of weakly and strictly dominated strategies leads both players to stop the game as soon as they have the possibility of doing so, even if this seems silly given that they can both get high payoffs by going on (450 and 1800 instead of 1 and 0).

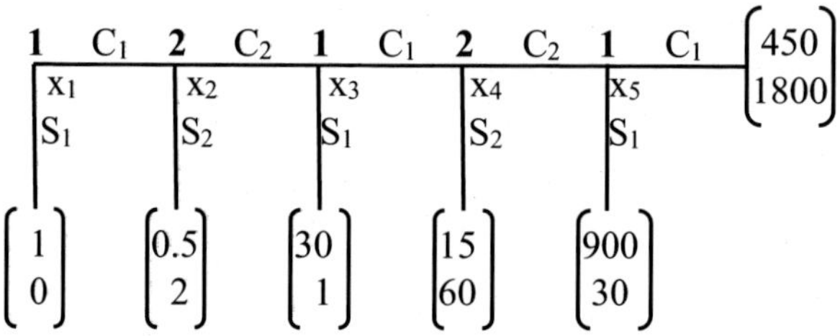

■ Figure E3.4

Questions

1. Show, by working on the extensive form game, that the only pure strategy Nash equilibria are such that player 1 stops at node x_1 and player 2 stops at node x_2. More generally show that in a pre-emption game, regardless of the (finite) length of the game, player 1 always stops the game at her first decision node on a Nash equilibrium path.
2. Find again the pure strategy Nash equilibria of the game in Figure E3.4, by working on the normal form game.
3. Comment by comparing with the ascending all pay auction/war of attrition game.

♣ EXERCISE 13 BERTRAND DUOPOLY

We come back to the Bertrand duopoly game *(see exercises 1 and 10 in chapters 1 and 2)*.

Question
Find the pure Nash equilibria in the discrete and in the continuous version of the game. Comment.

♣ ♣ EXERCISE 14 GUESSING GAME

Consider the guessing game *(see exercises 1 and 9 in chapters 1 and 2)*.

Question
Show that each player plays 0 in the Nash equilibrium of this game.

♣ EXERCISE 15 TRAVELLER'S DILEMMA (BASU)

Consider the traveller's dilemma, close to Basu's version, presented in Exercise 1 and 11 of chapters 1 and 2, with x_i an integer in [2,100], and the reimbursement rule:

- If $x_1 = x_2$, each traveller gets x_1
- If $x_i > x_j$ $i \neq j = 1,2$ then the traveller j gets $x_j + P$, where P is a fixed amount higher than 1, and the traveller i only gets max $(0, x_j - P)$, P being common knowledge.

Questions
1. Show that (2,2) is the only pure strategy Nash equilibrium. Does this result still hold when we replace max $(0, x_j - P)$ by $(x_j - P)$ (Basu's rule)?
2. Comment. Do real players play the Nash equilibrium? If so, why? If not, why?

♣ ♣ EXERCISE 16 TRAVELLER'S DILEMMA, STUDENTS' VERSION, P<49

Consider one of the students' versions of the traveller's dilemma *(see exercises 1 and 12 in chapters 1 and 2)* with x_i an integer in [2,100] and the reimbursement rule:

- If $x_1 = x_2$, each traveller gets x_1
- If $x_i > x_j$ $i \neq j = 1,2$ the traveller j gets $x_j + P$, where P is a fixed amount higher than 1, and the traveller i only gets $x_i - P$, P being common knowledge.

Suppose P<49

EXERCISES

Questions
1. Recall that x<100–2P can't be played and prove that there doesn't exist a Nash equilibrium in pure strategies.
2. Find the equilibrium in mixed strategies in the continuous version of the game (x is a real number in [2,100]). Comment.
3. Comment on the students' way of playing. See Figures E3.5a and E3.5b.

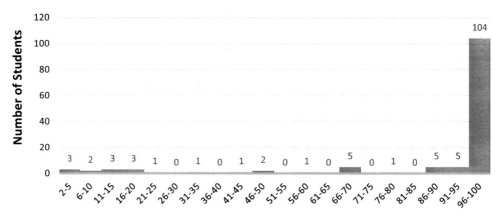

■ **Figure E3.5a** Traveller's dilemma, students' version, asked amounts for P=4

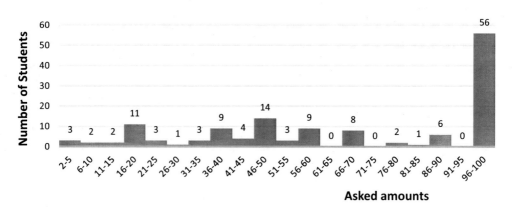

■ **Figure E3.5b** Traveller's dilemma, students' version, asked amounts for P=40

♣ EXERCISE 17 TRAVELLER'S DILEMMA, STUDENTS' VERSION, P>49, AND CAUTIOUS BEHAVIOUR

The equilibrium established in Exercise 16 for P<49, doesn't hold for P>49. The equilibria for P>49 are more difficult to establish. For example, for P=70, in the discrete version, it can be shown that there is a Nash equilibrium such that 2 is played with probability 41/70, and all the even integers in [44, 100] are played with probability 1/70. In other words, there is an atom on 2, and the remaining probability is uniformly distributed on {44,46,48,....96,98,100}.

This equilibrium is not so intuitive and my (137) L3 students do not seem to conform to it. Their asked amounts are given in the histogram in Figure E3.6

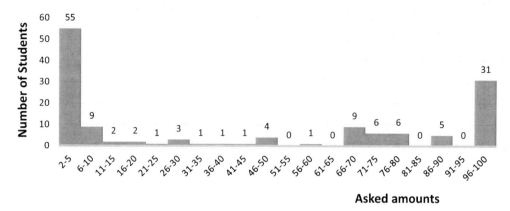

Figure E3.6 Traveller's dilemma, students' version, asked amounts for P=70

Question
Comment on the students' behaviour. What about a cautious strategy?

♣ EXERCISE 18 FOCAL POINT IN THE TRAVELLER'S DILEMMA

Question
We know that, when there are many Nash equilibria, like in a meeting game, then common history, shared moments of life, may help one equilibrium to become a focal point, and so may help the players to spontaneously converge to it. *But what about focal points in games where there is only one equilibrium, like in the traveller's dilemma studied in exercises 15, 16 and 17?* Focal points should not be necessary, should they?

♣ ♣ ♣ EXERCISE 19 ASYMMETRIC ALL PAY AUCTIONS

Consider the asymmetric first price all pay auction in Exercise 4 in chapter 2, with M=5, V_1=4, V_2=3, each player bidding x/2, where x is an integer from 0 to 10. The normal form of the game, after eliminating the weakly and strictly dominated strategies is given in Matrix E3.6a:

		Player 2		
		0	1	2
	0.5	(8.5,5)	(4.5,7)	(4.5,6)
Player 1	1.5	(7.5,5)	(7.5,4)	(3.5,6)
	2.5	(6.5,5)	(6.5,4)	(6.5,3)

Matrix E3.6a

Questions
1. Establish the mixed Nash equilibrium.
2. Suppose that the increment between 2 bids becomes 0.1. Matrix E3.6a becomes Matrix E3.6b.

EXERCISES

		Player 2				
		0	0.2	0.4	... 2.6	2.8
	0.1	(8.9,5)	(4.9,7.8)	(4.9,7.6)	(4.9,5.4)	(4.9,5.2)
Player 1	0.3	(8.7,5)	(8.7,4.8)	(4.7,7.6)	(4.7,5.4)	(4.7,5.2)
	...					
	2.7	(6.3,5)	(6.3,4.8)	(6.3,4.6)	(6.3,2.4)	(2.3,5.2)
	2.9	(6.1,5)	(6.1,4.8)	(6.1,4.6)	(6.1,2.4)	(6.1,2.2)

Matrix E3.6b

Comment on the new Nash equilibrium probabilities, and especially the one assigned to bid 0.
3. Imagine that player 1, respectively player 2, can play any bid in $[0,V_1]$, respectively in $[0,V_2]$. Find the new Nash equilibrium in this continuous setting.

♣♣ EXERCISE 20 WALLET GAME, FIRST PRICE AUCTION, WINNER'S CURSE AND A ROBUST EQUILIBRIUM

N players participate in a common value auction. There is an object of common value V. Each player i, i from 1 to N, has a partial information on V: he observes a signal x_i distributed on [0,1] according to a density function f_i (the functions f_i, i from 1 to N, are independent). The common value of the object is defined by: $V=V(x_1,x_2,...,x_N)$. The game goes as follows: after observing his private signal x_i, each player i makes an offer for the object. The winner is the player who makes the highest offer: he gets the object and pays the proposed bid (if K players play the highest bid, each gets and pays it with probability 1/K). The other players get and pay nothing.

In this exercise, we set N=2, we work with uniform distributions [0,1], and with the wallet game, the easiest common value auction, where V (the value of the wallet) is defined by: $V=x_1+x_2$.

Questions
1. In some experimental settings, people propose ½ in addition to their private signal. This behaviour is known as the winner's curse behaviour. Why do people behave in this way and why is it not a best way to play?
2. Find the symmetric Nash equilibrium of the wallet game.

♣♣ EXERCISE 21 WALLET GAME, FIRST PRICE AUCTION, NASH EQUILIBRIUM AND NEW STORIES

Consider the wallet game with N=2 and $V=x_1+x_2$, whose symmetric Nash equilibrium leads player 1 to bid x_1 and player 2 to bid x_2 (see Exercise 20).

Questions
1. The Nash equilibrium in a first price sealed bid auction, where player i values the object x_i i=1,2, x_i being uniformly distributed on [0,1], consists in bidding $x_i/2$. Find an easy rule of behaviour that fits with the Nash equilibrium in the wallet game.
2. Suppose that x_1 is uniformly distributed on [0,1] and that x_2 is uniformly distributed on [0, ½]. Show that the bid functions $b_1(x)=3x/4$ and $b_2(x)=3x/2$ constitute a Nash equilibrium. Find an easy rule of behaviour that fits with this equilibrium.

EXERCISES

♣ ♣ EXERCISE 22 TWO PLAYER WALLET GAME, SECOND PRICE AUCTION, A ROBUST SYMMETRIC EQUILIBRIUM

Consider the wallet game with N=2 and $V(x_1,x_2)=x_1+x_2$. The winner is again the player who makes the highest offer but he pays the price of the loser (second price auction) (if both make the same bid b, each gets the object and pays b with probability ½).

Questions
1. Find the symmetric Nash equilibrium. Comment.
2. Observe the no ex-post *regret property*.

♣ ♣ ♣ EXERCISE 23 TWO PLAYER WALLET GAME, SECOND PRICE AUCTION, ASYMMETRIC EQUILIBRIA

We now look for asymmetric Nash equilibria in the wallet game with second price auction.

Questions
1. Show that $b_1(x_1)=x_1+h(x_1)$ and $b_2(x_2)=x_2+h^{-1}(x_2)$, where h is a continuous function growing in x, constitute a Nash equilibrium. Take $h(x)=100x$. Comment. Do other asymmetric equilibria exist?
2. Show that the following profile of strategies is a Nash equilibrium.

   Player 1 bids $b_1(x_1) = x_1+1$ if $x_1 > c$ with $0 \leq c \leq 1$
   $ = x_1+d$ if $x_1 \leq c$ with $0 \leq d \leq 1$

   Player 2 bids $b_2(x_2) = x_2+c$ if $x_2 \geq d$
   $ = x_2$ if $x_2 < d$

   Comment.
3. What about dominated strategies in the two above equilibria and the role of the signal density functions?

♣ ♣ EXERCISE 24 N PLAYER WALLET GAME, SECOND PRICE AUCTION, MARGINAL APPROACH

In an N player wallet game with second price auction, and with $V(x_1,x_2,\ldots,x_N)=\sum_{i=1}^{N} x_i$ the Nash equilibria can be obtained *with a marginal approach*.

Questions
1. First set N=2. Suppose that player 1 switches from $b_1(x_1)=b$ to $b_1(x_1)=b+\varepsilon$. What additional payoff can she get? Deduce that, for player 1 to not be induced to play $b+\varepsilon$, this marginal additional payoff has to be null, and find again the symmetric Nash equilibrium. Why can't we take the same approach in a first price auction?
2. Take the same approach to find the symmetric Nash equilibrium in the N player wallet game when the signals are uniformly distributed on [0,1].

EXERCISES

♣♣♣ EXERCISE 25 SECOND PRICE ALL PAY AUCTIONS

Questions

1. Consider the second price sealed bid all pay auction, with M=5 and V=3, studied in chapter 1 (and chapter 6) and reproduced in Matrix E3.7:

   Check that $p_0=q_0=83/293=0.283$, $p_1=q_1=57/293=0.195$, $p_2=q_2=45/293=0.154$, $p_3=q_3=27/293=0.092$, $p_4=q_4=0$, $p_5=q_5=81/293=0.276$ (where p_0, p_1, p_2, p_3, p_4 and p_5, respectively q_0, q_1, q_2, q_3, q_4 and q_5 are the probabilities assigned by player 1, respectively player 2, to the bids 0, 1, 2, 3, 4 and 5) is a mixed Nash equilibrium of the game. Comment.

	Player 2					
	0	1	2	3	4	5
0	(6.5,6.5)	(5,8)	(5,8)	(5,8)	(5,8)	(5,8)
1	(8,5)	(5.5,5.5)	(4,7)	(4,7)	(4,7)	(4,7)
2	(8,5)	(7,4)	(4.5,4.5)	(3,6)	(3,6)	(3,6)
3	(8,5)	(7,4)	(6,3)	(3.5,3.5)	(2,5)	(2,5)
4	(8,5)	(7,4)	(6,3)	(5,2)	(2.5,2.5)	(1,4)
5	(8,5)	(7,4)	(6,3)	(5,2)	(4,1)	(1.5,1.5)

Player 1 rows labeled 0–5 on the left.

Matrix E3.7

2. Turn to the general case with V/2<M.
   a) Check that all the bids higher or equal to M–V/2 and lower than M are weakly dominated by M.
   b) Find the mixed Nash equilibrium of the continuous game (a bid is any real number in [0,M]). Comment.

♣ EXERCISE 26 SINGLE CROSSING IN A FIRST PRICE SEALED BID AUCTION

Consider the 2 player first price sealed bid auction studied in chapter 3 (section 3.1.2), where the value assigned to an object is uniformly distributed on [0,V] for each player. In the symmetric Nash equilibrium a player bids x/2, when he values the object x.

Question

The aim is to show that if the equilibrium bids are continuous in the value, then they are increasing in the value. In other words the aim is to prove the *single crossing property* according to which, if a player who values an object t weakly prefers a bid b' to a bid b, with b'>b, then a player who values the object t', with t'>t, prefers b' to b.

♣♣ EXERCISE 27 SINGLE CROSSING AND AKERLOF'S LEMON CAR

The studied context is an experience good model close to Akerlof's context, introduced in Exercise 7 of chapter 1. A seller wants to sell a car to a buyer. The car can be of n different qualities t_i, with $t_i<t_{i+1}$, i from 1 to n–1: we say that the seller is of type t. The seller's reservation price for a good (a car in Akerlof's model) of quality t_i is h_i, i from 1 to n. The seller sets a price for her good. The

buyer observes the price and accepts or refuses the transaction. The buyer's reservation price for a good of quality t_i is H_i, i from 1 to n. We set $h_i<h_{i+1}$ and $H_i<H_{i+1}$, for i from 1 to n–1. The buyer ignores the quality during the transaction, but has a prior probability distribution over the qualities, that is common knowledge of both players; it assigns probability ρ_i to the quality t_i, with $0<\rho_i<1$ for i from 1 to n and $\sum_{i=1}^{n} \rho_i = 1$. We suppose $H_i>h_i$ for any i from 1 to n, in order to make trade possible for any quality and we introduce Akerlof's assumption:

$$\frac{\sum_{i=1}^{j} \rho_i H_i}{\sum_{i=1}^{j} \rho_i} < h_j \text{ for j from 2 to n.} \quad (a)$$

Questions

1. The single crossing property says that if a seller of type t weakly prefers setting a price p' than a price p, with p'>p, then a seller of higher type t' prefers setting p' than p.

   Show that the single crossing property holds in this game.

2. Assumption (a) is central to Akerlof's well-known observation: 'if trade occurs, the car is sold at a unique price, regardless of its quality, because any type of seller wants to sell her car at the highest price. It follows that only the weakest quality can be sold on the market.' Justify Akerlof's conclusion.

3. Why do we not necessarily observe Akerlof's result at a Nash equilibrium?

♣ ♣ ♣ EXERCISE 28 DUTCH AUCTION AND FIRST PRICE SEALED BID AUCTION

The aim is to show, in a simple 2 player game, that the Nash equilibria in a Dutch auction are the same than in a first price sealed bid auction (FPSBA).

Questions

We work with a discrete game where the object has six possible values 0, 1, 2, 3, 4 or 5, each value occurring with probability 1/6. We say that a player is of type t if he values the object t.

1. Consider the symmetric profile of strategies defined by: a player of type 0 or 1 bids 0, a player of type 2 or 3 bids 1 and a player of type 4 or 5 bids 2. Show that this profile is a Nash equilibrium of the FPSBA game.

2. Switch to the Dutch auction. Consider the symmetric profile of strategies:

   A player of type 0 or 1 stays in the auction till the proposed price is 0, a player of type 2 or 3 only accepts the prices 1 and 0, and a player of type 4 or 5 only accepts the prices 0, 1 and 2. This amounts to saying that a player of type 4 and 5 accepts the price 2, a player of type 2 and 3 accepts the price 1, and a player of type 0 and 1 only accepts the price 0.

   Show that this profile of strategies is a Nash equilibrium. Comment.

Chapter 4

Backward induction and repeated games

INTRODUCTION

This chapter splits into two parts, backward induction and repeated games.

We first talk about backward induction, a way to solve games which is closely linked to Selten's concept of subgame perfect Nash equilibrium (SPNE). Subgame perfection and backward induction only apply to games in extensive form.

Once again, we start with much enthusiasm and we moderate this enthusiasm gradually as work proceeds. Backward induction solves a game by starting at the end of the game. At first look, this way of solving games may not seem very natural, given that in reality we start a game at its beginning. Yet, starting a game at the beginning does not stop you thinking about the end of the game. A player has to anticipate the future reactions to his decisions today to be able to choose the best current action. So he solves the future subgames before focusing on the decision to take at the beginning of the play: a clever reasoning requires backward induction. In fact, backward induction is a very old tool of reasoning[1] used in special games since 1912 (Zermelo 1912),[2] but also in optimization, especially in dynamic programming where it has never been contested. And we will see that backward induction can do a very nice job. In many contexts, it fits with strategies that are easier to justify than some Nash equilibria. Unfortunately, this is not always the case. Especially, backward induction suffers from a logical inconsistency: solving a subgame in a rational way when only an irrational behaviour can lead to it, is at least not meaningful.

In the first part of this chapter, we focus on SPNE and backward induction. In section 1, we expose the SPNE concept and its equivalence with the backward induction principle. We work on the ascending all pay auction. In the second section we provide other examples, namely the Ford Boyard sticks game and the negotiation games, where backward induction does a good job. In section 3 we turn to the limits of backward induction. We highlight the backward induction inconsistency, and we study the general sequential all pay auction, where backward induction leads to a strange result. In section 4 we turn to repeated games, with a stage game in normal or extensive form. We expose the classic results for finitely repeated games before turning to infinitely repeated games in section 5. We conclude with learning and backward induction.

BACKWARD INDUCTION AND REPEATED GAMES

1 SUBGAME PERFECT NASH EQUILIBRIUM AND BACKWARD INDUCTION

1.1 Subgame Perfect Nash equilibrium and Nash equilibrium

> **Definition 1: Subgame Perfect Nash equilibrium (SPNE)**
> A SPNE (Selten 1965[3]) is a Nash equilibrium that induces a Nash equilibrium in each subgame.

Comment: this concept impedes incredible behaviour in subgames that are out of the equilibrium path.

We show how it works in an ascending all pay auction, with V=3 and M=5, close to the game studied in the previous chapters, except that, at each step, both players simultaneously choose to stop or to overbid. The extensive form of the game is given in Figure 4.1a.

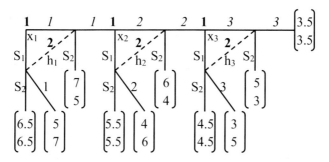

Figure 4.1a

We could of course pursue the game up to the bid 5, but for ease of explanation we only study a game with three steps: players 1 and 2 can bid 1, 2 or 3.

This game has three subgames; the one starting at x_1, the one starting at x_2, and the one starting at x_3, as represented in Figures 4.1a, 4.1c and 4.1b.

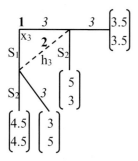

Figure 4.1b

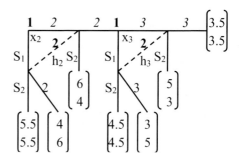

Figure 4.1c

151

BACKWARD INDUCTION AND REPEATED GAMES

> **STUDENTS' INSERT**
>
> We show that $((1/x_1, 2/x_2, 3/x_3), (S_2/h_1, S_2/h_2, 3/h_3))$ is a SPNE:
>
> To do so, we first show that it is a Nash equilibrium in the whole game (Figure 4.1a). As soon as player 2 stops at h_1, any strategy $(1/x_1, */x_2, */x_3)$ is a best response for player 1 in that it leads to the payoff 7, her highest payoff in the game (* means any available action). In the same way, given that player 1 goes on at every node, any strategy $(S_2/h_1, */h_2, */h_3)$ is a best response for player 2, given that it leads to the payoff 5, the best payoff player 2 can achieve given that player 1 never stops. Hence $((1/x_1, 2/x_2, 3/x_3), (S_2/h_1, S_2/h_2, 3/h_3))$ is a Nash equilibrium in the game of Figure 4.1a.
>
> Second we show that $((1/x_1, 2/x_2, 3/x_3), (S_2/h_1, S_2/h_2, 3/h_3))$ induces a Nash equilibrium in the subgame starting at x_3, i.e. that $(3/x_3, 3/h_3)$ is a Nash equilibrium of the game in Figure 4.1b. It is easy to check that this profile is the only Nash equilibrium of this game.
>
> Third, we show that $((1/x_1, 2/x_2, 3/x_3), (S_2/h_1, S_2/h_2, 3/h_3))$ induces a Nash equilibrium in the game in Figure 4.1c, i.e. that $((2/x_2, 3/x_3), (S_2/h_2, 3/h_3))$ is a Nash equilibrium of this game. Any strategy $(2/x_2, */x_3)$ is a best response for player 1 given that player 2 stops at h_2 (6>5.5). Reciprocally any strategy $(S_2/h_2, */h_3)$ is a best response for player 2, given that he gets 4 by doing so, whereas he gets less by bidding 2 (given that player 1 bids 2 at x_2 and 3 at x_3). Hence $((2/x_2, 3/x_3), (S_2/h_2, 3/h_3))$ is a Nash equilibrium of the game in Figure 4.1c.
>
> Hence $((1/x_1, 2/x_2, 3/x_3), (S_2/h_1, S_2/h_2, 3/h_3))$ is a SPNE.

It is not the only SPNE. The symmetric profile $((S_1/x_1, S_1/x_2, 3/x_3), (1/h_1, 2/h_2, 3/h_3))$ is also a SPNE. And there are three other SPNE:

$$((\pi_{1x1}(S_1)=1/3, \pi_{1x2}(S_1)=1/2, 3/x_3), (\pi_{2h1}(S_2)=1/3, \pi_{2h2}(S_2)=1/2, 3/h_3))$$

$$((S_1/x_1, \pi_{1x2}(S_1)=1/2, 3/x_3), (1/h_1, \pi_{2h2}(S_2)=1/2, 3/h_3))$$

and the symmetric profile $((1/x_1, \pi_{1x2}(S_1)=1/2, 3/x_3), (S_2/h_1, \pi_{2h2}(S_2)=1/2, 3/h_3))$ (see the exercises in this chapter).

This game is less artificial than the ascending all pay auction with V=3, M=5 studied in the previous chapters (and reproduced in Figure 4.2a): in this new game, both players are in a symmetric situation and no player has a structural advantage. So it is natural that the symmetric profile of each equilibrium is an equilibrium too. Let us stress the existence of an "always going on" equilibrium. This equilibrium is not due to the fact that we stopped the bids arbitrarily at 3. If, as in the asymmetric structure, we allow the players to bid up to 5, we again find an always going on SPNE where each player bids with positive probability up to 4, and bids 5 with probability 1 *(see the exercises)*. Consequently people overbid at equilibrium, even if they are losing money (bids 4 and 5), because it is better to go on when the other stops, and the other actually stops with enough positive probability.

We return to the game in Figure 4.1a. The five equilibria we proposed are the only SPNE. By contrast there exist other Nash equilibria. For example, $((1/x_1, 2/x_2, 3/x_3), (S_2/h_1, S_2/h_2, S_2/h_3))$ is a Nash equilibrium, in that it is best for player 2 to stop at h_1 given player 1's future behaviour, and it is best bidding 1 for player 1 given that player 2 stops at h_1. This Nash equilibrium is not a SPNE, because it does not induce a Nash equilibrium in subgame 4.1b, where bidding 3 is a strictly dominant action for both players. And it follows:

BACKWARD INDUCTION AND REPEATED GAMES

> **Property 1:** The set of SPNE of a game is included in the set of Nash equilibria.
>
> {SPNE} ⊆ {Nash equilibria}
>
> The set of SPNE is equal to the set of Nash equilibria if the game has no proper subgame. Each finite game has at least one SPNE.

To illustrate the second part of the property, we can mention the French plea bargaining game and the prisoners' disks game, which have no proper subgame. More generally, as soon as the game is simultaneous, or as soon as there is incomplete information (changed into imperfect information), the game often has no proper subgame; in all these games, the SPNE are the Nash equilibria of the game. In other words, subgame perfection is not a strong tool when there are few subgames.

Now observe that to establish that a profile is a SPNE, we have to check that it is a Nash equilibrium in a succession of games, from a smallest size to the size of the initial game, or from the largest size to the smallest one. This may be fastidious. That is why we usually take advantage of a structural property of the SPNE: at least in finite games, *the SPNE profiles can be found thanks to the backward induction principle*.

1.2 Backward induction

While the SPNE definition doesn't differ in finite and infinite games, the backward induction principle, which leads to solving a game by starting at the end, is clearly affected by the finiteness or the infiniteness of the game.

> **Definition 2:** When the game is a finite game, order the subgames from the smallest ones, starting from the end, to the largest (the game itself). Solving the game by **backward induction** means first selecting a Nash equilibrium in each smallest subgame at the end of the game, then jumping to the next subgames of the game (starting from the end), but replacing all the actions in the smallest subgames by the selected Nash equilibrium payoffs. Then you look for a Nash equilibrium of these new **reduced games,** and you jump to the next smallest subgames and proceed as before, and so on, until you reach the subgame which is the game itself, where you again proceed as before.

> ### ➤ STUDENTS' INSERT
>
> We first apply this principle to the finite game studied in the previous section.
> The smallest subgame is given in Figure 4.1b and (3.5, 3.5) are the only Nash equilibrium payoffs of this subgame.
>
> So we turn to the *reduced game* in Figure 4.1d: observe that we replaced the actions after (bid 2, bid 2) in the subgame of Figure 4.1c by the Nash payoffs (3.5, 3.5). One of the Nash equilibria in this reduced game is indeed ($2/x_2$, S_2/h_2) because 6>5.5 and 4>3.5.
>
> And we finally study the reduced game in Figure 4.1e (we replaced the actions after (bid 1, bid 1) in the game of Figure 4.1a by the above obtained Nash payoffs (6,4). And ($1/x_1$, S_2/h_1) is indeed a Nash equilibrium in this reduced game (7>6.5 and 5>4).

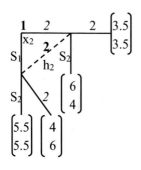

■ Figure 4.1d

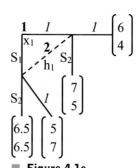

■ Figure 4.1e

So we proved that the SPNE $((1/x_1,2/x_2,3/x_3),(S_2/h_1,S_2/h_2,3/h_3))$ also resists the backward induction principle. This, as discussed above, is an expected result.

> **Property 2:** in a finite game, subgame perfection and the backward induction principle lead to the same selection of strategies.

This property implies that, in finite games, we almost systematically apply backward induction in order to find the SPNE, because backward induction makes us work on much easier games than subgame perfection: *it is easier to study the games 4.1b, 4.1d and 4.1e than the games 4.1b, 4.1c, and 4.1a.*

Things become more complicated when a game is infinite: backward induction needs to be redefined in that there no longer exists a smallest subgame starting from the end, given that there is no end (we will bypass this difficulty in section 4).

2 BACKWARD INDUCTION, DOMINANCE AND THE GOOD JOB OF BACKWARD INDUCTION/SUBGAME PERFECTION

2.1 Backward induction and dominance

Many times, especially in games with complete and perfect information, backward induction and iterative elimination of dominated strategies do the same or at least a similar job. But there is no equivalence.

We start outlining the similarity. To do so, we return to the two ascending all pay auction games studied in the previous chapters, reproduced in Figure 4.2a and Figure 4.3.

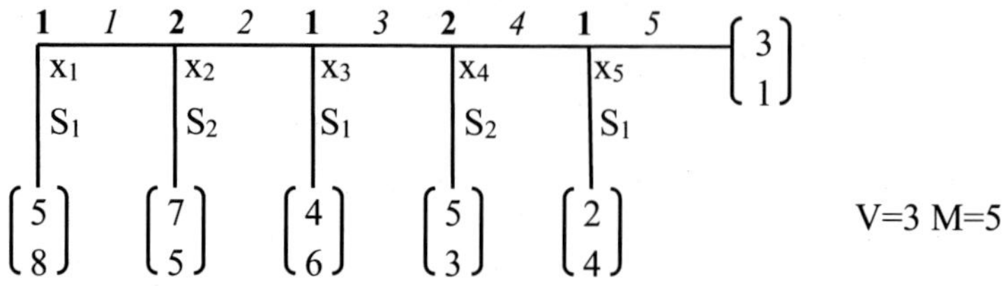

■ Figure 4.2a

BACKWARD INDUCTION AND REPEATED GAMES

We showed in chapter 2 that in the game in Figure 4.2a, the only behaviour resisting iterative elimination of dominated strategies leads player 1 to always go on and player 2 to always stop. These are also the only subgame perfect strategies of this game.

To see why, consider the succession of reduced games, reproduced in Figures 4.2b, 4.2c, 4.2d, 4.2e and 4.2f.

$$1 \xrightarrow{5}_{\substack{x_5 \\ S_1}} (3,1) \quad 2 \xrightarrow{4}_{\substack{x_4 \\ S_2}} (3,1) \quad 1 \xrightarrow{3}_{\substack{x_3 \\ S_1}} (5,3) \quad 2 \xrightarrow{2}_{\substack{x_2 \\ S_2}} (5,3) \quad 1 \xrightarrow{1}_{\substack{x_1 \\ S_1}} (7,5)$$

$$\begin{pmatrix}2\\4\end{pmatrix} \quad\quad \begin{pmatrix}5\\3\end{pmatrix} \quad\quad \begin{pmatrix}4\\6\end{pmatrix} \quad\quad \begin{pmatrix}7\\5\end{pmatrix} \quad\quad \begin{pmatrix}5\\8\end{pmatrix}$$

▪ **Figures 4.2b, 4.2c, 4.2d, 4.2e, 4.2f**

These reduced games are somewhat special in that only one player is playing; *looking for a Nash equilibrium amounts to looking for an optimal action of the unique decision taker.*

In the subgame in Figure 4.2b, player 1 bids 5 because 3>2.
In the reduced game in Figure 4.2c, player 2 stops because 3>1.
In the reduced game in Figure 4.2d, player 1 bids 3 because 5>4.
In the reduced game in Figure 4.2e, player 2 stops because 5>3.
In the reduced game in Figure 4.2f, player 1 bids 1 because 7>5.

Here backward induction *follows the same steps* as iterative elimination of dominated strategies.

In the other ascending all pay auction studied in chapter 2 (and reproduced in Figure 4.3), backward induction proceeds in the same way and again leads player 1 to always go on and player 2 to always stop. We could also eliminate dominated strategies by following the same steps, but in chapter 2 we proceeded in a more convincing way: we observed that, for player 1, always going on weakly dominates all her other strategies, which leads player 2 to stop at his first decision node. This approach has nothing to do with backward induction.

So backward induction can (partly) coincide with iterative elimination of dominated strategies; but there often exist ways to eliminate dominated strategies that do not coincide with the backward induction process.

In the game in Figure 4.3, the outcomes of these different processes were the same. But this is not always the case. So look at the game in Figure 4.4.

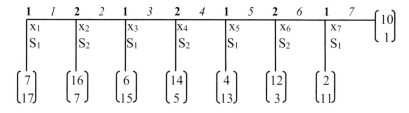

V=10 M=7

▪ **Figure 4.3**

155

BACKWARD INDUCTION AND REPEATED GAMES

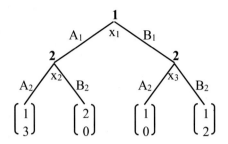

■ **Figure 4.4**

By backward induction, we first solve the two subgames at nodes x_2 and x_3: necessarily, player 2 plays A_2 at x_2 and B_2 at x_3. It follows that, at the reduced game starting at x_1, player 1 is indifferent between A_1 and B_1 so there are two SPNE (in pure strategies): $(A_1, A_2/x_2, B_2/x_3)$ and $(B_1, A_2/x_2, B_2/x_3)$).

By first eliminating player 2's weakly dominated strategies $(B_2/x_2, */x_3)$[4] and $(*/x_2, A_2/x_3)$, we also get the two equilibria $(A_1, A_2/x_2, B_2/x_3)$ and $(B_1, A_2/x_2, B_2/x_3)$.

Yet we can also first eliminate the weakly dominated strategy B_1 and then the strategies $(B_2/x_2, */x_3)$, which leads to the Nash equilibria $(A_1, A_2/x_2, B_2/x_3)$ and $(A_1, A_2/x_2, A_2/x_3)$, which are both more profitable to player 2 than $(B_1, A_2/x_2, B_2/x_3)$.

In some way, backward induction, by working only from the end of the game to its beginning, works too rigidly. Consequently it often coincides with only one way to eliminate dominated strategies. And, given that different iterative eliminations of weakly dominated strategies may lead to different outcomes, it is not astonishing that the SPNE are often a subset of the possible outcomes obtained with iterative elimination of dominated strategies.

2.2 The good job of backward induction/subgame perfection

In many games, subgame perfection rather does a good job in that it efficiently eliminates strange behaviours, sometimes called incredible threats, which are only possible because some subgames are not reached.

We discuss two examples, the sticks game and the negotiation game.

2.2.1 Backward induction and the Fort Boyard sticks game

We already know how to play this game in order to win, but let us prove the following result:

The profile such that:

$\forall k \in \mathbb{N}$, each player in front of

4k+1 sticks, takes 1, 2 or 3 sticks (2 and 3 are only possible if k>0)

4k+2 sticks, takes 1 stick

4k+3 sticks, takes 2 sticks

4k+4 sticks, takes 3 sticks

is a SPNE.

BACKWARD INDUCTION AND REPEATED GAMES

> **STUDENTS' INSERT**
>
> We first prove in a recursive way that a player in a reduced game where he is in front of 4k+1 sticks loses the game whatever he does, so that taking 1, taking 2 and taking 3 sticks are all best responses in the reduced game.
>
> It is obvious for k=0 and k=1.
>
> We suppose that this property is true for k and we prove that this implies that it is also true for k+1. So we prove that in a reduced game with 4(k+1)+1=4k+5 sticks, the player loses the game whatever he does, given the actions played in the following reduced games:
>
> - If he takes 1 stick, the opponent is in front of 4k+4 sticks, so he takes 3 sticks, which leaves the player in front of 4k+1 sticks, and he loses the game (by assumption).
> - If he takes 2 sticks, the opponent is in front of 4k+3 sticks, so he takes 2 sticks, which leaves the player in front of 4k+1 sticks, and he loses the game.
> - If he takes 3 sticks, the opponent is in front of 4k+2 sticks, so he takes 1 stick, which leaves the player in front of 4k+1 sticks, and he loses the game.
>
> So a player in front of 4k+1 sticks, given the actions (and payoffs) at the future reduced games, loses the game regardless of the number of sticks taken. Hence any behaviour in this reduced game is a best response.
>
> It automatically follows from this first proof that in a reduced game with 4k+2 sticks, taking one stick is an optimal action, given that it leaves the opponent in front of 4k+1 sticks. Let us more precisely prove that it is the only optimal action:
>
> If the player (say player i) takes 2 sticks, then the opponent is in front of 4k sticks (=4(k−1)+4), in which case he takes 3 sticks, and player i is in front of (4(k−1)+1) sticks and loses the game. And if player i takes 3 sticks, then the opponent is in front of 4(k−1)+3 sticks, in which case he takes 2 sticks, which leaves player i in front of 4(k−1)+1 sticks, so he loses the game.
>
> In a similar way, it automatically follows from the first proof that in a reduced game with 4k+3 sticks, taking 2 sticks is the only optimal action, and that in a reduced game with 4k+4 sticks, taking 3 sticks is the only optimal action.

Backward induction here is very natural. *Better than Nash equilibrium, it leads a player to always react best in a given situation, even if he should not be in this situation.* Backward induction is also more natural than elimination of weakly dominated strategies: as a matter of fact, remember that an equivalent strategy to a weakly dominant one cannot be eliminated in the sticks game even if it induces a strange behaviour in some unreached subgames.

Backward induction also requires that *each player plays rationally in each further subgame, even if he played irrationally in the past.* We'll later show that this may lead to a logical inconsistency. Yet, in our simple sticks game, expecting more rationality in the future is a reasonable assumption. Past silly actions are not associated with the fact that the player is really silly, but with *the fact that the game is too complicated at the beginning*: in front of many sticks (for example 50) players may be completely unable to discover the good way to play the game, but they may discover it when the number of sticks becomes smaller, for example when it reaches 9. *This has been observed experimentally for the companion race game (see the conclusion of chapter 2).*

BACKWARD INDUCTION AND REPEATED GAMES

2.2.2 Backward induction and negotiation games

Negotiation games gave rise to a rich literature, both in game theory, economics, law economics, political sciences and management. In this section we illustrate just one property of these games. *The power of the last player making an offer logically diminishes when the number of rounds of negotiation grows, and when time is costly.*

So let us suppose that two players can share a sum S that decreases over time.

We first suppose that the game only lasts one period and that S is equal to 10. Player 1 makes an offer x to player 2, with $x \in \{1,2,3,4,5,6,7,8,9\}$. Player 2 observes the offer and either accepts or refuses it. If he accepts, player 1 gets x and player 2 gets 10–x. If he refuses, both players get 0. This game is the *ultimatum game*, except that we work with integers and that we exclude the offers 0 and 10, which lead to a null payoff for one of the players (this avoids ties and multiple equilibria).

Given that player 2 is best off accepting each offer, player 1, at the unique SPNE of this game, proposes 9 and player 2 accepts, so player 1 gets 9 and player 2 gets 1.

We now suppose that the game lasts two periods. It starts as above but if player 2 refuses the offer x, the players switch to a second period. In this period, the sum to share is only 9. Player 2 makes a counteroffer y, with $y \in \{1,2,3,4,5,6,7,8\}$. Player 1 observes y and can accept or refuse it. If she accepts, she gets y and player 2 gets 9–y, if not, both get 0. This game is "scheduled" in Figure 4.5a (we just draw one branch x for player 1 (but x can take 9 values so we should draw 9 branches), and we just draw 1 branch for player 2's counteroffer y (but y can take 8 values).

We look for the SPNE of this game that satisfies the following property: if a player is indifferent between accepting and refusing an offer, he accepts it.[5]

Clearly, in the last subgames, player 1 accepts all the offers. It follows that player 2 proposes 1 in the next reduced games and is sure to get 8 in the second period. So, in the previous reduced games, player 2 refuses all the offers x>2 because he can get 8 by refusing. It derives that in the reduced game starting at the initial node, player 1 offers x=2 because player 2 accepts this offer and refuses any higher offer. So the SPNE path is such that player 1 proposes x=2 in period 1 and player 2 accepts it.

The SPNE is defined by: $((x=2, A_1/y \text{ y from 1 to 8}), (A_2/x \text{ if } x \leq 2, R_2/x \text{ if } x>2, y=1))$

We now suppose that there are 3 periods. The game starts as above, but if player 1 refuses the offer, a third period begins. In this period, the sum to share is only 8. Player 1 makes a counteroffer z, with $z \in \{1,2,3,4,5,6,7\}$. Player 2 observes z and can accept or refuse it. If he accepts it, he gets 8–z and player 1 gets z, if he refuses, both get 0. This game is "scheduled" in Figure 4.5b.

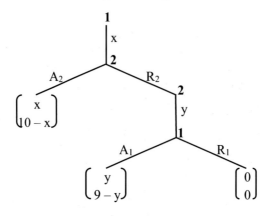

■ Figure 4.5a

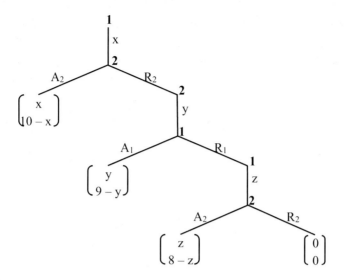

Figure 4.5b

The SPNE is obtained as follows: player 2 accepts all the offers z in the last subgames (because 8−z>0). It follows that player 1 proposes 7 in the previous reduced games and gets 7. So she refuses any offer lower than 7 in her reduced games in period 2. It derives that player 2 proposes 7 in his reduced games in period 2 and gets 2. So player 2 refuses each offer leading to a payoff lower than 2 in his reduced games in period 1. Finally player 1 proposes 8 in her reduced game in period 1. The SPNE is defined by: ((x=8, A_1/y if y≥7, R_1/y if y<7, z=7), (A_2/x if x≤8, R_2/x if x>8, y=7, A_2/z, z from 1 to 7)).

In the sequel, we will also work with 4 and 5 periods. The philosophy is still the same. At each additional round the sum to share decreases by 1 unity, the player who makes the counteroffer can propose an integer between 1 and the maximal value minus 1, and we suppose that, in case of indifference, a player always accepts.

Subgame perfection here clearly illustrates how the power of the player making an offer in the last period diminishes with time. So, first, the last player making an offer is more powerful: if there is one period, player 1 gets 9 and player 2 only 1, if there are two periods, player 2 gets 8 and player 1 only 2, if there are 3 periods player 1 gets 8 and player 2 only 2, if there are 4 periods (we can establish that) player 2 gets 7 and player 1 gets only 3, if there are 5 periods player 1 gets 7 and player 2 only 3, clearly 9>1, 8>2, and 7>3. But, second, the difference in payoffs between the player who makes the last offer and the other reduces as time goes on because the power linked to the last period is increasing in the last sum to share: 9−1>8−2>7−3.

The results are rather convincing. It is natural that the player making the last offer, especially if the number of periods is small (1,2,3), gets a high share of the sum because he can be sure to get it by waiting for the end of the game *(but see also the comments in the next section)*. And it is also natural to observe that the sharing in the first period becomes more egalitarian when the number of periods becomes large, because the advantage of the player making the last offer almost disappears when the sum to share at the last period goes to 0. Especially, we could show that if the game lasts 9 periods (in period 6, S=5, in period 7, S=4, in period 8, S=3, in period 9, S=2), player 1, in the ninth period, has no power any more (she can just propose 1), so the SPNE leads both players to share equally the sum 10 at the beginning of the play (see the exercises for 8 periods).

159

Yet, despite the results being rather intuitive, some experiments have established that real players do not always work in a backward induction way in negotiation games. Why is that? Is the game too long or too complicated? Johnson et al. (2002)[6] even showed that people did not use backward induction to solve a 3 step game. It may be that in real life, if the number of steps is high, people need some time to discover that the player making the last offer (say player i) has a special power. Or they may know this fact but fear that the opponent does not know it. So player i may fear that a low offer (to the opponent) will not be accepted at the beginning (because he fears that the opponent does not understand the rationality of his low offer); so he may prefer offering more than the SPNE offer in period 1 to be sure that his offer is accepted. For example, if the game lasts 5 periods, player 1, in the SPNE, should give 3 to player 2 at the beginning. But she may fear that player 2 does not understand the rationality of this low amount and subsequently refuses it. So player 1 may prefer giving 4 to player 2, which ensures her the payoff 6, *which is higher than the payoff she is sure to obtain by waiting for period 5 (payoff 5).*

Observe that in the above game the players agree immediately in period 1: this is logical because the sum (often called pie) to share is shrinking over time. Yet, in most experiments, people need some time to agree. So many authors turned to new bargaining games. Negotiation with *final offer arbitration* for example allow people to stop the negotiation and ask an arbitrator to share the pie; if the arbitrator chooses the compromise division that lies in between the most generous offers made by the two players (see for example Compte and Jehiel 2004),[7] nobody is best off being too generous at the beginning, given that a player benefits more from the compromise division if his most generous offer is *not* generous; this may lead players to gradualism, i.e. to only make concessions to the opponent gradually. A similar phenomenon may happen without final offer arbitration, when a player can't accept an offer that yields a lower (actualized) payoff than an earlier offer he has rejected in the past. This *commitment* model, for example studied by Li 2007,[8] leads to an impasse when the whole pie, once actualized, yields a lower payoff than a previous rejected offer. That is why in this model people also make concessions gradually. In other models, like the one studied by Fershtman and Seidmann (1993),[9] *last minute agreements* become possible. The presence of a last period negotiation, coupled with commitment (you can't accept at period t a share lower than a share you refused earlier), favours waiting for the last period if the discount factors are large. So arbitrators, deadline effects, gradual concessions and endogenous commitments often emerge in the more recent bargaining literature *(see the exercises for an approach of Li's model).*

3 WHEN THE JOB OF BACKWARD INDUCTION/SUBGAME PERFECTION BECOMES LESS GOOD

Up to now we studied games, especially the sticks game, where backward induction seemed quite natural. But is this approach always natural? We already mentioned that, in comparison to the different possible iterative processes of elimination of weakly dominated strategies, backward induction works in a rigid way. Let us examine some other flaws.

3.1 Backward induction, forward induction or thresholds?

I proposed to my students to play the two ascending all pay auctions in Figure 4.3 (M=7, V=10) and in Figure 4.6 (M=7, V=4). In both games, backward induction leads player 1 to always go on and player 2 to always stop.

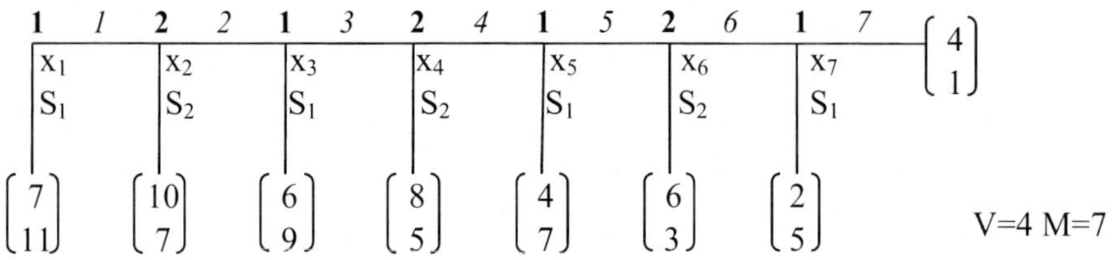

$$\begin{array}{cccccccc}
\underset{\begin{bmatrix}x_1\\S_1\end{bmatrix}}{1\quad\; 1} & \underset{\begin{bmatrix}x_2\\S_2\end{bmatrix}}{2\quad\; 2} & \underset{\begin{bmatrix}x_3\\S_1\end{bmatrix}}{1\quad\; 3} & \underset{\begin{bmatrix}x_4\\S_2\end{bmatrix}}{2\quad\; 4} & \underset{\begin{bmatrix}x_5\\S_1\end{bmatrix}}{1\quad\; 5} & \underset{\begin{bmatrix}x_6\\S_2\end{bmatrix}}{2\quad\; 6} & \underset{\begin{bmatrix}x_7\\S_1\end{bmatrix}}{1\quad\; 7} & \begin{bmatrix}4\\1\end{bmatrix}
\end{array}$$

$$\begin{bmatrix}7\\11\end{bmatrix}\quad \begin{bmatrix}10\\7\end{bmatrix}\quad \begin{bmatrix}6\\9\end{bmatrix}\quad \begin{bmatrix}8\\5\end{bmatrix}\quad \begin{bmatrix}4\\7\end{bmatrix}\quad \begin{bmatrix}6\\3\end{bmatrix}\quad \begin{bmatrix}2\\5\end{bmatrix} \qquad V=4\ M=7$$

Figure 4.6

The students were first asked to find the backward induction equilibria, something they did with no difficulty. But as I asked them to propose and justify their way of playing the game, they answered as follows:

In the game in Figure 4.3, they said that if they were player 1, they would always pursue the game, and if they were player 2, they would immediately stop, *not because of backward induction*, but because always going on weakly dominates any other strategy for player 1, and therefore stopping at the first decision node is player 2's best response. Given that these observations only need two steps of reasoning, each student reasonably expected that the opponent was able to follow the same reasoning and play accordingly. So the students got the same results as backward induction, *but not for backward induction reasons.*

With regard to the game in Figure 4.6, they had more difficulty choosing a way to behave, and once again, backward induction was of little help. The problem with backward induction is that the power it gives to player 1 rests on the fact that player 1 is the last potential player and that she overbids at her last potential decision node x_7. *But in reality, does being the last potential player really give you a strong power?* Not necessarily, because *the power is located at the end of the game, a location nobody wants to reach,* in that everybody gets a very low payoff. So can you reasonably threaten to always overbid, arguing that you will also do so at the end of the game? Well, your opponent may agree to stop at x_6, but he may also be quite sure that you will not lead him to x_6, because you get 6 which is lower than 7 (the payoff you get when you stop immediately) and which is not higher than the payoff you get when stopping at x_3. In other terms, *if your opponent thinks that you do not want to reach the end of the game, he has also no reason to take into account the power the end of the game gives you.*

Once the power of the last node becomes relative, *other arguments, completely ignored by backward induction*, become important. We already mentioned these arguments in chapter 1 (see section 1.2.2). One among them is forward induction. Given that each player is induced to go on each time the opponent stops next with probability q (with q>2/V), the question people have to answer once engaged in the race is who will throw in the towel first. Answering this question has little to do with backward induction. The feelings players have about their opponent necessarily depend on the bids they made in the past. So people play in a *forward induction way*. A consequence is that if the game is going on for a long time, i.e. if bids are large, it is difficult for you to expect that the other will throw in the towel at the next period because he did not do it before; so it is perhaps better to stop as soon as possible.

Thresholds may also be important. For example, as already mentioned in chapter 1, players may be reluctant to overbid, when the bids are higher than V/2 (because the seller of the object gets more than the value of the object), or when they are higher than V (in that case each player loses money), or when they reach M (there is no more money in the wallet).

BACKWARD INDUCTION AND REPEATED GAMES

I proposed this game to my (97) L3 students and to my M1 students (29 students), with M=100 and V=60. I asked them how long they would stay in the game as player 1 and as player 2. In the proposed game, the last player in the game tree is player 2 (he can propose 100). The results are given in the histograms in Figure 4.7a and 4.7b (already given in chapter 1).

First, despite no player can benefit from reaching the end node (where player 1 loses 99 and player 2 loses 40=100–60), player 2 benefits from being the last player to potentially play. As a matter of fact, the mean maximal bid my M1 students, respectively my L3 students, agree to bid is about 19 for player 1 and 36 for player 2, respectively 31 for player 1 and 52 for player 2.

Clearly, player 2 is more encouraged to bid than player 1: my students judge player 2 more powerful than player 1. There is clearly some backward induction behaviour. As a matter of fact, 34% of the L3 students advise player 1 to bid nothing or at most 1 (and only 12.4% advise him to bid till 99), whereas 35.1% of them advise player 2 to bid till 100 (and only 23.7% advise him to bid 2 or less). Moreover 11.3% of my L3 students almost play the backward induction path, given that they advise player 1 to bid 0 or 1 and player 2 to bid till 100.

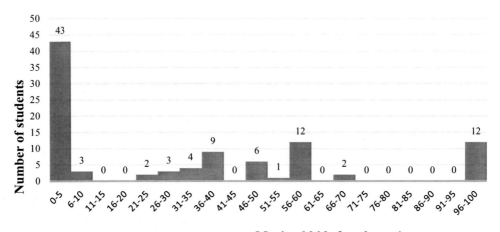

Maximal bids for player 1

■ **Figure 4.7a** Maximal bids for player 1 in the ascending all pay auction with M=100, V=60

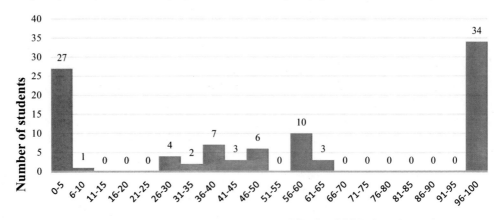

Maximal bids for player 2

■ **Figure 4.7b** Maximal bids for player 2 in the ascending all pay auction with M=100, V=60

Table 4.1 % of students playing threshold values in the ascending all pay auction (M=100, V=60)

Threshold values	Player 1		Player 2	
0 (not enter the game)	Max bid≤5	44.3%	Max bid≤5	27.8%
V/2	20≤max bid≤40	18.6%	20≤max bid≤40	13.4%
V	50≤max bid≤70	18.6%	50≤max bid≤70	19.6%
M	Max bid 99	12.4%	Max bid 100	35.1%
Total of these values		93.9%		95.9%

But it is worth noting that threshold points play also a key role, as highlighted in Table 4.1. In fact, the L3 students only propose few values for the bids, namely the threshold values, 0, V/2, V and M, as can be observed in Table 4.1. 37.2% of the students advise player 1 to invest up to an amount around V/2 or V, and 56.7% advise her either to stop fast or to go to the end. 33% advise player 2 to invest up to an amount around V/2 or V, and 62.9% advise him either to stop fast or to go to the end.

So the students' behaviour does not reject backward induction but it also includes threshold points and some cautiousness: as a matter of fact, this game is a dangerous one, in that you can lose a lot of money. *Many of my students don't like it*, so many of them refuse to enter the game, in that they bid nothing or a very low amount: this explains why, despite 35.1% of my L3 students advising player 2 to bid till 100 in accordance with backward induction, 27.8% of them advise him to stop at 5 or less (this probability is even much higher, 48%, for the M1 students).

It is worth noting that these results are in sharp contrast with the ones obtained for the companion game, the centipede game, where both players should stop immediately but where most of my students advise the players to go on, often till to the end of the game (see the exercises).

3.2 Inconsistency of backward induction, forward induction or momentary insanity?

Let us come back to our smaller ascending all pay auction in Figure 4.2a.

According to backward induction, player 2 should stop at both x_2 and x_4 and player 1 should go on at x_1, x_3 and x_5. But now imagine that she is actually at x_3. What should player 1 deduce from this fact and how should she react? If player 2 conformed to backward induction, he should have stopped at x_2 and player 1 should not be at x_3. But she is at x_3: so she has to observe that player 2 did not conform to backward induction up to now. So what should player 1 deduce for the future? Can she reasonably expect that player 2 will conform to backward induction in the future, i.e. can she expect that he will stop at x_4?

Backward induction is clearly based on rational actions in the future. Player 1's rational action at x_5 is to bid 5. That is why player 2 rationally stops at x_4. Knowing this rational action, player 1 goes on at x_3. And knowing this rational action, player 2 should rationally stop at x_2 (which in turn leads player 1 to rationally go on at x_1). So, given that player 2 overbids at x_2, player 1 is compelled to observe a failure in the chain of rational actions. So can she expect that player 2 will be rational at x_4? Clearly backward induction is confronted by the same problem as iterative elimination of dominated strategies (see chapter 2, section 3.3): given that player 2 doesn't stop at x_2, it becomes difficult for player 1 to anticipate player 2's future behaviour and therefore to know how to play at x_3.

BACKWARD INDUCTION AND REPEATED GAMES

This kind of problem arises each time a player reaches an information set he would not reach if the players conformed to backward induction in the past. In other words, it arises at each information set out of the (subgame perfect) equilibrium path.

How to solve this dilemma? Selten (1975)[10] solves it with the assumption of *temporary insanity*. He tells the following story: Normally player 2 should stop at x_2. But, because of a moment of *insanity*, he switched to bid 2 at x_2. But this insanity is only *temporary*. So player 1 can reasonably expect that player 2 will again stop at x_4. In other terms, if a behaviour does not conform to backward induction in the past, we should still expect that it conforms to backward induction in the future. So player 1 should go on at x_3.

Well, in the game at hand, this expectation is acceptable. It may be that player 2, at x_2, does not completely anticipate the evolution of the game (for example because the game is too complicated for him); when the game goes on, it becomes shorter and he understands it better. So he may understand that it is best for him to stop at x_4, which induces player 1 to go on at x_3.

But things become different in other games. A past action, which does not conform to backward induction, is not necessarily an error: it may signal a way to play in the future, in accordance with forward induction (see chapter 6 for more details).

For example look at the ascending all pay auction, with M=7, V=5, given in Figure 4.8.

In this game, if player 2 goes on at x_2 instead of stopping (as required by backward induction), player 1 may understand his deviation as a signal on future play. She may hold the following reasoning: *"Player 2 goes on because he expects to get more than 7. If I stop at x_3, he is happy because he gets 10, but if I go on, then he may go on at x_4 in order to get 8, which is also higher than 7. But if player 2 goes on at x_4, should I stop at x_3 (my payoff is 6) or should I overbid at x_3 (I can get 7 if I go on at x_5 and if player 2 stops at x_6, but I only get 5 if player 2 goes on at x_6)?"* It is difficult to answer this question, because it is difficult to know player 2's reaction once he discovers that he can no longer get more than 7, but one thing is sure: player 2's bid 2 at x_2 is not necessarily the result of "temporary insanity". *Past actions are not necessarily the best reactions to actions supposed to be played in the future, they may be signals that compel to reexamine the behaviour in the future.*

Nevertheless, temporary insanity is not necessarily an inappropriate assumption. Many games, like the sticks game, may, at the beginning, seem very complicated. So players can choose silly actions at the beginning and become more rational when the game goes on and comes to the end (because it becomes easier). In that case, expecting that players play rationally in the future, even if they did not in the past, is a reasonable assumption.

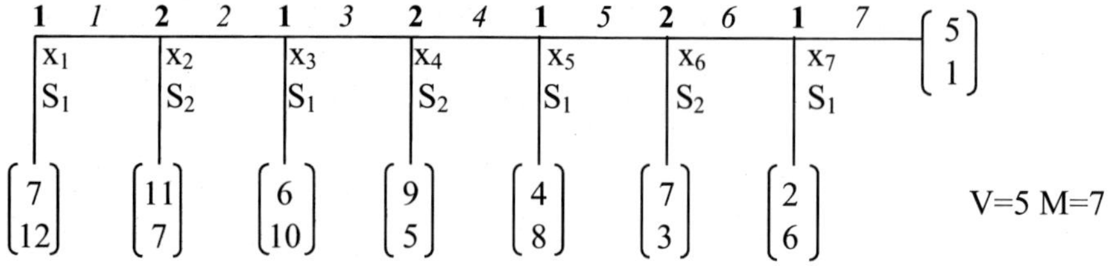

■ Figure 4.8

3.3 When backward induction leads to very strange results.

We return to the general non incremental ascending all pay auction, a game introduced in chapter 1 (section 1.2.2.) and depicted in Figure 4.9a, for M=5 and V=3. This section is rather technical: you can skip the proof and switch to the strange result.

We'll study this game for general values of V and M (with M>V). A similar game has been studied by O'Neill (1986)[11] and Leininger (1989),[12] both looking for special SPNE. To put it more precisely, our aim is to find the bid played by player 1 in the first round, in the SPNE path that seems to us the most natural, i.e. compatible with the following assumption: whenever a player is indifferent between stopping and overbidding M, he overbids.

We justify our assumption as follows. Suppose that at a decision node a player (say player 1) gets more (or the same amount) by playing M than by stopping, so M−b≤V where b is her previous bid. Bidding M is better than (or as good as) stopping. And bidding M is better than bidding any lower amount B, with b<b'<B<M, where b' is the opponent's (say player 2's) previous bid. As a matter of fact, if she bids B, the opponent will be called on to play, and he can and will choose an action *that leads him to win the auction*: given that he also gets more with M than by stopping (V≥M−b>M−b'), he will not stop in the future (because he plays a best response), so player 1 can only lose the auction and get at best M−B, which is lower than V, the payoff obtained with the bid M. So a player plays M if V>M−b (b being his previous bid) but he also plays M when he is indifferent between stopping and playing M (V=M−b). In the latter case, he plays M because this threat induces the opponent to stop at the previous round, given that overbidding only leads to lose more money. And given that an earlier stop of the opponent is the "best thing which may happen to a player", the player who has the possibility of playing M (and to get more or the same amount of money than by stopping) plays M.

Before establishing player 1's first SPNE bid, observe that player 1 has more strategic power than player 2: given that she bids first, she can be the first player being better off playing M than stopping. She necessarily takes advantage of this fact to win the auction. It derives that the equilibrium path will not go beyond the first round: either player 1 wins the auction and it would

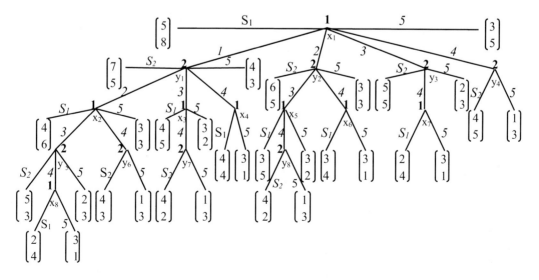

■ Figure 4.9a

be silly for player 2 to overbid in round 2, or player 1 immediately stops the game. It follows that the further bids in further rounds are only here to sustain the equilibrium action at the first round.

Let us make an additional observation. Being the first player able to justify the bid M ensures that the opponent throws in the towel, so each player is incited to play in a way to be the first to have the opportunity to optimally play M. And the first time a player can optimally play M only depends on *his* earlier bids, and not on the bids of the opponent, given that he is incited to play M as soon as M–b≤V, where b is his previous bid, a condition that does not rest on the opponent's bids.

Let us illustrate what will happen for given values of M and V, with V<M.

If player 1 plays V, the game stops, given that player 2 has no reason to overbid. If she bids b_1 lower than V, player 2 bids b_2 at the second round, with $b_1 < b_2 \leq V$.

Playing b_2 equal to V, i.e. the highest possible best response (playing more can only lead to losing money if winning the auction, so is not credible), allows player 2 to more quickly have the opportunity to get more money by bidding M than by stopping (because stopping induces a bigger loss). And what he plays has no impact on the first time player 1 is best off playing M (which only depends on player 1's earlier bids). So let us set that player 2 plays V.

In the third round player 1 gets M–b_1 if she stops the game. If she plays b_1', with V<b_1', she gets M–b_1'+V (if player 2 does not overbid), and so she is better off provided that M–b_1≤M–b_1'+V, i.e. b_1'≤V+b_1.

Now, either V+b_1≥M, or it is lower.

If V+b_1≥M, then player 1 plays M (given our previous comment).

Let us insist on the fact that this opportunity does not depend on player 2's action in the previous round (*regardless of player 2's bid*, player 1 gets M–b_1 by stopping and at least V by playing M). Player 2 knows this fact, so knows that player 1 plays M in the third round, and therefore optimally stops in the previous round: so player 1 wins the auction at price b_1. It follows that player 1's optimal bid b_1 in the first round is the lowest value such that V+b_1≥M. *Hence her optimal bid is M–V, each time M<2V (because b_1≤V)*. This bid ensures her winning the auction at the lowest possible price, given that she can credibly threaten to play M in the third round, compelling player 2 to stop the game at the second round.

And this threat does not work for a first bid b_1 lower than M–V. As a matter of fact, if she bids less than M–V in the first round, she will get more by stopping at the third round than by playing M, because stopping leads to more than M–(M–V)=V, whereas bidding M leads to the payoff V. It follows that either player 1 stops the game at the third round, leading player 2 to overbid in round 2, given that by doing so, he compels player 1 to stop at the first round, which is the best scenario for player 2. Or she bids b_1' higher than V but lower than M and player 2 is called on to play in a fourth round; he can choose a bid b_2', such that V<b_1'<b_2' and M–b_2'+V≥M–V (because stopping leads to M–V). Given that M<2V, player 2 can play M and will do so, and player 1 loses the auction and money.

It follows that, for player 1, playing b<M–V is not the best way to bid in the first round.

To summarize, **if M<2V**, player 1's optimal bid is M–V (followed by player 2's stopping action). You may be surprised by the fact that, sometimes we say that player 2 plays V in the second round, sometimes we say that he stops. Well, this is of no importance. Our aim is not to find player 2's optimal action at each round (we are not establishing the SPNE, but only its path). What matters is to observe that player 2 is able to win the auction when player 1 plays less than M–V, and that he is unable to, hence stops in round 2, when player 1 plays M–V.

BACKWARD INDUCTION AND REPEATED GAMES

If $M>2V$, $V+b_1$ is lower than M and it is best for player 1 to play $b_1'=V+b_1$, because bidding the highest possible amount helps her to more quickly have the opportunity to get more money by bidding M than by stopping.

Now it is player 2's turn to play. He bids b_2' with $V+b_1<b_2'$ and $M-V\leq M-b_2'+V$. Once again, let us suppose that player 2 plays the highest bid, i.e. 2V, in order to more quickly have the opportunity to play M. Given that $2V<M$, it's again player 1's turn to play (fifth round). She can play b_1'', with $2V<b_1''$ and $M-V-b_1\leq M-b_1''+V$, so $b_1''\leq 2V+b_1$.

Now, either $2V<M<3V$ or $M>3V$.

If $M<3V$, player 1's best first bid b_1 is $M-2V$. Indeed, it follows that $2V+b_1$ is equal to M. So player 1 can and will play M at the fifth round, which induces player 2 to stop at the fourth round (it is better to stop than to lose additional money). Hence playing $V+b_1$ in step 3 is a good action for player 1, in that it yields the same amount as stopping ($M-b_1-V+V=M-b_1$) and in that it induces player 2 to stop at round 2 (because overbidding only leads to losing money). And so player 1 wins the auction at price b_1 and gets a payoff $M-(M-2V)+V=3V>M$.

Player 1 cannot get more by switching to a bid $b_1<M-2V$, because, by so doing, she cannot play M at the fifth round ($V<M-(V+b_1)$), hence player 2 is called on to play at a sixth round, where he plays M, because $V\geq M-2V$ (so he compels player 1 to stop at round 5 and lose money).

If $M>3V$, it is best for player 1 to play $2V+b_1$ in round 5. Player 2 is called on to play in round 6, and he bids the highest bid b_2', with $b_2'\geq 2V+b_1$ and $M-2V\leq M-b_2'+V$, i.e. $b_2'=3V$. Given that $3V<M$, player 1 is again called on to play (in round 7). So, if $M<4V$, player 1's first best bid b_1 will be $M-3V$, because he can play M in the seventh round ($V=M-2V-b_1$), and so on.

As a result, we observe that player 1's best way to play in the first round is to bid (M–iV), when $iV<M<(i+1)V$ (i being an integer). This ensures that player 2 does not overbid in round 2, so we have established a SPNE path.

We make two observations:
- All the values, from 0 to V may be played, which is not strange. But what is strange, *very strange*, is that the chosen (winning) bid at the first round only depends on the link between M and a multiple of V!
- We have not tried to find out the optimal actions in all subgames. We have just established a SPNE path.

Examples:

If V=3 and M=5, *player 1 bids M–V=2* and wins the auction by doing so (she can threaten to play 5 in the third round). We illustrate this situation in the exercises.

If V=5 and M=9.9, *player 1 bids M–V=4.9, so her benefit is only 0.1*. She can't get a higher benefit, because if she bids less ($b_1<4.9$), player 2 can bid 5, player 1 can't play M=9.9 in the third round (because $5<9.9-b_1$), but player 2 can and will bid M=9.9 in the fourth round ($5>9.9-5$) (which leads player 1 to stop at the third round, justifies player 2's bid 5 in round 2 and compels player 1 to change her action in period 1).

By contrast, **if V=5 and M=10.1,** *then player 1 bids 0.1, wins the auction and gets the benefit 4.9!* As a matter of fact, regardless of player 2's bid in period 2 (necessarily ≤5), player 1 can bid 5.1 in period 3, because $M-5.1+5\geq M-0.1$, and regardless of player 2's overbid in period 4

167

(necessarily≤10), player 1 can bid M in period 5, because V=5≥M−5.1. So player 1 is sure to win the auction, which deters player 2 from overbidding in period 2.

So, switching from M=9.9 to M=10.1, for the same value V=5, drastically changes player 1's optimal bid, which is quite strange.

Once again, let us examine the assumption that a player overbids with M if he is indifferent between stopping and overbidding M. This assumption *has nothing to do with backward induction*, even if it seems quite natural in this game.

Let us illustrate this point on the game in Figure 4.9b. This game is part of the game in Figure 4.9a. Backward induction allows player 1 to stop (S_1) or bid 5 at z. When she bids 5, player 2 stops (S_2) at y because 5>2, so player 1 gets 6 (because she bids 2 at x). In other terms, by bidding 5 (=M), she compels player 2 to stop the game, which yields her the highest payoff 6 (the associated SPNE is ((2/x,5/z), S_2)). By contrast, if she stops at z, she allows player 2 to overbid (by bidding 3), because player 2 gets the same amount by overbidding and by stopping. It follows that player 1 only gets 3, which leads her to stop at x (5>3). So we get an additional SPNE ((S_1/x,S_1/z),3). Clearly, the first SPNE seems more natural than the second one. So, in some way, *backward induction does not systematically translate the "strategic thinking" a player may hold*. In other words, player 1 can convincingly threaten to bid 5, because she can bid 5 without losing money, and this threat, if understood, clearly gives her a higher payoff (6 instead of 5). Let us observe that, to our mind, player 1 has not even to *explicitly* threaten to bid 5 at z. Even without explicit threat, player 2 may hesitate overbidding because he may fear a kind of reprisal: he knows that, by overbidding, he hurts player 1 (who can only get the payoff 3 instead of the payoff 6 she gets when he stops), and he also knows that player 1 is indifferent between the strategy that hurts player 2 (bidding 5), and the other one (stopping). If I were player 2, *I would stop, even if backward induction allows player 1 to play S_1 at z*. I think that this game should be tested experimentally to see if players play in accordance with my expected behaviour, which isn't the only behaviour compatible with backward induction.

See the exercises in order to see the other SPNE paths and the detailed SPNE, according to the different possible behaviours in case of indifference.[13]

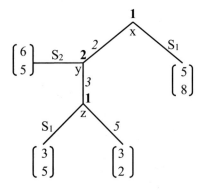

■ **Figure 4.9b**

4 FINITELY REPEATED GAMES

We now apply subgame perfection/backward induction to repeated games. A repeated game is a game with a special structure: at each period of play, the players play a same stage game which may be in normal or extensive form.

Repetition is a usual phenomenon in economics and social sciences in general. Transactions are repetitive, working activities are repetitive, and more generally most human relations and actions are repetitive. Repetition has many consequences, among them *the building of reputation, the possibility of learning, and also the broadening of the strategies' sets and the broadening of the strategic outcomes*, a point we will focus on.

4.1 Subgame Perfection in finitely repeated normal form games

4.1.1 New behaviour in finitely repeated normal form games

Consider the normal form game in Matrix 4.1:

		Player 2 A_2	B_2	C_2
	A_1	(7,1)	(1,2)	(3,0)
Player 1	B_1	(5,7)	(0.5,9)	(3,4)
	C_1	(0,0)	(0,1)	(4,5)

Matrix 4.1

This game, called *stage game*, has three Nash equilibria, (A_1, B_2), (C_1, C_2) and a mixed Nash equilibrium, where player 1 plays A_1 and C_1 with the probabilities $2/3$ and $1/3$ and player 2 plays B_2 and C_2 with probability $½$ each (in this equilibrium, player 1 and player 2's payoffs are 2 and 5/3).

> **Definition 3: finitely repeated games**
> We focus on repeated games with an observation of past actions before a new round. In this context, at each period the players play the stage game and get, at the end of the period, the payoffs assigned to the strategies played in the stage game. They observe all the played actions; then a new period begins: the players play again the stage game and so on, until the last period, after which the game stops.

What matters is that players, thanks to repetition, may choose different actions than the ones chosen in the stage game when it is only played once, *thanks to the possibility of introducing rewards and reprisals in future periods*.

To see why, suppose that the game in Matrix 4.1 is played two times, both players observing the actions played in period 1 before playing again.

First observe the *structure of this new game*. Given that the players observe the actions played in period 1 before playing period 2, there are nine subgames in period 2, and each of these subgames is the stage game. And the structure of the reduced game in period 1 is again the stage game, with regard to the actions and information structure.

BACKWARD INDUCTION AND REPEATED GAMES

And this is true regardless of the number of repetitions. At each period, the reduced games are the stage game, with regard to the available actions and information. This makes things rather easy in that each reduced game keeps an easy structure, that of the stage game.

Yet this does not mean that it is enough to study the stage game to establish the actions played in each period. As a matter of fact, the actions chosen in a period, say period t, may impact on the actions – and consequently the payoffs – in future periods, which in turn impact on the actions in period t. Yet there is one exception: the last period subgames. In the last period, there no longer exists a further period: consequently the actions chosen in the last period cannot induce rewards or reprisals in the future; they only impact on the payoffs obtained in this period, which are the stage game payoffs. It follows:

> **Property 3:** the Nash equilibria in the subgames of the last period are the Nash equilibria of the stage game.

Repetition becomes interesting when the stage game has several Nash equilibria that may be used to achieve a specific objective. For example, when the game in Matrix 4.1 is played twice, we may want both players in period 1 playing B_1 and A_2 because they get a high payoff by doing so. And the way to get this behaviour is quite simple. *The idea is to reward the players if they play as expected, and to punish them if they don't.*

> **➤ STUDENTS' INSERT**
>
> In period 1, player 1 may prefer playing A_1 if player 2 plays A_2, in order to get 7 instead of 5. But we can promise her in period 2 the Nash equilibrium (C_1,C_2) if she plays B_1 in period 1, and threaten her with the equilibrium (A_1,B_2) if she deviates to A_1. So by playing B_1, she gets 5+4=9, whereas if she plays A_1, she gets 7+1=8<9. So she is not induced to deviate.
>
> The same is true for player 2. He may prefer playing B_2 in period 1 if player 1 plays B_1, in order to get 9 instead of 7. But we can promise him, in period 2, the Nash equilibrium (C_1,C_2) if he plays A_2 in period 1, and threaten him with the equilibrium (A_1,B_2) if he deviates to B_2. So by playing A_2, he gets 7+5=12, whereas if he plays B_2 he gets 9+2=11<12. So he is not induced to deviate.
>
> In other words, we can build the following SPNE:
>
> Period 1: the two players play (B_1,A_2).
>
> Period 2: the two players play (C_1,C_2) if (B_1,A_2) has been played in period 1 or if both have deviated from (B_1,A_2).
>
> The two players play (A_1,B_2) if $(*,A_2)$ has been played in period 1, * being an action different from B_1, or if $(B_1,*)$ has been played in period 1, * being an action different from A_2.
>
> This is a SPNE because:
>
> A Nash equilibrium is played in each subgame of period 2 $((A_1,B_2)$ or $(C_1,C_2))$.
>
> And (B_1,A_2) is a Nash equilibrium in the reduced game of period 1. As a matter of fact, B_1 is player 1's best response, because it leads to the payoff 5+4=9, whereas A_1 leads to the payoff 7+1=8<9 and C_1 leads to the payoff 0+1=1<9. And A_2 is player 2's best response to B_1 because it leads to the payoff 7+5=12, whereas B_2 leads to the payoff 9+2=11<12 and C_2 leads to the payoff 4+2=6<12.

We reproduce the reduced period 1 game in Figure 4.10.

BACKWARD INDUCTION AND REPEATED GAMES

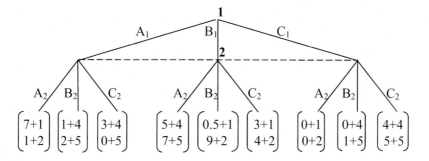

Figure 4.10

It immediately follows that, in a T period repeated game:

- All the players may get more than T times the best payoffs they get at a Nash equilibrium of the stage game. In other words the set of payoffs in the T played stage game may be larger than the set obtained by combinations of Nash equilibria of the stage game.
- The enlargement depends on the number and kind of Nash equilibria available in the stage game. Sometimes there is no enlargement. This may namely happen if there is just one Nash equilibrium in the stage game.

Let us justify these remarks.

In the studied game, player 1 gets 9 and player 2 gets 12, which are better than 8 and 10 (i.e. 4×2 and 5×2). So the repetition of the game allows the players to improve on two times the best payoffs they can get at equilibrium in the stage game.

This enlargement is possible because the structure of the stage game is such that it is both possible to punish both players, but also to reward them *simultaneously*. To understand this remark, look at the game in Matrix 4.2:

		Player 2		
		A_2	B_2	C_2
	A_1	(7,1)	(1,5)	(3,0)
Player 1	B_1	(5,7)	(0.5,9)	(3,4)
	C_1	(0,0)	(0,1)	(4,2)

Matrix 4.2

(A_1,B_2) and (C_1,C_2) are the two pure strategy Nash equilibria of this game. We again try to lead both players to play (B_1,A_2) in period 1.

It is again possible to induce player 1 to play B_1 in period 1, because we can reward her in period 2 with (C_1,C_2) if she plays B_1 and we can threaten her with (A_1,B_2) if she deviates, namely to A_1 (because 5+4>7+1).

And it is still possible to induce player 2 to play A_2 in period 1, because we can reward him in period 2 with (A_1,B_2) if he plays A_2 and we can threaten him with (C_1,C_2) if he deviates, namely to B_2 (because 9+2<7+5).

171

BACKWARD INDUCTION AND REPEATED GAMES

But there is a problem: what do we do when nobody deviates? The problem is that the equilibrium that rewards player 1 is the equilibrium that punishes player 2 and vice versa. In the previous game, (C_1,C_2) rewarded *both* players but this is no longer the case: the new mixed Nash equilibrium rewards no one, (A_1,B_2) rewards player 2 but punishes player 1 and (C_1,C_2) rewards player 1 but punishes player 2.

Well, there is also no equilibrium that punishes both players, *but this is not important*, due to the Nash unilateral deviation philosophy: we never study multilateral deviations, and therefore, regardless of the equilibrium we propose in case of multilateral deviations, it has no impact, so is of no importance.

Property 4

In an N player game, in order to sustain a profile that is not a Nash equilibrium of the stage game, there have to be at least *two* Nash equilibria in the stage game. Each player has to be sufficiently rewarded by one of the equilibria.

If the stage game has only one Nash equilibrium, then the only SPNE of the T times repeated game consists of playing, at each reduced game, the unique Nash equilibrium of the stage game, regardless of T.

We come back to the second observation in subsection 4.1.3.

What does "sufficiently rewarded" mean? In fact "sufficiently" is an adverb that is necessary if the game is only played twice. But its necessity diminishes with the number of repetitions.

Look at the following variant in Matrix 4.3:

		Player 2		
		A_2	B_2	C_2
Player 1	A_1	(7,1)	(1,2)	(3,0)
	B_1	(5,7)	(0.5,11)	(3,4)
	C_1	(0,0)	(0,1)	(4,5)

Matrix 4.3

When this game, which has the same equilibria as the game in Matrix 4.1, is only played twice, player 2 does no longer play A_2 in the first round, because he now gets at least 11+5/3 by deviating to B_2 and at most 7+5<11+5/3 by staying on A_2. So (C_1,C_2) *does not sufficiently reward* player 2.

But suppose now that the game is played three times. Consider the following profile:

Period 1: the players play (B_1,A_2)

Period 2 and period 3: if both played (B_1,A_2) in period 1, they play (C_1,C_2) in periods 2 and 3. If not, they play (A_1,B_2) in periods 2 and 3.

It is a SPNE because:

BACKWARD INDUCTION AND REPEATED GAMES

- It is easy to observe that a Nash equilibrium is played in each subgame of period 3. And a Nash equilibrium is played in each reduced game of period 2. This derives from the fact that, in each subgame of period 3, the played equilibrium does not depend on the actions chosen in period 2, so a Nash equilibrium of the stage game becomes a Nash equilibrium in each period 2 reduced game.
- A Nash equilibrium is played in the reduced game of period 1. As a matter of fact, player 1, by playing B_1, gets 5+4+4=13, while she gets at most 7+1+1=9 by deviating. And player 2 gets 7+5+5=17, while he gets at most 11+2+2=15 by deviating.

In other words, if the best deviation in period 1 leads to an additional payoff x, and if the difference of payoff between the rewarding equilibrium and the punishment equilibrium is y, you can induce a player to not deviate in the first round, as soon as the game is played T times with x<(T–1)y. In the game above, 3 periods are enough because 11–7<(3–1)(5–2). More generally, even if we suppose that the payoffs are discounted (given that the players are paid in each period) a sufficiently high number of periods T always exists, providing the discount factors are sufficiently large.

4.1.2 New behaviour, which links with the facts?

We showed in the previous section that to build a new behaviour in a repeated game context, it is sufficient to have an equilibrium that punishes everybody and one that (sufficiently) rewards everybody. In that case, like in our Matrix 4.1 game, in order to punish a player, we punish everybody. This is not problematic, even if the other players do not deviate, because the reprisal equilibrium *will never be played in reality*. Its only role is to prevent each player from deviating: thanks to its existence, the equilibrium path is such that only the rewarding stage game equilibrium is played.

But what about reality? In reality, *people are reluctant to threaten a player with a profile of actions that also harms people who did not deviate*. So compare the games in Matrices 4.1 and 4.4:

		Player 2			
		A_2	B_2	C_2	D_2
	A_1	(7,1)	(1,5)	(3,0)	(4,5)
Player 1	B_1	(5,7)	(0.5,9)	(3,4)	(0,0)
	C_1	(0,0)	(0,1)	(4,2)	(0,0)

Matrix 4.4

Suppose that both games are played twice. In both games (B_1,A_2) can be played in the first round.

Recall that in the game in Matrix 4.1, we threatened player 1 with (A_1,B_2), i.e. a threat that is also harmful for player 2 who is not supposed to deviate (when we study unilateral deviations of player 1). And the same is true for player 2.

By contrast, in the game in Matrix 4.4, if only player 1 deviates, we can punish her with the equilibrium (A_1,B_2) which penalizes player 1 but not player 2 (who did not deviate, and who gets the best Nash stage game equilibrium payoff 5). And if only player 2 deviates, we can punish him with the equilibrium (C_1,C_2) which penalizes player 2 but not player 1 (who did not deviate, and who gets the best Nash stage game equilibrium payoff 4). And it is possible, if nobody deviates, to reward both with the equilibrium (A_1,D_2). Finally, if both deviate, it seems natural to throw in the towel and play (A_1,D_2) too.

BACKWARD INDUCTION AND REPEATED GAMES

Most people will surely agree more with the threats and rewards in the game in Matrix 4.4 than with the threats and rewards in the game in Matrix 4.1. This should be tested experimentally.

In some ways, these remarks are linked to *renegotiation* (Benoît and Krishna)[14]: these authors claim that one should impose restrictions on the Nash equilibria chosen in the subgames. Given that people prefer getting more than less, they argue that only Pareto optimal Nash equilibria in subgames can be used to sustain an action profile in previous rounds. Their idea is that people may meet after a deviation and talk about the reprisals planned before the deviation, so they may renegotiate, namely when the reprisals are harmful for everybody. *But renegotiation raises a problem*. In the game of Matrix 4.1, after a unilateral deviation from (B_1,A_2), if the players talk together before playing again, they surely agree that the payoffs (4,5) are much better – than the payoffs (1,2) – *for both*, and so they may forgive the deviator, in order to get the payoffs (4,5) instead of the payoffs (1,2) in the second round. But of course, by doing so, they encourage the deviation in period 1, and the couple (B_1,A_2) can no longer be played in the first round. As a consequence, renegotiation is harmful for both players, who can no longer get the payoffs (5,7) in period 1.

4.1.3 Repetition and backward induction's inconsistency

We return to our first price sealed bid all pay auction game in Matrix 4.5.

		Player 2			
		0	1	2	3
Player 1	0	(6.5,6.5)	(5,7)	(5,6)	(5,5)
	1	(7,5)	(5.5,5.5)	(4,6)	(4,5)
	2	(6,5)	(6,4)	(4.5,4.5)	(3,5)
	3	(5,5)	(5,4)	(5,3)	(3.5,3.5)

Matrix 4.5

In this game there is a unique Nash equilibrium. So there is no possibility of punishing or rewarding a player (you need at least two Nash equilibria). As a consequence, regardless of the number of periods of play, the only SPNE leads everybody to play the unique Nash equilibrium in each round.

This strange result has often been commented on in game theory and in experimental studies, especially for the prisoner's dilemma game. In the prisoner's dilemma, with regard to subgame perfection, both players systematically denounce each other in all periods because there is no possibility of rewarding them or of punishing them. To put it more precisely:

In the last period, called period T, the two players denounce each other because double denouncement is the only Nash equilibrium of the stage game. It follows that in period T−1, each reduced game is strategically equivalent to the stage game, because, regardless of what players play in period T−1, they denounce in period T (so the payoffs to study in period T−1 are just the stage game ones, given that one adds the same constant – the payoff of double denouncement – to each of the stage game payoffs). It follows that the only Nash equilibrium in each T−1 reduced game is the only Nash equilibrium of the stage game, i.e. double denouncement.

In other words, given that everybody denounces in period T regardless of what is played in period T−1, a player is never incited to cooperate in period T−1, because he is neither rewarded for his cooperation nor punished for his denouncement. And so on. In period T–2, T–3..., until period 1, both players denounce because whatever they do, they will denounce in future periods.

In the same way, for the same reason, in the all pay auction game, players, in each period, play the mixed Nash equilibrium because whatever they do, they will play it in the next rounds.

Of course, this result is not intuitive, and the lack of intuition is again *linked to the backward induction's logical inconsistency.*

So imagine that the prisoner's dilemma is repeated 50 times, you are in period 11 and you observe that your opponent, instead of denouncing in the last 10 periods, systematically cooperated. So clearly he did not behave in accordance with backward induction: so why should he do so in the future? Can we just say that he was temporarily insane, and that he will behave in accordance with backward induction in the future? Obviously not: your opponent may try to inform you on his future behaviour. Perhaps he says: *"We can get a high utility if we both cooperate; regardless of the stability of this behaviour, let's play it because it is good for us. You see, I begin this peaceful and productive behaviour, and I even accept losing utility, in that I do not negatively react to your aggressive behaviour; but please, take my invitation to cooperate into account so that we both get a high utility. I cannot cooperate forever if you do not do so also."* Such implicit messages can be understood by players; this explains that cooperation is usually observed for a while in experimental repetitions of the prisoner's dilemma.

I would say that the backward induction's logical inconsistency is even more problematic in a repeated game than in a usual game: *a repetitive context better allows the transmission of a message*, because *you can repetitively play it* so that *it can no longer be interpreted as an error.*

We now stress a technical difficulty. When the unique Nash equilibrium of the game, like in the all pay auction game, is a mixed Nash equilibrium, then, each time you observe an action in support of the equilibrium, you cannot really know if the player deliberately chose this action (with probability 1), or if he chose it with the equilibrium probability. So, in the all pay auction, where both players play 0, 1 and 2 with probability $1/3$ at equilibrium, if your opponent plays 10 times 0, it is difficult to know if this 0 is the result of the equilibrium random draw or if it is a deliberate action, by which the opponent tries to signal that he incites you to play 0 too for a while, because the couple (0,0) is best for both.

With regard to theory, there is no difficulty: whatever your opponent played in the past, you play the unique Nash equilibrium in the future. And no player can be unilaterally better off by playing an equilibrium action with a probability different from the equilibrium one (by construction of the mixed Nash equilibrium). *But in reality, if the observed frequency is different from the equilibrium one, we may believe that the player sends a message: so, in real life, out-of-equilibrium but also equilibrium actions may become messages about future play.*

4.2 Subgame perfection in finitely repeated extensive form games

What about repetition when the stage game is in extensive form? We don't go into this topic much but would like to note two facts:

BACKWARD INDUCTION AND REPEATED GAMES

- First, you cannot change an extensive form stage game into a normal form game and work with this normal form game in the repeated context.
- Second, it is possible to play additional strategies in this context (in comparison with the stage game), providing there exist several **SPNE** in the stage game.

4.2.1 A forbidden transformation

Consider the *sequential* battle of the sexes game in Figure 4.11.

This game has a unique SPNE $(A_1, (A_2/A_1, B_2/B_1))$. The normal form of this game is given in Matrix 4.6.

		Player 2 $A_2/A_1\ A_2/B_1$	$A_2/A_1\ B_2/B_1$	$B_2/A_1\ A_2/B_1$	$B_2/A_1\ B_2/B_1$
Player 1	A_1	(3,1)	(3,1)	(0,0)	(0,0)
	B_1	(0,0)	(1,3)	(0,0)	(1,3)

Matrix 4.6

We suppose that the normal form game is played twice and look at the following profile:

Period 1: the players play $(B_1, (B_2/A_1, B_2/B_1))$

Period 2: the players play $(A_1, (A_2/A_1, A_2/B_1))$ regardless of the play in period 1

This profile is a SPNE of the *normal form game* played twice: $(A_1, (A_2/A_1, A_2/B_1))$ is a Nash equilibrium of each subgame in the second period and $(B_1, (B_2/A_1, B_2/B_1))$ is a Nash equilibrium of the reduced game in period 1, given that it is a Nash equilibrium of the stage game, and given that the actions played in period 2 do not depend on the actions played in period 1.

But this profile is not a SPNE of the sequential game played twice. Look at the second period: Player 2 is the player of the last subgames and he plays A_2 at each node following A_1 and B_2 at each node following B_1. It follows that player 1, in her reduced games in period 2, plays A_1. Now look at the first period. Player 2 is the player of the new reduced games and he plays A_2 at each node

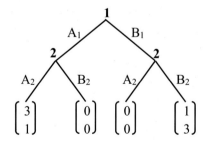

■ Figure 4.11

BACKWARD INDUCTION AND REPEATED GAMES

following A_1 and B_2 at each node following B_1 (these actions are not affected by the actions played in period 2 because, regardless of the actions chosen in period 1, player 1 and player 2 play A_1 and A_2 in the second round). It follows that player 1, in her reduced game in period 1, necessarily plays A_1. So the only SPNE of the sequential game played twice is such that player 1 plays A_1 followed by A_2 two times.

So we do not get the same behaviour when the stage game is in normal form and when it is in extensive form, and this is a general result. You can observe that the reason for a unique SPNE in the twice repeated extensive form stage game is that we have a unique SPNE in the extensive form stage game.

4.2.2 New behaviour in repeated extensive form games

We return to the Stackelberg first price all pay auction given in Figure 4.12 and suppose that it is repeated twice.

This game has three pure strategy SPNE (0, (1/0,2/1,3/2,0/3,0/4,0/5)), (2, (1/0,2/1,0/2,0/3, 0/4,0/5)), (3, (1/0,2/1,3/2,0/3,0/4,0/5)), that respectively lead to the payoffs (5,7) (6,5) and (5,5) (see the exercises). It follows that the following profile will be a SPNE of this game repeated twice:

Profile:

Period 1: Player 1 bids 0. Player 2 bids 0 if he observes 0, 2 if he observes 1, 3 if he observes 2, 0 if he observes 3, 4 and 5.

Period 2: player 1 plays 0 if both played (0,0) in period 1. In all other cases he bids 3. Player 2 bids 1 after 0, 2 after 1, 3 after 2, 0 after 3, 4 and 5.

This profile leads both players to bid 0 in the first round and to bid (0,1) in the second. So they get the interesting payoff profile ((6.5,6.5) (5,7)).

➤ STUDENTS' INSERT

We show that the above profile is a SPNE:

First examine the second period subgames. Clearly (1/0, 2/1, 3/2, 0/3, 0/4, 0/5) are best responses of player 2 in the (smallest) subgames following the bids 0, 1, 2, 3, 4 and 5. It immediately follows that bid 0 and bid 3 are the two best actions at the reduced games of player 1 in period 2.

Now examine the first period. Player 2 is called on to play. In each reduced game following a bid different from 0, regardless of player 2's action at this node, the profile of bids played in the second round is (3,0); consequently player 2 does not have to take the actions in period 2 into account and (2/1, 3/2, 0/3, 0/4, 0/5) are best responses. Now consider the reduced game following the bid 0. If he plays 0, he gets 6.5+7, because player 1 then plays 0 in the second round; if he makes another bid, he gets at most 7+5 (by playing 1 in the first round), because players 1 and 2 bid 3 and 0 in the second round. Consequently he bids 0. Now switch to player 1's reduced game. By bidding 0, she gets 6.5+5, given player 2's answer in period 1 and the actions played in period 2. If she makes another bid, then player 2 plays 2 after 1, 3 after 2, 0 after 3, 4 and 5 and in the second round, the played couple of bids is (3,0); so player 1 gets at most 5 in the first round and 5 in the second round. It follows that player 1 is best off bidding 0, because 6.5+5>5+5.

177

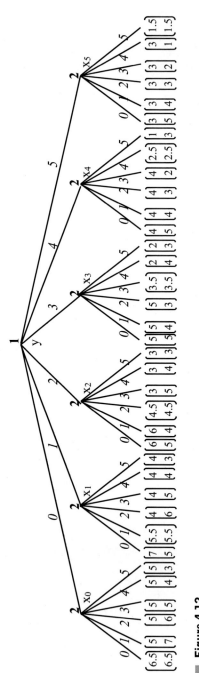

■ **Figure 4.12**

BACKWARD INDUCTION AND REPEATED GAMES

> In other words, the only player who may be incited to deviate is player 2 in period 1, when player 1 bids 0 (player 2 can get 7 by bidding 1 instead of 6.5 by bidding 0). So we have to punish him in the second round if he deviates in the first (with the SPNE (3, (1/0,2/1,3/2,0/3,0/4,0/5)), and reward him if he doesn't deviate (with the SPNE (0, (1/0,2/1,3/2,0/3,0/4,0/5)).

So it is possible to get new payoffs and actions when the stage game is in extensive form, providing several SPNE exist in the stage game, some of them punishing the deviators, and some of them rewarding the non deviators.

5 INFINITELY REPEATED GAMES

5.1 Adapted backward induction

The backward induction principle we applied up to now to get the SPNE, does not work when a game is infinitely repeated, given that the game has no end. Yet, despite the game being endless, the backward induction principle can be adapted, provided that the *payoffs are continuous at infinity*.

> **Definition 4:** the payoffs are **continuous at infinity** if the difference between the payoffs of two profiles of actions, identical up to a given level (period) t, goes to 0 when t goes to infinity.

The idea is that the actions played in the far future are of little importance, because they have a low impact on the payoffs. In economics, we are mostly in this context when we study repeated games because we usually introduce discount factors. Given that the players are paid after each period, it follows that the payoffs obtained in far future periods are multiplied by a high powered discount factor, so they go to 0. It follows that two stories of actions lead to almost the same payoffs when they are identical up to a period t, with t large.

For games whose payoffs are continuous at infinity, the backward induction principle generalizes as follows:

> **Property 5**: A profile of strategies is a SPNE if and only if it induces a Nash equilibrium in each reduced game, given the behaviour after this reduced game.

Thanks to this property, we will establish that turning to infinitely repeated games increases the number of subgame perfect Nash equilibrium outcomes of the games.

BACKWARD INDUCTION AND REPEATED GAMES

5.2 Infinitely repeated normal form games

5.2.1 New behaviour in infinitely repeated normal form games, a first approach

Look at the first price sealed bid all pay auction in normal form, with a smaller increment, given in Matrix 4.7.

		Player 2						
		0	0.5	1	1.5	2	2.5	3
	0	(6.5,6.5)	(5,7.5)	(5,7)	(5,6.5)	(5,6)	(5,5.5)	(5,5)
	0.5	(7.5,5)	(6,6)	(4.5,7)	(4.5,6.5)	(4.5,6)	(4.5,5.5)	(4.5,5)
	1	(7,5)	(7,4.5)	(5.5,5.5)	(4,6.5)	(4,6)	(4,5.5)	(4,5)
Player 1	1.5	(6.5,5)	(6.5,4.5)	(6.5,4)	(5,5)	(3.5,6)	(3.5,5.5)	(3.5,5)
	2	(6,5)	(6,4.5)	(6,4)	(6,3.5)	(4.5,4.5)	(3,5.5)	(3,5)
	2.5	(5.5,5)	(5.5,4.5)	(5.5,4)	(5.5,3.5)	(5.5,3)	(4,4)	(2.5,5)
	3	(5,5)	(5,4.5)	(5,4)	(5,3.5)	(5,3)	(5,2.5)	(3.5,3.5)

Matrix 4.7

Given that this stage game has a unique Nash equilibrium, where both players get 5.25 and play each bid from 0 to 2.5 with probability $1/6$, we know that it is not possible to get a SPNE in the finitely repeated game such that both players always bid 0.

Yet this becomes possible in the infinite repeated context. We introduce discount factors and propose the following profile of actions:

Period 1: the players bid (0,0)

Period t, with t from 2 to ∞: the players play (0,0) if (0,0) has been played in all previous periods, if not, they play the mixed Nash equilibrium of the stage game.

This profile yields the payoffs (6.5, 6.5) forever, by simply threatening to switch to the Nash equilibrium forever, as soon as at least one player deviated from the bid 0.

➤ STUDENTS' INSERT

To prove that the above profile is a SPNE of the infinitely repeated game, we add a "*state transition diagram*" (Figure 4.13) and we apply backward induction (as defined in property 5, for games with continuous payoffs at infinity).

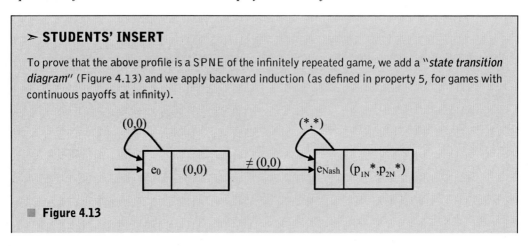

■ Figure 4.13

A "state transition diagram" explains how each player should play in each reduced game. It is compounded of states, actions and arrows. The state (here e_0 or e_{Nash}) is the type of reduced game the players are playing. The couple next to the state are the actions which the players should play in the state. So the actions to be played in e_0 are the bids 0, and the actions to be played in e_{Nash} are the mixed Nash equilibrium strategies p_N*. The arrows specify how one switches from one state to the same or to another, according to the actions chosen by the players. * means any action. The arrow without a starting state (that points toward e_0) indicates that the first played reduced game is an e_0 one.

A state transition diagram is just another way to expose a profile of strategies in an infinite repeated game.

A state often corresponds to a whole family of reduced games. Consequently the states partition the set of reduced games into as many sets as states. For example, in our studied game, a reduced game is associated with an e_0 state if the players only bid 0 before this reduced game or if it is the first reduced game (starting from the beginning). Every other reduced game is associated with an e_{Nash} state.

The arrows show that the game starts in an e_0 state, that the players stay in e_0 if both players bid 0, but switch to e_{Nash} as soon as another couple of actions is played. The arrows also show that, once in e_{Nash}, the game stays in e_{Nash} forever, regardless of the actions played in this state. So the arrows express the threats and rewards we want to express: the players start by playing cooperatively (both bid 0) and go on playing in this way as long as nobody chooses another bid. But if somebody makes another bid, they play the Nash equilibrium of the stage game forever.

According to the backward induction principle, we have to check if the strategies in each reduced game are a Nash equilibrium in the reduced game, given the behaviour of the players in the future. Given that all the reduced games corresponding to a same state are identical as regards the actions to be played and the consequences of all possible actions, it is enough to study one reduced game corresponding to each state.

Given the symmetry of the game, we only focus on the incentives of player 1.

In an e_0 state of period t, we observe the evolutions given in Table 4.2a (*called evolution table*).

Evolution Table 4.2a

period	t	t+1
Bid 0	e_0	e_0
	(0,0)	(0,0)
obtained payoff	6.5	$6.5\delta_1$
Bid a≠0	e_0	e_{Nash}
	(a,0)	$(p_{1N}*,p_{2N}*)$
obtained payoff	at most 7.5	$5.25\delta_1$

Table 4.2a examines the evolution of the states, the actions of both players and the payoff of player 1 if she bids 0 in period t (second line) and if she does not bid 0 in period t (third line).

δ_1 is player 1's discount factor. The arrow indicates that the game cycles on the state. More generally, we stop a table when the states in the second and in the third line are the same, or when the game cycles in both lines.

For example, if player 1 bids a≠0, the actions played in period t are (a,0) (because player 2 is supposed to play the expected bid 0), and the game switches to reduced games of type e_{Nash} in all the

BACKWARD INDUCTION AND REPEATED GAMES

> following periods (t+1,t+2,t+3,..) where both player are supposed to play p_N*. So player 1 gets at most 7.5 in period t (when she bids 0.5), then she gets the Nash payoff 5.25 forever. So player 1 is better off not deviating from the bid 0 if
>
> $6.5+6.5\delta_1/(1-\delta_1) \geq 7.5+5.25\delta_1/(1-\delta_1)$ i.e. $\delta_1 \geq 1/2.25 = 0.45$.
>
> Given the symmetry of the game, player 2 is better off not deviating in e_0 each time $\delta_2 \geq 0.45$, where δ_2 is his discount factor.
>
> In an e_{Nash} state, we observe the evolutions in Table 4.2b.
>
> ### ■ Evolution Table 4.2b
>
> | | t | t+1 |
> |---|---|---|
> | Bid p_{1Nash}* payoff | e_{Nash} $(p_{1N}*,p_{2N}*)$ 5.25 | e_{Nash} $(p_{1N}*,p_{2N}*)$ $5.25\delta_1$ |
> | Bid $a \neq p_{1Nash}$* payoff | e_{Nash} $(a,p_{2N}*)$ At most 5.25 | e_{Nash} $(p_{1N}*,p_{2N}*)$ $5.25\delta_1$ |
>
> It immediately follows that player 1 and (symmetrically) player 2 cannot be better off deviating from their Nash equilibrium strategy, regardless of the discount factor.
>
> So starting by both bidding 0, and keeping on bidding 0 till at least one player bids differently, in which case the Nash equilibrium of the stage game is played forever, is a SPNE of the infinite repeated game, provided that both players value enough the future ($\delta_1 \geq 0.45$ and $\delta_2 \geq 0.45$). In this SPNE, both players bid 0 forever and get 6.5 at each round.

There is a technical difficulty, already mentioned in section 4.1.3: given that only past *actions* are observed, nobody knows in period t+1 if the opponent played in accordance with the Nash equilibrium *probability distribution* in an e_{Nash} state in period t. This is not a problem here, in that the state after an e_{Nash} state does not depend on the played actions in this state. As a consequence, in an e_{Nash} reduced game, a player, to optimize his payoff, just has to maximize his payoff in the stage game, and so is best off not deviating from the Nash equilibrium. This explains the possibility of proposing the mixed Nash equilibrium profile in our state transition diagram. But in general, *as soon as the evolution of the states depends on the fact that the players do or don't play the planned probability distribution; a probability distribution can no longer be proposed as the strategy to be played in a state, given that only actions and not probabilities can be observed* (see further how to bypass this difficulty).

5.2.2. Minmax values, individually rational payoffs, folk theorem

A natural question is: which payoffs can the players expect to get in an infinite repetition of the game? The folk theorem answers this question. To present this theorem, we first introduce the notions of minmax value and individually rational payoff.

BACKWARD INDUCTION AND REPEATED GAMES

> **Definition 5 minmax values and individually rational payoffs.**
> Player i's **minmax value** of a stage game, V_i, is the payoff player i can achieve when all players fight against him, and he can react to their attack:
>
> So it is defined by:
>
> $V_i = \min_{p_{-i} \in P_{-i}} \max_{p_i \in P_i} u_i(p_i, p_{-i})$
>
> The **set of – strictly – individually rational payoffs** of the game is the set of payoffs such that each player gets at least – more than – his minmax value.

We illustrate these notions on the all pay auction game in Matrix 4.7. The minmax value for each player is 5 given that the opponent can impose a payoff lower or equal to 5 by bidding 3, but cannot impose a payoff lower than 5 because each player can get 5 by bidding 0.

So the set of feasible and individually rational strategies contains all possible couples of strategies such that each player gets at least 5. It is the set bounded by the points (5,5), (5,7.5), (6.5,6.5) and (7.5,5), represented in Figure 4.14.

The set of feasible and strictly individually rational strategies is the same set minus the couple (5,5) and the horizontal and vertical axes.

We now present the folk theorem.

> **Definition 6: Folk Theorem**[15]. Let $u^* = (u^*_1, \ldots, u^*_N)$ be a feasible and strictly individually rational N-tuple of payoffs of the stage game G. Suppose that, for each player i, i from 1 to N, there exists a profile of feasible and strictly individually rational payoffs $U(i) = (U_1(i) \ldots U_N(i))$, such that, for each player i, $u^*_i > U_i(i)$ and $U_i(j) > U_i(i)$ for any j, with $j \neq i$, j from 1 to N. Then there exists a discount factor $\underline{\delta}$, so that, for any $\delta > \underline{\delta}$, the infinitely repeated game G has a SPNE whose payoffs are u^* in each period of play.

According to this criterion, each feasible couple of payoffs in the strictly individually rational set can be obtained forever, provided that a rather loose condition is satisfied.

To prove this theorem, we choose to illustrate it on the game in Matrix 4.7.

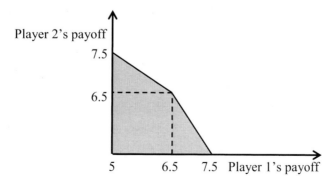

Figure 4.14

5.2.3 Building new behaviour with the folk theorem

We choose the strictly individually rational couple of payoffs u*=(6.5,6.5), obtained if both players bid 0, and we try to establish a minimal discount factor and a strategy profile that leads to bid 0 in each round and that rewards and punishes the players in the spirit of the folk theorem.

So we introduce the payoffs U(1)=(5.75, 7) and U(2)=(7, 5.75). These couples of strictly individually rational payoffs check: u_1*=6.5>U_1(1)=5.75 and U_1(2)=7>U_1(1)=5.75, u_2*=6.5> U_2(2)=5.75 and U_2(1)=7>U_2(2)=5.75. U(1) is obtained as follows: player 1 bids 0, whereas player 2 bids 0 with probability 0.5 and 0.5 with probability 0.5. U(2) is obtained in a symmetric way: player 2 bids 0, whereas player 1 bids 0 with probability 0.5 and 0.5 with probability 0.5.

Because probabilities are unobservable, we have to split U(1) and U(2) into two states: so U(1) will lead to two states, $\overline{e_{11}}$ and $\overline{e_{12}}$, one in which player 1 bids 0 and player 2 bids 0, and one in which player 1 bids 0 and player 2 bids 0.5: by switching from $\overline{e_{11}}$ to $\overline{e_{12}}$ and from $\overline{e_{12}}$ to $\overline{e_{11}}$, we ensure that player 2 bids 0 and 0.5 with probability 0.5. This trick does not exactly ensure the payoff 7 to player 2, given that he gets the payoff 6.5 one period and the payoff 7.5 the next period: so, due to the discount factor, he does not exactly get the mean payoff 7, but this is of no importance (discount factors are chosen to lead player 2 to conform to the strategy profile).

For the same reason, U(2) will lead to two states, $\overline{e_{21}}$ and $\overline{e_{22}}$, one in which player 2 plays 0 and player 1 plays 0, and one in which player 2 plays 0 and player 1 plays 0.5.

We now give the state transition diagram (Figure 4.15) which specifies a SPNE behaviour in the spirit of the folk theorem.

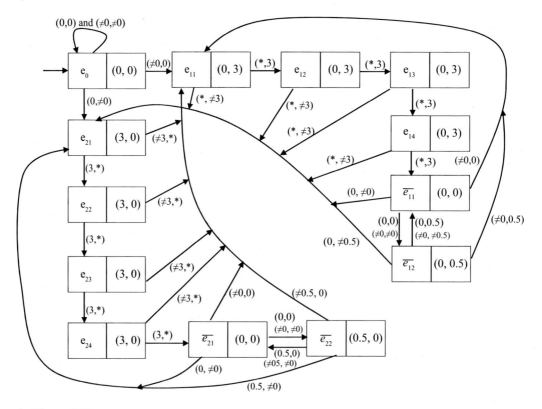

Figure 4.15 * means any available action

BACKWARD INDUCTION AND REPEATED GAMES

Well, this state transition diagram looks rather complicated. But let us describe it. The idea is:

- The game starts in e_0, both players bid 0 and stay in e_0 by doing so. It derives that if the bidders play as planned in the diagram, they stay in e_0 and bid 0 forever, as expected.
- Now suppose that player 1 deviates: in that case, first, she will be strongly punished four times, in that she gets four times her minmax payoff 5. That is why the diagram transits through the four states e_{11}, e_{12}, e_{13} and e_{14}. After that, the game reaches the two states $\overline{e_{11}}$ and $\overline{e_{12}}$, where it stays forever if the players play as expected: player 1 forever gets $(6.5+\delta_1 5)/2$ and player 2 forever gets $(6.5+7.5\delta_2)/2$, which are close to 5.75 and 7 at least for high discount factors.
- We return to the states e_{11}, e_{12}, e_{13} and e_{14}. The aim of these states is to strongly punish player 1 after her deviation in e_0 (so that she does not wish to deviate in e_0): to do so, the two players play four times the couple (0,3) (player 2's action ensures that player 1 cannot get more than 5).

The problem – and this is encountered often – is that in these states *the punisher, player 2, is as unhappy as the punished player, given that he also only gets 5. Moreover, by contrast to the punished player (who cannot get more by deviating from 0), the punisher, by switching from 3 to 0.5 can much increase his payoff. So it is necessary to punish the punisher, i.e. player 2, if he deviates from the punishing action 3, by punishing him in the folk spirit way*: he will get four times the minmax value 5 (in the four states e_{21}, e_{22}, e_{23} and e_{24}) forever followed by the payoff $(6.5+5\delta_2)/2$. The same punishment is applied if player 2 deviates from the planned behaviour in the states $\overline{e_{11}}$ and $\overline{e_{12}}$. This explains why the transition diagram switches to e_{21} if player 2 does not bid 3 in e_{11}, e_{12}, e_{13} and e_{14}, 0 in $\overline{e_{11}}$ and 0.5 in $\overline{e_{12}}$.

- What about player 1 if she deviates in the states e_{11}, e_{12}, e_{13} and e_{14} or in the states $\overline{e_{11}}$ and $\overline{e_{12}}$? She is not induced to deviate in e_{11}, e_{12}, e_{13} and e_{14}, so we do not punish her if she deviates (she punishes herself by deviating!). But she is strongly induced to deviate in $\overline{e_{11}}$ and $\overline{e_{12}}$: to prevent her from not bidding 0, the transition diagram switches to the state e_{11} if she bids more than 0, so that she gets four times her minmax payoff 5: the payoff she can get by deviating in $\overline{e_{11}}$ or $\overline{e_{12}}$ is therefore more than compensated by the loss due to the difference between 5 and the mean payoff in $\overline{e_{11}}$ and $\overline{e_{12}}$.
- Symmetric observations can be made for player 2.

We now show that the state transition diagram corresponds to a SPNE that leads everybody to always bid 0 for sufficiently high values of the discount factors.

Proof

We start in an e_0 state. Let us focus on player 1's behaviour (same reasoning for player 2). We get the evolution Table 4.3a:

BACKWARD INDUCTION AND REPEATED GAMES

■ Evolution Table 4.3a

	t	$t+1$	$t+2$	$t+3$	$t+4$	$t+5$	$t+6$
	e_0	e_0	e_0	e_0	e_0	e_0	e_0
Bid 0 payoff	(0,0) 6.5	(0,0) $6.5\delta_1$	(0,0) $6.5\delta_1^2$	(0,0) $6.5\delta_1^3$	(0,0) $6.5\delta_1^4$	(0,0) $6.5\delta_1^5$	(0,0) $6.5\delta_1^6$
	e_0	e_{11}	e_{12}	e_{13}	e_{14}	$\overline{e_{11}}$	$\overline{e_{12}}$
Bid a≠0 payoff	(a,0) At most 7.5	(0,3) $5\delta_1$	(0,3) $5\delta_1^2$	(0,3) $5\delta_1^3$	(0,3) $5\delta_1^4$	(0,0) $6.5\delta_1^5$	(0,0.5) $5\delta_1^6$

Player 1 will not deviate in e_0 if:

$$6.5+6.5\delta_1+6.5\delta_1^2+6.5\delta_1^3+6.5\delta_1^4+6.5\delta_1^5/(1-\delta_1^2)+6.5\delta_1^6/(1-\delta_1^2) \geq 7.5+5\delta_1+5\delta_1^2+5\delta_1^3+5\delta_1^4+6.5\delta_1^5/(1-\delta_1^2)+5\delta_1^6/(1-\delta_1^2)$$

i.e. $\delta_1 \geq 0.41$.

Symmetrically player 2 will not deviate in e_0 if $\delta_2 \geq 0.41$.

> We observe that $u_i^* > U_i(i))$ ensures, if player i does not deviate, an infinite sequence of 6.5, whereas a deviation in e_0 ensures an infinite sequence of 6.5 followed by 5; the infinity of these two sequences ensures that he is not incited to deviate in e_0, regardless of the number of punishing states e_{i*}, because the additional payoff due to the deviation (7.5–6.5) is necessarily more than compensated by the loss due to the payoff difference between the two infinite sequences, when the discount factor is sufficiently high.
>
> So it is the inequality $u_i^* > U_i(i)$ that ensures that player i does not deviate in e_0. The minmax payoffs V_i he gets a few periods after a deviation in e_0 help to avoid the deviation but are not essential.

We now consider an e_{11} state. Player 1 is not incited to deviate. Things are different for player 2 who can get 7.5 by not bidding 3. For player 2, we get the evolution Table 4.3b.

■ Evolution Table 4.3b

	t	$t+1$	$t+2$	$t+3$	$t+4$	$t+5$	$t+6$
	e_{11}	e_{12}	e_{13}	e_{14}	$\overline{e_{11}}$	$\overline{e_{12}}$	$\overline{e_{11}}$
Bid 3 payoff	(0,3) 5	(0,3) $5\delta_2$	(0,3) $5\delta_2^2$	(0,3) $5\delta_2^3$	(0,0) $6.5\delta_2^4$	(0,0.5) $7.5\delta_2^5$	(0,0) $6.5\delta_2^6$
	e_{11}	e_{21}	e_{22}	e_{23}	e_{24}	$\overline{e_{21}}$	$\overline{e_{22}}$
Bid a≠3 payoff	(0,a) at most 7.5	(3,0) $5\delta_2$	(3,0) $5\delta_2^2$	(3,0) $5\delta_2^3$	(3,0) $5\delta_2^4$	(0,0) $6.5\delta_2^5$	(0.5,0) $5\delta_2^6$

Player 2 does not deviate in e_{11} if:

$$5+5\delta_2+5\delta_2^2+5\delta_2^3+6.5\delta_2^4+7.5\delta_2^5/(1-\delta_2^2)+6.5\delta_2^6/(1-\delta_2^2) \geq 7.5+5\delta_2+5\delta_2^2+5\delta_2^3+5\delta_2^4+6.5\delta_2^5/(1-\delta_2^2)+5\delta_2^6/(1-\delta_2^2)$$ i.e. $\delta_2 \geq 0.80$.

BACKWARD INDUCTION AND REPEATED GAMES

> We observe that $U_2(1) > U_2(2)$) ensures that player 2 gets an infinite sequence of 6.5 followed by 7.5 by not deviating in a player 1 punishing state e_{11}, whereas he gets an infinite sequence of 6.5 followed by 5 by deviating; the infinity of these two sequences ensures that player 2 punishes player 1, even if he does not like doing that, because the additional payoff due to the deviation (7.5−5) is more than compensated by the loss due to the payoff difference between the two infinite sequences, when the discount factor is sufficiently high.
>
> So it is the inequality $U_j(i) > U_j(j)$ that ensures that player j imposes the minmax payoff to a player i who deviates in e_0, even if by doing so he also penalizes himself.

Now consider an e_{12} state. Player 1 is not incited to deviate and the same observation holds for e_{13} and e_{14} regardless of the discount factor.

For player 2, in e_{12}, we get the evolution Table 4.3c:

Evolution Table 4.3c

	t	t+1	t+2	t+3	t+4	t+5	t+6
	e_{12}	e_{13}	e_{14}	$\overline{e_{11}}$	$\overline{e_{12}}$	$\overline{e_{11}}$	$\overline{e_{12}}$
Bid 3 payoff	(0,3) 5	(0,3) $5\delta_2$	(0,3) $5\delta_2^2$	(0,0) $6.5\delta_2^3$	(0,0.5) $7.5\delta_2^4$	(0,0) $6.5\delta_2^5$	(0,0.5) $7.5\delta_2^6$
	e_{12}	e_{21}	e_{22}	e_{23}	e_{24}	$\overline{e_{21}}$	$\overline{e_{22}}$
Bid a≠3 payoff	(0,a) At most 7.5	(3,0) $5\delta_2$	(3,0) $5\delta_2^2$	(3,0) $5\delta_2^3$	(3,0) $5\delta_2^4$	(0,0) $6.5\delta_2^5$	(0.5,0) $5\delta_2^6$

Comparing Table 4.3c with Table 4.3b shows that player 2 is less incited to deviate in e_{12} than in e_{11}. Hence $\delta_2 \geq 0.80$ ensures that player 2 will not deviate from his expected behaviour in e_{12}. And the same observation holds for e_{13} and e_{14}.

We now switch to the states $\overline{e_{11}}$ and $\overline{e_{12}}$.

There is no problem for player 2, in that he is less induced to deviate in state $\overline{e_{11}}$ than in state e_{11}, as it easily derives from the evolution Table 4.3d (as compared to Table 4.3b).

Evolution Table 4.3d

	t	t+1	t+2	t+3	t+4	t+5	t+6
	$\overline{e_{11}}$	$\overline{e_{12}}$	$\overline{e_{11}}$	$\overline{e_{12}}$	$\overline{e_{11}}$	$\overline{e_{12}}$	$\overline{e_{11}}$
Bid 0 payoff	(0,0) 6.5	(0,0.5) $7.5\delta_2$	(0,0) $6.5\delta_2^2$	(0,0.5) $7.5\delta_2^3$	(0,0) $6.5\delta_2^4$	(0,0.5) $7.5\delta_2^5$	(0,0) $6.5\delta_2^6$
	$\overline{e_{11}}$	e_{21}	e_{22}	e_{23}	e_{24}	$\overline{e_{21}}$	$\overline{e_{22}}$
Bid a≠0 payoff	(0,a) At most 7.5	(3,0) $5\delta_2$	(3,0) $5\delta_2^2$	(3,0) $5\delta_2^3$	(3,0) $5\delta_2^4$	(0,0) $6.5\delta_2^5$	(0.5,0) $5\delta_2^6$

And it easily follows that he is not induced to deviate in state $\overline{e_{12}}$.

BACKWARD INDUCTION AND REPEATED GAMES

Hence $\delta_2 \geq 0.80$ ensures that player 2 will not deviate from his expected behaviour in the states $\overline{e_{11}}$ and $\overline{e_{12}}$.

Things are different for player 1. Let us first study the state $\overline{e_{11}}$ (evolution Table 4.3e):

■ **Evolution Table 4.3e**

	t	t+1	t+2	t+3	t+4	t+5	t+6
	$\overline{e_{11}}$	$\overline{e_{12}}$	$\overline{e_{11}}$	$\overline{e_{12}}$	$\overline{e_{11}}$	$\overline{e_{12}}$	$\overline{e_{11}}$
Bid 0	(0,0)	(0,0.5)	(0,0)	(0,0.5)	(0,0)	(0,0.5)	(0,0)
payoff	6.5	$5\delta_1$	$6.5\delta_1^2$	$5\delta_1^3$	$6.5\delta_1^4$	$5\delta_1^5$	$6.5\delta_1^6$
	$\overline{e_{11}}$	e_{11}	e_{12}	e_{13}	e_{14}	$\overline{e_{11}}$	$\overline{e_{12}}$
Bid a≠0	(a,0)	(0,3)	(0,3)	(0,3)	(0,3)	(0,0)	(0,0.5)
payoff	At most 7.5	$5\delta_1$	$5\delta_1^2$	$5\delta_1^3$	$5\delta_1^4$	$6.5\delta_1^5$	$5\delta_1^6$

Player 1 does not deviate if:

$$6.5+5\delta_1+6.5\delta_1^2+5\delta_1^3+6.5\delta_1^4+5\delta_1^5/(1-\delta_1^2)+6.5\delta_1^6/(1-\delta_1^2) \geq 7.5+5\delta_1+5\delta_1^2+5\delta_1^3+5\delta_1^4+6.5\delta_1^5/(1-\delta_1^2)+5\delta_1^6/(1-\delta_1^2)$$

i.e. $\delta_1 \geq 0.72$

> Here we observe the role of the minmax punishing states. Clearly, player 1 is not happy to face the infinite sequence of payoffs 6.5 followed by 5 (her $U_1(1)$ payoff). The minmax punishing states, which also help to avoid the deviation in the e_0 state, are mainly here to avoid the deviation from the U(1) states, $\overline{e_{11}}$ and $\overline{e_{12}}$. It can be checked, in our example, that less than three minmax punishment periods are not enough to induce player i to not deviate in the U(i) states.
>
> To put it more precisely, it is the inequality $U_i(i) > V_i$ that ensures that player i will not deviate in the U(i) states, providing the discount factor is sufficiently high: the difference between the $U_i(i)$ payoffs and the min max value, lost by player i in the e_{ik} states k=1 to 4, does more than outweigh the additional payoff obtained by deviating in a U(i) state.

We now study player 1's behaviour in a state $\overline{e_{12}}$ (evolution Table 4.3f):

■ **Evolution Table 4.3f**

	t	t+1	t+2	t+3	t+4	t+5
	$\overline{e_{12}}$	$\overline{e_{11}}$	$\overline{e_{12}}$	$\overline{e_{11}}$	$\overline{e_{12}}$	$\overline{e_{11}}$
Bid 0	(0,0.5)	(0,0)	(0,0.5)	(0,0)	(0,0.5)	(0,0)
payoff	5	$6.5\delta_1$	$5\delta_1^2$	$6.5\delta_1^3$	$5\delta_1^4$	$6.5\delta_1^5$
	$\overline{e_{12}}$	e_{11}	e_{12}	e_{13}	e_{14}	$\overline{e_{11}}$
Bid a≠0	(a,0.5)	(0,3)	(0,3)	(0,3)	(0,3)	(0,0)
payoff	At most 7	$5\delta_1$	$5\delta_1^2$	$5\delta_1^3$	$5\delta_1^4$	$6.5\delta_1^5$

■ 188

Player 1 will not deviate if:
$$5+6.5\delta_1+5\delta_1^2+6.5\delta_1^3+5\delta_1^4 \geq 7+5\delta_1+5\delta_1^2+5\delta_1^3+5\delta_1^4$$
i.e. $\delta_1 \geq 0.81$.

As above, it is again the inequality $U_i(i) > V_i$ that avoids the deviation.
The states e_{21}, e_{22}, e_{23} and e_{24}, $\overline{e_{21}}$ and $\overline{e_{22}}$ lead to a symmetric study.

> So, by looking for the strongest condition on the discount factors, we conclude that, providing both δ_1 and δ_2 are at least equal to 0.81, the proposed strategy profile, which is in the spirit of the folk theorem, is a SPNE, that leads both players to always bid 0.

Comments
- First, the comments in the boxes show that the folk theorem always works: the conditions $U_j(i) > U_i(i)$, $u^*_i > U_i(i)$, and $U_i(i) > V_i$ ensure the possibility of constructing a transition diagram that exploits these properties and that indeed leads all the players to play the actions associated with u* in each period, providing the discount factors are sufficiently high.
- Second, we observe that the necessary discount factors are high, and that it is not so easy to get lower conditions on these rates.
  For example, adding a (minmax) punishment period may – *strangely enough* – *induce more deviations*. So adding a minmax punishment period for player 1 helps player 1 to not deviate in e_0 and to not deviate in $\overline{e_{11}}$ and in $\overline{e_{12}}$, but it incites player 2 more to deviate in an e_{11} state because it postpones the coming back of high payoffs (you can check that we need $\delta_2 \geq 0.82$ for five punishment periods, instead $\delta_2 \geq 0.80$ for four periods).

5.2.4 Punishing and rewarding in practice

The folk theorem helps to know when outcomes can become SPNE outcomes, but the state transition diagram it leads to is rather complicated. What is more, it has two flaws: first it does not hesitate *to punish those who did not deviate* in the expected state e_0 (for example player 2 also suffers in the states e_{1k}, k from 1 to 4), and, second, it does not hesitate *to punish those who do not punish* (for example player 2 is punished if he does not respect the punishment programme in e_{1k}, k from 1 to 4).

In reality, people prefer strategy profiles that *only* punish the deviator in e_0.

And it is pleasant to observe that, in many cases, we can build strategy profiles that lead to nice payoffs and that only punish the deviators. And the cherry on the cake is that these profiles remain easy!

We will build such a profile step by step:

The first idea is that a deviation from the e_0 behaviour should lead to *a balancing* (the player who suffered from the deviation *should get more than in e_0 in order to compensate the loss*, and *the deviator who benefited from the deviation should get less*); after that, we should return to e_0, given that everybody appreciates this state. For example, during the balancing phase, the deviating player should play 0 whereas the other should play 0.5. So we propose the state transition diagram in Figure 4.16.

But the associated strategy profile is unfortunately not a SPNE, because, for example, player 1 in state e_1, after her deviation in e_0, is not motivated to accept the punishment.

BACKWARD INDUCTION AND REPEATED GAMES

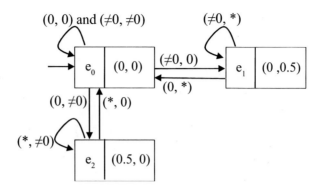

Figure 4.16 * means any available action.

As a matter of fact, we get the evolution Table 4.4:

Evolution Table 4.4

	t	$t+1$	$t+2$
Bid 0	e_1	e_0	e_0
	(0,0.5)	(0,0)	(0,0)
payoff	5	$6.5\delta_1$	$6.5\delta_1^2$
Bid $a\neq 0$	e_1	e_1	e_0
	(a,0.5)	(0,0.5)	(0,0)
payoff	At most 7	$5\delta_1$	$6.5\delta_1^2$

Player 1 doesn't deviate if $5+6.5\delta_1 \geq 7+5\delta_1$ which isn't possible. And adding a punishment period does not change anything because it leads to the condition $5+5\delta_1+6.5\delta_1^2 \geq 7+5\delta_1+5\delta_1^2$, which is even more difficult to fulfil.

The problem comes from the strong difference between 5 (the punishment payoff) and the potential payoff 7.

This leads to switching to a balancing behaviour such that deviating in the punishment phase leads to a payoff lower than 7 that prevents from deviating. This balancing behaviour will be still (but less) beneficial to the non deviating player: we propose that in e_1, player 2 bids 1.5 and player 1 bids 0. So player 1, by deviating in e_1, gets at best 6 by deviating (switching from bid 0 to bid 2). And we get $5+6.5\delta_1 \geq 6+5\delta_1$ which is possible. But the problem is now that player 2, in e_1, is induced to deviate to bid 0.5, in order to get 7.5 instead of 6.5. To prevent this deviation, we need to punish player 2 if he deviates.

Yet, given that player 2 is not the initial deviator (he is not the player who deviated in e_0), we like opting for a punishment that does not make player 1 better off than player 2. A way to do so is to switch to the Nash equilibrium for one period. So we get the state transition diagram in Figure 4.17.

BACKWARD INDUCTION AND REPEATED GAMES

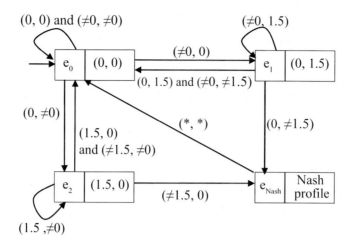

Figure 4.17 * means any action.

We check that the strategies in this state transition diagram constitute a SPNE:
In e_0, for player 1, we get the evolution Table 4.5a:

Evolution Table 4.5a

	t	t+1	t+2
	e_0	e_0	e_0
Bid 0	(0,0)	(0,0)	(0,0)
payoff	6.5	$6.5\delta_1$	$6.5\delta_1^2$
	e_0	e_1	e_0
Bid $a\neq 0$	(a,0)	(0,1.5)	(0,0)
payoff	At most 7.5	$5\delta_1$	$6.5\delta_1^2$

Player 1 does not deviate if $6.5+6.5\delta_1 \geq 7.5+5\delta_1$, i.e. $\delta_1 \geq 0.67$
The same condition holds for player 2, i.e. $\delta_2 \geq 0.67$.

In e_1, for player 1, we get the evolution Table 4.5b:

Evolution Table 4.5b

	t	t+1	t+2
	e_1	e_0	e_0
Bid 0	(0,1.5)	(0,0)	(0,0)
payoff	5	$6.5\delta_1$	$6.5\delta_1^2$
	e_1	e_1	e_0
Bid $a\neq 0$	(a,1.5)	(0,1.5)	(0,0)
payoff	At most 6	$5\delta_1$	$6.5\delta_1^2$

Player 1 does not deviate if $5+6.5\delta_1 \geq 6+5\delta_1$, which is possible for $\delta_1 \geq 0.67$.

BACKWARD INDUCTION AND REPEATED GAMES

For player 2 we get the evolution Table 4.5c:

▪ Evolution Table 4.5c

	t	t+1	t+2
Bid 1.5 payoff	e_1 (0,1.5) 6.5	e_0 (0,0) $6.5\delta_2$	e_0 (0,0) $6.5\delta_2^2$
Bid a≠1.5 payoff	e_1 (0,a) At most 7.5	e_{Nash} (p_{1N}^*, p_{2N}^*) $5.25\delta_2$	e_0 (0,0) $6.5\delta_2^2$

Player 2 does not deviate if $6.5+6.5\delta_2 \geq 7.5+5.25\delta_2$, which is possible for $\delta_2 \geq 1/1.25 = 0.8$.

The analysis is symmetric for the state e_2.

And in the state e_{Nash} nobody is induced to deviate because the played actions have no impact on the future (the players switch to e_0 regardless of the behaviour in e_{Nash}) and because the expected strategies are a Nash equilibrium of the stage game.

> In conclusion the strategy profile that corresponds to the state transition diagram in Figure 4.17 is a SPNE, provided that both discount factors are higher or equal to 0.8.

This is a rather nice equilibrium, because it partially balances the harmful deviations of the players. If player 1 deviates in state e_0, player 2 imposes on her the same loss than the one he incurred in e_0 (switch from the payoff 6.5 to 5), and he partially recovers his loss (he gets 6.5 but not 7.5); and, if he tries to get more than 6.5 in order to better recover his loss, he is punished, but not too strongly (he gets 5.25). Finally, if, for example, player 1 deviates in e_0 by playing 0.5, and if player 2 balances by playing 0.5 instead of 1.5 in e_1, both players get the same sequence of payoffs (7.5 then 5 then 5.25 then 6.5 for the deviating player 1, 5 then 7.5 then 5.25 then 6.5 for the balancing player 2). So there is some justice in this profile, even if, finally, player 2 accepts only getting 6.5 instead of 7.5 in e_1.

But the price to pay for this nice SPNE is a high discount factor. A way to get a smoother condition is to punish player 2 in e_1 more strongly when he does not play 1.5, simply by switching to e_2, as in the state transition diagram in Figure 4.18.

In this new state transition diagram, it can be shown that player 2 will not deviate in e_1, for $\delta_2 \geq 2/3$. And it follows that the strategy profile corresponding to the state transition diagram in Figure 4.18 is a SPNE provided that both discount factors are higher or equal to 2/3.

This profile has a nice property: it contains only three states. But the price to pay for the lower discount factors and the fewer number of states is a less ethical punishment (player 2 is too strongly punished if he tries to better balance by playing 0.5 in e_1 instead of 1.5, and the initial deviator (player 1) benefits from this punishment).

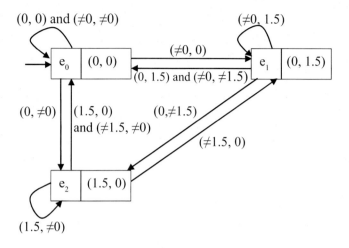

Figure 4.18

5.3 Infinitely repeated extensive form games

Up to now, in infinitely repeated games, we only worked with normal form stage games. What happens if the stage game is a game in extensive form? We already know that strategy profiles that are not Nash equilibria of the stage game may be played in a finite repetition context, if there are several SPNE in the stage game. When the game is repeated an infinite number of times, then it becomes even easier to justify from the stage game equilibrium behaviour (see for example Rubinstein and Wolinsky (1995)[16] for more detailed results). Many payoffs become achievable at each period, even if the game only contains a unique SPNE. However, this does not mean that the extensive form stage game and the associated normal form stage game lead to the same outcomes with the same justifications.

To illustrate these facts, we come back to the sequential battle of sexes stage game.

The infinite repetition of the associated normal form game for example yields a SPNE that leads the player to alternate between $(A_1, A_2/x_1A_2/x_2)$ and $(B_1, B_2/x_1B_2/x_2)$, and so to get the mean payoff 2. For example we can propose the simple state transition diagram in Figure 4.19.

It is easy to check that the profile corresponding to this state transition diagram is a SPNE regardless of the discount factors, mainly because the strategies in e_1 and e_2 are Nash equilibria of the stage game.

Things are different with the extensive form game. Given that player 2 observes player 1's action before playing, player 1 may be incited to play A_1 systematically, in order to compel player 2 to respond by A_2.

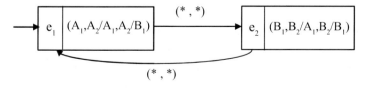

Figure 4.19 * means any available action

BACKWARD INDUCTION AND REPEATED GAMES

We again want both players to alternate (A_1 followed by A_2) and (B_1 followed by B_2). Yet, to punish player 1 when he does not play B_1 when it is time to play it, i.e. to prevent player 1 from deviating to A_1, player 2 has to play B_2 after A_1, an action that punishes player 1, *but also himself.* To constrain player 2 to play in this way, we punish him if he does not punish player 1. So we suggest that, in that case, both players play the path (A_1 followed by A_2) forever. The state transition diagram is given in Figure 4.20.

We prove that this strategy profile is a SPNE.

In each state only one player has to play. It is obvious that the behaviours are optimal in the states e_{11}, e_{12}, e_{22}, e_{41}, e_{42} and e_{43}, regardless of the discount factors, so we only focus on the states e_{21} and e_3.

In e_{21}, it is player 1's turn to play and we get the evolution Table 4.6a.

■ **Evolution Table 4.6a**

	t	t	t+1	t+1	t+2	t+2
	e_{21}	e_{22}	e_{11}	e_{12}	e_{21}	e_{22}
B_1 payoff	B_1	B_2 1	A_1	A_2 $3\delta_1$	B_1	B_2 δ_1^2
	e_{21}	e_3	e_{21}	e_{22}	e_{11}	e_{12}
A_1 payoff	A_1	B_2 0	B_1	B_2 δ_1	A_1	A_2 $3\delta_1^2$

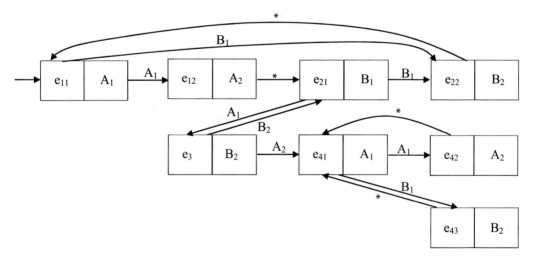

■ **Figure 4.20** * means any available action

Evolution Table 4.6b

	t	t+1	t+1	t+2	t+2
	e_3	e_{21} ←	e_{22}	e_{11}	e_{12}
B_2 payoff	B_2 0	B_1	B_2 $3\delta_2$	A_1	A_2 δ_2^2
	e_3	e_{41}	e_{42}	e_{41} ←	e_{42}
A_2 payoff	A_2 1	A_1	A_2 $1\delta_2$	A_1	A_2 $1\delta_2^2$

It follows that player 1 is best off playing B_1 regardless of the discount factor (playing A_1 yields a null payoff and compels him to play B_1 at the next round, so postpones the coming back of the actions A).

In e_3, it is player 2's turn to play. We get the evolution Table 4.6b.

Player 2 is best off playing B_2 if

$$0+3\delta_2/(1-\delta_2^2)+\delta_2^2/(1-\delta_2^2) > 1+\delta_2/(1-\delta_2^2)+\delta_2^2/(1-\delta_2^2), \text{ i.e. if } \delta_2 \geq 2^{1/2}-1$$

i.e. $\delta_2 \geq 0.42$

Player 2 has to sufficiently appreciate the payoff 3 associated with the actions (B_1, B_2) in the future to accept a null payoff in period t, in order to punish player 1.

> Conclusion: the proposed strategy profile is a SPNE provided that player 2's discount factor is higher or equal to 0.42. This is a loose condition, but it is stronger than in the normal form game (no constraint on the discount factors). This is logical given that, in the extensive form, it is necessary to prevent player 1 from always playing A_1, which requires that player 2 accepts punishing player 1, a punishment that is also harmful for himself.

CONCLUSION

Subgame perfection, or equivalently backward induction, is a quite powerful tool to study long or repeated games. If a game has many subgames (consequently many reduced games), it is easy to get the SPNE. This explains why, despite the fact that it leads to considering each past non expected behaviour as a temporary error, the criterion has been applied often, namely in industrial economics (we recall here the recent literature on negotiation which often refers to this concept and has been able to shine light on many characteristics of real negotiation).

Yet things become more complicated when a long game has no proper subgame, which is often the case when players have private information. In this case subgame perfection makes no selection among Nash equilibria, an issue we'll approach in the next chapter. Things change also when we cannot agree that past unexpected behaviours are errors. When past unexpected strategies are able to convey information on future behaviour, then backward induction is no longer the preferred tool to work with. We come back to this point in chapter 6.

To conclude, let me come back to two variants of Gneezy's and Dufwenberg's race game shown in the conclusion of chapter 2. I first proposed to my L3 students, as an exercise of their terminal exam, to study the easy game G(5,2)[17] with sticks. I first asked them to find the SPNE on the game tree (I gave the game tree), and then I asked them to check, without the game tree, that the first player, by taking 2 sticks at the beginning, is able to win the game: if player 1 takes 2 sticks, there remain 3 sticks to be taken; player 2 takes 1 or 2 sticks, which allows player 1 to complete to 5.

Then I asked how player 1 should play in order to be sure of winning the game G(7,2). Among the 180 students who solved the game G(5,2), 1/3 of them didn't solve the game G(7,2) (perhaps due to lack of understanding or lack of time (it was the last exercise of the exam)). My aim was to see if the students who solved the game used a backward induction reasoning. So I now focus on the students who solved the game.

Well, 15% of them just tried out the many possibilities of gathering 7 sticks and deduced the good way to play.

2/3 of the students who solved the game said that player 1 had to gather 4 sticks, so that player 2 is in front of the 3 last sticks to gather. So they partially solved the game backwards, in that they took into account that player 1 has to lead player 2 in front of a subgame with 3 remaining sticks. By continuing their backward reasoning, they should have observed that leading player 2 in front of a subgame with 6 remaining sticks is also optimal, because, regardless of the number of sticks player 2 takes, player 1 is able to lead him in front of a subgame with 3 remaining sticks. Yet very few students (4%) discovered this "3n" rule. Most of them just observed that the only sure way to gather 4 sticks is to start with 1 stick, and then, after player 2's choice, to complete to 4. Some of them, for example, showed that taking 2 sticks at the beginning is not a good way to play because player 2 can gather 4 sticks, so they deduced that the only sure way to get 4 sticks is to start with 1 stick.

I was pleasantly surprised by 10% of the other students, who used another backward argument. They took into account that a subgame with 5 remaining sticks is a winning game for the player who starts it. So they said: "player 1 should not take 2 sticks at the beginning, because if he takes 2 sticks, player 2 will be in front of 5 sticks and win the game". So they asked player 1 to start with one stick, and then they often switched to the previous backward result (a subgame with 3 remaining sticks is lost by the player who starts it) to get the good way to play. Finally, 9% of the students used a game tree to solve at least partially the game, so they used a backward induction argument.

So among the students who found the solution, many of them used, at least partially, backward induction arguments. But trial and error approaches were also proposed. And among the students who did not solve the game, but tried to solve it, some of them used statistics: they proposed many ways to gather 7 sticks and observed that the percentage of winning solutions for player 1 seems higher when he starts with one stick. Many students also focused on even and odd numbers (given that a player can take 1 or 2 sticks and given that 5 and 7 are odd). Some of them were convinced that if the number of sticks to be gathered is odd, then player 1 wins, and when it is even, player 2 wins (which is wrong). So people do not approach a problem in a unique way, and it is always important to see what may be focal points for them.

NOTES

1   For example, backward induction reasoning is omnipresent in military battles and historical sciences.
2   Zermelo, E., 1912. Uber eine Anwendung der Mengenlehre auf die Theorie des Schachspiels. In proceedings of the fifth International Congress of Mathematicians, Vol. II, pp. 501–504, Cambridge, Cambridge University Press.

3. Selten, R., 1965. Spieltheoretische Behandlung eines Oligopolmodells mit Nachfrageträgheit, Zeitschrift für die gesamte Staatswissenschaft, 121, 301–324 and 667–689.
4. * means any available action.
5. This assumption is linked to continuity. In a continuous setting, when a player refuses an offer in case of indifference, the opponent can propose him ε more and he will accept.
6. Johnson, E.J., Camerer, C., Sen, S. and Rymon, T., 2002. Detecting failures of backward induction: monitoring information search in sequential bargaining, *Journal of Economic Theory*, 104, 16–47.
7. Compte, O. and Jehiel, P., 2004. Gradualism in bargaining and contribution games, *Review of Economic Studies*, 71, 975–1000.
8. Li, D., 2007. Bargaining with history-dependent preferences, *Journal of Economic Theory*, 136, 695–708.
9. Fershtman, C. and Seidmann, D.J., 1993. Deadline effects and inefficient delay in bargaining with endogenous commitment, *Journal of Economic Theory*, 60, 306–321.
10. Selten, R., 1975. Reexamination of the perfectness concept for equilibrium points in extensive games, *International Journal of Game Theory*, 4, 25–55.
11. O'Neill, B. 1986. International escalation and the dollar auction, *Journal of Conflict Resolution*, 30 (1), 33–50.
12. Leininger, W. 1989. Escalation and cooperation in conflict situations, the dollar auction revisited, *Journal of Conflict Resolution*, 33 (2), 231–254.
13. We mention here that O'Neill (1986) and Leininger (1989) also looked for the SPNE of this game, by making different assumptions. O'Neill supposed that, whenever a player is indifferent between two or more bids, he chooses the smallest one, including the possibility of not bidding at all. Leininger, by contrast, supposed that, whenever indifferent between different bids, a player chooses the highest bid.
14. Benoît, J.P. and Krishna, V., 1993. Renegotiation in finitely repeated games, *Econometrica* 61, 303–323.
15. This version of the folk theorem is often attributed to Fudenberg, D. and Maskin, E. (1986). The folk theorem in repeated games with discounting or with incomplete information, *Econometrica*, 54 (3), 533–554.
16. Rubinstein, A. and Wolinsky, A., 1995. Remarks on infinitely repeated extensive-form games, *Games and Economic Behavior*, 9 (1), 110–115.
17. In the game G(M,2), played with sticks, two players, playing in turn, have to gather together M sticks by taking, in each round, 1 or 2 sticks. The winner is the first player gathering M sticks.

Chapter 4

Exercises

♣ EXERCISE 1 STACKELBERG ALL PAY AUCTION, BACKWARD INDUCTION AND DOMINANCE

Question
Find the Subgame Perfect Nash Equilibria (SPNE) in the Stackelberg first price all pay auction (chapter 4, section 4.2.2, Figure E4.1). Compare with the elimination of dominated strategies.

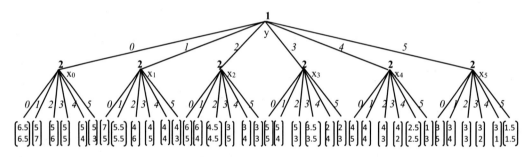

■ Figure E4.1

♣ EXERCISE 2 DUEL GUESSING GAME

Consider the guessing game in Figure E4.2 *(see exercises 6, 13 and 11 in chapters 1, 2 and 3)*

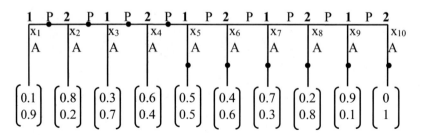

■ Figure E4.2

Questions
1. Find the SPNE of this game.

■ 198

EXERCISES

2. Compare the backward induction procedure with the elimination of dominated strategies (Exercise 13 in chapter 2) and with the way the Nash equilibrium is obtained (Exercise 11 in chapter 3).

♣ EXERCISE 3 CENTIPEDE PRE-EMPTION GAME, BACKWARD INDUCTION AND THE STUDENTS' WAY OF PLAYING

Questions
1. Find the SPNE of the centipede pre-emption game (Figure E4.3a) (*see the exercises 5, 8 and 12 in chapters 1,2 and 3*).

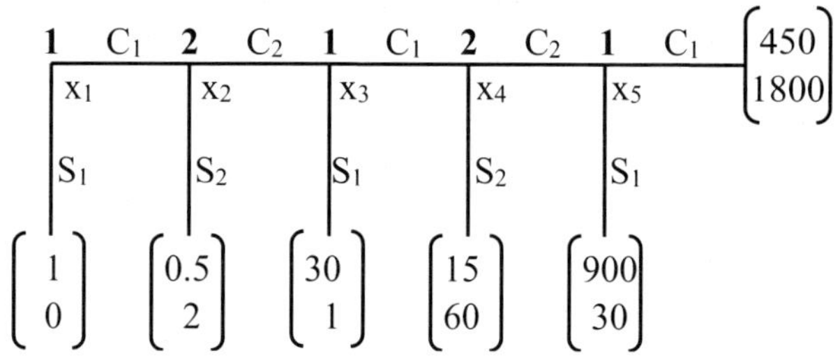

```
   1  C₁   2  C₂   1  C₁   2  C₂   1  C₁  ⎡450 ⎤
   │x₁     │x₂     │x₃     │x₄     │x₅    ⎣1800⎦
   │S₁     │S₂     │S₁     │S₂     │S₁
   ⎡1⎤    ⎡0.5⎤   ⎡30⎤   ⎡15⎤   ⎡900⎤
   ⎣0⎦    ⎣ 2 ⎦   ⎣ 1⎦   ⎣60⎦   ⎣ 30⎦
```

Figure E4.3a

2. *I proposed to my (122) L3 students the centipede game in Figure E4.3b. I asked them when they would stop the game (if called on to play), as player 1 and as player 2. Their answers are given in Table E4.1 and in Figures E4.3c and E4.3d. Comment. Compare with the (same) students' behaviour in the ascending all pay auction (chapter 4, section 3.1).*

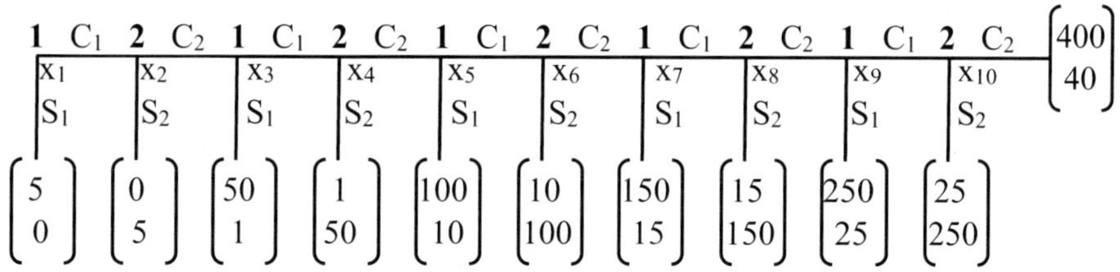

```
 1 C₁ 2 C₂ 1 C₁ 2 C₂ 1 C₁ 2 C₂ 1 C₁ 2 C₂ 1 C₁ 2 C₂ ⎡400⎤
 │x₁  │x₂  │x₃  │x₄  │x₅  │x₆  │x₇  │x₈  │x₉  │x₁₀ ⎣ 40⎦
 │S₁  │S₂  │S₁  │S₂  │S₁  │S₂  │S₁  │S₂  │S₁  │S₂
 ⎡5⎤  ⎡0⎤ ⎡50⎤ ⎡ 1⎤ ⎡100⎤ ⎡ 10⎤ ⎡150⎤ ⎡ 15⎤ ⎡250⎤ ⎡ 25⎤
 ⎣0⎦  ⎣5⎦ ⎣ 1⎦ ⎣50⎦ ⎣ 10⎦ ⎣100⎦ ⎣ 15⎦ ⎣150⎦ ⎣ 25⎦ ⎣250⎦
```

Figure E4.3b

Table E4.1

	Player 1		Player 2
Stops at x_1	11.5%	Stops at x_2	13.1%
Stops at x_3	3.3%	Stops at x_4	6.6%
Stops at x_5	9.8%	Stops at x_6	11.5%
Stops at x_7	10.7%	Stops at x_8	13.1%
Stops at x_9	29.5%	Stops at x_{10}	39.3%
Never stops	35.2%	Never stops	16.4%

EXERCISES

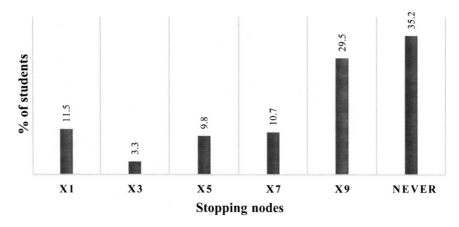

■ **Figure E4.3c** Centipede game, stopping nodes for player 1

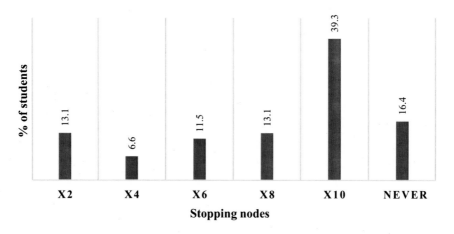

■ **Figure E4.3d** Centipede game, stopping nodes for player 2

♣♣ EXERCISE 4 HOW TO SHARE A SHRINKING PIE

Consider the game studied in chapter 4 (section 2.2), but with 8 periods instead of 2 or 3. Two players share a pie that shrinks with time, or, in other words, share a sum S that decreases over time. In period 1, S=10, in period 2, S=9, in period i, S=11–i, with i from 1 to 8. In the odd periods j, player 1 makes an (integer) offer x to player 2, with $x \in \{1,..., 10-j\}$. Player 2 observes the offer and accepts or refuses it. If he accepts, player 1 gets x and player 2 gets 11–j–x. If he refuses, the game switches to period j+1. In the even periods j, player 2 makes an offer x to player 1, with $x \in \{1,..., 10-j\}$. Player 1 observes the offer and accepts or refuses it. If she accepts, player 1 gets x and player 2 gets 11–j–x. If she refuses, both players get 0 if j=8. If j≠8, the game switches to period j+1.

Questions:
1. Write the scheduled game.
2. Find the SPNE of the game, by supposing that each player accepts an offer when s/he is indifferent between accepting and refusing.

EXERCISES

♣♣ EXERCISE 5 ENGLISH ALL PAY AUCTION

Consider the English all pay auction (Figure E4.4) studied in chapter 4 (section 1).

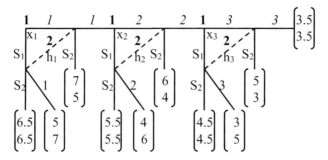

Figure E4.4

Questions:
1. Show that the three profiles $((\pi_{1x1}(S_1)=1/3, \pi_{1x2}(S_1)=1/2, 3/x_3), (\pi_{2h1}(S_2)=1/3, \pi_{2h2}(S_2)=1/2, 3/h_3))$, $((S_1/x_1, \pi_{1x2}(S_1)=1/2, 3/x_3), (1/h_1, \pi_{2h2}(S_2)=1/2, 3/h_3))$ and $((1/x_1, \pi_{1x2}(S_1)=1/2, 3/x_3), (S_2/h_1, \pi_{2h2}(S_2)=1/2, 3/h_3))$ are SPNE.
2. Suppose that the players are allowed to bid 4 and 5. Give the new extensive form and look for a SPNE such that both players always overbid with positive probability. Comment.

♣♣ EXERCISE 6 GENERAL SEQUENTIAL ALL PAY AUCTION, INVEST A MAX IF YOU CAN

Look at the game in Figure E4.5 (studied in chapter 4, section 3.3)

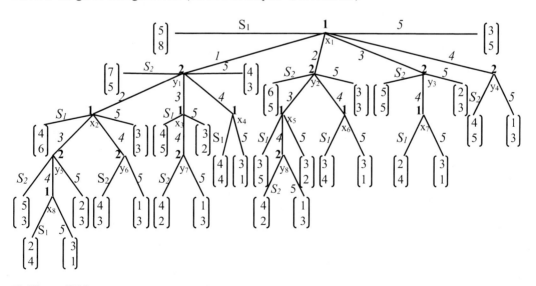

Figure E4.5

EXERCISES

Questions
1. Find the SPNE in which a player, when he is indifferent between stopping and overbidding 5 (=M), chooses to overbid. Show that this SPNE is in accordance with the reasoning given in chapter 4.
2. Find the SPNE where a player, when he is indifferent between two bids (respectively between bidding and stopping), chooses the lowest one (respectively chooses to stop), (O'Neill's (1989) assumption).
3. Find the SPNE where player 1 chooses the lowest bid (stops), and player 2 the largest one, when s/he is indifferent between two bids (between stopping and bidding).
4. Find the SPNE paths of this game.

♣♣ EXERCISE 7 GIFT EXCHANGE GAME

Consider an extensive form gift exchange game we talked about in chapter 1 (section 3.3) and in Exercise 9 of chapter 3: the employer first offers a wage, w_i, i from 1 to 3, with $w_3 > w_2 > w_1$, the employee observes the wage and then offers a certain level of effort, E_i, i from 1 to 3, with $E_3 > E_2 > E_1$. The payoffs of both players, according to the couples of wage and effort, are given in Table E4.2.

■ **Table E4.2** The first, respectively the second coordinate of each couple is the employee's, respectively the employer's payoff.

		employer		
		w_1	w_2	w_3
employee	E_1	(2,2)	(4,1)	(5.5,0)
	E_2	(1,4)	(3,3)	(4.5,2.5)
	E_3	(0,6)	(2,5)	(3.5,4)

Questions
1. Show that there exists a unique SPNE.
2. Change the structure of the game as follows: the employee commits himself to an effort for each wage (once he gets the wage, he must provide the effort he promised). Then the employer proposes a wage. Comment on the structure of the game (give the strategy sets of the players) and find the SPNE path in pure strategies.
3. Now reverse the roles. Suppose that the employer commits herself to a wage for each effort (once she observes the effort, she pays the promised associated wage). Then the employee chooses an effort. Comment again on the structure of the game and find the SPNE path in pure strategies. Comment.

EXERCISES

♣ EXERCISE 8 REPEATED GAMES AND STRICT DOMINANCE

Consider the normal form stage game given in Matrix E4.1.

		Player 2 A_2	B_2
Player 1	A_1	(3,5)	(15,15)
	B_1	(6,6)	(17,2)
	C_1	(8,1)	(16,1)

Matrix E4.1

Questions

1. Find all the Nash equilibria of this game
2. Suppose that the stage game is repeated twice (players observe the actions played in period 1 before they play in period 2). Show that it is possible to build a SPNE in which player 2 totalizes a payoff equal to 16 on the two periods. Give the extensive form of the reduced game in period 1.
3. Discuss the possibility of playing a strictly dominated action of the stage game in a repeated context. Discuss also the number and the particularities of the Nash equilibria of the stage game necessary to get the SPNE in question 2.

♣ EXERCISE 9 THREE REPETITIONS ARE BETTER THAN TWO

Consider the stage game in normal form given in Matrix E4.2

		A_2	Player 2 B_2	C_2
Player 1	A_1	(7,4)	(9,3)	(0,0)
	B_1	(0,0)	(10,10)	(0,14)
	C_1	(1,0)	(15,0)	(1,1)

Matrix E4.2

This game has two pure strategy Nash equilibria (A_1,A_2) and (C_1,C_2), and a mixed strategy Nash equilibrium where player 1 plays A_1 with probability 1/5 and C_1 with probability 4/5, and player 2 plays A_2 with probability 1/7 and C_2 with probability 6/7 (player 1's and player 2's expected payoffs are 1 and 0.8).

Questions

1. Suppose that the stage game is repeated twice, the players observing the actions played in period 1 before playing in period 2. Explain why it is not possible to construct a SPNE in which both players get 10 in the first period. Propose a change in the payoffs that would allow getting a SPNE with a payoff 10 in period 1.
2. Show that it is possible to get a SPNE in which both players get 10 in the first period if the game in Matrix E4.2 is played three times. Comment.

203

EXERCISES

♣ EXERCISE 10 ALTERNATE REWARDS IN REPEATED GAMES

Consider the normal form stage game in Matrix E4.3.

		Player 2		
		A_2	B_2	C_2
Player 1	A_1	(7,1)	(1,5)	(3,0)
	B_1	(5,7)	(0.5,9)	(3,4)
	C_1	(0,0)	(0,1)	(4,2)

Matrix E4.3

This game has two pure strategy Nash equilibria (A_1,B_2) and (C_1,C_2) and a mixed strategy Nash equilibrium in which player 1 plays A_1 and C_1 with probability 1/6 and 5/6, and player 2 plays B_2 and C_2, each with probability ½ (player 1 gets 2, player 2 gets 5/3).

Questions
1. Suppose that the stage game is repeated twice, each player observing the past actions before playing in period 2. Explain why it is not possible to construct a SPNE in which the players play (B_1,A_2) in the first period.
2. Show that it is possible to construct such a SPNE if the game is played three times. Comment.

♣♣ EXERCISE 11 SUBGAME PERFECTION IN RUBINSTEIN'S[1] FINITE BARGAINING GAME

Consider a bargaining game, where two players try to share a pie of unit size. The game is sequential and lasts at most T periods. In each odd (respectively even) period, player 1 (respectively player 2) offers a division x of the pie. Player 2 (respectively player 1) accepts the offer, in which case player 1 gets the share x and player 2 the share 1–x, or he/she refuses the offer. In this case, the game proceeds to the next period.

The game ends if an offer is accepted or after period T. If the players fail to agree in period T, both get 0.

Player 1 has a discount factor δ_1, $0<\delta_1<1$, player 2 has a discount factor δ_2, $0<\delta_2<1$.

Question
Find the SPNE of this bargaining game.

♣♣ EXERCISE 12 SUBGAME PERFECTION IN RUBINSTEIN'S INFINITE BARGAINING GAME

Consider the game in Exercise 11, but suppose now that bargaining may be infinite (T is infinite).

Question
Find the SPNE of the game.

♣ ♣ ♣ EXERCISE 13 GRADUALISM AND ENDOGENOUS OFFERS IN A BARGAINING GAME, LI'S INSIGHTS

In all the negotiation/bargaining games studied up to now, the negotiation stopped in the first period: player 1 makes an offer that is immediately accepted by player 2. And this result is logical, given that the pie is shrinking over time. *Yet it doesn't fit with reality.*

In Li's model (2007), two players again bargain over a pie of unit size according to the infinite alternating offers procedure. In each period they can wait or make a concession to the other player, which can be accepted or refused. The point is that a player is committed to never accepting at period t an offer that gives him a lower discounted payoff than the payoff associated with an earlier refused offer. This leads to *gradualism*, i.e. to make offers that slowly grow in time, which postpones the date of transaction.

An impasse is a situation such that the sum of the two shares the players can accept is higher than 1. In that case, both players get 0. We introduce a same discount factor δ for both players. We call z_{it} the highest (at period t discounted) payoff player i has rejected till period t, i=1,2.

Questions
1. Suppose that, at period t, it is player i's turn to make an offer. Show that in a SPNE, player j accepts any offer equal or higher than $\max(\delta-z_{it}, z_{jt})$ and rejects any lower offer; $\max(\delta-z_{it}, z_{jt})$ is called the *clinching offer*.
2. Fix $\delta=0.7$. Given that it is not rational to end with an impasse, the game necessarily ends with a clinching offer. Show that player 1 cannot prefer making the clinching offer in period 1, if player 2 makes it in period 2. Then show that it is best for player 1 to make an offer refused in period 1, but that leads player 2 to make the clinching offer in period 2. Give the offers on the SPNE path.

♣ ♣ EXERCISE 14 INFINITE REPETITION OF THE TRAVELLER'S DILEMMA GAME

Let us consider the traveller's dilemma *(see the exercises 1, 11 and 15 in chapters 1, 2 and 3)* where player i gets max $(0, x_j-P)$ if $x_i>x_j$ i≠j=1,2 (version close to Basu). We suppose that the prime P is lower or equal to 96.

Question
Suppose that this game is repeated an infinite number of times.

Find different SPNE such that both players ask for 100 at each period. First propose profiles that include the Nash equilibrium, then propose a profile close to the folk theorem, and finally propose a profile that benefits more the players who do not deviate.

♣ EXERCISE 15 INFINITE REPETITION OF THE GIFT EXCHANGE GAME

Consider again the extensive form gift exchange game in Exercise 7. The employer first offers a wage w_i, i from 1 to 3, with $w_3>w_2>w_1$, and then the employee, after observing the wage, offers a certain level of effort E_i, i from 1 to 3, with $E_3>E_2>E_1$. The payoffs of both players, according to the salaries and efforts, are given in Table E4.2 (see Exercise 7).

EXERCISES

We know (see Exercise 7) that the unique SPNE of this game leads the employee to offer the lowest effort regardless of the wage, compelling the employer to offer the lowest wage w_1. Yet of course (w_1, E_1) is a very bad couple of actions: (w_3, E_3) looks much better (it is namely a fair Pareto optimal couple).

Question

The game is repeated an infinite number of times. Find a SPNE such that, in each period, the employer offers w_3 and the employee offers the effort E_3.

NOTE

1   Rubinstein, A., 1982. Perfect equilibrium in a bargaining model, *Econometrica*, 50, 97–110.

Chapter 5

Trembles in a game

INTRODUCTION

Let's (a little) shake the game! Many things in real life justify trembles in a game. On the one hand, players may incorrectly know the structure of a game and/or they may wrongly optimize; common knowledge of the structure of the game may be non existent and the ability to calculate a best response pure utopia. Trembles in the structure or in the behaviour may help to cope with these imperfections. On the other hand, in many games, trembles in strategies ensure that the game explores all possible directions. So they prevent incredible threats (nobody wants to play), because they ensure that each information set is reached, so that the players are compelled to play with positive probability the strategy they proposed to play. This chapter will focus on trembles and their consequences, and the core concept developed is Selten's (1975) trembling-hand equilibrium.

We proceed as follows: In the first section we develop Selten's perfect equilibrium/alias trembling-hand equilibrium. Selten doesn't shake the whole game: he only perturbs strategies. His idea is that we push on a button to validate our strategy, and that with a small probability because of distraction or temporary insanity, or a trembling hand, we push on a wrong button. Selten claims that a strategy profile should resist at least some small trembles. In section 2, we talk about Selten's closest related concepts, Kreps and Wilson's (1982) sequential equilibrium and Harsanyi's perfect Bayesian equilibrium. These concepts develop the new notion of beliefs. In this section we also wonder if a strategy profile has to resist all sorts of trembles, or if some trembles are more natural than others. This leads us to Myerson's (1978) proper equilibrium. In section 3 we go further, and shake more: instead of only perturbing the strategies, we slightly perturb the structure of the game. This can be done in many ways; we just propose four nice examples that show in a progressive way when very small changes have no impact and when, by contrast, they can destroy upper hemicontinuity and lower hemicontinuity of equilibrium sets. We work with a carrier pigeon (Myerson's game), and we study successively an ascending all pay auction with unknown budgets, negotiation games with an unknown number of negotiation rounds, and Rubinstein's electronic mail game. In section 4, we perturb payoffs with Jackson et al. (2012) and the concept of quantal response (McKelvey and Palfrey (1995)). We finally study replicator equations in order to link them to Selten's initial approach before concluding.

TREMBLES IN A GAME

1 SELTEN'S PERFECT EQUILIBRIUM

1.1 Selten's horse, what's the impact of perturbing strategies?

We start with Selten's[1] famous horse game, depicted in Figure 5.1.

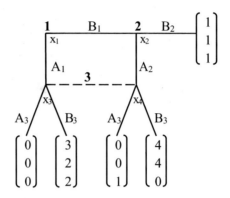

■ Figure 5.1

This game has two pure strategy Nash equilibria, (A_1,B_2,B_3) and (B_1,B_2,A_3), that share a common point: one of the players is not called on to play, player 2 in the first equilibrium, player 3 in the second. Despite this common point, the first equilibrium seems more fragile than the second one. Why?

Suppose that players choose their actions by pushing on a button and may slightly tremble while doing so. So, when a player wants to play A, he plays A with probability $1-\varepsilon$, and B with probability ε, where ε goes to 0 (the same is true if he wants to play B).

Now observe the equilibrium (A_1,B_2,B_3). Player 2 is called on to play, because player 1 plays B_1 with a small positive probability ε. So, knowing that player 3 plays B_3 with a probability close to 1 (because player 3 also slightly trembles), he should play A_2 and not B_2, because he gets a payoff close to 4 by playing A_2 in contrast to the payoff 1 obtained with B_2.

Now study the equilibrium (B_1,B_2,A_3). Again, if players 1 and 2 slightly tremble, player 3 is called on to play. But he can still play A_3 (with a probability going to 1) in this perturbed environment because he may imagine that player 2 trembled more often than player 1. So he may imagine that he is at x_4 with a probability higher than two times the probability to be at x_3, which leads him to play A_3 (with a probability going to 1).

This way of doing is linked to subgame perfection, which requires that players behave optimally (in conformity with the Nash equilibrium) in all subgames. But many games have no subgame, like Selten's horse, leaving subgame perfection without bite. Introducing trembles ensures that all the subgames and *all the information sets* are reached. Consequently players will have to behave in a rational way anywhere in the game tree, not only in subgames. *A natural consequence of this fact is that each perfect equilibrium is a SPNE but not the reverse.*

1.2 Selten's perfect/trembling hand equilibrium

Selten (1975) introduces the notion of a *perturbed game*:

> **Definition 1: perturbed game (Selten 1975).** Let G be a game in extensive form with perfect recall. A perturbed game $\hat{G}^k$ of G is the couple (G, η^k) where η^k is a function that, at each information set h, assigns a minimal positive probability to each action a of the set of actions at h, A_h, the sum of the minimal probabilities at h being lower than 1:
>
> $$\forall a \in A_h, \eta^k(a) > 0 \text{ and } \sum_{a \in A_h} \eta^k(a) < 1$$

$\eta^k(a)$ is the minimal probability with which action a is played in the perturbed game. The constraint $\sum_{a \in A_h} \eta^k(a) < 1$ is a natural one, given that Selten works with minimal probabilities going to 0.

It follows that a Nash equilibrium in a perturbed game $\hat{G}^k$ is a profile of behavioural strategies π^k, such that, for each player i, π_i^k is a best answer to π_{-i}^k and checks the property: $\forall h_i, \forall a \in A_{hi}, \pi_{ihi}^k(a) \geq \eta^k(a)$.

> **Definition 2: test sequence of perturbed games:** A sequence $\hat{G}^1 \hat{G}^2 \ldots \hat{G}^k$ is a test sequence of perturbed games, if $\forall h, \forall a \in A_h, \eta^k(a) \to 0$ when k→∞.

For Selten, a profile of behavioural strategies π^* is a perfect equilibrium if and only if it resists the introduction of perturbations:

> **Definition 3: perfect equilibrium.** π^* is a perfect equilibrium of the game G if there exists a test sequence of perturbed games of G, $\hat{G}^1 \hat{G}^2 \ldots \hat{G}^k$, and a sequence of Nash equilibria of these perturbed games, $\pi^1 \pi^2 \ldots \pi^k$, such that $\pi^k \to \pi^*$ when k→∞.

Comments

The main comment is that Selten's concept is not so demanding. As a matter of fact it imposes no constraint on the minimal probabilities except that they converge to 0 when k goes to infinity. So, to prove that a profile of behavioural strategies of the game G is perfect, it is enough to find a sequence of minimal perturbations going to 0, such that one sequence of associated Nash equilibria converges to the tested profile. The way we choose the perturbations is completely open. As a consequence, when a profile is not perfect, *it does not resist any perturbations in the strategies of the opponents, and is really not very robust*. So it seems reasonable to only accept perfect equilibria.

But how do we justify the introduction of these perturbations? Selten describes the minimal probabilities as the expression of trembles or temporary insanity. That is why Selten's concept is

TREMBLES IN A GAME

also called *the trembling hand concept*. But to my mind, the best way to justify the concept is to talk about *minimal robustness*. As a matter of fact, as said above, if a profile is not perfect, then it disappears as soon as one shakes the strategies a little: so it doesn't have many chances to be observed in reality given that many small errors or perturbations affect the behaviour in reality.

Properties of perfect equilibria

> **Property 1: perfection, subgame perfection and Nash equilibrium**
> Given that a perfect equilibrium is the limit of a sequence of Nash equilibria in perturbed games, given that the expected payoffs are continuous functions of the strategies, and given that the strategy sets go the original sets of strategies when the perturbations go to 0, it follows that *a perfect equilibrium necessarily is a Nash equilibrium.*
>
> And given that, in a perturbed game, each subgame is reached with positive probability, a perfect equilibrium necessarily is a SPNE.

So we have {Perfect Equilibria}∈{SPNE}∈{Nash Equilibria}, and it derives the question of existence. Selten proves the existence by using the existence of Nash equilibria in special normal form games.

> **Property 2: existence of perfect equilibria**
> In each extensive form finite game with perfect recall there exists at least one perfect equilibrium.

A useful property is: given that, in a perturbed game, each information set is reached with positive probability, a behavioural strategy is a best response in the perturbed game if and only if each local strategy is a best response to the remaining strategies. So we get:

> **Property 3:**
> To check, in a perturbed game $\hat{G}^k$, if a profile of behavioural strategies π^k is a Nash equilibrium of $\hat{G}^k$, it is necessary and sufficient to check if, for each player i at each of his information set h, the local strategy π_{ih}^k is a best response to π_{-ih}^k.

1.3 Applications and properties of the perfect equilibrium

1.3.1 Selten's horse, trembles and strictly dominated strategies

We apply the trembling hand concept to Selten's horse.

TREMBLES IN A GAME

> **STUDENTS' INSERT**

We first prove that the first equilibrium $E_1=(A_1,B_2,B_3)$ is not perfect.

We have to show that, regardless of the minimal perturbations introduced, (A_1,B_2,B_3) cannot be the limit of a sequence of Nash equilibria π^k of a test sequence of perturbed games. Suppose the contrary and look for player 2's behaviour in π^k. By playing A_2, player 2 gets the payoff $4\pi_1^k(B_1)\pi_3^k(B_3)$, by playing B_2 he gets $1\pi_1^k(B_1)$. Given that π^k is supposed to converge to the tested equilibrium when k goes to ∞, $\pi_3^k(B_3)$ goes to 1 when k goes to ∞, hence $4\pi_1^k(B_1)\pi_3^k(B_3)>\pi_1^k(B_1)$ when k goes to ∞. Consequently player 2 has to play B_2 with the minimal probability, which goes to 0 when k goes to ∞. So π_2^k can't converge to player 2's strategy in E_1, regardless of the introduced minimal perturbations.

We now prove that the second equilibrium $E_2=(B_1,B_2,A_3)$ is perfect. It is enough to find minimal perturbations such that a sequence of Nash equilibria π^k of the sequence of perturbed games goes to E_2 when k goes to infinity. We call ε_1^k, ε_2^k, ε_3^k the minimal probabilities assigned to A_1, A_2 and B_3, and propose the profile π^k defined by: $\pi_1^k(A_1)=\varepsilon_1^k$, $\pi_1^k(B_1)=1-\varepsilon_1^k$, $\pi_2^k(A_2)=\varepsilon_2^k$, $\pi_2^k(B_2)=1-\varepsilon_2^k$, $\pi_3^k(A_3)=1-\varepsilon_3^k$, $\pi_3^k(B_3)=\varepsilon_3^k$. By construction π^k goes to E_2 when k goes to ∞. So it remains to show that π^k is a Nash equilibrium of the perturbed game. To do so, we check if each player's behaviour is a best response in the perturbed environment:

Player 1: A_1 leads to the payoff $3\pi_3^k(B_3)=3\varepsilon_3^k$. B_1 leads to the payoff: $\pi_2^k(B_2)+4(1-\pi_2^k(B_2))\pi_3^k(B_3)=1-\varepsilon_2^k+4\varepsilon_2^k\varepsilon_3^k$. It follows that B_1 leads to a higher payoff than A_1, when k goes to ∞. So player 1's best response in the perturbed environment consists of assigning to B_1 the highest possible probability, $1-\varepsilon_1^k$, i.e. 1 minus the minimal probability ε_1^k she has to assign to A_1.

Player 2: A_2 leads to the payoff $4\pi_1^k(B_1)\pi_3^k(B_3)=4(1-\varepsilon_1^k)\varepsilon_3^k$ whereas B_2 leads to the payoff $\pi_1^k(B_1)=1-\varepsilon_1^k$. B_2 leads to a higher payoff than A_2, when k goes to ∞. So player 2's best response in the perturbed environment consists to assign to B_2 the highest possible probability, $1-\varepsilon_2^k$, i.e. 1 minus the minimal probability ε_2^k he has to assign to A_2.

Player 3: A_3 leads to the payoff $1\pi_1^k(B_1)\pi_2^k(A_2)=(1-\varepsilon_1^k)\varepsilon_2^k$. B_3 leads to the payoff $2\pi_1^k(A_1)=2\varepsilon_1^k$. We want that player 3 gets more with A_3 than with B_3. So we need: $(1-\varepsilon_1^k)\varepsilon_2^k \geq 2\varepsilon_1^k$. If this condition is fulfilled, player 3's best response in the perturbed environment indeed consists of assigning to A_3 the highest possible probability, $1-\varepsilon_3^k$, i.e. 1 minus the minimal probability ε_3^k he has to assign to B_3. Can the condition be fulfilled, i.e. is it possible to set: $\varepsilon_2^k \geq 2\varepsilon_1^k/(1-\varepsilon_1^k)$? The answer is yes, in that it is possible to have ε_2^k going to 0 when k goes to ∞, because $2\varepsilon_1^k/(1-\varepsilon_1^k)$ goes to 0 when k goes to ∞.

So by choosing minimal probabilities such that $\varepsilon_2^k \geq 2\varepsilon_1^k/(1-\varepsilon_1^k)$, π^k is a Nash equilibrium of the perturbed game. And, given that π^k goes to E_2 when k goes to ∞, we conclude that E_2 is a perfect equilibrium.

We will comment again on the unrestricted choice we have in selecting the minimal probabilities: in order to justify player 3's action A_3, player 2's trembles are supposed to be more numerous than player 1's, so as to counterbalance the payoff 2 player 3 can obtain with B_3 with the payoff 1 obtained with A_3.

Trembles are far from any rationality. So the players namely also tremble towards strictly dominated strategies which leads to this unexpected result:

TREMBLES IN A GAME

> **Property 4:** Eliminating strictly dominated strategies can suppress a perfect equilibrium in a finite extensive form game with perfect recall.

We prove this result with the game in Figure 5.2.

> ### ➢ STUDENTS' INSERT
>
> (C_1, C_2) is a perfect equilibrium. To prove it, we call ε_1^k, e_1^k, ε_2^k, e_2^k the minimal probabilities assigned to A_1, B_1, A_2 and B_2, and propose the profile: $\pi_1^k(A_1)=\varepsilon_1^k$, $\pi_1^k(B_1)=e_1^k$, $\pi_1^k(C_1)=1-\varepsilon_1^k-e_1^k$ and $\pi_2^k(A_2)=\varepsilon_2^k$, $\pi_2^k(B_2)=e_2^k$, $\pi_2^k(C_2)=1-\varepsilon_2^k-e_2^k$.
>
> By construction π^k goes to (C_1,C_2) when k goes to ∞. We now show that π^k is a Nash equilibrium of the perturbed game.
>
> Player 1: A_1 leads to the payoff $2\pi_2^k(A_2)+\pi_2^k(B_2)=2\varepsilon_2^k+e_2^k$, B_1 leads to the payoff $4\pi_2^k(A_2)+2\pi_2^k(B_2)+\pi_2^k(C_2)=4\varepsilon_2^k+2e_2^k+1-\varepsilon_2^k-e_2^k=3\varepsilon_2^k+e_2^k+1$ and C_1 leads to the payoff 2. C_1 is the best response, so player 1's best response in the perturbed game is to assign the minimal probabilities, ε_1^k and e_1^k, to A_1 and B_1, and to assign to C_1 the remaining probability $1-\varepsilon_1^k-e_1^k$.
>
> Player 2: A_2 leads to the payoff $2\pi_1^k(B_1)=2e_1^k$, B_2 leads to the payoff 0, and C_2 leads to the payoff $2\pi_1^k(A_1)=2\varepsilon_1^k$. C_2 is the best response if $\varepsilon_1^k \geq e_1^k$.
>
> So by choosing minimal probabilities that check $\varepsilon_1^k \geq e_1^k$, π^k is a Nash equilibrium of the perturbed game, and given that it converges to (C_1,C_2) when k goes to ∞, (C_1,C_2) is a perfect equilibrium.
>
> Yet A_1 is strictly dominated by B_1. If we suppress A_1, the only SPNE (and consequently the only perfect equilibrium) becomes (B_1,A_2). So (C_1,C_2) does not resist the elimination of strictly dominated strategies. This is due to the fact that the perfection of (C_1,C_2) requires that player 1 trembles more toward A_1 (the strictly dominated strategy) than toward B_1 (the strategy that strictly dominates A_1); this illustrates that trembles are not affected by payoffs.

Of course this rather strange result, which is opposite to the one obtained for Nash equilibria – *no Nash equilibrium can be eliminated by the suppression of a strictly dominated strategy* – clearly highlights that Selten's trembles are neutral, free, and have, in contrast with best responses, no link with the players' payoffs.

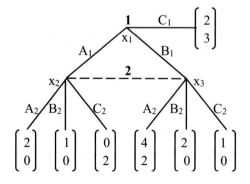

■ Figure 5.2

1.3.2 Trembles and completely mixed behavioural strategies

An additional remark concerns the link between trembling hand perfection and completely mixed equilibria. We get:

> **Property 5:** A completely mixed behavioural Nash equilibrium is automatically perfect.

The proof of this property is obvious. Consider a completely mixed behavioural Nash equilibrium π^*, a test sequence $\hat{G}^1, \hat{G}^2 \ldots \hat{G}^k$ in which each player plays each action with at least the probability ε^k, with $\varepsilon^k < \min_{a \in A} \pi^*(a)$ (where A is the set of actions in the game), which is possible, given that π^* is completely mixed. Now consider a sequence of Nash equilibria of these perturbed games, π^1 $\pi^2 \ldots \pi^k$ all equal to π^*, which is possible given our definition of ε^k. So we automatically have $\pi^k \to \pi^*$, and π^k is automatically a Nash equilibrium of $\hat{G}^k$, given that each player best responds to the others strategies (because π^* is a Nash equilibrium) and each player plays each of his actions with a probability higher than the minimal one ε^k.

So completely mixed behavioural Nash equilibria *are perfect by construction. To my mind, this is surely a bad point of the trembling hand concept*. I am not at all shocked by the fact that shaking can lead to assigning higher minimal probabilities to strictly dominated actions than to undominated ones, given that shaking is neutral. But I find quite annoying that, by construction, mixed behavioural strategies *will in fact not been shaken.* Let me illustrate my claim with a small variant of the horse game given in Figure 5.3 (the only change is player 1's payoff after (B_1, B_2); it switches from 1 to ½).

This game has a completely mixed behavioural Nash equilibrium in which player 1 plays A_1 with probability 1/5, player 2 plays A_2 with probability ½, and player 3 plays A_3 with probability ¾. Given our proof above, this equilibrium is perfect (it is enough to fix $\varepsilon^k < 1/5$ in any perturbed game). But, to my mind, shaking the game should lead to slightly shake the probabilities with which the 3 players play their actions. So, in π^k, player 1 should for example play A_1 with probability $1/5 + \varepsilon_1^k$ (and B_1 with the complementary probability $4/5 - \varepsilon_1^k$), player 2 should play A_2 with probability $1/2 + \varepsilon_2^k$ (and B_2 with the complementary probability $1/2 - \varepsilon_2^k$), and player 3 should play A_3 with probability $3/4 + \varepsilon_3^k$ (and B_3 with the complementary probability $1/4 - \varepsilon_3^k$), where $\varepsilon_1^k, \varepsilon_2^k, \varepsilon_3^k$ are positive or negative and going to 0. Yet, clearly, given that none of these probabilities is the

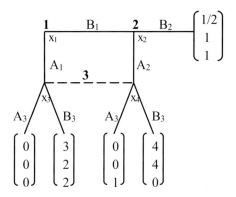

■ Figure 5.3

TREMBLES IN A GAME

minimal probability to play the action (none of these probabilities goes to 0), these are only acceptable if A and B give the same payoff in the perturbed environment. So we should have $3(1/4-\varepsilon_3^k)=\frac{1}{2}(1/2-\varepsilon_2^k)+4(1/2+\varepsilon_2^k)(1/4-\varepsilon_3^k)$ for player 1, $1=4(1/4-\varepsilon_3^k)$ for player 2 and $2(1/5+\varepsilon_1^k)=(4/5-\varepsilon_1^k)(1/2+\varepsilon_2^k)$ for player 3, which is only possible if $\varepsilon_1^k=\varepsilon_2^k=\varepsilon_3^k=0$. So, clearly, we cannot, in this example, converge to the completely mixed equilibrium as soon as we shake each probability a little. In other terms, really shaking all the probabilities, even with minimal probabilities going to 0, would impede the completely mixed behavioural Nash equilibrium from being robust! And I would prefer this conclusion (which is not the one Selten makes) because, to my mind, mixed strategies in Nash equilibria have a fragility: given that two actions can be played with a positive probability only if they yield the same payoff, shaking the strategies leads to the equality of the two payoffs disappearing and destroys the mixed strategy.

1.3.3 Trembles and weakly dominated strategies

Selten's trembling hand concept is particularly efficient with regard to the elimination of local strategies whose optimality derives only from not being on the equilibrium path. It also often eliminates weakly dominated strategies, but not systematically.

We illustrate these facts on two games.

Look at the game in Figure 5.4.

> ### ➢ STUDENTS' INSERT
>
> (A_1,A_2,A_3) is a SPNE that is not perfect. Suppose the contrary, i.e. that there exists a sequence of Nash equilibria π^k in a test sequence of perturbed games that goes to (A_1,A_2,A_3) and consider player 1 at x_1. By playing C_1, she gets 2, by playing A_1, she gets $\pi_2^k(A_2)(2\pi_3^k(A_3)+1-\pi_3^k(A_3))+1-\pi_2^k(A_2)=\pi_2^k(A_2)\pi_3^k(A_3)+1<2$. Consequently player 1 has to play A_1 with the minimal probability, so π^k can't converge to (A_1,A_2,A_3).

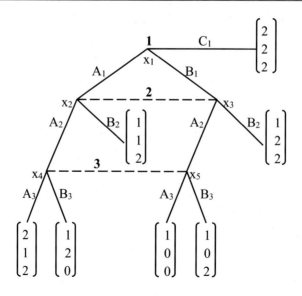

■ Figure 5.4

TREMBLES IN A GAME

Clearly, (A_1, A_2, A_3) is not perfect because A_1 is weakly dominated by C_1; given that, in a perturbed game, all the strategies are played with positive probability, it automatically follows that A_1 necessarily leads to a lower payoff than C_1 (by definition of a weakly dominated strategy) which leads to the result.

Yet, despite this fact, *not all the weakly dominated strategies are eliminated.* When a player is called on to play several times in a game, some weakly dominated strategies can still be played in a perfect equilibrium.

To see why, consider a slight variant of the ascending all pay auction game, with imperfect information on player 1's budget, which we already presented in chapter 1 (Figure 5.5).

Consider the SPNE E where player 1 stops at both X_1 and x_1, plays 3 at X_2, and player 2 overbids at h_1. Clearly, player 1 plays a weakly dominated strategy, in that $(S_1/X_1, */x_1, 3/X_2)$ is weakly dominated by $(1/X_1, */x_1, 3/X_2)$, $*$ being any same action at x_1. Yet E is perfect for the following reason:

Call ε_1^k e_1^k ε_2^k ε_3^k the minimal probabilities on the bids 1 at X_1, 1 at x_1, S_2 at h_1 and S_1 at X_2 and consider the perturbed profile π^k that goes to E, defined by:

$$\pi_{1\times 1}^k(S_1) = 1-\varepsilon_1^k, \ \pi_{1\times 1}^k(S_1) = 1-e_1^k, \ \pi_2^k(2) = 1-\varepsilon_2^k, \ \pi_{1\times 2}^k(3) = 1-\varepsilon_3^k.$$

Clearly player 1 plays optimally at X_2. Player 2, at h_1, gets $2[\rho\varepsilon_1^k + (1-\rho)e_1^k]$ by playing S_2 and he gets $3\rho\varepsilon_1^k \varepsilon_3^k + 3(1-\rho)e_1^k$ by bidding 2. If $e_1^k \geq \rho\varepsilon_1^k(2-3\varepsilon_3^k)/(1-\rho)$, which is always possible when k goes to ∞, player 2 indeed plays his best strategy by bidding 2 with the maximal probability. Player 1 at x_1 also plays her best response because S_1 leads to $(1-\rho)2$, whereas bidding 1 leads to the lower payoff $(1-\rho)(4\varepsilon_2^k + (1-\varepsilon_2^k))$ (when k goes to ∞). We now focus on the interesting decision node X_1. By playing S_1, player 1 gets $\rho 3$. By playing 1, she gets: $\rho(5\varepsilon_2^k + (1-\varepsilon_2^k)(2\varepsilon_3^k + 3(1-\varepsilon_3^k))) = \rho(5\varepsilon_2^k + (1-\varepsilon_2^k)(3-\varepsilon_3^k))$. $\rho 3$ is larger than $\rho(5\varepsilon_2^k + (1-\varepsilon_2^k)(3-\varepsilon_3^k))$ as soon as $\varepsilon_3^k \geq 2\varepsilon_2^k/(1-\varepsilon_2^k)$, which is possible because $2\varepsilon_2^k/(1-\varepsilon_2^k)$ goes to 0 when k goes to ∞. So player 1, for perturbations that check $\varepsilon_3^k \geq 2\varepsilon_2^k/(1-\varepsilon_2^k)$ can play S_1 at X_1 with the maximal probability $1-\varepsilon_1^k$. π^k is a Nash equilibrium of the perturbed game (for $e_1^k \geq \rho\varepsilon_1^k(2-3\varepsilon_3^k)/(1-\rho)$ and $\varepsilon_3^k \geq 2\varepsilon_2^k/(1-\varepsilon_2^k)$) and therefore E is perfect.

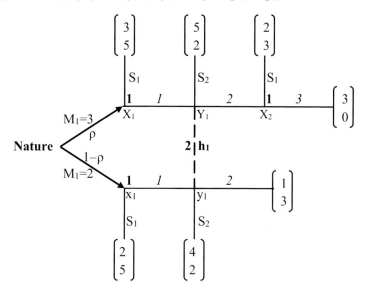

■ **Figure 5.5**

How can this happen? Simply because player 1 trembles at X_1 and X_2. So, when choosing her action at X_1 *she can't be sure that she will play 3 at X_2*! If she fears that she trembles at X_2 much more than player 2 trembles at h_1 (in accordance with the condition on the perturbations), it is better for her to play S_1 at X_1, because, by bidding 1, the loss due to her tremble (payoff 2) may more than compensate the advantage (payoff 5) due to the tremble of player 2!

Well, is that really natural? It is, in the spirit of trembles. *Trembles constrain each player to consider local strategies, one by one, introducing trembles on all, including his future local strategies.* In the spirit of trembles you can't be sure to play a given action tomorrow – which is not a silly assumption – and this changes the impact of weak dominance when behavioural strategies contain more than one local strategy.

2 SELTEN'S CLOSEST RELATIVES, KREPS, WILSON, HARSANY AND MYERSON: SEQUENTIAL EQUILIBRIUM, PERFECT BAYESIAN EQUILIBRIUM AND PROPER EQUILIBRIUM

2.1 Kreps and Wilson's sequential equilibrium: the introduction of beliefs

Selten's well-known concept led to two variants: the sequential equilibrium and the perfect Bayesian equilibrium. Kreps and Wilson's[2] sequential equilibrium, published in 1982, seven years after Selten's concept, is very similar to the perfect equilibrium, but presents a main procedural innovation: *it explicitly introduces beliefs* at the information sets.

2.1.1 Beliefs and strategies: consistency and sequential rationality

Now a player, at an information set, has beliefs about the nodes of the information set; as a consequence, when he is looking for his best local strategy, he takes these beliefs (as well as the strategies played further in the game) into account. So Kreps and Wilson do not test a profile of behavioural strategies, but they test couples (μ,π),[3] where π is a profile of behavioural strategies and μ is a system of beliefs, i.e. a function that goes from each non terminal node x to [0,1], such that, for any information set h, $\sum_{x \in h} \mu(x) = 1$.

The concept of belief is natural in games: generally we play in a given way because we believe that others have played and will play in a given way. But, in Kreps and Wilson's approach, beliefs are a limited concept: they directly stem from perturbations which are similar to those of Selten.

> **Definition 4: sequential equilibrium.** A couple (μ,π) is a sequential equilibrium if it is consistent and sequentially rational.

> **Definition 5:** a couple (μ,π) is **sequentially rational** if, for any player i, at any information set h of this player, the local strategy π_{ih} is player i's best response given the beliefs μ at h, $\mu(h)$, and given the local strategies of π played after the information set h.

TREMBLES IN A GAME

> **Definition 6:** let Ψ be the set of couples (μ^k, π^k) such that $\pi^k \in \widetilde{\Pi}$, where $\widetilde{\Pi}$ is the set of behavioural strategies that assign a positive probability to any action available in the game tree, and μ^k is the system of beliefs associated with π^k via Bayes rule. A couple (μ, π) is **consistent** if it is the limit of a sequence (μ^k, π^k) in Ψ, when k goes to ∞.

In other words, consistency of the couple (μ, π) requires that we find a sequence of completely perturbed strategies π^k, such that, on the one hand, the sequence of perturbed strategies π^k goes to the tested profile π, as for the perfect equilibrium, on the other hand, the beliefs μ are the limit of the beliefs μ^k obtained with the Bayes rule applied to the completely perturbed strategies π^k. *So the beliefs are far from free; they directly derive from the perturbed strategies.*

The change, when we compare the sequential equilibrium and the perfect equilibrium concept, lies in the fact that we no longer require that π^k is a Nash equilibrium of the perturbed game. Sequential rationality just requires that players locally best reply at each information set, working with their local beliefs $\mu(h)$ – and not the beliefs $\mu^k(h)$ – and the π's – and not π^k's – local strategies in the sequel of the game. *The perturbations are only here to restrict the liberty of beliefs, they do not intervene in the calculation of the best responses.* This will explain why some profiles, namely those with weakly dominated strategies, may be sequential even if not perfect, because the other players may not play in π the strategies that weaken the weakly dominated strategies. So we get a first property:

> **Property 6:** some sequential equilibria are not perfect, *but each perfect equilibrium is sequential.*

As a matter of fact, if π^* is perfect, there exists a sequence of Nash equilibria π^k in a test sequence of perturbed games that converges to π^*. π^k belongs to $\widetilde{\Pi}$. Take the sequence (μ^k, π^k), where μ^k is obtained via Bayes's rule applied to π^k. The sequence (μ^k, π^k) goes to (μ, π^*), with μ being the limit of μ^k, when k goes to ∞ (this limit exists given that no denominator is null). So (μ, π^*) is consistent. We know that π^k is a Nash equilibrium in the perturbed game. Given that, at any information set h of any player i, $\mu^k(h)$ is a summarized way to express the behaviours above the information set h, this means that the local strategy π_{ih}^k is the best response given the beliefs μ^k and the local perturbed strategies of π^k further in the game. By continuity of the expected payoff functions, if an inequation is true for $\mu^k(h)$, π_{ih}^k and π^k, when k goes to infinity, it is also true for the limits of $\mu^k(h)$, π_{ih}^k and π^k, i.e. for $\mu(h)$, π_{ih}^* and π^*. It follows that π_{ih}^* is the best local response given $\mu(h)$ and the further local strategies of π^*. So (μ, π^*) is sequentially rational.

> So we have:
>
> {Perfect Equilibria} $\subseteq$ {Sequential Equilibria} $\subseteq$ {SPNE} $\subseteq$ {Nash Equilibria}
>
> Due to property 2, it follows that in each extensive form finite game with perfect recall there exists at least one sequential equilibrium.

2.1.2 Applications of the sequential equilibrium

We now illustrate how the sequential equilibrium works, the meaning of consistent beliefs and the links with the perfect equilibrium.

> ### ➤ STUDENTS' INSERT
>
> *To do so we first come back to the game in Figure 5.4. We call π_E the profile (A_1,A_2,A_3) and we look for a system of beliefs μ, such that (μ,π_E) is a sequential equilibrium.*
>
> *We choose a completely perturbed strategy profile π^k that goes to π_E, we calculate via Bayes's rule the associated beliefs μ^k and we establish μ, the limits of μ^k. We get:*
>
> *$\mu^k(x_1)=\mu(x_1)=1$ (because there is only one node in the information set) and $\mu^k(x_2)=\pi_1^k(A_1)/(\pi_1^k(A_1)+\pi_1^k(B_1))$. Given that $\pi_1^k(A_1)$ goes to 1 and $\pi_1^k(B_1)$ goes to 0, $\mu^k(x_2)$ goes to 1 (and, by complementarity $\mu^k(x_3)$ goes to 0). So $\mu(x_2)=1$ and $\mu(x_3)=0$. We also have $\mu^k(x_4)=\pi_1^k(A_1)\pi_2^k(A_2)/(\pi_1^k(A_1)\pi_2^k(A_2)+\pi_1^k(B_1)\pi_2^k(A_2))=\mu^k(x_2)$. So $\mu^k(x_4)$ goes to 1 (and $\mu^k(x_5)$ goes to 0) and we get $\mu(x_4)=1$ and $\mu(x_5)=0$. Consequently (μ, π_E) where μ is defined as above is consistent.*
>
> *We now prove that (μ, π_E) is sequentially rational. Given that player 3 assigns the belief 1 to the node x_4, he indeed best replies by playing A_3 (2>0). Given that player 2 assigns the belief 1 to the node x_2, and given that player 3 plays A_3, player 2 compares the payoff 1 he obtains with A_2, with the payoff 1 he obtains with B_2, hence A_2 is a best response. Finally, player 1 gets 2 with C_1, 1 with B_1 and 2 with A_1 (given that players 2 and 3 play A_2 and A_3). Hence A_1 is a best reply, and (μ, π_E) is sequentially rational.*
>
> *Given that (μ, π_E) is consistent and sequentially rational, it is a sequential equilibrium.*

This study leads to three remarks.

First we observe that (A_1,A_2,A_3), when associated with the above beliefs, is a sequential equilibrium, whereas it is not a perfect equilibrium. This is due to the observation made above: given that sequential rationality is checked on unperturbed strategies and beliefs, some weakly dominated strategies can be played at a sequential equilibrium but not in a perfect equilibrium.

Second, applying Bayes's rule to π_E leads to $\mu(x_2)=\pi_{E1}(A_1)/(\pi_{E1}(A_1)+\pi_{E1}(B_1))=1/(1+0)=1$, and $\mu(x_4)=\pi_{E1}(A_1)\pi_{E2}(A_2)/(\pi_{E1}(A_1)\pi_{E2}(A_2)+\pi_{E1}(B_1)\pi_{E2}(A_2))=1$. So we have the property:

> **Property 7:** when the Bayes's rule can be applied directly to the tested profile of strategies (i.e. the denominator is not null), it automatically leads to the consistent beliefs.

Third, the calculi linked to the sequential rationality are easier than the ones linked to perfection (there are no epsilons!). So, if both concepts are really close, we are tempted to switch to the sequential concept. But, are the concepts really close? Up to now, we just know that each perfect equilibrium is sequential and that some sequential equilibria that contain a weakly dominated strategy may fail to be perfect (as (A_1, A_2, A_3) in Figure 5.4). But Kreps and Wilson establish a more precise link:

> **Property 8:** Kreps and Wilson call **strict sequential equilibrium** a sequential equilibrium such that any action that does as well as another action in the equilibrium's support (i.e. the set of actions played with positive probability at equilibrium), is also in the equilibrium's support. The **sequential equilibrium** is called **weak** in the complementary case. Kreps and Wilson establish that for almost all payoffs u (i.e. all payoffs u except a set whose closure is of Lesbegue measure 0), every strict sequential equilibrium is perfect.

Especially (A_1, A_2, A_3) is not a strict sequential equilibrium because C_1 leads to the same payoff as A_1 but is not in the equilibrium's support.

So, given that in general strict sequential equilibria and perfect equilibria coincide, given that the calculi are more friendly for the sequential equilibrium concept, we can first focus on the strict sequential equilibria of a game, and then examine more cautiously its weak sequential equilibria.

We now examine other consequences on beliefs implied by the consistency property.

> ### ➤ STUDENTS' INSERT
>
> To this aim, we show that the SPNE $E=(C_1,A_2,B_3)$ is not a sequential equilibrium of the game in Figure 5.4. To do so, we prove that we can't find beliefs μ so that (μ,E) is sequentially rational and consistent.
>
> For (μ,E) to be sequentially rational, μ has to be chosen such that B_3 is player 3's best response given μ. A_3 leads to the payoff $2\mu(x_4)$ and B_3 leads to the payoff $2\mu(x_5)$. So B_3 is player 3's best response if and only if $\mu(x_4) \leq 1/2$.
>
> A_2 has to be player 2's best response given μ and player 3's action B_3. A_2 leads to the payoff $2\mu(x_2)$ whereas B_2 leads to the payoff $\mu(x_2)+2\mu(x_3)$. For A_2 to be the best response, we need $\mu(x_2) \geq 2/3$.
>
> Yet (μ,E), with the beliefs μ defined as above, is not consistent. To see why, consider any completely perturbed profile of strategies π^k that goes to E, and calculate $\mu^k(x_2)$ and $\mu^k(x_4)$.
>
> We already observed, in the previous students' insert, that $\mu^k(x_2)=\pi_1^k(A_1)/(\pi_1^k(A_1)+\pi_1^k(B_1))$ and $\mu^k(x_4)=\pi_1^k(A_1)\pi_2^k(A_2)/(\pi_1^k(A_1)\pi_2^k(A_2)+\pi_1^k(B_1)\pi_2^k(A_2))=\pi_1^k(A_1)/(\pi_1^k(A_1)+\pi_1^k(B_1))=\mu^k(x_2)$, so $\mu(x_2)=\mu(x_4)$. It is therefore impossible to both have $\mu(x_2) \geq 2/3$ and $\mu(x_4) \leq 1/2$.

This example highlights the links between beliefs at some information sets, and beliefs at other information sets ($\mu(x_2)=\mu(x_4)$). Kreps and Wilson illustrate these links with stories of behaviours.

> **Kreps and Wilson's stories of behaviours (or lexicographic consistency)**
>
> Kreps and Wilson interpret $\mu(x_2)=\mu(x_4)$ in the following way. They say that beliefs translate *a sequence of stories that the players have in mind*. The first story is the expected profile of strategies. As long as nobody deviates from this profile, people believe in the first story and beliefs translate it: in other words, beliefs respect Bayes's rule as long as we can apply it (we already showed above that this automatically follows from the consistency property). But now imagine that a player is compelled to observe that somebody deviated from the expected profile. For example, in the game in Figure 5.4, player 2, when called on to play, is compelled to observe that player 1 deviated from (C_1,A_2,B_3). If so he switches to the second story in the

sequence of stories, i.e. a new story that is compatible with the fact that he is called on to play, and that justifies his playing A_2. This new story is necessarily such that player 1 plays A_1 at least 2 times more often than B_1, so $\mu(x_2) \geq 2/3$. But consider now player 3. Player 3, when called on to play, has to observe that player 1 deviated from C_1 but he can still believe that player 2 played A_2. This coincides with Kreps and Wilson's *inertia* in beliefs: as long as you can believe in a story, you have to believe in it. You switch to the following story in the sequence of stories only if you have the proof that the story in which you believe is wrong. But, what is more, *the sequence of stories is the same for player 3 as for player 2*. So player 3 has to believe that player 2 plays A_2 (first story), and that player 1 plays A_1 at least two times more often than B_1 (second story, common to all the players). But then, of course, he assigns at least probability 2/3 to x_4 which prevents him from playing B_3.

So, to summarize, each player builds his beliefs thanks to a sequence of stories of behaviours that favours inertia (each time the observed actions are compatible with the story t, no player is allowed to switch to story t+1), and is the same for all the players in the game. Well, of course, this story of stories is somewhat artificial. Clearly, inertia in beliefs and the fact that the story sequence is the same for each player immediately follow from the consistency requirement: *as long as all the players calculate beliefs with the same perturbed strategies, they respect inertia, and they necessarily have the same sequence of stories*!

But Kreps and Wilson, with their story of stories, *highlight some limits of the perturbations. As long as people calculate with the same perturbations, they can't have different interpretations of the evolution of a game. So, if you want to really build a theory of beliefs evolution, such that people may have different beliefs that may or not converge, like in true life, you surely can't stick to a unique system of perturbations for all the players.*

To explain things differently, we return to the profile (C_1, A_2, B_3) in Figure 5.4. To sustain it, player 2 has to suppose that player 1 deviated two times more to A_1 than to B_1. Player 3, by contrast, has to suppose that player 1 deviated more to B_1 than to A_1. These beliefs are incompatible in game theory, *beliefs have to be right*, so can't be different among players: players can't agree to disagree.[4] Yet, in our case, no player suffers from disagreeing in beliefs: in fact, as long as player 1 plays C_1, players 2 and 3 can't discover that they disagree. So why should they aim to agree on beliefs, when disagreeing is possible and harms nobody? I think that in real life, beliefs among different people may be inconsistent for a very long time, if nobody suffers from the situation: as a matter of fact, only bad payoffs, only the impression that one can be better off by changing things, motivate the evolution of a situation, and hence of beliefs.

We finally apply the sequential equilibrium concept to Selten's horse.

> **STUDENTS' INSERT**
>
> $E_1 = (A_1, B_2, B_3)$ can't be part of a sequential equilibrium (μ, E_1), because no couple (μ, E_1) is sequentially rational. As a matter of fact, necessarily, $\mu(x_2)=1$, so player 2's best response, given that player 3 plays B_3, is A_2 and not B_2.
>
> $E_2 = (B_1, B_2, A_3)$ can be part of a sequential equilibrium. Let us construct adequate beliefs μ. Given that the information sets $\{x_1\}$ and $\{x_2\}$ are singletons, $\mu(x_1) = \mu(x_2) = 1$. For (μ, E_2) to be sequentially

rational, A_3 has to be player 3's best response, so, given that A_3 leads to the payoff $\mu(x_4)$ whereas B_3 leads to the payoff $2\mu(x_3)$, we need $\mu(x_4) \geq 2\mu(x_3)$, i.e. $\mu(x_4) \geq 2/3$. An easy way to get such beliefs in the limit is to build π^k as follows: $\pi_1^k(A_1)=\varepsilon^k$, $\pi_1^k(B_1)=1-\varepsilon^k$, $\pi_2^k(A_2)=3\varepsilon^k$, $\pi_2^k(B_2)=1-3\varepsilon^k$, $\pi_3^k(A_3)=1-\varepsilon^k$, $\pi_3^k(B_3)=\varepsilon^k$, with $\varepsilon^k \to 0$ when $k \to \infty$. It follows that $\mu^k(x_4)=(1-\varepsilon^k)3\varepsilon^k/[\varepsilon^k+(1-\varepsilon^k)3\varepsilon^k]$ which goes to ¾>2/3 (and of course π^k goes to (the probabilities assigned to the actions in) E_2).

And it is obvious that B_1 and B_2 are sequentially rational (1>0), so (μ, E_2) is a sequential equilibrium.

2.2 Harsanyi's perfect Bayesian equilibrium

2.2.1 Definition, first application and links with the sequential equilibrium

Perfect Bayesian equilibria, usually attributed to Harsanyi,[5] are almost redundant with sequential equilibria. They are developed for a special class of games, two player signalling games. We give the perfect Bayesian equilibrium's (PBE) definition and some terminology associated with this concept, before outlining the links between PBE and sequential equilibria.

Let (μ, π_E) be a couple constituted of a system of beliefs μ and a profile of strategies π_E in a signalling game. π_E gives an action (or a probability distribution on actions) for each type of player 1 and for player 2 after each possible sent message by player 1. Given that player 1's information sets are singletons, only player 2's beliefs matter.

Definition 7: Perfect Bayesian equilibrium. (μ, π_E) is a perfect Bayesian equilibrium if it is sequentially rational and if μ is obtained via Bayes's rule applied to π_E, whenever Bayes's rule is applicable.

Definition 8: Separating PBE. A PBE is **separating (or perfectly revealing)** if each type of player 1 plays different messages. It follows that after each message on the equilibrium path, player 2 perfectly knows who sent the message (his beliefs are 0 or 1).

Definition 9: Pooling PBE. A PBE is **pooling (or unrevealing)** if all types of player 1 play the same message – or the same messages with the same probabilities. It follows that after each message on the equilibrium path, player 2 gets no more information than the prior information (so his beliefs are the prior ones).

Definition 10: Semi separating PBE. A PBE is **semi separating (semi revealing)** if it is neither pooling nor separating.

TREMBLES IN A GAME

Let us compare the PBE with the sequential equilibrium. By definition, a PBE is sequentially rational. But it is less demanding than the sequential equilibrium with regard to consistency. Beliefs have to check Bayes's rule whenever it applies, i.e. at each information set on the equilibrium path. But elsewhere *beliefs are free*. The only requirement is that there exist beliefs that justify the actions. Yet, given the special structure of signalling games, we later show that this smoother requirement reveals to be identical to consistency, providing the number of types and the number of messages per type is finite.

We illustrate the definitions and the above property on the following signalling game in Figure 5.6.

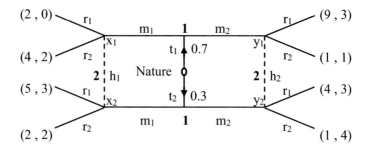

■ **Figure 5.6**

> **STUDENTS' INSERT**

This game has a separating PBE: $((m_2/t_1, m_1/t_2), (r_1/m_1, r_1/m_2), \mu(x_1)=\mu(y_2)=0, \mu(x_2)=\mu(y_1)=1)$

We prove it this way: Given player 2's answers, player 1 plays optimally, when she is of type t_1 (she compares 9, obtained with m_2 followed by r_1, with 2 obtained with m_1 followed by r_1), and when she is of type t_2 (she compares 5, obtained with m_1 followed by r_1, with 4 obtained with m_2 followed by r_1). Player 2 behaves optimally given his beliefs (at h_1 he gets 3 with r_1 and only 2 with r_2, given that his beliefs are concentrated on x_2, at h_2 he gets 3 with r_1 and only 1 with r_2, given that his beliefs are concentrated on y_1). Player 2's beliefs have to satisfy Bayes's rule at each of his information sets, because both sets are on the equilibrium path. Bayes's rule is satisfied: $\mu(x_2)=p(x_2)/(p(x_1)+p(x_2))$ where $p(x)$ means the probability to reach x on the equilibrium path, hence $\mu(x_2)=0.3\times1/(0.7\times0+0.3\times1)=1$. In the same way, we get $\mu(y_1)=p(y_1)/(p(y_1)+p(y_2))=0.7\times1/(0.7\times1+0.3\times0)=1$ And the complementary probabilities are equal to 0. More quickly, given that only t_1 plays m_2, and only t_2 plays m_1, player 2's beliefs are logically focused on x_2 after observing m_1 and on y_1 after observing m_2.

This game has two pooling PBE: $((m_1/t_1, m_1/t_2), (r_2/m_1, r_2/m_2), \mu(x_1)=0.7, \mu(x_2)=0.3, \mu(y_1)=0, \mu(y_2)=1)$ and $((m_2/t_1, m_2/t_2), (r_2/m_1, r_1/m_2), \mu(x_1)=1, \mu(x_2)=0, \mu(y_1)=0.7, \mu(y_2)=0.3)$.

We prove the first equilibrium: Given player 2's answers, player 1 plays optimally, when she is of type t_1 (she compares 4, obtained with m_1 followed by r_2, with 1 obtained with m_2 followed by r_2), and when she is of type t_2 (she compares 2, obtained with m_1 followed by r_2, with 1 obtained with m_2 followed by r_2). Player 2 behaves optimally given his beliefs (at h_1 he gets $3\times0.3=0.9$ with r_1 and 2 with r_2, given that his beliefs are the prior ones, at h_2 he gets 3 with r_1 and 4 with r_2, given that his beliefs are concentrated on y_2). Player 2's beliefs have only to satisfy Bayes's rule at h_1 because m_1 is on the equilibrium path. Bayes's rule is satisfied: $\mu(x_1)=p(x_1)/(p(x_1)+p(x_2))=0.7\times1/(0.7\times1+0.3\times1)=0.7$, consequently $\mu(x_2)=0.3$. By contrast, m_2 is not played at equilibrium, so

> player 2 is free with regard to his beliefs at h_2: he can assign m_2 to t_2. We observe, pedagogically speaking, that the aim at each set which is not on the equilibrium path, is always to choose beliefs that sustain the equilibrium. So assigning m_2 to t_2 leads to play r_2 after m_2, so both types of player 1 only get 1 by playing m_2, which keeps them from deviating from m_1.
>
> This game has also a semi separating PBE: (($\pi_{1t1}(m_2)=p=3/14$, m_2/t_2), (r_2/m_1, $\pi_{2h2}(r_1)=q=3/8$), $\mu(x_1)=1$, $\mu(x_2)=0$, $\mu(y_1)=1/3$, $\mu(y_2)=2/3$)[6]
>
> We prove it this way. Given player 2's answers, player 1 plays optimally: t_1 compares 4, obtained with m_1 followed by r_2 with $9q+1-q=4$, obtained with m_2, so she can play both messages with any probability at equilibrium, and t_2 compares 2 obtained with m_1 followed by r_2 with $4q+1-q=17/8$ obtained with m_2, so m_2 is his optimal action. Player 2 behaves optimally given his beliefs (at h_1 he gets 0 with r_1 and 2 with r_2, given that his beliefs are focused on x_1, at h_2 he gets 3 with r_1 and $\mu(y_1)+4\mu(y_2)=3$, so he can play both responses with any probability). Player 2's beliefs have to satisfy Bayes's rule at h_1 and h_2, given that both messages are played with positive probability. This is indeed the case: $\mu(x_1)=1$, given that only t_1 plays m_1, and $\mu(y_1)=0.7p/(0.7p+0.3)=1/3$ for $p=3/14$).

We observe that all these equilibria are sequential equilibria. This is obvious for the separating equilibrium and for the semi separating one (putting an ε on the non played messages does not change anything on the limit probabilities obtained with Bayes's rule). With regard to the (studied) pooling equilibrium, adding perturbations to player 1's strategies does not change the limit probabilities after observing m_1. But they have an impact after observing the non equilibrium message m_2. Yet, given that we want $\mu(y_2)=1$, it is enough to fix $\pi_{1t1}(m_2)=\varepsilon^{k2}$ and $\pi_{1t2}(m_2)=\varepsilon^k$, with ε^k going to 0 when k goes to ∞. We immediately get $\mu^k(y_2)=0.3\varepsilon^k/(0.3\varepsilon^k+0.7\varepsilon^{k2})$ that goes to 1 when k goes to ∞.

So, more generally, a possible difference with the sequential equilibrium, if there is any, may only appear after a non equilibrium message. But there will be no difference between PBE and sequential equilibria when the signalling game is a finite game.

> **Property 9:** In a finite signalling game, each PBE is a sequential equilibrium.

Proof

We just have to show that, for any distribution of beliefs $\mu(h)$, we can find a system of perturbations such that $\mu^k(h)\to\mu(h)$. Given that slight perturbations have almost no impact after messages that belong to the equilibrium path, it is enough to focus on the beliefs after messages out to the equilibrium path. Let m be such a message. Call D the set of types d, such that $\mu(d/m)=0$, and F the types f such that $\mu(f/m)=c_f\neq 0$. We fix, for any d of D, $\pi_{1d}^k(m)=\varepsilon^{k2}$ whereas, for any f or F, we fix $\pi_{1f}^k(m)=K_f\varepsilon^k$ where K_f is a constant. It automatically follows that $\mu^k(d/m)\to 0=\mu(d/m)$ when k goes to ∞. And, for any f of F, $\mu^k(f/m)=\rho_f K_f \varepsilon^k/\Sigma_{j\in F} \rho_j K_j\varepsilon^k=\rho_f K_f/\Sigma_{j\in F} \rho_j K_j$ an expression we want equal to c_f. This system of Card F equations and Card F unknowns (K_j) is easy to solve because it is enough to fix $K_j=c_j/\rho_j$ for any j in F.

It automatically follows that, in finite signalling games, given that the PBE is the easiest concept to apply (among perfect, sequential and perfect Bayesian equilibrium), people usually work with the PBE concept!

TREMBLES IN A GAME

2.2.2 French plea-bargaining and perfect Bayesian equilibria

We briefly return to the French plea-bargaining game in Figure 5.7 (this game is detailed in chapter 1, sections 3.2 and 3.3): this game has two PBE that differ according to the value of ρ.

> **STUDENTS' INSERT**
>
> Regardless of ρ, there is a pooling PBE, such that both types of player 1 plead guilty, because the jury convicts at trial with probability 1, given that he can assign the action "don't plead guilty" to the guilty defendant (with probability 1).
>
> If ρ≤½ there is a second pooling equilibrium where both types of defendant go to trial and are released at trial; this is due to the fact that the jury's beliefs at the trial are the prior ones, and 1−ρ≥ρ, which justifies the action "release".
>
> If ρ≥½ there exists a semi separating equilibrium defined as follows.
>
> The guilty defendant goes to trial with probability (1−ρ)/ρ, pleads guilty with probability (2ρ−1)/ρ, the innocent defendant only goes to trial, the jury releases with probability u, and convicts with probability 1−u. Let us justify this last equilibrium. The guilty defendant is indifferent between going to trial and pleading guilty because she gets u with both actions, so she can go to trial with probability (1−ρ)/ρ. The innocent defendant prefers going to trial to pleading guilty because he gets v by pleading guilty and u>v when going to trial. The jury's beliefs are ½ on both types (because [ρ(1−ρ)/ρ]/[ρ(1−ρ)/ρ+1−ρ]=½) so he is indifferent between releasing and convicting, so he can release with probability u.

We observe a good property of this game: regardless of the values of the parameters, there always exists a PBE such as the innocent defendant goes to trial with probability 1. So the possibility of pleading guilty does not lead an innocent defendant to necessarily plead guilty, even if her record is bad (large value of ρ) and the prosecutor is generous. *Consequently game theory leads to an ethical*

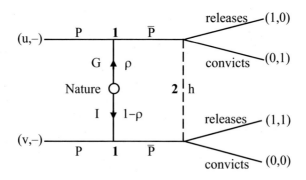

■ **Figure 5.7**

result which was not necessarily intuitive. As a matter of fact, many people fear that the plea bargaining procedure leads innocent defendants to plead guilty: game theory does not say that this will never happen, but it says that innocent defendants can rationally go to trial, i.e. that there always exist equilibria where their best action consists of going to trial.

2.3 Any perturbations or only a selection of some of them? Myerson's proper equilibrium

Selten has introduced robustness, but in quite a loose way. We just have to find *one sequence of perturbations* that sustains the tested equilibrium for this equilibrium to be perfect. But, what about the other perturbations? If we really tremble by choosing our actions, we can tremble in any way. So shouldn't we be more demanding? Shouldn't we require that an equilibrium resists any sequence of perturbations?

Well, in many games, there exist equilibria such that, for *any sequence* of perturbations going to 0, there is a sequence of perturbed equilibria that goes to the tested equilibrium. For example, in the game of Figure 5.2, (B_1,A_2) resists any introduced perturbations. In the game in Figure 5.6, the separating PBE$((m_2/t_1,m_1/t_2),(r_1/m_1,r_1/m_2)$, $\mu(x_1)=\mu(y_2)=0$, $\mu(x_2)=\mu(y_1)=1)$ also resists any introduced perturbations. More generally, as soon as the equilibrium path crosses all the information sets of the game and if each player gets, at each of his information sets, more by playing the equilibrium action than another one, then ε perturbations can neither affect the beliefs, nor affect the play of the players.

Completely mixed behavioural Nash equilibria also, by construction, resist any sequence of perturbations in Selten's spirit.

But requiring that a profile of strategies resists any sequence of small perturbations is too restrictive. Do we really want such a requirement? If we come back to Selten's horse game (Figure 5.1), none of the two examined equilibria resists any sequence of perturbations (recall that we had restrictions on the perturbations to get (B_1,B_2,A_3) perfect). In Selten's horse game, there is a third perfect equilibrium profile[7] that is neither completely mixed nor reaches all the information sets: $(B_1,B_2,\pi_3(A_3)=¾)$. Let us call ε_1^k player 1's minimal probability on A_1, ε_2^k player 2's minimal probability on A_2, and ε_3^k player 3's minimal probability on both actions (no impact). Clearly, for player 3 to be able to play A_3 and B_3 with probability ¾ and ¼, one requires that $2\varepsilon_1^k=(1-\varepsilon_1^k)\pi_2^k(A_2)$. Hence $\pi_2^k(A_2)=2\varepsilon_1^k/(1-\varepsilon_1^k)$. Given that player 2 is indifferent between B_2 and A_2 (as long as in the perturbed game player 3 plays as in the unperturbed game, which is possible because ¼ and ¾ are higher than the minimal probability), player 2 can indeed play A_2 with this probability as soon as it is higher than the minimal probability ε_2^k. But if $\varepsilon_2^k>2\varepsilon_1^k/(1-\varepsilon_1^k)$, this is again not possible.

So clearly, in Selten's horse game, no equilibrium is the limit of a sequence of equilibria for any chosen perturbations.

Well, what should we deduce?

First, to my mind, we should deduce that it is stupid to reject an equilibrium because it does not resist all profiles of small perturbations. Requiring that an equilibrium resists all perturbations means that it has, at unreached information sets, *to resist any beliefs*. This simply does not make sense. Actions that are optimal for any beliefs do not often exist. Things would be too easy…and completely uninteresting.

Second, if it seems inappropriate to require that an equilibrium resists any sequence of perturbations, *this does not mean that we should accept any sequence of perturbations to justify a profile*. Once again, Selten's requirement is a minimal one: a profile that is not perfect does not resist any system of perturbations, so it has no robustness and should be rejected. But this does not mean that a perfect equilibrium is a convincing equilibrium.

TREMBLES IN A GAME

Selten did not want to colour his perturbations: perturbations are just trembles. But this is not the case for many authors after him. Myerson (1978),[8] with his proper equilibrium concept, was one of the first game theorists who wanted to restrict the set of acceptable perturbations. Roughly speaking, Myerson says that is easier to tremble toward an action that can lead to a high payoff, than toward an action that can only lead to a low one. In other words, Myerson links the perturbations to the payoffs.

Let us present his concept. Myerson works on normal form games and so departs from Selten and Kreps and Wilson's extensive form game approach. As explained in chapter 1 (section 2.3) this is significant, given that perturbations in the normal form correlate the actions of a same player at different decision rounds. This however doesn't happen in our studied example.

> **Definition 11: Proper equilibrium**
> A completely mixed profile in a game with N players $(p_1^k,\ldots,p_N^k)$ is an ε^k proper equilibrium if: $\forall i \in \mathcal{N}$, $\forall s_i$ and $s_i' \in S_i$, if $U_i(s_i/p_{-ik}) < U_i(s_i'/p_{-ik})$ then $p_{ik}(s_i) \leq \varepsilon_k p_{ik}(s_i')$.
>
> A proper equilibrium is a profile that is a limit of a sequence of ε^k proper equilibria, when k goes to ∞ and ε^k goes to 0.

We apply this concept to the game in Figure 5.2 (where each player has only one information set, consequently only plays once, which impedes the correlation problem).

The normal form of this game is given in Matrix 5.1.

		Player 2 A_2	B_2	C_2
	A_1	(2,0)	(1,0)	(0,2)
Player 1	B_1	(4,2)	(2,0)	(1,0)
	C_1	(2,3)	(2,3)	(2,3)

Matrix 5.1

In this game (C_1,C_2) is not a proper equilibrium.

As a matter of fact, in an ε^k proper equilibrium, given that A_1 is strictly dominated by B_1, we have $p_1^k(A_1) \leq \varepsilon^k p_1^k(B_1)$. Given that C_2 leads to the payoff $2p_1^k(A_1)+3p_1^k(C_1)$, whereas A_2 leads to the higher payoff $2p_1^k(B_1)+3p_1^k(C_1)$, we have $p_2^k(C_2) \leq \varepsilon^k p_2^k(A_2)$. Given that B_2 leads to a lower payoff than C_2 (because B_2 is weakly dominated by C_2), we also have $p_2^k(B_2) \leq \varepsilon^k p_2^k(C_2) \leq \varepsilon^{k2} p_2^k(A_2)$.

It follows that B_1 leads to the payoff: $4p_2^k(A_2)+2p_2^k(B_2)+p_2^k(C_2)=4(1-p_2^k(B_2)-p_2^k(C_2))+2p_2^k(B_2)+p_2^k(C_2)=4-2p_2^k(B_2)-3p_2^k(C_2)$ that goes to 4, given that $p_2^k(B_2) \leq \varepsilon^{k2}$ and $p_2^k(C_2) \leq \varepsilon^k$. Given that C_1 leads to the payoff 2, we get:

$p_1^k(C_1) \leq \varepsilon^k p_1^k(B_1)$.

So an ε^k proper equilibrium checks:

$p_1^k(A_1) \leq \varepsilon^k p_1^k(B_1)$ and $p_1^k(C_1) \leq \varepsilon^k p_1^k(B_1)$.
$p_2^k(C_2) \leq \varepsilon_k p_2^k(A_2)$ and $p_2^k(B_2) \leq \varepsilon^{k2} p_2^k(A_2)$.

TREMBLES IN A GAME

It derives that the probability assigned to B_1 and to A_2 necessarily goes to 1 in the limit. Hence only (B_1, A_2) is a proper equilibrium.

Myerson establishes the existence of proper equilibria in normal form games. Indirectly, Myerson works on the set of possible beliefs. For example, the restrictions on the strategies he requires in the game in Matrix 5.1 induce that player 2's beliefs in the game in Figure 5.2 are such that $\mu^k(x_2)$ goes to 0, hence $\mu(x_2)=0$ and $\mu(x_3)=1$ (because $p_1^k(A_1) \leq \varepsilon^k p_1^k(B_1)$); player 2 is not allowed to believe that player 1 plays a strictly dominated strategy. Myerson's work is pursued by many authors, namely those working on forward induction (see chapter 6).

3 SHAKING OF THE GAME STRUCTURE

Up to now, we introduced perturbations on strategies (Selten, Kreps and Wilson, Myerson), so we shook the game only a little. But we can perturb a game in other ways. We may, for example, suppose that the players, with a small probability, have in mind a wrong game, so that, with an ε probability, they behave optimally but in a game that is not the game they really play.

In this section we work with this kind of perturbation. We show that a small change in the structure of the game may have no impact or a small impact, but can also completely disrupt the way of playing. This section is also the occasion to work with a carrier pigeon and his enemies, e-mails; i.e. to present some new stimulating story games, Myerson's carrier pigeon game and Rubinstein's electronic mail game. And we will also come back to the ascending all pay auction with unknown budgets and negotiation games with an unknown number of periods of negotiation.

3.1 When only a strong structural change matters, Myerson's carrier pigeon game

Myerson's[9] carrier pigeon game goes as follows. Consider the following signalling game, in Figure 5.8a.

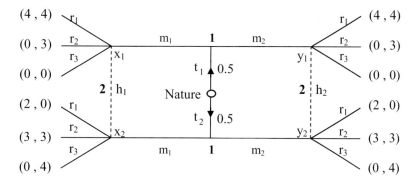

■ **Figure 5.8a**

In this game, player 1 says "I am of type t_1" – message m_1 – or she says nothing – message m_2. m_1 and m_2 have no economic meaning – they are cheap talk – so the payoffs after m_1 and m_2 are the same. This game has a specificity: *player 1 of type t_1 (we later simply say t_1) would be very happy to be recognized by player 2, but player 1 of type t_2 (we later simply say t_2) would not.* As a matter of fact, if player 2 recognizes t_1, he plays r_1 (regardless of the sent message) and both players get

227

TREMBLES IN A GAME

the highest payoff 4, and if player 2 recognizes t_2, he plays r_3 (regardless of the sent message), and t_2 gets the worse payoff 0. It follows that there is no separating PBE. As a matter of fact, if t_1 plays m_i and t_2 plays m_j $j \neq i$, then player 2 plays r_1 after m_i and r_3 after m_j, which leads t_2 to (copy) switch to m_i. There only exist pooling equilibria that give a 0 payoff to t_1 (because player 2 plays r_2 after the played message).

To bypass this difficulty, *Myerson introduces a carrier pigeon*. The bird's work goes as follows: each time player 1 sends the message m_1, the carrier pigeon either brings the message to player 2, with probability $(1-\rho)$, or drops it, with probability ρ. We therefore get the game in Figure 5.8b.

Clearly when ρ, who measures the structural change, goes to 0, the game structure is not strategically modified: behaviours will not change, so the game is stable to small modifications. The new game becomes interesting when it becomes beneficial for t_1 to send m_1 and for t_2 to send m_2. And this separating equilibrium is possible if player 2 is able to switch from r_3 to r_2 after m_2, that is to say if $3\mu(y_1)+3\mu(y_3) \geq 4\mu(y_3)$, i.e. if $\mu(y_1) \geq 1/4$, i.e. $0.5\rho/(0.5\rho+0.5) \geq 1/4$, so $\rho \geq 1/3$. In that case, t_1 chooses to send m_1 whereas t_2 chooses m_2 (t_1 gets $(1-\rho)4+\rho 0$, which is better than 0, t_2 gets 3, which is better than the payoff she gets with m_1). Everybody is better off – except for t_2 who gets the same payoff; even player 2 gets more: $(0.5)[4(1-\rho)+\rho 3]+0.5 \times 3 = 3+0.5(1-\rho) > 3$.

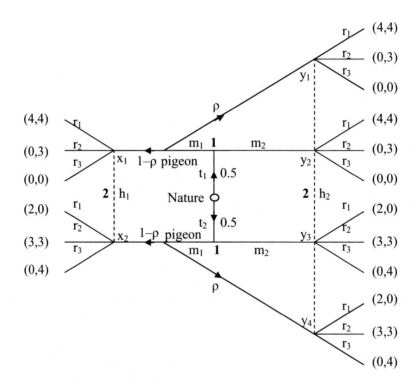

■ **Figure 5.8b**

So it is because the carrier pigeon drops the message enough (he drops the message at least 1/3 of the time) that everybody is happy. So Myerson proved that *bringing errors into a game may help all the players*, by changing the incentives to mimic other players. Observe that *the best carrier pigeon is one that drops the message with probability 1/3*!

3.2 Small change, a lack of upper hemicontinuity in equilibrium strategies

We return to the ascending all pay auction in Figure 5.5. We showed earlier that (($S_1/X_1,S_1/x_1$, $3/X_2$),$2/h_1$) is a perfect equilibrium. Recall also that if $M_1=M_2=2$, and if this fact is common knowledge, the game reduces to the lower branch, and the only SPNE is such that player 1 stops the game and player 2 bids 2 if called on to play.

The game in Figure 5.5 has another sequential equilibrium E=(($1/X_1$, p=$2\rho/(1-\rho)$, $3/X_2$), q=$2/3$) where p is player 1's probability to bid 1 at x_1 and q is player 2's probability to bid 2 at h_1.

As a matter of fact the associated beliefs are immediate (given by Bayes's rule applied to E): $\mu(X_1)=\mu(x_1)=\mu(X_2)=1$, $\mu(Y_1)=\rho.1/(\rho 1.+(1-\rho)2\rho/(1-\rho))=1/3$, $\mu(y_1)=2/3$. Sequential rationality is also satisfied. Player 1 behaves optimally at X_2 (obvious), at X_1 (because $3<5/3+3\times 2/3$), and at x_1, because $2=4/3+1\times 2/3$). Player 2, given his beliefs, gets 2 by stopping and $0\times 1/3+3\times 2/3=2$ by bidding 2, hence his answer is optimal.

And this equilibrium holds for any positive ρ, checking $2\rho/(1-\rho)\leq 1$, i.e. $\rho\leq 1/3$.

Well, we clearly observe here a lack of upper hemicontinuity. Even for ρ going to 0, player 2, at his information set, stops the game with probability 1/3, whereas, when $\rho=0$, he never stops the game. *So a very slight change in the structure of the game may induce strong changes in behaviours.*

But, despite the fact that player 2 has a different behaviour when $\rho=0$ and when ρ goes to 0, in this example, the equilibrium path is not really modified. As a matter of fact, player 1, when $M_1=2$, bids 1 only with probability $2\rho/(1-\rho)$ that goes to 0 when ρ goes to 0. It follows that player 2's change in behaviour has only a very small impact on the equilibrium path, and so on the equilibrium payoffs. So we get a lack of upper hemicontinuity in strategies, but not in payoffs.

But things can be much different, even in this game (see next section).

3.3 Large changes, a lack of upper hemicontinuity in equilibrium payoffs

First we switch to negotiation, but with imperfect information on the number of steps of negotiation. Remember that an offer from player i, i=1,2, is an amount offered to player 1. Recall (see chapter 4, section 2.2.2) that, if there are two periods of negotiation, the only SPNE is such that player 1 offers 2 in the first step and accepts all the offers in period 2, player 2 only accepts the offers 1 and 2 in period 1, and offers 1 in period 2. This equilibrium is profitable for player 2, given that he gets 8 at equilibrium, player 1 getting only 2.

If there are three periods of negotiation, the only SPNE is such that player 1 offers 8 in the first period, only accepts the offers higher or equal to 7 in period 2, and offers 7 in period 3; player 2 only accepts the offers lower or equal to 8 in period 1, offers 7 in period 2, and accepts all the offers in period 3. This equilibrium is profitable for player 1, given that she gets 8 at equilibrium, player 2 getting only 2.

But now suppose that player 2 ignores if there are two periods or three periods of negotiation. More precisely, he is quite sure that there are three periods (probability 1–ρ), but he thinks that there is a small probability ρ that there are only two periods of negotiation.

The synthetic extensive form of this game is given in Figure 5.9. The prior distribution is common knowledge of both players. And player 1 knows if there are two or three periods, this knowledge also being common knowledge.

TREMBLES IN A GAME

This game is rather complicated and the reader can switch to the conclusion, to see how a small uncertainty can induce a strong change in payoffs, if s/he doesn't want to participate in the construction of the equilibrium.

We first work on the structure of the game in that we can eliminate many strategies that will not be played at equilibrium. In fact player 2, in period 2, only offers 1 or 7, because, if there are three periods, player 1 will never accept less than 7 (because she can get 7 by waiting for the third period), and she accepts 1 if there are only two periods. To put it more precisely, for player 2, in the second period, the offer 8 is weakly dominated by the offer 7 and any offer from 2 to 6 is weakly dominated by the offer 1.

So, with just two offers p and p' for player 1 in period 1, we get the game tree in Figure 5.10.

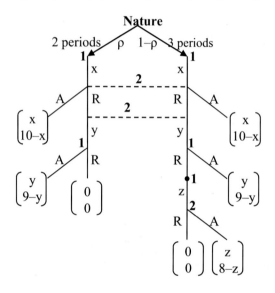

■ **Figure 5.9**

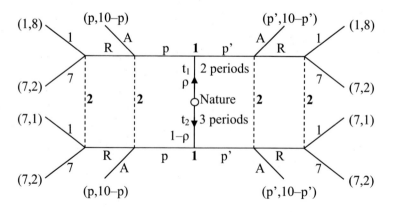

■ **Figure 5.10**

We explain the players' payoffs after player 2's offers 1 and 7: if player 2 offers 1 and if there are only two periods, player 1 accepts 1, hence the payoffs (1,8). If there are three periods, player 1 refuses and offers 7 in period 3, an offer that will be accepted by player 2, hence the payoffs (7,1). If player 2 offers 7, the offer is always accepted, hence the payoffs (7,2).

Now recall that player 1, if there are three periods, never plays less than 7 in period 1 given that she is sure to get 7 (by waiting for period 3 and offering 7 in this period). And player 2 gets 2 for sure by offering 7 in period 2, so he will never accept an offer higher than 8 in period 1, whether the game lasts two or three periods. It follows that player 1, if there are three periods, only plays 7 or 8 in period 1. So player 2 understands any other offer as a signal of a two period game, so refuses any offer different from 7 or 8 and higher than 2. In the sequel, we assume that he accepts 2 (in that he cannot get more than 8). It follows that, at equilibrium, player 1, if there are two periods of play, at most plays 3 offers, 2, 7 or 8, in period 1.

We say that player 1 is of type t_1, respectively t_2, if there are two, respectively three periods of negotiation. We'll construct a sequential equilibrium where both types of player 1 offer 7 in period 1. To help this equilibrium to emerge, we suppose (if possible) that player 2 is right assigning offers 2 and 8 to t_1. So player 2 accepts 2 and refuses 8 (and any offer different from 7). Consequently player 1 is not induced to deviate from 7 if 7 is accepted.

We represent the part of the game following the offer 7 in Figure 5.11.

First we study player 2's behaviour at h_2 after refusing offer 7 in period 1. Given that both t_1 and t_2 play 7 in period 1, Bayes's rule leads to $\mu(y_1)=\rho$ and $\mu(y_2)=1-\rho$. Player 2 compares 2 to $8\mu(y_1)+\mu(y_2)$ $=8\rho+(1-\rho)$ and offers 7 if $\rho\leq1/7$. So, at h_1, after observing the offer 7, he accepts it because $3>2$.

If $\rho>1/7$, he plays 1, and therefore, after observing 7, he compares 3 with $8\mu(x_1)+\mu(x_2)=8\rho+(1-\rho)$, and he accepts 7 if $\rho\leq2/7$.

It remains to justify that player 2 refuses any offer from 3 to 8, different from 7, in period 1. We simply suppose that, in the perturbed game, t_1 plays any offer different from 7 with probability ε^k (and the offer 7 with the complementary probability), and t_2 plays any offer different from 7 with probability ε^{k2}. It follows that player 2 assigns an offer different from 7 to t_1 with probability $\rho\varepsilon^k/(\rho\varepsilon^k+(1-\rho)\varepsilon^{k2})=\rho/(\rho+(1-\rho)\varepsilon^k)$ which goes to 1. So player 2 refuses any offer different from 1 and 2 (after this refusal, he proposes 1 in period 2, which will be accepted by t_1 and refused by t_2). And these perturbations have no impact on $\mu(x_1)$ and $\mu(y_1)$ (and perturbing player 1's actions in period 2 and 3 and player 2's actions in period 1 and 2 has no impact on player 1's information sets because they are singletons).

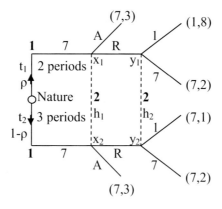

Figure 5.11

TREMBLES IN A GAME

> Conclusion: for $\rho \leq 2/7$, we get a sequential equilibrium (with an offer 7 immediately accepted) which is interesting for player 1 of type t_1 (she gets 7 instead of 2 with complete information), less favourable to player 1 of type t_2 (she gets 7 instead of 8 with complete information), and in between for player 2. He gets 3, which is better than 2, the payoff he gets when facing a player 1 of type t_2, but worse than 8, the payoff he gets when facing a player 1 of type t_1.

So, for very smooth changes – ρ goes to 0 – we get a completely new equilibrium, with new equilibrium payoffs, that does not exist for $\rho=0$: in this new equilibrium player 1 no longer gets 8 but only 7, and player 2 gets 3 instead of 2.

We mention that the equilibrium for $\rho=0$ still exists when ρ is small but different from 0 (player 1 plays 8 in period 1 regardless of type, and her offer is accepted). So the introduction of a very small ρ does not suppress the old equilibrium, but it allows a new equilibrium to emerge. This illustrates the lack of upper hemicontinuity in equilibria and equilibrium payoffs.

In some games, this lack of upper continuity may coincide with the emergence of new equilibria that are completely opposite to the old one. Let us show it with the ascending all pay auction in Figure 5.5.

Imagine that the starting game is the one in which player 2 is sure that $M_1=3$. In this game, the only SPNE is such that player 1 bids 1, player 2 stops the game, and player 1 bids 3 if player 2 bids 2.

Yet we showed in section 1.3.3. that in the game of Figure 5.5 there exists a sequential equilibrium where player 1 stops at both X_1 and x_1, plays 3 at X_2, and player 2 overbids at h_1, whatever the value of ρ. When ρ goes to 1 (hence $1-\rho$ goes to 0), we are arbitrarily close to the starting game but we have a diametrically opposed equilibrium. In the game without uncertainty, player 1 goes on and player 2 stops, whereas, in the game with very limited uncertainty, player 1 can stop the game and player 2 can go on.

3.4 Very large changes, a lack of lower hemicontinuity in equilibrium behaviours and payoffs, Rubinstein's e-mail game

In some cases, small structural changes have even stronger effects. They simply lead to the suppression of some equilibria of the game before the change. That's the case in the e-mail game, *another story game*.

The e-mail game is the modern version of Rubinstein's mail game,[10] an old story with a military battle, hills, forests, and so on. Let us present it.

The starting game is the following one. There are two states of nature, x and y, and only player 1 is informed on the state of nature. Player 2 has a prior distribution on these states, that assigns a probability ρ to x, and $1-\rho$ to y, with $\rho>0.5$. This prior distribution is common knowledge of both players. Now suppose, in this starting game, that player 1, if the state is y, sends a message m to player 2, and that player 2 observes this message before the play. After that, both players play a simultaneous game: each player chooses one among two actions, X or Y.

The payoffs depend on the state of nature, as illustrated in the Matrices 5.2:

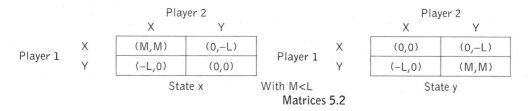

Matrices 5.2

The starting game is represented in extensive form in Figure 5.12.

Given that player 2 gets the message m when the state is y, all happens as if he knows the state of nature. So ((X/x,Y/y), (X/x,Y/y)) is an obvious SPNE (perfect) equilibrium of the game.

Now bring in the small uncertainty introduced by Rubinstein. Suppose that both players have a computer; when they get a message in their mail box, this box automatically replies to the message received. Suppose however that the computers have some bugs so that each message or reply is only sent with probability 1–ε. With probability ε, the process stops, with ε very small, close to 0.

So this game is very close to the previous one. For example, when m is sent, player 2 gets the message with probability 1–ε, almost one. More, given the infinite replying process, after a certain number of replies, both players know y, they know that the opponent knows y, that the opponent knows that they know y..., but there is still a very small uncertainty, given that they are never sure that the last reply they sent arrived at its destination. And this is enough to impede the equilibrium of the starting game from being an equilibrium of the new game. *Moreover, the only equilibrium in this new game is such that, at any information set of the game, each player plays X!*

Rubinstein proved this fact by focusing on the information sets of both players.

We write t=j, when player 1 has j messages or replies on her computer, T=j, when player 2 has j messages or replies on his computer. For example, t=1 means that player 1 sent the message but got no reply. So player 1 does not know if the message never came to player 2 (T=0), or if it came to player 2 (T=1), but the reply of the message did not come back. So the two events (t=1,T=0) and (t=1,T=1) are in the same information set of player 1. More generally, the set of information sets of player 1 can be written:

$$H_1=\{\{t=0\},\{(t=1,T=0),(t=1,T=1)\},\{(t=2,T=1), (t=2,T=2)\}\ldots\{(t=j,T=j-1),(t=j,T=j)\}\ldots\}$$

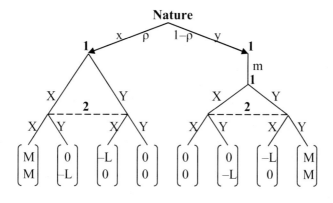

Figure 5.12

TREMBLES IN A GAME

For player 2, when T=0, this means that either player 1 did not send a message (t=0), or she sent it but the message never arrived (t=1). When T=1, this means that player 1 sent the message and did not get a reply (t=1), or that she sent the message, got a reply but her own reply did not arrive to player 2's computer (t=2). So, more generally, the set of information sets of player 2 can be written:

$$H_2 = \{\{(T=0,t=0),(T=0,t=1)\}, \{(T=1,t=1), (T=1,t=2)\} \ldots \{(T=j,t=j),(T=j,t=j+1)\} \ldots\}$$

These information sets also appear clearly on the (partial) extensive form of the game given in Figure 5.13.

What is interesting in this game is its resolution. Whether we work on the information sets or on the game tree, we solve the game in a very precise order, by starting from the beginning and switching from one information set to the next. (This approach is not unlike the one used to solve the prisoners' disks game in chapter 2, section 4).

Given that we would like to come close to the equilibrium ((X/x,Y/y), (X/x,Y/y)), we first fix that player 1 plays X when the state is x. And now we turn to player 2 at the information set that follows this action, i.e $\{\{T=0,t=0),(T=0,t=1)\}$, h_1 in the game tree.

If player 2 plays X, he gets M if he is at the first part of his information set (T=0,t=0) – i.e. the node x_1 in the game tree – and he gets 0 at the second part (T=0,t=1), i.e the nodes x_3 and x_4 in the game tree (player 1's action at node y_1 has no impact).

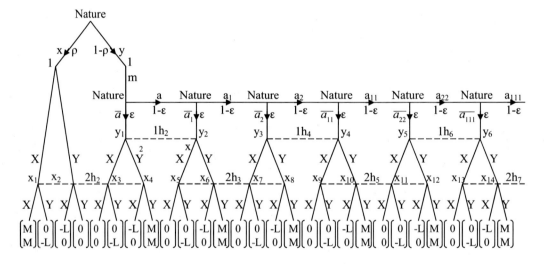

Figure 5.13 A steam engine for an e-mail game, strange, isn't it?

a=the message arrives to player 2, $\bar{a}$ it doesn't arrive
a_1=the reply arrives to player 1, $\bar{a}_1$ it doesn't arrive
a_2=the reply to a_1 arrives to player 2, $\bar{a}_2$ it doesn't arrive
a_{11}=the reply to a_2 arrives to player 1, $\bar{a}_{11}$ it doesn't arrive
a_{22}=the reply to a_{11} arrives to player 2, $\bar{a}_{22}$ it doesn't arrive
a_{111}=the reply to a_{22} arrives to player 1, $\bar{a}_{111}$ it doesn't arrive
and so on.

If he plays Y, he gets –L at the first part, and at best M at the second part. Yet the probability of reaching the first part is ρ, whereas the probability of reaching the second part is (1–ρ)ε. So the sequential beliefs are ρ/(ρ+(1–ρ)ε) and (1–ρ)ε/(ρ+(1–ρ)ε)), and (–Lρ+M(1–ρ)ε)/(ρ+(1–ρ)ε)<0 given that ε is close to 0. Consequently player 2 plays X at the set {(T=0,t=0),(T=0,t=1)} – at information set h_1 on the game tree.

Now we turn to player 1's information set {(t=1,T=0),(t=1,T=1)}–h_2 on the game tree. She knows that when t=1 and T=0 (node y_1 on the game tree), player 2 plays X. The probability of reaching y_1; i.e. (t=1,T=0), is (1–ρ)(ε), whereas the probability of reaching y_2, (t=1,T=1), is (1–ρ)(1–ε)ε. So the consistent sequential beliefs at h_2 are (1–ρ)ε/((1–ρ)(1–ε)ε+(1–ρ)ε) and (1–ρ)(1–ε)ε/ ((1–ρ)(1–ε)ε+(1–ρ)ε); i.e. 1/(2–ε) and (1–ε)/(2–ε). So the probability of reaching y_1, (t=1, T=0), is higher than ½.

Consequently, if player 1 plays X, she gets 0 (regardless of player 2's action at h_3), whereas if she plays Y, she gets at best (–L)/(2–ε)+M(1–ε)/(2–ε)<0. So she plays X.

Now we switch again to player 2, at information set {(T=1,t=1), (T=1,t=2)}, h_3 on the game tree. The probability of being at (T=1,t=1), i.e. at x_5 in the game tree (given that player 1 plays X at h_2), is (1–ρ)(1–ε)ε, whereas the probability of being at (T=1,t=2), i.e. at the nodes x_7 or x_8, is (1–ρ)(1–ε) ×(1–ε)ε. It follows that the consistent beliefs are 1/(2–ε)(>1/2) on x_5 (T=1,t=1), and only (1–ε)/(2–ε) at (T=1,t=2) (dispatched on x_7 and x_8, on an (up to now) unknown way). So, by playing X, player 2 gets 0 (regardless of being at x_7 or x_8), by playing Y, he gets at best (–L)/(2–ε)+M(1–ε)/(2–ε)<0. So he plays X.

And so on. The next step leads us to h_4 on the game tree, i.e. {(t=2,T=1), (t=2,T=2)}. The belief of assigning to y_3, i.e. (t=2,T=1), is again 1/(2–ε) higher than ½, which justifies to play X at this information set (because X leads again to a payoff 0, whereas Y leads again at best to (–L)/(2–ε)+M(1–ε)/(2–ε)<0).

So the iterative resolution of the game, whether on the extensive form (switching from h_1 to h_2, then to h_3, then to h_4, then to h_5, then to h_6…), or on the graph below, clearly leads to establish progressively that the game has only one sequential equilibrium where each player plays X at each of his/her information set.

H_1={{t=0},{(t=1,T=0),(t=1,T=1)},{(t=2,T=1), (t=2,T=2)}…{(t=j,T=j–1), (t=j,T=j)}…}

H_2={{(T=0,t=0),(T=0,t=1)},{(T=1,t=1), (T=1,t=2)}………………….{(T=j,t=j),(T=j,t=j+1)}..}

Well, this funny game shows that very small changes in a game tree (an infinite small break in the common knowledge of the state y is sufficient) can lead to the disappearance of an equilibrium. So in this game, a small change induces a lack of lower hemicontinuity in the behaviours and payoffs.

4 PERTURBING PAYOFFS AND BEST RESPONSES IN A GIVEN WAY

Myerson was among the first game theorists who modified Selten's uncoloured approach of perturbations by linking perturbations to payoffs. Many theorists now agree with such an approach. Replicator equations (which have been developed independently and before the game theory work on perturbations) belong to this trend of literature. Quantal responses directly perturb payoffs. We will not go deeply into these trends of literature; we just give some insights in order to show strong differences with Selten's approach. But before that, we turn to a more neutral way to perturb payoffs and strategies, proposed by Jackson et al. (2012).

4.1 Trembling*-hand perfection

We discussed earlier that is does not make sense to require that an equilibrium resists all kind of perturbations. Some authors however, among them Jackson et al. (2012),[11] require this kind of robustness, yet they smoothen the requirement by shaking also the payoffs.

Let us outline their approach.

They first define the notion of ex ante or interim ε equilibrium, that are identical in a game where each player has no private information (chosen by Nature). Given that we will focus on Selten's horse (without moves of Nature), both definitions are identical. We adapt here their definition to our notations:

> **Definition 12:** a profile π is an (interim) ε equilibrium of the game, for some $\varepsilon \geq 0$, if for all i from 1 to N: $\forall \pi_i' \in \Pi_i : u_i(\pi) \geq u_i(\pi_i', \pi_{-i}) - \varepsilon$.

They define a new trembling hand concept in the following way:

> **Definition 13:** a profile π is a trembling*-hand perfect equilibrium if for any sequence of perturbed games, there exists a sequence of interim ε – equilibria that converge to π.

So, on the one hand, Jackson et al. require that an equilibrium resists any sequence of perturbations going to 0, but, on the other hand, they allow best responses to not exactly be best responses, in that a small loss in payoffs is allowed.

They justify these "almost" best answers by the fact that payoffs might be slightly mis-specified. And they are right: *why should we tremble by playing and have no doubt about payoffs?*

Yet Jackson et al. observe that the trembling*-hand perfect concept makes no selection between the Nash equilibria.

To see why, return once more to Selten's horse and to the SPNE (A_1, B_2, B_3). This equilibrium is not perfect because, as soon as player 2 is called on to play (because of a small tremble of player 1) player 2 can only play A_2 because player 3 plays B_3 with a probability going to 1. But it is a trembling*-hand perfect equilibrium. As a matter of fact, if player 1 plays B_1 with probability ε_1^k and player 3 plays B_3 with probability $1-\varepsilon_3^k$, player 2 gets ε_1^k by playing B_2 and $4\varepsilon_1^k(1-\varepsilon_3^k)$ by playing A_2. So by playing B_2 he only loses $4\varepsilon_1^k(1-\varepsilon_3^k)-\varepsilon_1^k=3\varepsilon_1^k-4\varepsilon_1^k\varepsilon_3^k$. So just define ε^k by $\varepsilon^k= 3\varepsilon_1^k-4\varepsilon_1^k\varepsilon_3^k$. It follows that $(\pi_1^k(A_1)=1-\varepsilon_1^k, \pi_2^k(B_2)=1-\varepsilon_2^k, \pi_3^k(B_3)=1-\varepsilon_3^k)$ is an interim ε^k equilibrium, and given that π^k goes to (A_1, B_2, B_3) when k goes to ∞, (A_1, B_2, B_3) is a trembling*-hand perfect equilibrium.

Clearly, the trembling*-hand perfect equilibrium concept leaves player 2's behaviour without impact. Player 2 is only called on to play with probability ε_1^k, which multiplies all his payoffs by ε_1^k. Consequently, automatically, any action can only make him lose a very low amount of payoffs. So,

TREMBLES IN A GAME

provided we add a small amount to the payoff associated with the action B_2, B_2 automatically becomes a best response. In fact, the impact of actions out of the equilibrium path is neutralized, so we come back to the Nash spirit, where actions out of the equilibrium path are automatically optimal.

But this reasoning, to my mind, does not really make sense when we directly work on an extensive form game. With regard to Selten's horse, all happens as if Jackson et al. work with its normal form given in Matrices 5.3, or with an extensive form game out of this normal form, given in Figure 5.14:

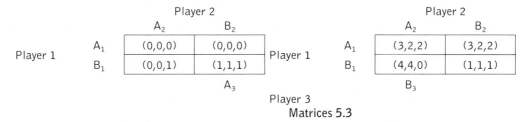

Matrices 5.3

In this game, it shocks nobody to study player 2's payoffs by assuming that player 1 plays B_1 with probability ε_1^k and taking this probability into account in the payoffs.

But clearly, Selten's horse game has nothing to do with the game in Figure 5.14, despite the fact that (A_1, B_2, B_3) is not perfect in both games. The best way to see that is to turn to the sequential equilibrium concept: (A_1, B_2, B_3) is not a sequential equilibrium in the game in Figure 5.1 (Selten's horse game) but it is a sequential equilibrium in the game in Figure 5.14. As a matter of fact, in Selten's horse, when working with the sequential rationality, you put a probability 1 *on the unique decision node* of player 2, so that he compares the payoffs 1 and 4 and plays A_2. By contrast, in the game of Figure 5.14, player 2, by consistency, puts probability 1 on x_1 and 0 on x_2, and so he can rationally play B_2, which leads to the same payoff than A_2.

Kreps and Wilson, by changing the actions above an information set into beliefs at this information set, give these actions much more weight in terms of payoffs. Whereas, in Selten's horse game (Figure 5.1), with Selten's approach, player 2 compares ε_1^k to $4\varepsilon_1^k (1-\varepsilon_3^k)$, i.e. a very low payoff to another very low one – so that the introduction of small changes in payoffs can change the optimal action – with Kreps and Wilson's approach, player 2 compares 1 to 4; by doing so, adding small payoffs doesn't change the conclusion that he has to play A_2. By working with Kreps and Wilson's beliefs and sequential rationality, in Selten's horse, a profile that resists all perturbations doesn't exist, even if we add low additional payoffs. Notably, even (B_1, B_2, A_3) does

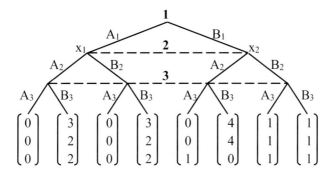

Figure 5.14

not resist all perturbations: player 3's beliefs can switch from {0 on x_3, 1 on x_4} to {1 on x_3, 0 on x_4} and, of course, for beliefs higher than 1/3 (on x_3), for example ½, player 3 compares (½×2) to (½×1) and prefers B_3, regardless of the small additional payoffs we introduce. So we cannot systematically counterbalance small strategy perturbations by small payoff perturbations and, as a consequence, in many games, an equilibrium that resists all kind of small strategy perturbations will not exist, even if we add ε perturbations in payoffs.

4.2 Quantal responses

Quantal responses are often presented as a way to perturb best responses (see Turocy 2005),[12] but the philosophy of quantal responses is quite original.

In this section we present the philosophy of this approach as it was proposed by its authors, McKelvey and Palfrey (1995).[13]

Quantal responses rest on a *perturbation of the payoffs linked to the strategies*. The idea is that players play best responses, but that they imperfectly know the payoffs assigned to different profile of strategies.

McKelvey and Palfrey work on normal form games. Call p the mixed profile played by the players. Player i, when he plays the pure strategy s_{ij}, gets the payoff:

$$\underline{u}_i(s_{ij}, p_{-i}) = U_i(s_{ij}, p_{-i}) + \varepsilon_{ij}$$

where U_i are the expected payoffs in the matrix game and ε_{ij} is a random error. So the true payoff of a player depends on a random error vector. More precisely, an error ε_{ij} is assigned to each strategy s_{ij} of player i, and it affects the payoff linked to strategy s_{ij} (by being added to it). So each player i has an error vector $\varepsilon_i = (\varepsilon_{i1} \ldots \varepsilon_{iCardSi})$ that is distributed according to a joint distribution with density function $f_i(\varepsilon_i)$, that checks $E(\varepsilon_i) = 0$. And the profile of distributions of errors $(f_1 \ldots f_N)$ is called *admissible* if each $f_i(.)$ checks $E(\varepsilon_i) = 0$, i from 1 to N.

McKelvey and Palfrey require that players play best responses in this *statistically* perturbed environment. They define R_{ij}, the set of perturbations such that s_{ij} is player i's best response:

$$R_{ij}(\underline{u}_i(s_{ij}, p_{-i})) = \{\varepsilon_{ij} / \forall k \neq j \text{ from 1 to Card } S_i, \ U_i(s_{ij}, p_{-i}) + \varepsilon_{ij} \geq U_i(s_{ik}, p_{-i}) + \varepsilon_{ik}\}.$$

According to McKelvey and Palfrey, player i selects s_{ij} each time it does better than any other pure strategy of her strategy set S_i, i.e. each time ε_{ij} is in $R_{ij}(\underline{u}_i(s_{ij}, p_{-i}))$. So he plays s_{ij} with the probability

$$\int_{R_{ij}(u_i(s_{ij}, p_{-i}))} f_i(\varepsilon_{ij}) d\varepsilon_{ij}$$

. This probability is written $\underline{p}_{ij}(\underline{u}_i(s_{ij}, p_{-i}))$. *This approach is quite original and it has some links with the best reply matching behaviour we will see in chapter 6: in both approaches, you don't look for the strategy that is the best response in the mean context, but you play a strategy as many times as it is a best reply.*

A quantal response equilibrium is a profile of mixed strategies that check the above property.

TREMBLES IN A GAME

Definition 14: let G be a game in normal form and f an admissible profile of distributions of errors. A **quantal response equilibrium** is any mixed profile p such that

$\forall i \in \mathcal{N}$, $\forall j$ from 1 to Card S_i, $p_{ij} = \underline{p}_{ij}(\underline{u}_i(s_{ij}, p_{-i}))$

McKelvey and Palfrey prove that, if, for each player i, f_i *is a log Weibull distribution,* then a quantal response equilibrium checks:

$\forall i \in \mathcal{N}$, $\forall j$ from 1 to Card S_i, $p_{ij} = e^{\lambda U_i(s_{ij}, p_{-i})} \Big/ \sum_{k=1}^{CardS_i} e^{\lambda U_k(s_{ik}, p_{-i})}$

where λ is a parameter that can be chosen from 0 to ∞.

In that case, the quantal response equilibrium is called a **logit equilibrium**.

Clearly, when $\lambda=0$, each strategy is chosen with the same probability 1/CardS$_i$ (this strategy is called the centroid), but when λ goes to ∞, then the logit equilibrium is a Nash equilibrium.

Let us show this last result.

Suppose that $U_i(s_{ij}, p_{-i}) > U_i(s_{ik}, p_{-i})$.

It follows:

$$p_{ik} / p_i = e^{\lambda U_i(s_{ik}, p_{-i})} / e^{\lambda U_i(s_{ij}, p_{-i})} = e^{\lambda(U_i(s_{ik}, p_{-i}) - U_i(s_{ij}, p_{-i}))}$$

The last term goes to 0 when λ goes to ∞. It derives that $p_{ik}=0$.

It follows that the *logit equilibrium correspondence*, that associates to each λ its logit equilibrium, goes from the centroid to a Nash equilibrium, when λ goes from 0 to ∞.

Let us outline this correspondence for Selten's horse (studied as a normal form game). For each λ, we have to solve:

$\ln(p_{A1}/p_{B1}) = \lambda(3p_{B3} - p_{B2} - (1-p_{B2})4p_{B3})$ and $p_{A1} + p_{B1} = 1$
$\ln(p_{A2}/p_{B2}) = \lambda(p_{B1}4p_{B3} - p_{B1})$ and $p_{A2} + p_{B2} = 1$
$\ln(p_{A3}/p_{B3}) = \lambda(p_{B1}(1-p_{B2}) - 2(1-p_{B1}))$ and $p_{A3} + p_{B3} = 1$

so

$\ln(1-p_{B1}) - \ln(p_{B1}) = \lambda(-p_{B3} - p_{B2} + 4p_{B2}p_{B3})$
$\ln(1-p_{B2}) - \ln(p_{B2}) = \lambda p_{B1}(4p_{B3} - 1)$
$\ln(1-p_{B3}) - \ln(p_{B3}) = \lambda(3p_{B1} - p_{B1}p_{B2} - 2)$

where p_{Ai} and p_{Bi} are player i's probabilities of playing A_i and B_i, i=1 to 3.

For each λ we solve this system of three equations and three unknowns (see Turocy (2005) for an illustration). When λ goes to ∞, the system is solved for p_{B1} and p_{B2} going to 1, p_{B3} going to 0, provided $3p_{B1} - p_{B1}p_{B2} - 2 > 0$, which, for $p_{B1} = 1 - \varepsilon_1$ and $p_{B2} = 1 - \varepsilon_2$ *induces the same restrictions on* ε_1 *and* ε_2 *than Selten's perfection criterion.*

McKelvey and Palfrey more generally show that the logit equilibrium operates a selection between the Nash equilibria, which can be different from the one in the perfect equilibrium selection.

239

4.3 Replicator equations

Quantal responses are linked to replicator equations (see Turocy (2005) who presents quantal responses as a mix of replicator equations and best responses. What are replicator equations?

Well, these are very old equations, called Lotka and Volterra's[14] equations, first established in biology. They are used in evolutionary game theory where they represent an adaptive way of behaviour.

The idea is quite simple. Suppose that, at the beginning, each possible strategy is played with positive probability. Some strategies, at a given time t, do better than the mean, some do worse. The strategies that do better will be played, at time t+dt, with a higher probability, and those which do worse will be played with a lower probability. This directly derives from biology: the species which are better off than the mean will replicate (reproduce) more often than those which are worse off.

In this paragraph we do not aim to go into evolutionary game theory, we just aim to propose some links and some main differences between a replicator approach and Selten's trembling hand concept, by studying Selten's horse with replicator equations.

We start in a context where each action is played with positive probability, as in Selten's perturbed environment approach. Then we fix $\pi_{A1}(t)=\varepsilon_1(t)$, $\pi_{A1}^{\cdot}(t)=d\pi_{A1}(t)/dt=\dot{\varepsilon}_1(t)$, $\pi_{A2}(t)=\varepsilon_2(t)$, $\pi_{A2}^{\cdot}(t)=d\pi_{A2}(t)/dt=\dot{\varepsilon}_2(t)$, $\pi_{B3}(t)=\varepsilon_3(t)$, $\pi_{B3}^{\cdot}(t)=d\pi_{B3}(t)/dt=\dot{\varepsilon}_3(t)$. So we spontaneously define a sequence of minimal probabilities ε_1^k, ε_2^k, ε_3^k, where k designs time and goes to ∞.

The difference between Selten's approach and the replicator equations is that we cannot choose the sequence of minimal probabilities as we want. The sequence $\varepsilon_i(t)$, by contrast to a free sequence ε_i^k, i from 1 to 3, depends on the payoffs obtained with A_1, A_2 and B_3 at time t. To put it more precisely, the replication rate of a strategy is linked to the difference between the payoff it yields and the mean payoff of all available strategies. We get:

$\dot{\varepsilon}_1(t)/\varepsilon_1(t)=(U_{1A1}(t)-U_1(t))$
$\dot{\varepsilon}_2(t)/\varepsilon_2(t)=(U_{2A2}(t)-U_2(t))$
$\dot{\varepsilon}_3(t)/\varepsilon_3(t)=(U_{3B3}(t)-U_3(t))$

Where $U_i(t)$ is player i's mean payoff at time t and $U_{iX}(t)$ is player i' payoff by playing action X at time t.

It follows:

$\dot{\varepsilon}_1(t)=\varepsilon_1(t)(U_{1A1}(t)-\varepsilon_1(t)U_{1A1}(t)-(1-\varepsilon_1(t))U_{1B1}(t))$
    $=\varepsilon_1(t)(1-\varepsilon_1(t))(U_{1A1}(t)-U_{1B1}(t))$
$\dot{\varepsilon}_2(t)=\varepsilon_2(t)(1-\varepsilon_2(t))(U_{2A2}(t)-U_{2B2}(t))$
$\dot{\varepsilon}_3(t)=\varepsilon_3(t)(1-\varepsilon_3(t))(U_{3B3}(t)-U_{3A3}(t))$

These equations, in Selten's horse game, become:

$\dot{\varepsilon}_1(t)=\varepsilon_1(t)(1-\varepsilon_1(t))(3\varepsilon_3(t)-(1-\varepsilon_2(t))-4\varepsilon_2(t)\varepsilon_3(t))$
    $=\varepsilon_1(t)(1-\varepsilon_1(t))(3\varepsilon_3(t)-1+\varepsilon_2(t)-4\varepsilon_2(t)\varepsilon_3(t))$
$\dot{\varepsilon}_2(t)=\varepsilon_2(t)(1-\varepsilon_2(t))(4(1-\varepsilon_1(t))\varepsilon_3(t)-(1-\varepsilon_1(t)))$
    $=\varepsilon_2(t)(1-\varepsilon_2(t))(1-\varepsilon_1(t))(4\varepsilon_3(t)-1)$
$\dot{\varepsilon}_3(t)=\varepsilon_3(t)(1-\varepsilon_3(t))(2\varepsilon_1(t)-(1-\varepsilon_1(t))\varepsilon_2(t))$
    $=\varepsilon_3(t)(1-\varepsilon_3(t))(2\varepsilon_1(t)-\varepsilon_2(t)+\varepsilon_1(t)\varepsilon_2(t))$

Clearly, if $\varepsilon_i(t)$, i from 1 to 3, are close to 0, then $\dot{\varepsilon}_1(t)<0$ and $\dot{\varepsilon}_2(t)<0$, and, if $2\varepsilon_1(t)-\varepsilon_2(t)+\varepsilon_1(t)\varepsilon_2(t)<0$, i.e. $\varepsilon_2(t)>2\varepsilon_1(t)/(1-\varepsilon_1(t))$, then $\dot{\varepsilon}_3(t)<0$ too and we get three sequences $\varepsilon_1(t)$, $\varepsilon_2(t)$ and $\varepsilon_3(t)$ that go to 0 and can represent an original, payoff motivated, way to choose Selten's sequences of minimal probabilities. And given that $\varepsilon_1(t)$, $\varepsilon_2(t)$ and $\varepsilon_3(t)$ are the probabilities to play A_1, A_2 and B_3 the sequence of perturbed strategies automatically converges to the equilibrium (B_1,B_2,A_3).

Yet, is it possible to construct $\varepsilon_1(t)$, $\varepsilon_2(t)$ and $\varepsilon_3(t)$ so that $\varepsilon_2(t)>2\varepsilon_1(t)/(1-\varepsilon_1(t))$ regardless of t? Let us suppose that, at a given time $\underline{t}$ (for example the beginning of the game), we choose $\varepsilon_1(\underline{t})$, $\varepsilon_2(\underline{t})$ so that $\varepsilon_2(\underline{t})>2\varepsilon_1(\underline{t})/(1-\varepsilon_1(\underline{t}))$ (hence $\dot{\varepsilon}_1(\underline{t})<0$, $\dot{\varepsilon}_2(\underline{t})<0$ and $\dot{\varepsilon}_3(\underline{t})<0$). The only thing we have to check is that $\dot{\varepsilon}_3(t)$ keeps negative for any $t>\underline{t}$ (given the -1 in the other two equations, $\dot{\varepsilon}_1(t)$ and $\dot{\varepsilon}_2(t)$ keep negative without difficulty).

Observe that if $\dot{\varepsilon}_2(t)/\varepsilon_2(t)>\dot{\varepsilon}_1(t)/\varepsilon_1(t)$ then $(\varepsilon_2(t+dt)-\varepsilon_2(t))/(dt\varepsilon_2(t))>(\varepsilon_1(t+dt)-\varepsilon_1(t))/(dt\varepsilon_1(t))$, i.e. $\varepsilon_2(t+dt)>\varepsilon_1(t+dt)\varepsilon_2(t)/\varepsilon_1(t)$

It follows that $2\varepsilon_1(t+dt)-\varepsilon_2(t+dt)+\varepsilon_1(t+dt)\varepsilon_2(t+dt)=2\varepsilon_1(t+dt)-\varepsilon_2(t+dt)(1-\varepsilon_1(t+dt))$
$<2\varepsilon_1(t+dt)-\varepsilon_1(t+dt)(1-\varepsilon_1(t+dt))\varepsilon_2(t)/\varepsilon_1(t)$
$<\varepsilon_1(t+dt)(2\varepsilon_1(t)-\varepsilon_2(t)(1-\varepsilon_1(t+dt)))/\varepsilon_1(t))$

So if $(2\varepsilon_1(t)-\varepsilon_2(t)+\varepsilon_1(t)\varepsilon_2(t))<0$, then $2\varepsilon_1(t+dt)-\varepsilon_2(t+dt)+\varepsilon_1(t+dt)\varepsilon_2(t+dt)$ is also lower than 0, because $\varepsilon_1(t+dt)<\varepsilon_1(t)$.

It follows that $\dot{\varepsilon}_3(t)$, if negative for $\underline{t}$, keeps negative for $t>\underline{t}$.

So it remains to show that $\dot{\varepsilon}_2(t)/\varepsilon_2(t)>\dot{\varepsilon}_1(t)/\varepsilon_1(t)$ for any $t>\underline{t}$.
$\dot{\varepsilon}_2(t)/\varepsilon_2(t)=(1-\varepsilon_2(t))(1-\varepsilon_1(t))(4\varepsilon_3(t)-1)$ and
$\dot{\varepsilon}_1(t)/\varepsilon_1(t)=(1-\varepsilon_1(t))(3\varepsilon_3(t)-1+\varepsilon_2(t)-4\varepsilon_2(t)\varepsilon_3(t)$
$(1-\varepsilon_2(t))(4\varepsilon_3(t)-1)=4\varepsilon_3(t)-1-4\varepsilon_2(t)\varepsilon_3(t)+\varepsilon_2(t)>3\varepsilon_3(t)-1+\varepsilon_2(t)-4\varepsilon_2(t)\varepsilon_3(t)$
because $4\varepsilon_3(t)>3\varepsilon_3(t)$. So $\dot{\varepsilon}_2(t)/\varepsilon_2(t)>\dot{\varepsilon}_1(t)/\varepsilon_1(t)$ for any $t>\underline{t}$.

So we have found three payoff motivated sequences $\varepsilon_1(t)$, $\varepsilon_2(t)$ and $\varepsilon_3(t)$ that go to 0 and that can represent three possible Selten's sequences that justify the perfectness of the equilibrium (B_1,B_2,A_3).

This is a good result. But will that always be the case? The response is negative. The proximity between the replicator equations and Selten's approach rests on the payoffs in Selten's horse. In some ways, player 2 is more induced to deviate to A than player 1 (he may switch from the payoff 1 to the payoff 4, whereas player 1 can only switch from 1 to 3). Things would be different if we replaced the payoffs (3,2,2) by (6,2,2) ((B_1,B_2,A_3) is still a Perfect equilibrium)

In that case, $\dot{\varepsilon}_1(t)/\varepsilon_1(t)$ becomes $(1-\varepsilon_1(t)(6\varepsilon_3(t)-1+\varepsilon_2(t)-4\varepsilon_2(t)\varepsilon_3(t))$, $\dot{\varepsilon}_2(t)/\varepsilon_2(t)$ being unchanged, so that $\dot{\varepsilon}_2(t)/\varepsilon_2(t)<\dot{\varepsilon}_1(t)/\varepsilon_1(t)$, because $6\varepsilon_3(t)>4\varepsilon_3(t)$. So $\varepsilon_3(t)$ grows in time after a while and our equations do no longer converge to (B_1,B_2,A_3).

Here we are close to a literature that comes nearer to forward induction (see chapter 6), in that we look for which payoffs a player rather prefers deviating than another. We will later see that (B_1,B_2,A_3) checks Kohlberg's forward induction requirement in Selten's horse, but that it no longer checks this requirement when we switch from the payoffs (3,2,2) to the payoffs (6,2,2).

CONCLUSION

Errors (trembles) are quite natural in games and it is not surprising to see them in strategies, payoffs and game trees. Sometimes, errors are unstructured, like in Selten, Kreps and Wilson, Jackson et al. Sometimes they have to fit minimal properties (proper equilibrium, quantal response equilibria, replicator equations).

Structured trembles will have a role to play in the next chapter, where they will be linked to the potential payoffs they may allow to be achieved (forward induction). So they will lose their nature of "error". Trembles will become intentional deviations in order to achieve higher payoffs.

NOTES

1. Selten, R., 1975. Reexamination of the perfectness concept for equilibrium points in extensive games, *International Journal of Game Theory*, 4, 25–55.
2. Kreps, D. and Wilson, R., 1982. Sequential Equilibria, *Econometrica*, 50, 863–894.
3. Kreps and Wilson call (μ,π) a "pair" or an "assessment"; I prefer couple because of the mathematical writing.
4. For a well-known paper on this topic, see Aumann, R. J., 1976. Agreeing to disagree, *The Annals of Statistics*, 4 (6), 1236–1239.
5. Harsanyi, J.C., 1967–1968. Games with incomplete information played by Bayesian players, I–III *Management Science*, 14 and 15, 159–182, 320–334, 486–502.
6. For signalling games, we often switch to more intuitive notations: we write $\pi_1(m/t)$ for $\pi_{1t}(m)$, $\pi_2(r/h)$ for $\pi_{2h}(r)$ or even $\pi_2(r/m)$ when h is the information set reached when m is played. We simply write m/t when $\pi_{1t}(m)=1$ and r/m when $\pi_2(r/m)=1$. We also write $\mu(t/m)$ for the belief to be in front of type t when observing m and we sometimes write $u_1(t,m,,r)$, respectively $u_2(t,m,r)$, for player 1's utility when he is of type t, plays m followed by r, respectively for player 2's utility when he faces a player 1 of type t who plays m, followed by r.
7. We should add the equilibria where players 1 and 2 play B and player 3 plays A_3 with a probability between 3/4 and 1, which are also perfect but do not change our conclusion.
8. Myerson, R.B. 1978. Refinements of the Nash equilibrium concept, *International Journal of Game Theory*, 7(2), 73–80.
9. Myerson, R.B., 1991. *Game Theory*, Cambridge, Massachusetts, Harvard University Press.
10. Rubinstein, A., 1989. The electronic mail game: strategic behaviour under almost common knowledge, *American Economic Review*, 79 (3), 385–391.
11. Jackson, M.O., Rodriguez-Barraquer, T. and Tan, X., 2012. Epsilon-equilibria of perturbed games, *Games and Economic Behaviour*, 75, 198–216.
12. Turocy, T.L., 2005. A dynamic homotopy interpretation of the logistic quantal response equilibrium correspondence, *Games and Economic Behaviour*, 51, 243–263.
13. McKelvey, R.D. and Palfrey T.R., 1995. Quantal response equilibria for normal form games, *Games and Economic Behaviour*, 10, 6–38.
14. Volterra, V., 1931. *Leçon sur la théorie mathématique de la lutte pour la vie*, Paris, Gauthier-Villars.

Chapter 5

Exercises

♣ EXERCISE 1 PERFECT AND PROPER EQUILIBRIUM

Consider the extensive form game in Figure E5.1.

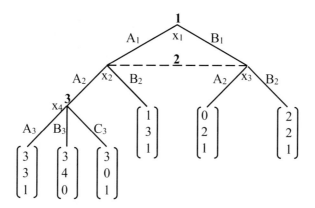

Figure E5.1

Questions
1. Show that (A_1, A_2, A_3) is a subgame perfect Nash equilibrium. Then show that (A_1, A_2, A_3) is a perfect equilibrium.
2. Is (A_1, A_2, A_3) still a perfect equilibrium if you eliminate the weakly dominated strategy B_3?
3. Explain why (A_1, A_2, A_3) is not a proper equilibrium.

♣ EXERCISE 2 PERFECT EQUILIBRIUM WITH WEAKLY DOMINATED STRATEGIES

Consider the extensive form game in Figure E5.2.

EXERCISES

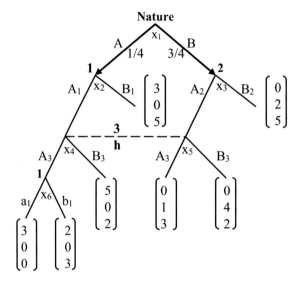

Figure E5.2

Question

Check that $((B_1/x_2, a_1/x_6), B_2/x_3, A_3/h)$ is a SPNE and that the strategy $(B_1/x_2, a_1/x_6)$ is weakly dominated. Then show that $((B_1/x_2, a_1/x_6), B_2/x_3, A_3/h\ )$ is a perfect equilibrium. Comment.
Hint: Nature is not a true player, don't perturb its strategy.

♣♣ EXERCISE 3 PERFECT EQUILIBRIUM AND INCOMPATIBLE PERTURBATIONS

Consider the extensive form game in Figure E5.3.

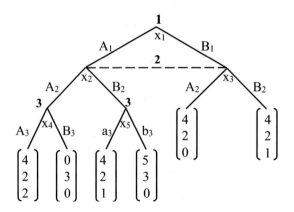

Figure E5.3

Question

Check that $(A_1, A_2, (A_3/x_4, a_3/x_5))$ is a SPNE, but that it is not a perfect equilibrium
Hint: try to show that it is perfect and discover what impedes this perfection.

EXERCISES

♣ EXERCISE 4 STRICT SEQUENTIAL EQUILIUBRIUM

Consider again the extensive form game in Figure E5.3.

Question
Show that $(A_1, A_2, (A_3/x_4, a_3/x_5))$ is a sequential equilibrium, but that it is not a strict sequential equilibrium. Comment.

♣ EXERCISE 5 SEQUENTIAL EQUILIBRIUM, INERTIA IN THE PLAYERS' BELIEFS

Consider the extensive form game in Figure E5.4.

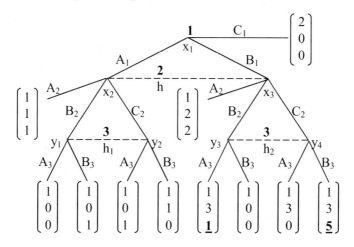

Figure E5.4

Questions
1. Check that $(C_1, A_2, (A_3/h_1, A_3/h_2))$ is a SPNE, then show that it is possible to build beliefs which form sequentially rational couples with $(C_1, A_2, (A_3/h_1, A_3/h_2))$.
2. Show that none of these couples is consistent, so that it is not possible to build a sequential equilibrium with $(C_1, A_2, (A_3/h_1, A_3/h_2))$.
3. What happens if we interchange the bold underlined payoffs 1 and 5? Build the sequential equilibrium if possible.

EXERCISES

♣♣ EXERCISE 6 CONSTRUCT THE SET OF SEQUENTIAL EQUILIBRIA

Consider the extensive form game in Figure E5.5.

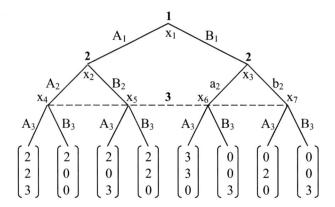

Figure E5.5

Questions

1. Check that $(A_1, (A_2/x_2, b_2/x_3), A_3)$ is a SPNE, then show that if player 1 plays A_1, consistency requires that player 3 assign a belief 0 to the nodes x_6 and x_7. Deduce player 3's sequentially rational strategy. Then deduce player 2's sequentially rational action at x_3. Is it possible to build a sequential equilibrium in which player 1 plays A_1?
2. Is it possible to build a sequential equilibrium in which player 1 plays B_1?
3. Find the unique sequential equilibrium of this game.

♣♣ EXERCISE 7 PERFECT EQUILIBRIUM, WHY THE NORMAL FORM IS INADEQUATE, A LINK TO THE TREMBLING*-HAND EQUILIBRIUM

Consider the extensive form game in Figure E5.6.

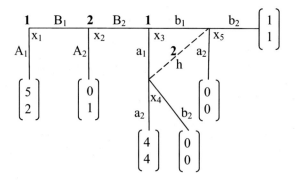

Figure E5.6

Questions
1. Check that (A_1a_1, A_2a_2) is not a perfect equilibrium; then write the game in normal form. Call this game G and try to adapt the perfect equilibrium concept to the normal form game. What happens? Comment.
2. Link the analysis to Jackson et al.'s conception of error.

♣ EXERCISE 8 PERFECT BAYESIAN EQUILIBRIUM

Consider the signalling game in Figure E5.7.

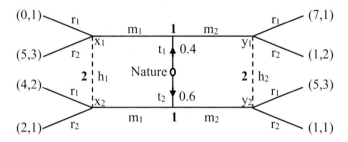

Figure E5.7

Questions
1. Find all the pure strategies perfect Bayesian equilibria (PBE) of this game.
2. Show that there exists a semi separating PBE such that player 1 of type t_1 only plays m_2, whereas player 1 of type t_2 plays both m_1 and m_2 with positive probability.

♣♣ EXERCISE 9 PERFECT BAYESIAN EQUILIBRIUM, A COMPLEX SEMI SEPARATING EQUILIBRIUM

Consider the signalling game in Figure E5.8, out of Umbhauer.[1]

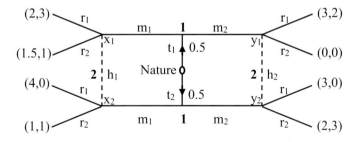

Figure E5.8

Question
Find all the PBE.

EXERCISES

♣♣ EXERCISE 10 LEMON CARS, EXPERIENCE GOOD MARKET WITH TWO QUALITIES

Consider the experience good market, defined in chapter 1 (Exercise 7).
We here work with the values $h_1=40$, $H_1=60$, $h_2=100$, $H_2=120$, $\rho_1=\rho_2=0.5$.
A symbolized version of the game with two prices 60 and P is given in Figure E5.9.

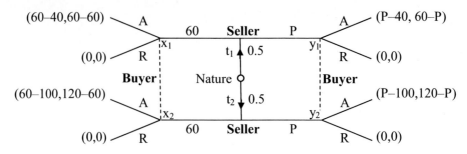

■ **Figure E5.9** The first and second coordinates of each payoff vector are respectively the payoff of the seller and the buyer.

Questions

1. Prove that there exists a semi separating PBE such that the seller of type t_1 plays 60 and P, with $100<P<120$, the seller of type t_2 only plays P, and the consumer buys the good at price 60 with probability 1.
2. Prove that there exists a separating PBE such that the seller of type t_1 plays $H_1=60$ and the seller of type t_2 plays $H_2=120$, and the consumer buys the good at price 60 with probability 1. What about Akerlof's remark (see chapter 3, Exercise 27)?

♣♣♣ EXERCISE 11 LEMON CARS, EXPERIENCE GOOD MARKET WITH n QUALITIES

Let us come back to the experience good model close to Akerlof's context (the generalization of the game above) exposed in chapter 3 (Exercise 27). There are different qualities t_i, with $t_i<t_{i+1}$ for i from 1 to n–1. t_i is the type of the seller. We suppose here that the seller's reservation prices h_i and the buyer's reservation prices H_i, check

$$\frac{\sum_{i=k}^{j} \rho_i H_i}{\sum_{i=k}^{j} \rho_i} < h_j \text{ for j from 2 to n and k from 1 to j–1} \quad (a)$$

and $h_i<H_i<h_{i+1}<H_{i+1}$ for any i from 1 to n–1. (b)

The consumer (buyer) and the seller payoffs are respectively $(H_i-p)q$ and $(p-h_i)q$ when the good is sold at price p and bought with probability q.

We write $\pi_i(p)$ the probability that a seller of type t_i sets price p and $q(p)$ the probability that a the buyer agrees to buy at price p.

EXERCISES

Questions

1. Consider a PBE in which each type of seller gets a positive payoff. Show that:
   - each type of seller plays at most three prices with a positive probability;
   - if t_i plays three prices p, p', and p", with p<p'<p", then p'=H_i;
   - if t_i plays a price p different from H_i, then p is also played with positive probability by the adjacent type t_{i-1} or t_{i+1}.

2. Show that there exists an infinite number of PBE, in which the seller of type t_i, i from 1 to n–1, plays the prices p_i* and p_{i+1}* with positive probabilities and in which the seller of type t_n plays the price p_n* with probability 1, with: p_1*=H_1 and h_i<p_i*<H_i for i from 2 to n (and therefore p_i*<p_{i+1}* for i from 1 to n–1).

The buyer accepts p_1* with probability 1 and accepts each price p_i*, i from 2 to n, with probability q(p_i*).

Show that $\pi_i(p_i$*), i from 2 to n, and q(p_i*), i from 1 to n, are defined by:

$\pi_{i-1}(p_i$*)=$\rho_i\, \pi_i(p_i$*)(H_i–p_i*)/[$\rho_{i-1}(p_i$*–H_{i-1})]

q(p_i*)=(p_{i-1}*–h_{i-1})q(p_{i-1}*)/(p_i*–h_{i-1}).

You can suppose that the buyer assigns each price p differently to the equilibrium prices, with $H_{i-1} \leq H_i$, to t_{i-2}, for i from 3 to n, and each price p, with p<H_2, to t_1. Hence he refuses the trade at each non equilibrium price higher than H_1. He accepts all the out of equilibrium prices lower than H_1.

♣ ♣ EXERCISE 12 PERFECT BAYESIAN EQUILIBRIA IN ALTERNATE NEGOTIATION WITH INCOMPLETE INFORMATION

Consider the game with incomplete information on the number of rounds of negotiation, either 2 or 3, studied in chapter 5 (section 3.3).

Question
We already established that for $\rho \leq 2/7$, there is a pooling equilibrium where both types of player 1 propose 7 and player 2 accepts this offer. Find all the other PBE.

NOTE

1 Umbhauer, G., 1991. Information asymétrique et principe d'induction, *Revue Economique*, 42 (6), 1089–1110.

Chapter 6

Solving a game differently

INTRODUCTION

Well, there are still games that resist the topics and tools we developed up to now. The envelope game, for example, is one of these games which give one a headache: Nash equilibria say that we should not exchange amounts higher than 1, but we know perfectly well that we are ready to exchange much higher amounts.

That's why we turn, in this chapter, to some alternative ways of coping with games. The first one comes back to mixed Nash equilibria and the strange meaning of their probabilities, which often does not fit with intuition and common behaviour. Best-reply matching, developed by Kosfeld, Droste and Voorneveld (2002), provides a new approach of mixed strategies, which better fits with the usual way of working with probabilities. The second approach, regret minimization, completely departs from an equilibrium approach. Halpern and Pass (2012) exploit the idea that people try not to regret a decision, so opt for a strategy that minimizes the potential regrets. They apply their concept to a new, very stimulating story game, the traveller's dilemma, a story where an airline company destroys luggage but is not in a hurry to reimburse. Halpern and Pass's concept also gives new insights on the envelope game. The third approach, level-k reasoning, exploits the idea that the ability to run a reasoning is usually limited and highly heterogeneous among players. Some players (level-0 players) are unable to run a reasoning, so may play any strategy at hand with the same probability. Other players (level-1 players) best react to level-0 players, and more generally, level-k players best react to level k-1 players. The fourth approach, forward induction, is very different: it focuses on incentives and on the meaning of out of equilibrium actions. The general idea is that an unexpected action is not necessarily a tremble or an error, but can signal future actions.

Section 1 is about best reply matching, section 2 discusses regret minimization, section 3 explores level-k reasoning and section 4 introduces forward induction.

1 BEST REPLY MATCHING

I have been teaching game theory for more than 20 years and I still observe that my students don't readily accept Nash equilibria in mixed strategies. The main difficulty they encounter is the meaning of the probabilities. As we discussed in chapter 3, when a player plays two actions A and B with probabilities ¼ and ¾ in a Nash equilibrium, these probabilities have no meaning for him: as soon as he plays two actions with a positive probability, these two actions lead to the same payoff. So he could play A and B with any probability, and, if he chooses ¼ and ¾ at equilibrium, it is only because these probabilities optimize the other players' strategies.

Yet, in reality, when we play in a mixed way, when we assign the probabilities ¼ and ¾ to actions A and B, so when we play B three times more than A, we do this because *we think that B is more often the best strategy (three times more than A)*. This common perception of equilibrium probabilities is at the basis of the concept of Best Reply Matching (BRM) developed by Kosfeld, Droste and Voorneveld (2002).[1]

1.1 Definitions and links with correlation and the Nash equilibrium

1.1.1 Best reply matching, a new way to work with mixed strategies

---

Definition 1 (Kosfeld et al. 2002): Normal form BRM equilibrium
Let G=($\mathcal{N}$, S_i, $\succ_i$, i∈$\mathcal{N}$) be a game in normal form. A mixed strategy p is a BRM equilibrium if for every player i ∈ $\mathcal{N}$ and for every pure strategy $s_i \in S_i$:

$$p_i(s_i) = \sum_{s_{-i} \in B_i^{-1}(s_i)} \frac{1}{\text{Card } B_i(s_{-i})} p_{-i}(s_{-i})$$

In a BRM equilibrium, the probability assigned to a pure strategy is linked to the number of times the opponents play the strategies to which this pure strategy is a best reply: if player i's opponents play s_{-i} with probability $p_{-i}(s_{-i})$, and if the set of player i's best responses to s_{-i} is the subset of pure strategies $B_i(s_{-i})$, then each strategy of this subset is played with the probability $p_{-i}(s_{-i})$ divided by the cardinal of $B_i(s_{-i})$.

This concept has a link with the intuitive concept of rationalizability developed by Bernheim (1984)[2] and Pearce (1984),[3] a strategy s_i being rationalizable if a pure strategy profile s_{-i} exists to which s_i is a best response. Kosfeld, Droste and Voorneveld just go further: they observe that, if the opponents often play s_{-i}, then s_i often becomes the best response, and therefore they argue that it is rational (rationalizable) for player i to often play s_i. More precisely, they require that, if s_{-i} is played with probability $p_{-i}(s_{-i})$, s_i is played with the same probability if s_i is the only best reply to s_{-i}, with probability $p_{-i}(s_{-i})/K$ if there are K best responses to s_{-i}. Given that the same condition is checked for each pure strategy, each player's probability distribution (on pure strategies) is justified by the opponents' probability distributions, which ensures a strong consistency.

Given that, in the case of several best replies to a profile s_{-i}, there is no real motivation to play each best reply with the same probability, it seems natural to generalize Kosfeld et al.'s criterion by allowing the playing of the different best replies with different probabilities as follows:

Definition 2: Generalized BRM equilibrium in normal form games (Umbhauer 2007).[4]
Let G=($\mathcal{N}$, S_i, $\succ_i$, i∈$\mathcal{N}$) be a game in normal form. A mixed strategy p is a generalized BRM (GBRM) equilibrium if for every player i ∈ $\mathcal{N}$ and for every pure strategy $s_i \in S_i$:

$$p_i(s_i) = \sum_{s_{-i} \in B_i^{-1}(s_i)} \delta_{s_i} p_{-i}(s_{-i})$$

SOLVING A GAME DIFFERENTLY

with $\delta_{s_i} \in [0, 1]$ for any s_i belonging to $B_i(s_{-i})$ and $\sum_{s_i \in B_i(s_{-i})} \delta_{s_i} = 1$

The consistency of best reply matching is quite different from the consistency of mixed Nash equilibria. We illustrate this difference on the game in Matrix 6.1a:

Player 2

		A_2	B_2			A_2	B_2
Player 1	A_1	(4,4)	(1,3)		A_1	$b_1 b_2$	
	B_1	(1,1)	(8,10)		B_1		$b_1 b_2$

Matrix 6.1a Matrix 6.1b

This game has a mixed strategy Nash equilibrium where player 1 plays A_1 with probability 0.9 and B_1 with probability 0.1, and player 2 plays A_2 with probability 0.7 and B_2 with probability 0.3. Player 1 gets the mean expected payoff 3.1 and player 2 gets the mean expected payoff 3.7.

Many students – and not only they – have problems with this equilibrium. *Why not play B_1 and B_2 with higher probabilities?* Why does player 1 play A_1, the strategy that only leads to the payoffs 1 and 4, with probability 0.9, whereas B_1 leads to the payoffs 1 and 8? And why does player 2 play A_2, the strategy that only leads to the payoffs 1 and 4, with probability 0.7, whereas B_2 leads to the payoffs 3 and 10? The reason for this strange behaviour is that player 1 has to put weight on A_1 *because it helps player 2* to get the same payoff with A_2 and B_2 (player 2's payoffs with A_2 and B_2 are closer when player 1 plays A_1 than when she plays B_1–4 and 3 are closer than 1 and 10). Similarly, player 2 puts weight on A_2 *because it helps player 1* to get the same payoff with A_1 and B_1 (player 1's payoffs with A_1 and B_1 are closer when player 2 plays A_2 than when he plays B_2–4 and 1 are closer than 1 and 8). *But this is really not very natural.*

> ## ➤ STUDENTS' INSERT
>
> We turn to BRM. Matrix 6.1b is the best reply matrix: b_1, respectively b_2, means that the corresponding action of player 1, respectively player 2, is a best answer to the corresponding action of player 2, respectively player 1 (for example, the upper b_1 means that A_1 is a best response to A_2). So, by noting p_1 and p_2, respectively q_1 and q_2, the probabilities assigned to A_1 and B_1, respectively A_2 and B_2, we get:
>
> For player 1: $p_1=q_1$ (because A_1 is the best response to A_2, so is played as many times as is played A_2) and $p_2=q_2$ (because B_1 is the best response to B_2, so is played as many times as is played B_2)
>
> For player 2: $q_1=p_1$ (because A_2 is the best response to A_1, so is played as many times as is played A_1) and $q_2=p_2$ (because B_2 is the best response to B_1, so is played as many times as is played B_1)
>
> So we have an infinite number of BRM equilibria, each characterized by the fact that A_1 and A_2, on the one side, and B_1 and B_2, on the other side, are played with the same probability.

The BRM family of equilibria is intuitive: in this game, it is good playing A each time the other player plays A and it is good playing B each time the other player plays B. Hence, playing A as many times as the opponent plays A, and playing B as many times as he plays B is quite intuitive.

SOLVING A GAME DIFFERENTLY

What is more, we can get BRM equilibria with high payoffs for both players: for example, if both players play A with probability 0.1 and B with probability 0.9, player 1 gets: 4×0.01+1×0.09+1×0.09+8×0.81=6.7 and player 2 gets: 4×0.01+3×0.09+1×0.09+10×0.81=8.5. These payoffs are more attractive than the mixed strategy Nash equilibrium ones.

1.1.2 Link between best reply matching equilibria and Nash equilibria

The above set of BRM equilibria contains the two pure strategy Nash equilibria (A_1,A_2) and (B_1,B_2) of the game but not its mixed strategy Nash equilibrium. More generally the set of GBRM equilibria has a nice non empty intersection with Nash equilibria.

> **Property 1:** Each pure strategy Nash equilibrium is a GBRM equilibrium.

Proof: if s* is a pure strategy Nash equilibrium, then, for any player i from 1 to N, s_i* is a best response to s_{-i}*. So s_i* is played with the probability assigned to s_{-i}* – more generally can be played with the probability assigned to s_{-i}* if there are several best replies to s_{-i}*, i.e. 1, which is the probability assigned to it in the Nash equilibrium.

1.1.3 Link between best reply matching equilibria and correlated equilibria

BRM has a link with correlation. Let us first give the definition of a correlated equilibrium.

> **Definition 3: Correlated equilibrium**
> A correlated equilibrium is a probability distribution p(.) over XS_i, i from 1 to N, such that, for every player i and every pure strategy s_i in the support of p(.),
>
> $$\forall s_i' \in S_i, \sum_{s_{-i}} u_i(s_i,s_{-i})p(s_i,s_{-i}) \geq \sum_{s_{-i}} u_i(s_i',s_{-i})p(s_i,s_{-i})$$

In a correlated equilibrium, an external player, Nature, plays a lottery over the different profiles of pure strategies, observes the outcome, and proposes to each player the associated pure strategy. The lottery, which is common knowledge of all the players, is a correlated equilibrium if each player is always induced to play the proposed action, given the information he can infer on the actions played by the others.

So for example, the lottery that puts probability 0.1 on (A_1,A_2), 0 on (A_1,B_2), 0 on (B_1,A_2) and 0.9 on (B_1,B_2) is a correlated equilibrium. As a matter of fact, when the lottery selects (B_1,B_2), player 1 is called on to play B_1 and player 2 is called on to play B_2, and each player is happy to play B given that he can deduce from the lottery that the other player also plays B. And similarly when the lottery selects (A_1,A_2) (just replace B by A).

In the BRM equilibrium, where each player plays A with probability 0.1 and B with probability 0.9, there is a some coordination between the player's actions, given that each player plays A, respectively B, as many times as the other player plays A, respectively B. But this does not mean

253

SOLVING A GAME DIFFERENTLY

that player 1 plays A_1 each time player 2 plays A_2. So, when player 1 plays B_1, what she does with probability 0.9, player 2 may play B_2 – he does it with probability 0.9 – but he may also play A_2 – he does it with probability 0.1. In other words, there is a *strong coordination on the probabilities, but not on the actions*.

In our correlated equilibrium the coordination among the actions is quite stronger: each time player 1 plays A_1, player 2 plays A_2, and each time player 1 plays B_1, player 2 plays B_2. That is why, in the BRM equilibrium, player 1 and player 2 only get the payoffs 6.7 and 8.5 whereas they get in the correlated equilibrium the higher payoffs 4×0.1+8×0.9=7.6 and 4×0.1+10×0.9=9.4. But correlation, by contrast to BRM, requires a common lottery which may become difficult to define when the game is more conflicting.

So there is a link between BRM equilibria and correlated equilibria in that both concepts try to coordinate the players' actions but the coordination is stronger in the correlated equilibrium than in the best reply matching.

1.2 Auction games, best reply matching fits with intuition

BRM often fits well with strategic phenomena.

We consider first in Matrix 6.3a, the first price sealed-bid all pay auction, with M=5 and V=3, studied in the previous chapters.

		Player 2					
		0	1	2	3	4	5
	0	(6.5,6.5)	(5,7)	(5,6)	(5,5)	(5,4)	(5,3)
	1	(7,5)	(5.5,5.5)	(4,6)	(4,5)	(4,4)	(4,3)
Player 1	2	(6,5)	(6,4)	(4.5,4.5)	(3,5)	(3,4)	(3,3)
	3	(5,5)	(5,4)	(5,3)	(3.5,3.5)	(2,4)	(2,3)
	4	(4,5)	(4,4)	(4,3)	(4,2)	(2.5,2.5)	(1,3)
	5	(3,5)	(3,4)	(3,3)	(3,2)	(3,1)	(1.5,1.5)

Matrix 6.3a

The associated best reply matrix is Matrix 6.3b. It highlights a strategic fact, the push toward a higher bid: if a player bids x, it is better for the opponent to bid x+1 (providing x≤2). So, by noting p_0, p_1, p_2, p_3, p_4, p_5 and q_0, q_1, q_2, q_3, q_4, q_5 the probabilities assigned to the bids 0, 1, 2, 3, 4 and 5 by player 1 and player 2,[5] the BRM concept leads to:

$p_0=q_2/2+q_3+q_4+q_5$ (because bidding 0 is one of the two best responses to the opponent's bid 2, so should be played half of the time player 2 bids 2, and bidding 0 is the best response when the opponent bids 3, 4 or 5, so should be played as many times as player 2 plays these bids)

$p_1=q_0$
$p_2=q_1$
$p_3=q_2/2$
$p_4=p_5=0$
$p_0+p_1+p_2+p_3+p_4+p_5=1$

and

$q_0=p_2/2+p_3+p_4+p_5$
$q_1=p_0$
$q_2=p_1$
$q_3=p_2/2$
$q_4=q_5=0$
$q_0+q_1+q_2+q_3+q_4+q_5=1$

SOLVING A GAME DIFFERENTLY

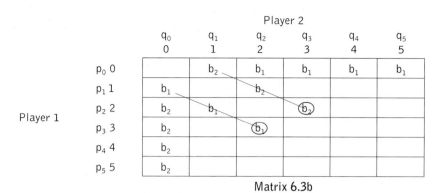

Matrix 6.3b

The only solution of this set of equations is:

$p_0=p_1=p_2=q_0=q_1=q_2=2/7$ and $p_3=q_3=1/7 \; p_4=q_4=p_5=q_5=0$

This result is quite close to the mixed strategy Nash equilibrium. We can even strengthen this similarity with the GBRM criterion, by suppressing the circled b_1 and b_2 (player 1 and player 2 only choose to bid 0 when the opponent bids 2) so that we get:

$p_0=q_2+q_3+q_4+q_5$ $\qquad\qquad$ $q_0=p_2+p_3+p_4+p_5$
$p_1=q_0$ $\qquad\qquad$ $q_1=p_0$
$p_2=q_1$ $\qquad$ and $\qquad$ $q_2=p_1$
$p_3=p_4=p_5=0$ $\qquad\qquad$ $q_3=q_4=q_5=0$
$p_0+p_1+p_2+p_3+p_4+p_5=1$ $\qquad$ $q_0+q_1+q_2+q_3+q_4+q_5=1$

The only solution of this set of equations is: $p_0=p_1=p_2=q_0=q_1=q_2=1/3$ and $p_3=q_3=p_4=q_4=p_5=q_5=0$, which is exactly the mixed strategy Nash equilibrium.

So the mixed Nash equilibrium and the above GBRM equilibrium are the same. Yet the philosophy of both equilibria is quite different. Whereas the Nash equilibrium is obtained by requiring that each bid leads to the same payoff for the bidder, the GBRM equilibrium derives from the following reasoning:

Bid 1 is the best response as often as the opponent bids 0
Bid 2 is the best response as often as the opponent bids 1
Bid 0 is the best response as often as the opponent bids 2
(given that 3, 4 and 5 are never played).

So given the symmetry of their status (each strategy is a best response to one of the two others) each bid has to be played with the same probability. *It seems to us that this motivation is much more intuitive than the equalization of the payoffs.*

We now consider the second price sealed bid all pay auction game (Matrix 6.4a), with M=5 and V=3, studied in earlier chapters.

255

SOLVING A GAME DIFFERENTLY

Player 2

	0	1	2	3	4	5
0	(6.5,6.5)	(5,8)	(5,8)	(5,8)	(5,8)	(5,8)
1	(8,5)	(5.5,5.5)	(4,7)	(4,7)	(4,7)	(4,7)
2	(8,5)	(7,4)	(4.5,4.5)	(3,6)	(3,6)	(3,6)
3	(8,5)	(7,4)	(6,3)	(3.5,3.5)	(2,5)	(2,5)
4	(8,5)	(7,4)	(6,3)	(5,2)	(2.5,2.5)	(1,4)
5	(8,5)	(7,4)	(6,3)	(5,2)	(4,1)	(1.5,1.5)

Player 1 (rows)

Matrix 6.4a

This game has six pure strategy Nash equilibria (0, 3), (0, 4), (0, 5), (3, 0), (4, 0) and (5, 0).

It has also a mixed strategy Nash equilibrium defined by: $p_0=q_0=83/293=28.3\%$, $q_1=p_1=57/293=19.5\%, p_2=q_2=45/293=15.4\%, p_3=q_3=27/293=9.2\%, p_4=q_4=0, p_5=q_5=81/293=27.6\%$: the equilibrium probabilities decrease in the bids from bid 1 to bid 4 but there is a high positive probability on bid 5. The structure of this equilibrium is the one obtained in a continuous context. More precisely, we establish *(Exercise 25 in chapter 3)* that in a continuous setting, x from 0 to M–V/2 is bid with probability $dx(e^{-x/V})/V$, x between M–V/2 and M (excluded) is bid with probability 0, and M is bid with probability $e^{(V/2-M)/V}$. So there is an atom on bid M, and the probability to bid x exponentially decreases in x, from bid 0 up to a certain threshold bid (M–V/2) (then it is equal to 0).

Yet my L3 students clearly *did not play* in this way: 38% bid 0, 9% bid 1.5% bid 2, 20.5% bid 3, 16% bid 4 and 15% bid 5. In other words, about 2/5 of my students played 0, few of them played 1 and 2, one-fifth of them played 3 and 31% almost equally shared on 4 and 5.

What says BRM? The best replies are given in Matrix 6.4b:

Player 2

		q_0 0	q_1 1	q_2 2	q_3 3	q_4 4	q_5 5
p_0	0		b_2	b_2	b_1b_2	b_1b_2	b_1b_2
p_1	1	b_1		b_2	b_2	b_2	b_2
p_2	2	b_1	b_1		b_2	b_2	b_2
p_3	3	b_1b_2	b_1	b_1		b_2	b_2
p_4	4	b_1b_2	b_1	b_1	b_1		
p_5	5	b_1b_2	b_1	b_1	b_1		

Player 1 (rows)

Matrix 6.4b

The symmetric repartition around the diagonal with doublings on the 0 line and the 0 column leads to the system of equations:

$p_0=q_3/3+q_4+q_5$ $q_0=p_3/3+p_4+p_5$
$p_1=q_0/5$ $q_1=p_0/5$
$p_2=q_0/5+q_1/4$ $q_2=p_0/5+p_1/4$
$p_3=q_0/5+q_1/4+q_2/3$ $q_3=p_0/5+p_1/4+p_2/3$
$p_4=q_0/5+q_1/4+q_2/3+q_3/3$ $q_4=p_0/5+p_1/4+p_2/3+p_3/3$

$p_5=q_0/5+q_1/4+q_2/3+q_3/3=p_4$
$q_5=p_0/5+p_1/4+p_2/3+p_3/3=q_4$
$p_0+p_1+p_2+p_3+p_4+p_5=1$
$q_0+q_1+q_2+q_3+q_4+q_5=1$

The solution of this system is:

$p_0=q_0=180/481=37.4\%$, $p_1=q_1=p_0/5=7.5\%$, $p_2=q_2=p_0/4=9,4\%$, $p_3=q_3=p_0/3=12.5\%$, $p_4=p_5=q_4=q_5=4p_0/9=16.6\%$

Clearly, these probabilities are far from the Nash equilibrium ones and they better fit with the students' probabilities, except for p_2 (higher) and p_3 (lower). BRM, as well as the students, copes with the fact that bids 1 and 2 are seldom best responses to the bids of the opponent: bid 1 is a best response only if player 2 plays 0 (and in this case, bids 2, 3, 4 and 5 are also best responses), bid 2 is a best response only if player 2 bids 0 or 1 and in each of these two cases, many other bids are also best responses. And BRM, as well as the students, nicely copes with the fact that 0, 3, 4 and 5 are often best replies.

1.3 Best reply matching in ascending all pay auctions

We now turn to the ascending all pay auction (Figure 6.1) studied in the previous chapters. We recall that one Nash equilibrium path leads player 1 to stop immediately, whereas the second one – the backward induction equilibrium path – leads player 1 to bid 1 at x_1 and player 2 to stop at x_2.

BRM, applied to the normal form of this game (Matrix 6.5a), leads to the best reply Matrix 6.5b.

	Player 2			
	$(S_2/x_2 S_2/x_4)$	$(S_2/x_2 4/x_4)$	$(2/x_2 S_2/x_4)$	$(2/x_2 4/x_4)$
$(S_1/x_1 S_1/x_3 S_1/x_5)$	(5,8)	(5,8)	(5,8)	(5,8)
$(S_1/x_1 S_1/x_3\ 5/x_5)$	(5,8)	(5,8)	(5,8)	(5,8)
$(S_1/x_1 3/x_3 S_1/x_5)$	(5,8)	(5,8)	(5,8)	(5,8)
$(S_1/x_1 3/x_3\ 5/x_5)$	(5,8)	(5,8)	(5,8)	(5,8)
$(1/x_1 S_1/x_3 S_1/x_5)$	(7,5)	(7,5)	(4,6)	(4,6)
$(1/x_1 S_1/x_3\ 5/x_5)$	(7,5)	(7,5)	(4,6)	(4,6)
$(1/x_1\ 3/x_3 S_1/x_5)$	(7,5)	(7,5)	(5,3)	(2,4)
$(1/x_1\ 3/x_3\ 5/x_5)$	(7,5)	(7,5)	(5,3)	(3,1)

Player 1

Matrix 6.5a

$$\begin{array}{c|c|c|c|c|c|c}1 & 1 & 2 & 2 & 1 & 3 & 2 & 4 & 1 & 5\\\hline x_1 & x_2 & x_3 & x_4 & x_5\\ S_1 & S_2 & S_1 & S_2 & S_1\end{array}\begin{pmatrix}3\\1\end{pmatrix}$$

$\begin{pmatrix}5\\8\end{pmatrix}$ $\begin{pmatrix}7\\5\end{pmatrix}$ $\begin{pmatrix}4\\6\end{pmatrix}$ $\begin{pmatrix}5\\3\end{pmatrix}$ $\begin{pmatrix}2\\4\end{pmatrix}$ V=3 M=5

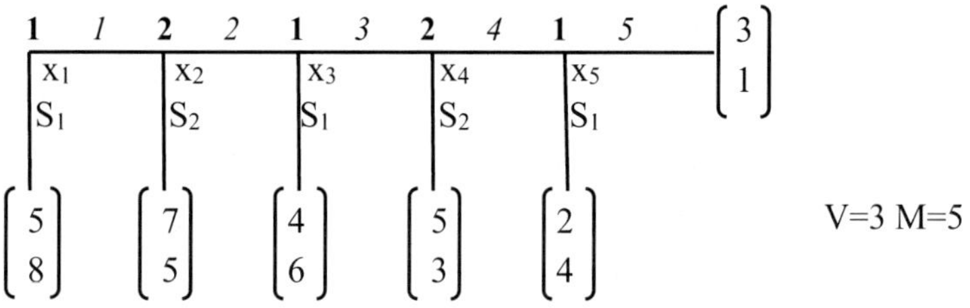

Figure 6.1

SOLVING A GAME DIFFERENTLY

Player 2

	q_1 $(S_2/x_2 S_2/x_4)$	q_2 $(S_2/x_2 4/x_4)$	q_3 $(2/x_2 S_2/x_4)$	q_4 $(2/x_2 4/x_4)$
p_1 $(S_1/x_1 S_1/x_3 S_1/x_5)$	b_2	b_2	$b_1 b_2$	$b_1 b_2$
p_2 $(S_1/x_1 S_1/x_3 5/x_5)$	b_2	b_2	$b_1 b_2$	$b_1 b_2$
p_3 $(S_1/x_1 3/x_3 S_1/x_5)$	b_2	b_2	$b_1 b_2$	$b_1 b_2$
p_4 $(S_1/x_1 3/x_3 5/x_5)$	b_2	b_2	$b_1 b_2$	$b_1 b_2$
p_5 $(1/x_1 S_1/x_3 S_1/x_5)$	b_1	b_1	$\underline{b_2}$	$\underline{b_2}$
p_6 $(1/x_1 S_1/x_3 5/x_5)$	b_1	b_1	$\underline{b_2}$	$\underline{b_2}$
p_7 $(1/x_1 3/x_3 S_1/x_5)$	$b_1 b_2$	$b_1 b_2$	*b_1*	
p_8 $(1/x_1 3/x_3 5/x_5)$	$b_1 b_2$	$b_1 b_2$	*b_1*	

Player 1 labels the rows.

Matrix 6.5b

The system of equations:

$p_1=p_2=p_3=p_4=q_3/6+q_4/4$ $\quad q_1=q_2=p_1/4+p_2/4+p_3/4+p_4/4+p_7/2+p_8/2$
$p_5=p_6=q_1/4+q_2/4$ $\quad q_3=q_4=p_1/4+p_2/4+p_3/4+p_4/4+p_5/2+p_6/2$
$p_7=p_8=q_1/4+q_2/4+q_3/6$
$p_1+p_2+p_3+p_4+p_5+p_6+p_7+p_8=1$ $\quad q_1+q_2+q_3+q_4=1$

has a unique solution: $p_1=p_2=p_3=p_4=2.5/26$, $p_5=p_6=3.5/26$, $p_7=p_8=4.5/26$, $q_1=q_2=7/26$, $q_3=q_4=6/26$

Consequently, BRM leads player 1 to stop immediately – only – with probability $p_1+p_2+p_3+p_4=5/13=38.5\%$; she stops at x_3 with probability $p_5+p_6=3.5/13=26.9\%$, she stops at the last decision node x_5 with probability $2.25/13=17.3\%$ and she never stops with probability $2.25/13=17.3\%$.

Player 2 stops at his first decision node with the higher probability $q_1+q_2=7/13=53.8\%$, he stops at x_4 with probability $3/13=23.1\%$, and he never stops with probability $3/13=23.1\%$. These probabilities give a slight advantage to player 1: she goes on at her first decision node with probability $8/13=61.5\%$ and player 2 stops after this first move with probability $7/13=53.8\%$. So *BRM behaviour is somewhere between Nash and backward induction behaviour with regard to the first stopping decisions, but it also allows the players to always overbid. This fits with the real behaviour in some contexts (see my L3 students for the game with M=100 and V=60 and Shubik's grab the dollar game, chapter 1, section 1.2.2 and chapter 4, section 3.1).*

It is interesting to observe that BRM exploits the fact that it is worth overbidding when the opponent stops at the next round. For example, the bold underlined b_2 expresses that it is best for player 2 to overbid at x_2 when he expects that player 1 stops at x_3.

Yet some probabilities are difficult to understand a priori, notably player 1's positive probability of stopping at x_5. In fact, $(1/x_1, 3/x_3, S_1/x_5)$ for example, and therefore stopping at x_5 after overbidding at earlier nodes, is a best response each time player 2 stops at or before x_4, *so that x_5 is not reached* (see the b_1 in italics in Matrix 6.5b). We see here a limit of the normal form approach: an action A, for example S_1/x_5, is part of a best response just/only because it will not be played due to the other player's behaviour. So, in the BRM equilibrium, A will be played with positive probability, but it will be played with the probability with which the other players play the strategies that leave A without bite.

SOLVING A GAME DIFFERENTLY

1.4 Limits of best reply matching

There is another problem with BRM. In some ways, BRM is an asymmetric criterion: it is only interested in how many times an action is the best response, it does not take into account how many times a strategy is the worst, or at least a bad response.

To explain this fact, we return to the second price all pay auction in Matrix 6.4a and calculate the payoff a player gets with each strategy when confronted with a player who plays in the BRM way.

Bidding 0 leads to the payoff 6.5×0.374+5×0.075+5×0.094+5×0.125+5×0.166+5×0.166 =5.561

Bidding 1 leads to the payoff 8×0.374+5.5×0.075+4×0.094+4×0.125+4×0.166+4×0.166 =5.6085

Bidding 2 leads to the payoff 8×0.374+7×0.075+4.5×0.094+3×0.125+3×0.166+3×0.166 =5.311

Bidding 3 leads to the payoff 8×0.374+7×0.075+6×0.094+3.5×0.125+2×0.166+2×0.166 =5.1825

Bidding 4 leads to the payoff 8×0.374+7×0.075+6×0.094+5×0.125+2.5×0.166+1×0.166 =5.287

Bidding 5 leads to the payoff 8×0.374+7×0.075+6×0.094+5×0.125+4×0.166+1.5×0.166 =5.619

Observe that bid 3, which is played with a higher probability than bid 1 and 2 in the BRM equilibrium, does not lead to a high payoff. The reason for this is that bid 3 is a bad answer to bids 3, 4 and 5 and that BRM does not take this fact into account. In some way, only dealing with when an action is a best response is perhaps too restrictive. For example, bidding 1 is a best reply only one time, whereas bidding 3 is a best reply three times, but bidding 1 is better than bidding 3 three times (when the opponent plays 3, 4 or 5) and only worse two times (when the opponent plays 1 or 2). But how should we take into account the fact that an action is the worst or a bad answer to a profile of strategies? This is not obvious.

Finally let us stress a problem, common to BRM and to the pure strategy Nash philosophy: when we look for best responses, we don't take into account small and big differences in payoffs. The comparison is ordinal, not cardinal.

So look at the game given in Matrix 6.6 below:

		Player 2	
		A_2	B_2
Player 1	A_1	(4.01, 4.01)	(1, 3.99)
	B_1	(3.99, 1)	(79, 79)

Matrix 6.6

In this game, A is the best reply to A (and the worst reply to B), and B is the best reply to B (and the worst reply to A), so the difficulty mentioned above has no impact – each action is one time the worst response. So BRM (rightly) leads to $p_1=q_1$ and $p_2=q_2$, where p_1, p_2 and q_1, q_2, are the probabilities assigned to A_1, B_1 and A_2, B_2.

SOLVING A GAME DIFFERENTLY

It follows an infinite family of BRM equilibria where both players play the same action with the same probability. Yet, if you ask people about the way to play this game, they answer that they play B. This difference stems from the fact that a real player takes cardinal differences into account, in contrast to BRM (and pure strategy Nash equilibria) which only order the payoffs: so a real player observes that the possible loss linked to the play of B, 4.01–3.99, is much lower than the loss linked to the play of A, 79–1.

It may be interesting to mix the BRM approach with a cardinal approach, by weighting the probability assigned to an action by its weighted payoff contribution. We could proceed in the following way:

$p_1 = 4.01 q_1/(4.01+3.99) + 1 q_2/(1+79)$
$p_2 = 3.99 q_1/(4.01+3.99) + 79 q_2/(1+79)$
$q_1 = 4.01 p_1/(4.01+3.99) + 1 p_2/(1+79)$
$q_2 = 3.99 p_1/(4.01+3.99) + 79 p_2/(1+79)$
$p_1 + p_2 = 1 \quad q_1 + q_2 = 1$

It follows that $p_1 = q_1 = 1/40.9$ and $p_2 = q_2 = 39.9/40.9$, which seems much closer to a real behaviour.

2 HALPERN AND PASS'S REGRET MINIMIZATION

2.1 A new story game, the traveller's dilemma

Best reply matching is unfortunately of minimal help in games with special best reply structures like the game we now present: the traveller's dilemma. This game (due to Basu 1994[6] and studied by Halpern and Pass 2012[7]) is very stimulating from a game theoretic point of view and to our mind, should become *a new story game*.

Two travellers have identical luggage (they paid the same price for it and they value it in the same way) that unfortunately is damaged by an airline. The company agrees to pay for the damage and proposes the following game: first each traveller i=1,2 asks for a given amount x_i (without knowing the amount proposed by the other traveller), where x_i has to be an integer in [2,100]. Second the company observes the two offers and proposes to reimburse the travellers in the following way:

- If $x_1 = x_2$, each traveller gets x_1.
- If $x_i > x_j$ i≠j=1,2 the traveller j gets $x_j + P$, where P is a fixed amount higher than 1, and the traveller i only gets $x_j - P$. So the company wants to reward the traveller who asks for the lowest amount and to "punish" the one who asks "too much". Consequently its aim is to induce the two travellers not to ask for too high an amount.

To my students, who were shocked by the fact that $x_j - P$ can be negative – which means that you have to pay the company which damaged your luggage! – I proposed a slightly different version: if $x_i > x_j$ i≠j=1,2 the traveller j gets $x_j + P$, but the traveller i gets max $(0, x_j - P)$.

And with regard to Nash equilibrium and iterative elimination of weakly dominated strategies *(see Exercise 11 in chapter 2 and Exercise 15 in chapter 3)*, the company achieves its aim, both with Basu's (Halpern and Pass) rule and my slight variation: both travellers ask for 2, regardless of

SOLVING A GAME DIFFERENTLY

P>1. The reason for this is close to the one that justifies the equality between the price and the marginal cost in a Bertrand price competition. As in Bertrand's competition *(see Exercise 10 in chapter 2 and Exercise 13 in chapter 3)*, each traveller is better off asking one unit less than the opponent's demand x (because x–1+P>x>x–P; i.e. it is better to ask one unit less rather than the same amount, and rather than a higher amount). It follows that no couple (x_1,x_2) with $x_1 > x_2$, but also no couple (x_1,x_2) with $x_1 = x_2 > 2$ can be a Nash equilibrium. So the demands converge to the only Nash equilibrium (2,2).

BRM also leads to the same outcome (regardless of the version). Given that x–1 is the best reply to x, except for 2 which is the best answer to 2, it automatically follows that the whole probability will be on 2 *(see Exercise 2 in this chapter)*.

Yet, even more than Bertrand's equilibrium, this result is quite unnatural. Halpern and Pass (2012) claim that people who played this game, even game theorists, do not play 2, especially if P is low, for example P=2. Becker (2005)[8] showed that among the 45 members of the Game Theory Society who played a pure strategy, 31 proposed an amount higher than or equal to 96, 7 proposed an amount higher than 90 (and lower than 96) and only 3 of them proposed the Nash equilibrium strategy 2.

What is more, by contrast to what happens in the Bertrand game, the behaviour of the players varies with the value of P. Capra et al. (1999)[9] show that people tend to play high values for low values of P, even when the game is repeated, whereas for high values of P, they start with much lower demands and quickly converge to the Nash solution after a few repetitions of the game.

The results obtained with my students confirm Becker and Capra's results. I proposed this game many times. So I just talk here about the results I obtained with (102) L3 students faced to P=5 and P=30, and about the results I got with other (112) L3 students faced to P=4 and to P=40 *(see also Exercise 15 in chapter 3)*.

The results are summarized in Table 6.1.

Table 6.1

	P=4	P=5	P=30	P=40
% of players playing 2	2.7	4.9	6.9	18.8
% of players playing less than 20	7.1	11.8	13.7	30.4
% of players playing less than 50	16.7	24.5	57.8	78.6
% of players playing less than 80	31.3	41.2	80.4	92
% of players playing 100	31.3	24.5	13.7	4.5
Mean played value	76.7	71	46.1	33.3
Median value	90	89.5	42.5	30

It clearly derives from Table 6.1 that my students play differently when faced to low and large values of P. Whereas about 60% or more of them ask for amounts higher than 80 when P is equal to 4 or 5 (about 25% or more ask for 100), about 60% or more of them ask for amounts lower than 50 when P is equal to 30 or 40; for P=40, about 1/5 of them ask for P=2. And, as a consequence, the mean value of the asked amount decreases with P.

You can also see on the histograms in Figures 6.2a, 6.2b, 6.2c, 6.2d, that there are classes of amounts which are not played by the students, and that these classes differ for low and large values of P. For example, for P=4, no student asks for an amount between 11 and 25, whereas, for P=30,

SOLVING A GAME DIFFERENTLY

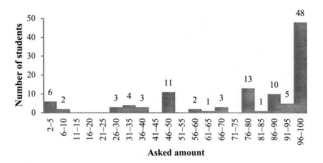

■ **Figures 6.2a** Traveller's dilemma, 'Basu's' version, asked amounts for P=4

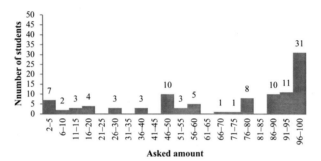

■ **Figures 6.2b** Traveller's dilemma, 'Basu's' version, asked amounts for P=5

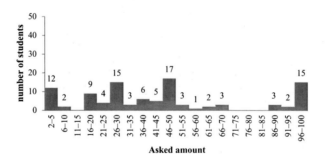

■ **Figures 6.2c** Traveller's dilemma, 'Basu's' version, asked amounts for P=30

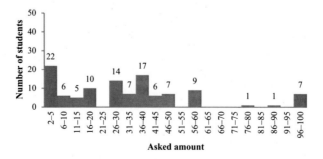

■ **Figure 6.2d** Traveller's dilemma, 'Basu's' version, asked amounts for P=40

no student asks for an amount between 71 and 85 (for P=40, only 1 student asks for an amount between 61 and 85). Clearly, the students focus on large amounts when P is low and on low amounts when P is large; this is confirmed by the median value which is very large (much larger than the mean value) for low values of P (90 and 89.5 for P=4 and P=5), and which is low (lower than the mean value) for large values of P (42.5 for P=30 and 30 for P=40).

And this seems quite natural. If P is low (2, 3, 4, 5 for example) when you ask for a high amount you do not risk a lot. You risk much less than if you ask for a low amount! For example, if you ask for 95, even if your opponent only asks for 80, you get 80–P, i.e. a high amount. But if you ask for 40, and if your opponent asks for 80, you get 40+P, which is much lower than 80–P! 95 does not dominate 40 (because you can get more with 40 when the opponent plays between 40 and 40+2P–1), yet you get more with 95 than with 40 each time the opponent plays more than 40+2P (and there are much more values higher than 40+2P, for P low, than between 40 and 40+2P–1).

But when P is high, for example P=50, then of course it becomes dangerous to play high values: if you still ask for 95 and the opponent asks for 80, you only get 80–50=30, a much lower amount than if you play 40 (in which case you get 40+50=90). More precisely, if P=50, you get more with 40 than with 95 each time the opponent asks for an amount between 40 and 94, whereas you get more with 95 only if he plays between 95 and 100! Given that a player can now get high amounts when asking for a low one (you get at least 52 when playing 2, the opponent playing more), and can get very low payoffs by asking for high amounts, he is strategically induced to propose rather low values for high values of P, and so to come close to the Nash equilibrium (observe that among one-fifth of my L3 students play 2 when P=40).

To summarize, the traveller's dilemma, by contrast to the Bertrand game, exhibits the following *asymmetry* taken into account by real players, but not taken into account by the Nash equilibrium concept nor by the iterative elimination of dominated strategies:

when P is low, you can get high payoffs with high demands and low payoffs with low demands, but when P is high, you can get low payoffs with high demands and high payoffs with low demands.

2.2 Regret minimization

2.2.1 Definition

Halpern and Pass (2012) proposed an original method to cope with these facts, i.e. they proposed to *minimize maximal regret*. Their concept directly derives from regret minimization in decision theory (Savage, 1951,[10] Niehans 1948[11]). The idea is the following: Suppose that player i chooses a strategy p_i in a context he does not manage. Suppose that his utility depends on p_i and a state of nature s which belongs to a set S. So he gets $u_i(p_i, s)$. Call p_i^* the best strategy in state s. The regret of p_i in state s is the difference between the highest utility he can get in this state and the utility he gets by playing p_i:

$regret_u(p_i,s) = u_i(p_i^*,s) - u_i(p_i,s)$.

Call $regret_u(p_i) = \max_{s \in S} regret_u(p_i,s)$. $Regret_u(p_i)$ is the maximal regret of a player playing p_i. The *minimax regret decision rule* simply asks player i to play the strategy p_i that minimizes $regret_u(p_i)$.

Halpern and Pass adapt this decision rule to games. The states of nature become the profiles of strategies played by the other players, and the decision theory definitions become:

SOLVING A GAME DIFFERENTLY

> **Definition 4: Minimizing regret strategy**
> regret$_u$(s_i,s_{-i})=u_i(s_i*,s_{-i})−u_i(s_i,s_{-i}), where s_i* is player i's best response when the other players play s_{-i}. regret$_u$(s_i)=max$_{s-i \in S-i}$ regret$_u$(s_i,s_{-i}). Player i's minmax regret strategy is the strategy $\hat{s}_i$ that minimizes regret$_u$(s_i). Halpern and Pass say that $\hat{s}_i$ minimizes regret.

So the idea, like in decision theory, is to take the strategy that yields the minimal regret in the worst situation, i.e. that minimizes the maximal possible regret. Observe that Halpern and Pass implicitly work in pure strategies.

We apply their criterion first to the traveller's dilemma and then to the envelope game.

2.2.2 Regret minimization and the traveller's dilemma

Suppose that traveller 1 chooses an amount x_1.

If traveller 2 chooses x_2 lower than x_1, the best decision would be to choose x_2−1 if x_2>2, x_2 if x_2=2. With x_1, traveller 1 gets x_2−P, with x_2−1, she gets x_2−1+P. So her maximal loss is x_2−1+P−x_2+P=2P−1 if x_2>2, 2−2+P=P if x_2=2. It follows that the maximal regret is 2P−1.

If player 2 chooses x_2>x_1, the best decision would again be x_2−1. But this time player 1 gets x_1+P with x_1 and x_2−1+P with x_2−1. So the regret is x_2−1−x_1 and the maximal regret is obtained for x_2=100, and is equal to 99−x_1. If traveller 2 asks for the same amount, the maximal regret is only P−1.

So the maximal regret is max (2P−1, 99−x_1) =2P−1 if 2P−1>99−x_1 i.e. x_1>100−2P
=99−x_1 if x_1<100−2P

For x_1=100−2P, the regret is 2P−1=99−x_1. Clearly the lowest maximal regret is 2P−1 and obtained for any value x between 100−2P and 100.[12]

So, for P=2, the best is to choose a value between 96 and 100. But for P=30 for example, the best is to choose x_1 between 40 and 100. These results are consistent with real players' behaviour.

I have here to mention that the slight variant I proposed to my students (if x_j>x_i, j gets max (0, x_i−P) instead of x_i−P) does not lead to the same result if P>101/3 *(see Exercise 7 in this chapter)*. In that case the strategy that minimizes regret is (101−P)/2. So, for P=40, these strategies are the prices 30 and 31 and the associated regret is 69, much lower than 2P−1=79. This is due to the fact that low values of the opponent can no longer lead to negative payoffs. Observe that 30 is the median value of the amounts asked by my students for P=40.

Halpern and Pass do not stop their analysis at this level. They proceed to an iterative regret minimization. Given that both travellers now play an integer in [100−2P, 100], the best decision becomes 100−2P+1. The main reason for this is that, if you play 100−2P+1, your opponent, if he plays a lower price, can only play 100−2P, so you only lose (100−2P)−(100−2P−P)=P. And if he plays more, then your maximal regret is 99−(100−2P+1)=2P−2. And max (2P−2, P) is necessarily lower than the maximal regret 2P−1 you get if you ask for a different amount. So, for P=2, each player plays 97, for P=30, each player plays 41, which are quite intuitive values.

But personally I do not appreciate the iterative process, even if it leads to a good result because to apply regret minimization iteratively, you have to assume that other players also minimize their regret. Yet, generally, when you choose regret minimization, you do so because you have no idea of the way other players are playing: so how could you suppose that they behave as regret minimizers?

SOLVING A GAME DIFFERENTLY

2.2.3 Regret minimization and the envelope game

As you know, iterative elimination of weakly dominated strategies, as well as the Nash equilibrium concept, are of no help in the envelope game, where they lead to the highly improbable result that a player should not exchange an envelope that contains more than $1.

We now apply Halpern and Pass's criterion for a given value K (we suppose K even). If a player exchanges a value x, the worst-case scenario is an opponent who only exchanges 1 and has this value in his envelope. So the player loses x–1. If he doesn't exchange x, the worst-case scenario is an opponent who exchanges K and has this value in his envelope. So the player loses K–x.[13] The player compares x–1 and K–x.

So, if x≤K/2, the best is to exchange, because the maximal loss by doing so, x–1, is lower than the maximal loss in case of no exchange, K–x. If x≥K/2+1, the best is to not exchange, because the maximal loss by doing so, K–x, is lower than the maximal loss in case of exchange, x–1. So a player should exchange up to K/2 and not exchange more, a behaviour we often observe in reality, but only when K is not too large. So for K=100 (see the histogram 6.3a) clearly many students play K/2=50 (about 1/3 of the students play this value). And the mean value of the maximal exchanged value is 48, close to K/2. Things are different as you can see in the histogram 6.3b, for K=100,000. K/2=50,000 is still the most played value, but only by 1/5 of the students. The mean value of the maximal exchanged value is 33,256, much lower than K/2 and 1/5 of the students exchange their envelope only if the amount is lower or equal to 5,000=K/20!

In some way, I wonder if we shouldn't take into account the following observation: generally you regret missing the opportunity to get more money but you *much more* regret losing an amount of money you *had*. In other words, you regret to not have exchanged your envelope with 1,000$ with an envelope that contains 31,000$, but you much more regret to have exchanged your envelope with 31,000$ with an envelope that contains 1,000$. Losing 30,000 dollars *you had* is worse than not getting 30,000 additional dollars.

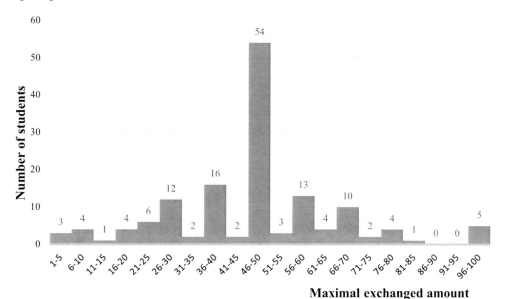

Figure 6.3a Envelope game, maximal amount exchanged for K=100

SOLVING A GAME DIFFERENTLY

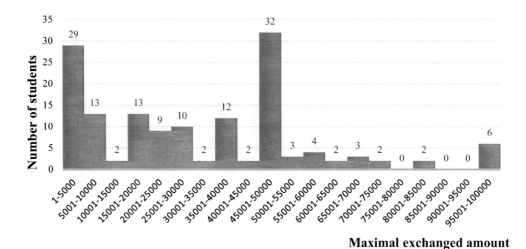

■ **Figure 6.3b** Envelope game, maximal amount exchanged for K=100,000

"A bird in the hand is worth two in the bush": so we could say that if a player exchanges a value x (the bird in the hand), he values $2(x-1)$ his maximal regret $(x-1)$. And if he doesn't exchange x (so he doesn't try to get a bird in the bush), he just values K–x his maximal regret (K–x). So he minimizes regret by exchanging as long as $2(x-1) \leq K-x$, i.e. $x \leq (K+2)/3$. For K=100,000, (K+2)/3=33,334, which much better fits with the students maximal mean exchanged number, 33,256.

Yet of course, if we begin evaluating different regrets differently, we may find an evaluation that fits with any observation, which is not credible. *Nevertheless, in some contexts, such a way of proceeding seems natural.*

For another interesting example of regret minimization, see Bertrand's game in Exercise 5.

2.3 Regret minimization in the ascending all pay auction/war of attrition: a switch from normal form to extensive form

Once more we come back to the ascending all pay auction in Figure 6.1. Does regret minimization lead to an interesting result?

We first study regret minimization in the normal form (Matrix 6.7a) and then we adapt regret minimization to the extensive form (Figure 6.4).

		Player 2			
		$S_2/x_2 S_2/x_4$	$S_2/x_2 4/x_4$	$2/x_2 S_2/x_4$	$2/x_2 4/x_4$
Player 1	$(S_1/x_1 S_1/x_3 S_1/x_5)$	(5,8)	(5,8)	(5,8)	(5,8)
	$(S_1/x_1 S_1/x_3 5/x_5)$	(5,8)	(5,8)	(5,8)	(5,8)
	$(S_1/x_1 3/x_3 S_1/x_5)$	(5,8)	(5,8)	(5,8)	(5,8)
	$(S_1/x_1 3/x_3 5/x_5)$	(5,8)	(5,8)	(5,8)	(5,8)
	$(1/x_1 S_1/x_3 S_1/x_5)$	(7,5)	(7,5)	(4,6)	(4,6)
	$(1/x_1 S_1/x_3 5/x_5)$	(7,5)	(7,5)	(4,6)	(4,6)
	$(1/x_1 3/x_3 S_1/x_5)$	(7,5)	(7,5)	(5,3)	(2,4)
	$(1/x_1 3/x_3 5/x_5)$	(7,5)	(7,5)	(5,3)	(3,1)

Matrix 6.7a

SOLVING A GAME DIFFERENTLY

In the normal form game, when player 1 plays $(S_1/x_1, S_1/x_3, S_1/x_5)$, his maximal regret is *2* because she could get 7 instead of 5 by playing $(1/x_1, S_1/x_3, S_1/x_5)$ if player 2 plays $(S_2/x_2, S_2/x_4)$ for example (see the bold italic terms). The same is true if player 1 plays $(S_1/x_1, S_1/x_3, 5/x_5)$, $(S_1/x_1, 3/x_3, S_1/x_5)$ or $(S_1/x_1, 3/x_3, 5/x_5)$. When player 1 plays $(1/x_1, S_1/x_3, S_1/x_5)$, her maximal regret is *1* because she could get 5 instead of 4 by playing $(S_1/x_1, S_1/x_3, S_1/x_5)$ if player 2 plays $(2/x_2, S_2/x_4)$ (see the underlined terms). The same is true for $(1/x_1, S_1/x_3, 5/x_5)$. If player 1 plays $(1/x_1, 3/x_3, S_1/x_5)$, her maximal regret is *3* because she could get 5 instead of 2 by playing $(S_1/x_1, S_1/x_3, S_1/x_5)$ if player 2 plays $(2/x_2, 4/x_4)$ (see the bold underlined terms). And if player 1 plays $(1/x_1, 3/x_3, 5/x_5)$, her maximal regret is *2* because she could get 5 instead of 3 by playing $(S_1/x_1, S_1/x_3, S_1/x_5)$ if player 2 plays $(2/x_2, 4/x_4)$ (see the bold underlined terms). *So the strategy that minimizes her regret consists of bidding 1 at x_1 and stopping at x_3.*

When player 2 plays $(S_2/x_2, S_2/x_4)$, his maximal regret is *1*, given that he could get 6 instead of 5 by playing $(2/x_2, S_2/x_4)$ when player 1 plays $(1/x_1, S_1/x_3, S_1/x_5)$ (see the bold italic terms). The same is true if player 2 plays $(S_2/x_2, 4/x_4)$. When player 2 plays $(2/x_2, S_2/x_4)$, his maximal regret is *2*, given that he could get 5 instead of 3 by playing $(S_2/x_2, S_2/x_4)$ when player 1 plays $(1/x_1, 3/x_3, S_1/x_5)$ (see the underlined terms). When player 2 plays $(2/x_2, 4/x_4)$, his maximal regret is *4*, given that he could get 5 instead of 1 by playing $(S_2/x_2, S_2/x_4)$ when player 1 plays $(1/x_1, 3/x_3, 5/x_5)$ (see the bold underlined terms). *So the strategy that minimizes his regret is to stop at his first decision node.*

Well, with regret minimization, the game ends with the backward induction payoffs (7,5), but player 1 would stop at x_3 if called on to play at this node.

By the way, we observe that Halpern and Pass's criterion gives the same result when working on the reduced normal form (Matrix 6.7b) in that the bold, italic, underlined and bold underlined terms are still present.

		Player 2		
		(S_2/x_2)	$(2/x_2 S_2/x_4)$	$(2/x_2 4/x_4)$
Player 1	(S_1/x_1)	(**5**,8)	(5,8)	(**5**,8)
	$(1/x_1 S_1/x_3)$	(*7*,*5*)	(4,6)	(4,6)
	$(1/x_1 3/x_3 S_1/x_5)$	(7,5)	(5,3)	(**2**,4)
	$(1/x_1 3/x_3 5/x_5)$	(7,5)	(5,3)	(**3**,**1**)

Matrix 6.7b

We now try *to adapt Halpern and Pass's reasoning directly to the extensive form* (Figure 6.4)

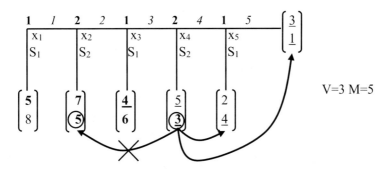

Figure 6.4

SOLVING A GAME DIFFERENTLY

Observe player 1 at x_1: when she stops, her maximal regret is *2* because she could get 7 by going one step further instead of 5 obtained by stopping. When she bids 1, her maximal regret is *1*, because she can get 4 (if player 2 goes on and if she stops at x_3), instead of 5 obtained by stopping. So we say that, given that 2>1, going on (bidding 1) *locally minimizes her regret at x_1* (see her bold payoffs in Figure 6.4).

At x_3, when she stops, her maximal regret is *1* because she could get 5 by going one step further instead of 4 obtained by stopping. When she bids 3, her maximal regret is *1*, because she can get 3 (if player 2 goes on and she bids 5 at x_5), instead of 4 obtained by stopping. So, given that 1=1, we say that going on (bidding 3) and stopping *locally minimize her regret at x_3* (see her underlined payoffs in Figure 6.4).

At x_5, when she stops, her maximal regret is 1 because she could get 3 by bidding 5 instead of 2 obtained by stopping. When she bids 5, her maximal regret is 0, because it's her best decision. So her *local minimizing regret decision* is to bid 5 at node x_5.

We now study player 2's behaviour. At x_2, when he stops, his maximal regret is *1* because he could get 6 by bidding 2 instead of 5 obtained by stopping. When he bids 2, his maximal regret is 2, because he can get 3 (if player 1 goes on and if he stops at x_4), instead of 5 obtained by stopping. So stopping *locally minimizes his regret at x_2* (see his bold payoffs).

At x_4, when he stops, his maximal regret is 1, because he could get 4 by bidding 4 instead of 3 obtained by stopping. When he bids 4, his maximal regret is 2, because he can get 1 (if player 1 bids 5), instead of 3 obtained by stopping. So stopping *locally minimizes his regret at x_4* (see his underlined payoffs).

This result contrasts with the one obtained for the normal form, as concerns player 1's decision (player 2 stops at his first decision node regardless of the chosen form). Player 1, in the extensive form, goes on at x_1 and x_5 and can also go on at x_3, whereas she necessarily stops at x_3 (and has no constraint at x_5) in the normal form. So in the extensive form approach we are, in this game, quite close to the backward induction path.

Why do we get different results in the normal and in the extensive form game? Look at player 2 for example. In the normal form we assign regret 4 to the strategy $(2/x_2,4/x_4)$ because it leads to the payoff 1, whereas the strategy $(S_2/x_2,S/x_4)$ leads to the payoff 5, when player 1 plays $(1/x_1,3/x_3,5/x_5)$. There is no regret 4 in the extensive form, because when player 2 decides to stop or to bid 4 at x_4, he is at x_4, and so he can no longer get the payoff 5, *because he cannot come back to an earlier decision node* (see the circled numbers and the impossible arrow in Figure 6.4). So he just has to compare 3 to 4 and 3 to 1 (the possible arrows), which leads to maximal regrets 1 and 2 and consequently to stop at x_4.

2.4 The limits of regret minimization

2.4.1 A deeper look into all pay auctions/wars of attrition

In the previous game, we saw that the way of estimating regrets is clearly different in the normal and extensive forms. Should we work in normal or extensive form? Aren't the results somewhat artificial?

At least in the extensive form approach, observe that the local regrets highly depend on the bid increment and on the value of the object: in the game in Figure 6.4, it is because the increment of the bids (2 for a player) is higher than the potential additional win (V–2=1) that player 2 always stops.

Consequently, in games with large values of V, everybody may always go on in the extensive form game, which leads us far from backward induction, *but also far from realism*.

We illustrate this remark with the game below in Figure 6.5, with V=5 and M=9.

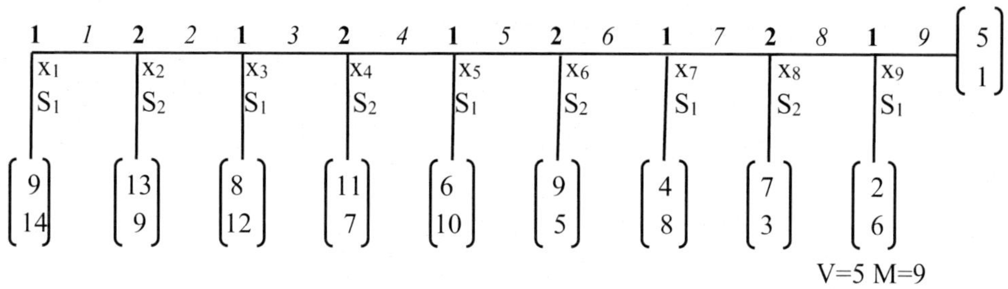

SOLVING A GAME DIFFERENTLY

```
 1  1   2  2   1  3   2  4   1  5   2  6   1  7   2  8   1  9    ⎡5⎤
|x₁    |x₂    |x₃    |x₄    |x₅    |x₆    |x₇    |x₈    |x₉       ⎣1⎦
|S₁    |S₂    |S₁    |S₂    |S₁    |S₂    |S₁    |S₂    |S₁

⎡9 ⎤   ⎡13⎤   ⎡8 ⎤   ⎡11⎤   ⎡6 ⎤   ⎡9⎤   ⎡4⎤   ⎡7⎤   ⎡2⎤
⎣14⎦   ⎣9 ⎦   ⎣12⎦   ⎣7 ⎦   ⎣10⎦   ⎣5⎦   ⎣8⎦   ⎣3⎦   ⎣6⎦
                                                V=5 M=9
```

Figure 6.5

At x_1, player 1's maximal regret is 4 (=13–9) when she stops and 1 when she goes on (=9–8), so she goes on. At x_3, her maximal regret is 3 (=11–8) when she stops and 2 when she goes on (=8–6), so she goes on. At x_5, her maximal regret is 3 (=9–6) when she stops and 1 when she goes on (=6–5), so she goes on. At x_7, her maximal regret is 3 (=7–4) when she stops and 0 when she goes on (because she can at least get 5 instead of 4 by stopping), so she goes on. At x_9, she goes on (best response, no regret). *So player 1 goes on at each decision node (as in the backward induction path).*

Yet things, as conjectured, are similar for player 2. At x_2, his maximal regret is 3 (=12–9) when he stops and 2 (=9–7) when he goes on, so he goes on. At x_4, his maximal regret is 3 (=10–7) when he stops and 2(=7–5) when he goes on, so he goes on. At x_6, his maximal regret is 3 (=8–5) when he stops and 2 when he goes on (=5–3), so he goes on. At x_8, his maximal regret is 3 (=6–3) when he stops and 2 when he goes on (=3–1), so he goes on. *It derives that player 2 goes on at each decision node, in contrast to his backward induction stopping behaviour. The least we can say is that player 2's behaviour at x_8 is rather strange because it is difficult to expect player 1 stopping at x_9!*

Clearly, the reason of this continual behaviour is that V–2=3 (the value of the object minus the two additional units necessary to overbid) is higher than 2 (the loss incurred by stopping at the next decision round rather than at the present one, because of the two additional units of bid).

We do not get the same comparisons with the reduced normal form (it leads to the same results as the normal form) given in Matrix 6.8a.

		Player 2			
	(S_2/x_2)	$(2/x_2$ $S_2/x_4)$	$(2/x_2$ $4/x_4 S_2/x_6)$	$(2/x_2 4/x_4$ $6/x_6 S_2/x_8)$	$(2/x_2 4/x_4$ $6/x_6 8/x_8)$
(S_1/x_1)	(9,14)	(9,14)	(9,14)	(9,14)	(9,14)
$(1/x_1 S_1/x_3)$	(13,9)	(8,12)	(8,12)	(8,12)	(8,12)
$(1/x_1 3/x_3 S_1/x_5)$	(13,9)	(11,7)	(6,10)	(6,10)	(6,10)
$(1/x_1 3/x_3 5/x_5 S_1/x_7)$	(13,9)	(11,7)	(9,5)	(4,8)	(4,8)
$(1/x_1 3/x_3 5/x_5 7/x_7 S_1/x_9)$	(13,9)	(11,7)	(9,5)	(7,3)	(2,6)
$1/x_1 3/x_3 5/x_5 7/x_7 9/x_9$	(13,9)	(11,7)	(9,5)	(7,3)	(5,1)

Player 1 labels the rows.

Matrix 6.8a

SOLVING A GAME DIFFERENTLY

You can check that (S_1/x_1) leads to the maximal regret 4, $(1/x_1S_1/x_3)$ and $(1/x_13/x_3S_1/x_5)$ lead to the maximal regret 3, $(1/x_13/x_35/x_5S_1/x_7)$ leads to the maximal regret 5, $(1/x_1,3/x_3,5/x_5,7/x_7,S_1/x_9)$ leads to the maximal regret 7, and $(1/x_1, 3/x_3, 5/x_5, 7/x_7, 9/x_9)$ leads to the maximal regret 4. So bidding 1 and stopping thereafter, or overbidding 3 and stopping thereafter minimize player 1's regret.

(S_2/x_2) and $(2/x_2S_2/x_4)$ lead to the maximal regret 3, $(2/x_24/x_4S_2/x_6)$ leads to the maximal regret 4, $(2/x_2,4/x_4,6/x_6,S_2/x_8)$ leads to the maximal regret 6, and $(2/x_2,4/x_4,6/x_6,8/x_8)$ leads to the maximal regret 8. So stopping immediately or bidding 2 and stopping thereafter minimize player 2's regret.

So the game stops with the payoffs (13,9), (8,12) or (11,7).

Whether we work on the normal or on the extensive form, we give much weight to V–2=3, given that stopping, in comparison to overbidding one time leads to this loss if the opponent stops at the next round (this appears in bold (for player 1), in bold and italic (for player 2) on the diagonal in Matrix 6.8a). Yet, in the normal form, a strategy that stops at x_i is also compared to the decision to go out immediately, so, when deciding to stop or not at x_i, you also compare M–i+2 to M and this comparison leads to a higher regret as soon as i–2>V–2, hence i>V (see the comparisons of underlined numbers). So a player minimizes regret by stopping at x_{i-2} with i the first number higher than V, which means that he overbids till at most i–4, with i the first number higher than V; so he bids up to V–3 or V–2 (in our game, i–4=7–4=3 for player 1, i–4=6–4=2 for player 2).

Why V–2 or V–3? Well, if you bid V–2 and if the opponent *does not stop* at the next decision node, you will have to bid V if you still want to get the object at the next decision node, hoping that your opponent will not overbid. So the behaviour expresses *a limited cautious* behaviour in that the player keeps the possibility of investing one more time without losing money.

2.4.2 Regret minimization: a too limited rationality in easy games

In the ascending auction game, the minimizing regret concept seems more convincing in the normal form than in the extensive form. But, whatever form we work with, this concept has a flaw: it assumes a too limited rationality.

Consider again the extensive form game in Figure 6.5 and player 2 at node x_8; even very limited rationality is enough to understand that player 1 will bid 9 at x_9, and so player 2 should stop and not bid 8. He should have no regret stopping because it is completely inappropriate to expect an additional win of 3 by bidding 8.

So let's study the easier game that begins at x_8. Its normal form is given in Matrix 6.8b.

		Player 2	
		S_2/x_8	$8/x_8$
Player 1	S_1/x_9	(7,3)	(2,6)
	$9/x_9$	(7,3)	(5,1)

Matrix 6.8b

For this easy game, normal and extensive forms lead to the same comparisons of payoffs. Player 2 plays $8/x_8$ which yields the maximal regret 2, whereas (S_2/x_8) yields the maximal regret 3. Yet, given that player 1 never plays S_1/x_9, it is unreasonable to fear a regret 3 that will never happen! So in easy games, the lack of rationality behind regret minimization becomes obvious.

This leads us back to the origin of regret minimization. At the beginning, it is a concept for decision theory, where a player is confronted by different states of nature. Halpern and Pass adapt it to game theory, by replacing the states of nature by the profiles of strategies of the other players. Yet there is a big difference between states of nature and strategies. Whereas it is reasonable to expect any state of nature (because Nature is neutral), it may be silly to expect some profiles of strategies, because they will never be played if players are rational!

In other words, Halpern and Pass's criterion better suits games where each possible opponents' profile of strategies may be rational: the traveller's dilemma belongs to these games as each amount x is the best answer to the amount x+1 (and 100 is a good amount if the opponent plays 100 too, even if it is not the best answer).

To conclude, we add the following remark. In the previous section, we said that best reply matching exhibits a kind of asymmetry, because it only leads to counting the situations where a strategy is the best response. Minimizing regret leads to *the opposite asymmetry,* given that we only focus on the worst payoff differences: why not also take into account the payoff differences in favour of a strategy?

3 LEVEL–*K* REASONING

Level-*k* reasoning, developed by many authors, among them Stahl and Wilson (1994)[14] and Nagel (1995),[15] is an old idea linked to limited heterogeneous rationality and conjectures. Put simply, the idea is that in an N-player game players are heterogeneous with regard to their ability to run a reasoning. Some of the players are of level-0; i.e., they choose randomly, so play all the strategies at hand with the same probability. Other players are of level-1; i.e. they assume that the other players are of level-0 and they best react to them. And so on.

> **Definition 5: level-*k* reasoning**
> Players of level-0 usually randomize uniformly on their strategy set.[16] Players of level-*k* best respond to players of level k-1, randomizing uniformly in case of indifference.

This definition is sometimes referred to as 'original level-*k*': it is often generalized, giving rise to *cognitive hierarchy theory* (see Camerer Ho and Chong (2004)[17]), according to which a level-*k* player best responds to a probability mix of level-0, level-1,...and level-*k*-1 strategies. But we mainly work with definition 5.

3.1 Basic level-*k* reasoning in guessing games and other games

3.1.1 Basic level-k reasoning in guessing games

Level-*k* reasoning has mainly been illustrated on the guessing game (first exposed in Moulin 1986[18] and also studied by Nagel 1995), *another story game.*

SOLVING A GAME DIFFERENTLY

> **Definition 6: the guessing game**
> In the N player guessing game, each player has to choose a real number in [0,100]. The winner is the player closest to p times the average of all chosen numbers, where p belongs to]0,1[. In case of a tie, all the winners share equally the amount of the win.
> Very often p=2/3, and we will keep this value.

We first recall *(see Exercise 14 in chapter 3)* that the only Nash equilibrium of this game leads all players to play 0. The reason for this is that a player is always better off when he proposes less than the mean of the proposed values, down to 0 in theory.

But in practice, many players reason as follows: They understand the need to propose a rather small number, some of them are even able to run the reasoning that ends on 0, but they fear that other players are unable to do so. And of course, if it is fine proposing a low number when others also propose low numbers it would be inappropriate to propose 0 when the others all propose 60 for example (60 is closer than 0 to 40 (≈60×2/3)).

So, the idea is that each player is able to run the downward reasoning a certain number of steps (levels) but that this number of steps differs among the players.

Especially, level-0 players are not able at all to run a reasoning and they play randomly: so the mean value they are supposed to choose is 50. Level-1 players play as if the other players were level-0 players, so propose a number close to 2/3×50, i.e. 33. Agranov et al. (2012)[19] looked for how many people are able to run this level-1 reasoning by asking a player to play against a computer, supposed to randomize on [0,100]. Agranov et al. observed that 49% of the subjects chose numbers between 30 and 37, which led them to conclude that about 50% of the players are able to run one step of reasoning (which, according to Agranov, coincides here with the optimal behaviour).

Now suppose that all your opponents run (exactly) one step of reasoning: so they are all level-1 players and play a value around 33. If you best reply in this context (you are a level-2 player) then you have to propose a value around 2/3×33, i.e. around 22. Agranov et al. (2012) tested the presence of level-2 reasoning by asking players to play against other players, which they presented as graduate students used to playing this game. They also proposed playing against a group of seven opponents, x of them being computers, 7–x of them being graduate experienced students, so as to observe if the number of level-2 players grows with the proportion of graduate students. Agranov et al. observed that the distribution of choices shifts to the left when the proportion of graduate students rises (some of them even propose 0), but that the mean proposed number is still high, around 31, even in the presence of seven graduate students.

I mention here the results I got with a class of (38) M1 students; I asked them to think the game over for a week, propose a number (between 0 and 100) and explain to me why they chose it. I mention that my students never studied level-*k* reasoning before, but that most of them studied iterative elimination of dominated strategies in a previous game theory course. The results are: 10 students proposed a number but gave no explanation, 4 looked for focal numbers or intuition, 5 of them discovered the Nash value 0 and played low numbers, 4 worked with iterative dominance, but 15 of them clearly made a level-*k* reasoning. Eight students, i.e. 21% of the students, hold the level-1 reasoning and played around 33, 2 students, i.e. 5%, hold the level-2 reasoning and played 22.22, and *5 students (13%) explicitly hold a level-3 reasoning* and played around 14,80. So clearly 39% of my students hold a level-*k* reasoning, despite never having heard about such a reasoning in previous lectures.

3.1.2 Basic level-k reasoning in the envelope game and in the traveller's dilemma game

We return to two other games where level-*k* reasoning may seem natural.

We first focus on the envelope game, with K=100 (to avoid the presence of too many risk adverse players). In this two player game, 37.7% of my students proposed to exchange their envelope only if it contains an amount lower than 50, 33.6% proposed exchanging it up to the value 50, and 28.7% proposed exchanging it even for values higher than 50.

Many of my students confessed that they played as if the opponent always accepts exchanging his envelope. And they also confessed that they supposed that the values are uniformly distributed. So clearly, the expected amount in the opponent's envelope is 50, and it is optimal to exchange up to 50 (maximization of mean payoff), which explains the high percentage of players ready to exchange up to 50 (see the peak on 50 in Figure 6.3a).

Well, what is a level-0 player in this game? Can we say that it is an opponent that is ready to exchange all his envelopes (as supposed by many of my students), or is it a player that exchanges up to the mean value 50? This question is not so easy to answer and raises a first problem of level-*k* reasoning: *what is the level-0 reasoning in a general game*, not as trivial as the guessing game (see Heap, Arjona and Sudgen 2014[20] for a study on level-0 behaviour)? Let us suppose that the level-0 player is one that exchanges all his envelopes, so the students who exchange up to the value 50 are level-1 players (so it seems that 1/3 of my students played like level-1 players). What is a level-2 player in this game? If a player expects that the opponent exchanges his envelope up to 50, then the mean amount contained in his envelope is 25, and the player should be ready to exchange up to 25. And a level-3 player should be ready to exchange up to 12, because the mean value exchanged by a level-2 player is 12.5. And so on, until a level-6 player should be ready to exchange it up to 1. So, rather quickly, in fewer steps than with the iterative elimination of weakly dominated strategies, the level-*k* reasoning leads to the conclusion that a player should not exchange an envelope unless it contains the number 1.

In passing, let us observe (see Figure 6.3a) that it is not clear that, for K=100, many of my students are level-2 players. There is a small peak around 25 (14.4% of the students exchange up to a value in [20,30]) but there are also other small peaks higher than 50 (17.1% of the students exchange up to a value in [60,70])!

The data conform more to level-2 behaviour for K=100,000, in that we have a peak around 25,000 (21.2% of the students exchange up to a value in [20,000, 30,000] and we have no peak for values higher than 50,000. We have also about 1/5 of the students that exchange only up to 5,000; this may be due to higher degrees of level-*k* reasoning (a level-5 reasoning leads to exchange at most 3,125) but surely not exclusively.

As a matter of fact, for K=100,000, the notion of risk aversion is difficult to omit. So, even a level-1 player can play around 25,000. As a matter of fact, even if a player supposes that the opponent exchanges all his envelopes, and so expects a mean value of 50,000, he may be ready to exchange only envelopes that contain less than 25,000 because he is risk adverse and not because he runs a higher level-2 reasoning.

We now return to the traveller's dilemma studied in the previous section. What is a level-0 player in this game? Somebody who randomizes on [2,100]? Let us assume it, even if it is not very natural here: *in the guessing game, when you are the winner, you always get the prize, say 1, but in the traveller's dilemma the amount you get depends on what is played; if you play 2 you get at most 2+P, which is low if P is low, whereas you can get 80+P when you play 80*. So, even somebody

SOLVING A GAME DIFFERENTLY

that is not able to run a reasoning may hesitate to ask for very low amounts when P is low. So the level-0 player is not easy to define. But suppose that it is somebody who plays any value between 2 and 100. Hence a level-1 player a priori proposes 50, because the opponent's mean expected demand is 51, and 50 is the best reply to 51. Then a level-2 player plays 49, because 49 is the best answer to 50, and we come back to a process which seems similar to iterative elimination of dominated strategies, except that it starts at 51 instead of 100. This is however not exactly true. A level-1 player does not necessarily best react to the demand 51: *maximizing against a player that uniformly randomizes on a set of values is not equivalent to maximizing against a player that plays the mean value of the set*. This is illustrated in the next section and in Exercise 9 in this chapter. But the iterative decreasing process happens regardless of the level-1 demand.

3.2 A more sophisticated level-*k* reasoning in guessing games

The problem with level-*k* reasoning in the guessing game is that it is more complicated than presented above, as observed by Breitmoser (2012).[21] So imagine that the other players are level-0 players and that they play 50 on average. Suppose you play x lower than 50. Then you win each time x is closer to 2/3 times the average, that is to say if:

$$((N-1)50+x)2/(3N)-x<50-((N-1)50+x)2/(3N)$$

If N is large, this inequation can be approximated by: 200/3–50<x, i.e. x>50/3. That is to say, a level-1 player can play any number between 50/3 and 50 and not only 100/3. In other words, Agranov et al. (2012) were wrong as they claimed that only subjects playing between 30 and 37 best respond. They should have added all the players playing between 16.67 and 30, and those playing between 37 and 50, which gives a much higher percentage, close to 100%!

Now there is another problem. What is a level-2 player if a level-1 player can reasonably play between 50/3 and 50? In accordance with the definition, given that the players are indifferent between all the values between 50/3 and 50, level-1 players randomize equally on these values so the mean value they play is (50/3+50)/2=100/3, which is Agranov's unique best response.

Now imagine that you are a level-2 player and that you play x. You win each time:

$$((N-1)100/3+x)2/3N-x<100/3-((N-1)100/3+x)2/3N$$

i.e. x>100/9. So a level-2 player can play any value between 100/9 and 100/3, i.e. between 11.11 and 33.33, and not only the mean value 200/9.

So, first a level-*k* player has no reason to only play one value: there is a range of best responses, and 2y/3, where y if the mean number played by the level k-1 players, is only the mean value played by level-*k* players, if these players randomize equally between all the winning values, i.e. the range]y/3,y[.

Second, as a consequence, observing a mode at 33 (level-1 players), or around 22 (level-2 players), is rather an annoying fact, in that it does not coincide with the assumption "they randomize equally between all best-responses" in definition 5. In some way a more adequate definition of a level-*k* player, that fits with Agranov's mode observations and with their treatment of best responses would be:

SOLVING A GAME DIFFERENTLY

> **Definition 7: mean level-k behaviour**
> A level-k player best responds to a level $k-1$ player; in case of indifference, he only plays the mean of the best responses.

But why should a player behave in this way? If we keep definition 5's randomizing behaviour, another consequence is that it becomes very difficult to distinguish a level-1 player from a level-2 player because the values between 16.67 and 33.33 can be played by both types of players!

Third, there is an additional problem. As already mentioned above, playing against level-0 players means playing against players that play the uniform distribution on [0,100]. This is not identical to playing against players that play the mean value 50.

We illustrate this point for N small, N=3.

According to the basic level-k reasoning a level-1 player best reacts to the value 50, and plays $x=2/3\times(2\times50+x)/3$, so $x=200/7=28.57$ (we cannot write $x=2/3\times50=33.33$ because N is too small and the value played by the third player has a strong impact on the mean). According to the previous remark, a level-1 player, playing x, wins as soon as:

$(100+x)2/9 - x < 50 - (100+x)2/9$, which is true for any number lower than 50.

Yet, neither playing only 28.57, nor playing any number lower than 50 is the adequate level-1 behaviour *because best reacting to the uniform distribution on [0, 100] is not equivalent to best reacting to the mean value 50.* Following Breitmoser, let us establish the right level-1 behaviour for N=3. Suppose that player 1 is the level-1 player and that players 2 and 3 are the level-0 players.

W.l.o.g, we set $x_2 < x_3$. Player 1 wins:

- each time $x_1 < x_2 < x_3$ and $2(x_1+x_2+x_3)/9 - x_1 < x_2 - 2(x_1+x_2+x_3)/9$, i.e. $x_3 < (5x_1+5x_2)/4$ (observe that $(5x_1+5x_2)/4 < 100$ if $x_2 < 80-x_1$)
- and each time $x_2 < x_1 < x_3$ and $x_1 - 2(x_1+x_2+x_3)/9 < 2(x_1+x_2+x_3)/9 - x_2$, i.e. $x_3 > (5x_1+5x_2)/4$ (observe that if $x_1 < 40$, then $x_2 < 40$ and $(5x_1+5x_2)/4 < 100$)

Given that x_1's optimal value will be lower than 40, player 1 maximizes on x_1 the following sum of integrals:

$$\int_{x_1}^{80-x_1} (\int_{x_2}^{\frac{5x_1+5x_2}{4}} f(x_3)dx_3) f(x_2)dx_2 + \int_{80-x_1}^{100} (\int_{x_2}^{100} f(x_3)dx_3) f(x_2)dx_2 + \int_{0}^{x_1} (\int_{\frac{5x_1+5x_2}{4}}^{100} f(x_3)dx_3) f(x_2)dx_2$$

This maximization leads to $x_1=25.81$ which is both quite far from 28.57 and quite far from the interval [0,50]!

Now imagine that player 1 is a level-2 player and players 2 and 3 are level-1 players, hence play 25.81. It is easy to observe that 0 is one among player 1's best answer given that: $(2/9) (2\times25.81) - 0 < 25.81 - (2/9) (2\times25.81)$. *So the players can play the Nash equilibrium in only two steps.*

Finally, why should a level-k player imagine only being confronted with level $k-1$ players? Let us for a moment come back to cognitive hierarchy theory and to N large. Level-0 players are supposed to play the field and level-1 players are supposed to play $50\times2/3$ (we here omit Breitmoser's observation). According to cognitive hierarchy theory, a level-2 player can suppose

275

that the population is compounded of a proportion p of level-0 players and a proportion 1–p of level-1 players. In that case he plays (p50+(1–p)100/3)2/3=(100p+200)/9, a value that goes from 200/9 to 100/3 when p goes from 0 to 1.

So clearly, when we add all the remarks in this paragraph, we can almost justify each behaviour observed in an experimental setting. So level-k reasoning does not really help us to make a prediction on the way people will behave. Only one result seems to emerge. When N is low, then level-k behaviours can fast converge to 0, especially if we eliminate the assumption that in case of indifference all the numbers are played with the same probability.

3.3 Level-*k* reasoning versus iterated dominance, some limits of level-*k* reasoning

3.3.1 Risky behaviour and decreasing wins

We will adopt, in this paragraph, definition 7 in order to better cope with Agranov's modes, and return to the envelope game (with K=100). Again suppose that, like in the guessing game, a level-0 player exchanges all his envelopes. So the mean amount in the envelope is 50, a level-1 player is ready to exchange all his envelopes up to 100(1/2) and a level-2 player is ready to exchange his envelope up to 25=100(1/2)².

This is similar to the guessing game, where the mean number played by level-0 players is also 50, the mean number played by a level-1 player is 50 (2/3), and the mean number played by a level-2 player is 50 (2/3)².

Yet, despite this similarity, there are two big differences between the two games.

In the guessing game, when you are a level-*k* player (with definition 7) and the others are level-*k*-1 players, then you are sure to win the game.

This is not the case in the envelope game. When you are for example a level-2 player and you exchange up to 25, given that the level-1 player exchanges his envelope up to 50, you can either win or lose. On the one hand, when there is a 24 in your envelope and a 2 in your opponent's, you lose money by exchanging and you regret your decision. On the other hand, when there is a 27 in your envelope and a 45 in your opponent's, you do not exchange (because you stop exchanging at 25) and this is a pity because you could have won money by exchanging. So, whereas a level-*k* player (as defined in definition 7) in the guessing game never regrets his decision, a level-*k* player in the envelope game can regret his decision. This also explains why the notion of risk we talked about in a previous paragraph can't be forgotten in the envelope game.

We point out another particularity of guessing games. In these games you win the prize, say 1, each time you propose the number that is closest to 2/3 times the average proposition. So you get 1 if you are a level-*k* player and the others are of level *k*-1: even if you play 50 (2/3)^k, which may be close to 0, you get the (maximal) win 1.

This is not true in the envelope game. If you are a level-*k* player and the other is a level *k*-1 player, you never exchange your envelope when its amount is higher than $100/2^k$ and your opponent only exchanges it when it is lower than $100/2^{k-1}$. So you get at best $100/2^{k-1}$; for example you get at best 6 by exchanging when k=5! By contrast, if the opponent is a level-0 player and you are a level-1 player, then you can get up to 100. So the wins differ with the ability to run a multilevel reasoning. Whereas, in the guessing game, the level-*k* player (opposed to level *k*-1 players) gets the sure (maximal) win 1 regardless of k, in the envelope game, the level-*k* player (opposed to a level *k*-1 player) is not sure to win and his win is strictly decreasing in k. *In other words, in the envelope* game, there is a phenomenon not observed in the guessing game: *a player hopes that the opponent runs only a very few steps of reasoning.*

The traveller's dilemma is intermediary between the envelope game and the guessing game. We suppose that a level-0 player asks for any amount between 2 and 100 with the same probability, so the mean asked amount is 51. So let's assume that the level-1 player asks for 50^{22}, so that the level-k player asks for 51–k. *As in the guessing game (with definition 7), a level-k player will always win against a level k-1 player (because he asks for a lower amount) but, as in the envelope game, his win is strictly decreasing in k, because he gets 51–k+P.* So, as in the envelope game, a player hopes that the opponent runs only a very few steps of reasoning.

3.3.2 Level-k reasoning is riskier than iterated dominance

What is the link between level-k reasoning and iterative elimination of dominated strategies (IEDS)?

I would say that it is difficult (risky) to establish a level-k strategy, in that it depends on the level-0 strategy, which is often difficult to define. This difficulty does not arise with iterative elimination of dominated strategies.

Let us briefly recall this elimination in the guessing game *(see Exercise 9 in chapter 2)*.

In the guessing game, at the first step, a player only eliminates the numbers higher than 100×2/3. As a matter of fact, the only thing he is sure about, is that the opponents do not play more than 100. So their mean value is at most 100, and he can only be better off playing 100×2/3=66.66 than playing more. If rationality is common knowledge, everybody is able to do the same reasoning, hence does not play more than 66.66; a player knows this fact, so does not play more than $100 \times (2/3)^2$. Following this reasoning, we get a decreasing sequence of values, 66.66, 44.44, $100(2/3)^k$, which is different from the level-k one, but which ends with the same value 0.

But the motivations of the sequence are different. The calculi do not depend on the behaviour of a strange level-0 player. So your behaviour is not risky as long as you do not overestimate the rationality of the other players. So imagine that you are sure that the others are able to deduce that they should not play more than 66.66. If you are not sure that they can run the reasoning further, you can choose not to propose more than 66.66×2/3=44.44. By doing so, you are not sure to win, but you are sure to be in the good range of values [0, 44.44]. More generally, if you are sure that the others have k-1 crossed levels of rationality, then it is good proposing a number not higher than $100(2/3)^k$.

The same observation can be made for other games.

In the envelope game, for example, in the first step, you only eliminate the strategies that consist of exchanging the envelope with the number K (let's say 100). In the second step, you suppose that nobody exchanges 100, so you have no reason to exchange 99. In the third step, you suppose that everybody is sufficiently clever not to exchange 100 and 99, and so you don't exchange 98. And so on: if you think that the opponent has k-1 levels of crossed rationality, you do not exchange envelopes that contain more than 100–k. This does not mean that you are sure to get a lot of money, and *this does not mean that you will exchange values lower than 100–k*. It only means that *you will not exchange numbers higher than 100–k,* and that you will not regret this decision.

3.3.3 Something odd with basic level-k reasoning?

I would say that there is something odd with basic level-*k* reasoning. If you play against a (randomizing) computer, then you are right starting with a level-0 opponent. But, if you suppose that there are only level *k*-1 opponents, which justifies your level-*k* behaviour, then *there are no level-0 players in the population, but the whole reasoning is based on their existence.*

Let me be more precise: even if only level *k*-1 opponents and you exist (you are a level-*k* player), you optimize your behaviour against the level *k*-1 players, who *optimize their behaviour against non existing level-k-2 players*, …who optimize … against non existing level-0 players.[23] Isn't this strange? This behaviour does not occur with IEDS in that additional levels of rationality do not presuppose the existence of a stupid reasoning. In the guessing game for example, with IEDS, when I do not play more than 44.44, it is because I think that my opponents are able to understand that it is not clever to play more than 66.66. And if they play in this way it is because they indeed realize that no player can play more than 100, and so that it is not clever to play higher values than 66.66. *They do not base their reasoning on the existence of less clever players.* What is more, when I decide not to play more than 44.44, I am not even sure that there will not exist players that react like myself or even that run additional steps of reasoning. I just make the assumption that players are clever enough to not play more than 66.66.

Well, to my mind, this way of playing is more logical, isn't it?

4 FORWARD INDUCTION

In the two previous sections, we coped with limited rationality. Forward induction belongs to another trend, in that it supposes that people are able to interpret signals sent by other players, sometimes in a sophisticated way. So, because of this sophisticated ability to interpret actions, *forward induction has not got wind in its sails*: it is rather an old topic, mainly developed in the years 1985–1995. Yet, at the beginning, forward induction was a very intuitive concept: when a player chooses an action (expected or not), he may be trying to signal something to the others, who therefore have to react to this signal in order to play in an appropriate way. Forward induction is a sound reaction to backward induction: whereas, by backward induction, only the future actions influence the current ones, by forward induction, the present actions influence the future ones.

But there are different (at least two) ways to proceed.

Information can emerge from unexpected actions. For example, players play the equilibrium path they always play, but suddenly they observe that somebody deviates from this path: in accordance with forward induction, the players try to interpret the deviation in a rational way; i.e. they interpret the player's deviation as a signal about his future behaviour, assuming that he surely deviates in order to get a higher payoff. So we here adopt a sequential approach, in that the deviations have to be observed in order to be analysed. But this approach still gives much weight to traditional game equilibrium concepts (usually backward induction ones) in that a backward induction equilibrium is usually the starting point of the forward induction reasoning (we study actions out of the given backward induction equilibrium path). Yet we know that Nash equilibria and backward induction criteria are not the panacea, so building forward induction on that basis is perhaps not the best way to proceed.

SOLVING A GAME DIFFERENTLY

Another approach consists of constructing a new normal form equilibrium concept that contains forward induction arguments. That is the idea of Kohlberg and Mertens (1986) in their stable equilibrium concept. We will start with this concept.

4.1 Kohlberg and Mertens' stable equilibrium set concept

Forward induction supposes that past actions convey information on future ones. *So past and future actions are linked.* That is why Kohlberg and Mertens (1986)[24] claim that forward induction should be studied on normal form games, given that the normal form spontaneously links the actions at different stages of decisions – this contrasts with the other trend of forward induction based on the analysis of deviations.

More precisely, Kohlberg and Mertens work on the reduced normal form of a game and propose the following criterion:

Definition 8: Kohlberg and Mertens' (1986) Stable Equilibrium Set
A set A of equilibria is stable if it is minimal with respect to the property: $\forall \varepsilon > 0, \exists \delta_0 > 0 / \forall p' \in$ set of completely mixed strategies, and $\forall (\delta_1 \ldots \delta_N)$ with $0 < \delta_i < \delta_0$ for any player i from 1 to N, the perturbed game where each mixed strategy p_i of a player i is replaced by the strategy $(1-\delta_i)p_i + \delta_i p_i'$ leads to a Nash equilibrium that is ε close to A.

Well, we again face a robustness concept. The idea is this time to perturb strategies in any direction and to still get an equilibrium that is ε close to the tested set. By contrast to Selten's concept, this criterion requires robustness against any small perturbations of the strategies, but it does not test the robustness of one equilibrium, but *the robustness of a set of equilibria.* Moreover, Kohlberg and Mertens' notion of perturbation is not exactly Selten's one, given that minimal probabilities in the extensive form and in the normal form of a game are not the same *(see Exercise 11 in this chapter).*

Let us show how stability works on Selten's horse, recalled in Figure 6.6. We showed in chapter 5 that is was not possible to find a Nash equilibrium that resists all kinds of perturbations.

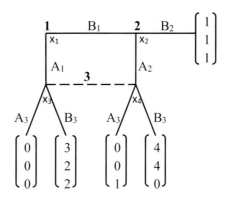

▨ **FIGURE 6.6**

279 ▨

SOLVING A GAME DIFFERENTLY

Now examine the reduced normal form of this game, in Matrices 6.9.

		Player 2	
		A_2	B_2
Player 1	A_1	(0,0,0)	(0,0,0)
	B_1	(0,0,1)	(1,1,1)

A_3
Player 3

		Player 2	
		A_2	B_2
Player 1	A_1	(3,2,2)	(3,2,2)
	B_1	(4,4,0)	(1,1,1)

B_3

Matrices 6.9

We show that a minimal set of equilibria leading to the outcome (1,1,1), hence to the path (B_1,B_2) is stable. We proceed in this way:

> **STUDENTS' INSERT**

Suppose that player 1 plays A_1 with probability p_1, player 2 plays A_2 with probability p_2, and player 3 plays A_3 with probability p_3.

- Player 1 is best off playing B_1 as soon as $(1-p_2)+4p_2(1-p_3) \geq 3(1-p_3)$
- Player 2 is best off playing B_2 as soon as $(1-p_1) \geq 4(1-p_1)(1-p_3)$, i.e. $p_3 \geq \frac{3}{4}$.

When $p_3 \geq \frac{3}{4}$ and p_2 goes to 0, then $(1-p_2)+4p_2(1-p_3) \geq 3(1-p_3)$ is fulfilled.

- Player 3 can play $p_3 \geq \frac{3}{4}$ if $(1-p_1)p_2 \geq 2p_1$ i.e. $p_2 \geq 2p_1/(1-p_1)$

Now introduce perturbations: ε_1 is any small probability to play A_1, ε_2 any small probability to play A_2, ε_3 any small probability to play B_3 (this probability has no impact on the players).

- When $\varepsilon_2 \geq 2\varepsilon_1/(1-\varepsilon_1)$, we set $p_1=\varepsilon_1$, $p_2=\varepsilon_2$ and $1-p_3=\varepsilon_3$, the conditions above are verified, and we get an equilibrium ε close to the path (B_1,B_2)
- When $\varepsilon_2 < 2\varepsilon_1/(1-\varepsilon_1)$, we set $p_2=2p_1/(1-p_1)=2\varepsilon_1/(1-\varepsilon_1) > \varepsilon_2$, which is possible only if player 2 is indifferent between his two actions, i.e. if $p_3 = \frac{3}{4}$. And this is possible given that $p_2=2p_1/(1-p_1)$. So we get again an equilibrium ε close to the path (B_1,B_2)
- So, regardless of the perturbations introduced, the payoffs (1,1,1) are stable: the minimal set of equilibria is $\{(B_1, B_2, A_3), (B_1, B_2, p_3(A_3)=3/4)\}$.

To see what this means in terms of forward induction we present the criterion in more literary terms.

> **STUDENTS' INSERT**

When player 2 trembles more than player 1, $\varepsilon_2 \geq 2\varepsilon_1/(1-\varepsilon_1)$, player 3 is induced to play A_3 which deters player 1 and 2 from deviating. But, if player 2 trembles less than player 1, $\varepsilon_2 < 2\varepsilon_1/(1-\varepsilon_1)$, then player 3 is induced to play B_3, which destroys the outcome (1,1,1), unless player 2 can become indifferent between his both actions, so that he can play $p_2=2p_1/(1-p_1)=2\varepsilon_1/(1-\varepsilon_1)$. This is possible as soon as $p_3=\frac{3}{4}$. For this probability player 2 is indifferent between his two actions and he can play $p_2=2p_1/(1-p_1)=2\varepsilon_1/(1-\varepsilon_1) > \varepsilon_2$. And player 1 is still not induced to deviate, which sustains the outcome (1,1,1).

SOLVING A GAME DIFFERENTLY

It is interesting at this level to come back to our remarks on perturbations in chapter 5. We noticed in chapter 5 that for Selten's horse *a Nash equilibrium* that resists any kind of small perturbations doesn't exist. We have now shown that there exists *a Nash equilibrium outcome*, (1,1,1), that resists any kind of small perturbations: the reason for this is that sometimes (1,1,1) is associated with the Nash equilibrium (B_1, B_2, A_3); sometimes it is associated with the Nash equilibrium $((B_1, B_2, \pi_3(A_3)=\frac{3}{4})$.

What is interesting in the justification of the stability of the outcome (1,1,1) is that player 1 is not induced to deviate when player 2 is indifferent. Things would be different if the payoffs (3,2,2) were replaced by (6,2,2) $((B_1,B_2,A_3)$ is still a perfect equilibrium with these payoffs). In that case, to sustain the outcome (1,1,1), the conditions become:

- Player 1 is best off playing B_1 as soon as $(1-p_2)+4p_2(1-p_3)) \geq 6(1-p_3)$
- Player 2 is best off playing B_2 as soon as $(1-p_1) \geq 4(1-p_1)(1-p_3)$, i.e. $p_3 \geq \frac{3}{4}$.

Yet, when p_2 goes to 0, player 1's condition becomes $p_3 \geq 5/6 > 3/4$.

- Player 3 can play $p_3 \geq 5/6$ if $(1-p_1)p_2 \geq 2p_1$ i.e. $p_2 \geq 2p_1/(1-p_1)$

But this time, when we introduce perturbations such as $\varepsilon_2 < 2\varepsilon_1/(1-\varepsilon_1)$, for player 3's condition to be fulfilled, we need $p_2 \geq 2p_1/(1-p_1) \geq 2\varepsilon_1/(1-\varepsilon_1) > \varepsilon_2$, which is possible only if player 2 can be indifferent between his two actions, i.e. if $p_3=\frac{3}{4}$. But this is not possible because $p_3 \geq 5/6 > \frac{3}{4}$. And so the outcome (1,1,1) does not resist all perturbations.

In literary terms, when player 2 trembles less than player 1, $\varepsilon_2 < 2\varepsilon_1/(1-\varepsilon_1)$, then player 3 is induced to play B_3, which destroys the outcome, unless player 2 can become indifferent between both actions so that he can play $p_2=2p_1/(1-p_1)=2\varepsilon_1/(1-\varepsilon_1)$. But this is only possible if $p_3=\frac{3}{4}$, a condition that cannot be fulfilled because player 1 deviates from B for this value of p_3.

In other words, to sustain the outcome (1,1,1), player 3 needs to put enough beliefs on player 2. When the trembles respect this condition there is no problem. If they do not, in order to still be able to put enough beliefs on player 2, it is necessary that player 2 plays the deviating action more often than required by the trembles. So he has to get the same payoff with his equilibrium action, 1, than with the deviating action, $(1-p_3)$ 4. This is possible in the first case but not in the second, because in the second case player 1 compares 1 and $(1-p_3)$ 6, and so deviates when player 2's payoffs are equalized. *In other words, there is a forward induction content in the Stable equilibrium concept. In the first case, player 2 seems more induced to deviate than player 1, because $(1-p_3)$ $3 \geq 1 \Rightarrow (1-p_3)$ $4 > 1$. Hence a deviation is more easily attributed to player 2 than to player 1, which sustains the equilibrium. But in the second case, player 1 is more induced to deviate than player 2, because $(1-p_3)$ $4 \geq 1 \Rightarrow (1-p_3)$ $6 > 1$. Yet, if player 3 more easily assigns the deviation to player 1, then he plays B_3, which encourages the deviation.*

By the way, also observe that we made a similar observation with replicator equations in chapter 5. We also showed that it was possible to converge to (B_1,B_2,A_3) for the payoffs (3,2,2) but that this was not possible with the payoffs (6,2,2), because of the difference between the player 1 and 2's payoffs (3 and 4, and 6 and 4).

Kohlberg and Mertens show the following property:

SOLVING A GAME DIFFERENTLY

> **Property**: for any game in extensive form, such that no player gets two times the same payoff, there always exists a stable set with a unique outcome that corresponds to a Nash equilibrium outcome.

So Kohlberg and Mertens' criterion combines good properties: robustness against any kind of small perturbations, some forward induction ideas, and an existence result.

4.2 The large family of forward induction criteria with starting points

We now turn to the extensive form approach of forward induction, the analysis of deviations. We do not want to be exhaustive, so we just develop inferior strategies, the interest to deviate and consistent deviations.

Most of the criteria are developed for signalling games, but Kohlberg's (1990) *self-enforcing concept* applies to all extensive form games. So we start with this concept, applied to Selten's horse. Then we turn to signalling games in order to describe other ways to approach forward induction.

4.2.1 Selten's horse and Kohlberg's self-enforcing concept

The self-enforcing concept is close in spirit to Kohlberg and Mertens' stable equilibrium concept, but it perhaps better explains why some players are more induced to deviate.

> **Definition 9: inferior strategy and self-enforcing outcome Kohlberg (1990)[25]**
> Consider an outcome V, i.e. a distribution of probabilities on the vectors of payoffs of a game. A strategy of player i is an ***inferior strategy***, if, at any possible – admissible – Nash equilibrium corresponding to V, player i gets less with this strategy than with V. This strategy is inferior, in that it is never a best response in any admissible Nash equilibrium corresponding to V.
>
> An outcome is ***self-enforcing*** if it corresponds to a perfect equilibrium, not only in the game under study, but also in the game obtained after iterative elimination of inferior, and also strictly and weakly dominated strategies.

Kohlberg starts with a perfect equilibrium outcome. So let us focus on Selten's horse and on the probability distribution that assigns probability 1 to the payoffs (1,1,1), that follow players 1 and 2's action B.

Kohlberg starts by looking for any possible – called *admissible* – Nash equilibrium that sustains this outcome. Clearly player 1 and player 2 have to play B, and player 3, to prevent players 1 and 2 from deviating to A, has to play A_3 with probability p_3 such that:

$3(1-p_3) \geq 1$ and $4(1-p_3) \geq 1$.
Yet $4(1-p_3) \geq 1 \Rightarrow 3(1-p_3) < 1$

It follows that A_1 is never a best response for player 1, because it leads to a lower payoff than the equilibrium one at any admissible equilibrium leading to the payoffs (1,1,1). So A_1 is an inferior strategy for player 1. A probability distribution on outcomes is self-enforcing, if it still corresponds to a perfect equilibrium after the deletion of the inferior strategies. If we suppress A_1 from the game tree, then player 3 plays A_3 when called on to play (because he necessarily plays after player 2's action A_2). It follows that players 1 and 2 play B; so the payoffs (1,1,1) are still a perfect equilibrium outcome, even after suppression of inferior strategies. So the outcome (1,1,1) is self-enforcing. Things are different when the payoffs (3,2,2) are replaced by the payoffs (6,2,2). The probability distribution that assigns 1 to the payoffs (1,1,1), still corresponds to a perfect equilibrium. Player 1 and player 2 still play B, but player 3, to prevent them from deviating to A, now has to play A_3 with probability p_3 such that:

$6(1-p_3) \geq 1$ and $4(1-p_3) \geq 1$.
Yet $6(1-p_3) \geq 1 \Rightarrow 4(1-p_3) < 1$

So this time A_2 is never a best response *for player 2* (because it leads to a lower payoff than the equilibrium one, at any admissible equilibrium leading to the payoffs (1,1,1)). So A_2 is an inferior strategy for player 2. But if we suppress A_2 from the game, player 3, when called on to play, plays B_3 (because he can only be called on to play after A_1), so the unique perfect equilibrium outcome of this new game is (6,2,2) (because player 1 necessarily plays A_1 followed by B_3). Hence, in this case, the probability distribution that assigns 1 to the payoffs (1,1,1) is not self-enforcing.

Observe that Kohlberg uses the same equations as Kohlberg and Mertens, even if he does not introduce perturbations. So it is not surprising that both criteria lead to the same conclusions (see Kohlberg for more links between the two criteria).

4.2.2 Local versus global interpretations of actions

We now study a signalling game in Figure 6.7, in order to present two additional forward induction criteria.

Consider the PBE E in which player 1 plays m_1 regardless of type, followed by r_1, m_2 being followed by r_1, being for example assigned to t_1 with probability 0.6. This equilibrium (its outcome), is neither stable nor self-enforcing. Let us show it:

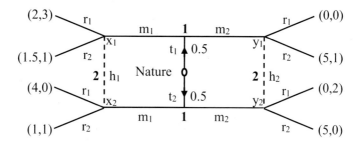

■ Figure 6.7

SOLVING A GAME DIFFERENTLY

> **STUDENTS' INSERT**
>
> E and its associated payoffs do not constitute a stable set; call p_1, respectively p_2, the probability assigned by t_1, respectively t_2, to m_2, and call q the probability assigned by player 2 to r_1 after m_2. For E to be a Nash equilibrium, we need $5(1-q) \leq 2$ and $5(1-q) \leq 4$, hence $q \geq 3/5$. So player 2 has to play r_1 with positive probability which is only possible if $2\mu(y_2) \geq \mu(y_1)$, hence $2p_2 \geq p_1$. Now we introduce small perturbations so that t_1, respectively t_2, plays m_2 with at least probability ε_1, respectively ε_2. If $2\varepsilon_2 \geq \varepsilon_1$, then the equilibrium is automatically sustained. But if $2\varepsilon_2 < \varepsilon_1$, we need, for the equilibrium to be sustained, $2p_2 \geq p_1 \geq \varepsilon_1 > 2\varepsilon_2$. So t_2 has to play m_2 with a probability higher than the minimal one, which requires $5(1-q)=4$, which is incompatible with the condition $5(1-q) \leq 2$, so the equilibrium is not stable.
>
> E is also not self-enforcing. As a matter of fact $5(1-q) \geq 2 \Rightarrow 5(1-q) < 4$. So m_2 is an inferior strategy for t_2. If one deletes this strategy, then m_2 is automatically assigned to t_1, hence followed by r_2 and the equilibrium is destroyed.
>
> So, both the stable equilibrium concept and the self-enforcing one highlight that t_1 is more incited to deviate to m_2 than t_2 (5 is farther from 2 than from 4): they argue that forward induction should take this fact into account and therefore lead player 2 to play r_2 after m_2 which leads t_1 to deviate.

We now propose Banks and Sobel's (1987)[26] ***D criterion***, which only applies to signalling games. Whereas Kohlberg and Kohlberg and Mertens only focus on actions sustaining an outcome, Banks and Sobel look for ***the wish to deviate*** from an equilibrium. So they start with a given equilibrium (usually a PBE) E, and they establish beliefs after each deviating action, these beliefs fitting with the fact that some players are more induced to deviate than others.

We first define Banks and Sobel's D criterion:

> **Definition 10: Banks and Sobel's beliefs construction**
> Consider the function $\mu(t,r)$ defined by:
>
> $\mu(t,r)=1$ if $u_1(t,m,r) > u_1^*(t)$ where $u_1^*(t)$ is type t's payoff at equilibrium E
>
> $\mu(t,r) \in [0,1]$ if $u_1(t,m,r) = u_1^*(t)$
>
> $\mu(t,r)=0$ if $u_1(t,m,r) < u_1^*(t)$
>
> $\Gamma(r) = \{\gamma \in \text{set of beliefs systems}/ \exists \mu(.,r) \text{ and } \exists c \in \mathbb{R}^{+*}/ \forall t \in T, \gamma(t) = c\mu(t,r)\rho(t)\}$

Definition 11: Banks and Sobel's D criterion

(1) For any out of equilibrium E message m, define the following iterative process:

Γ_0=initial complete set of beliefs systems and R_0=set of all possible responses after m

$\forall n>0$ (integer) $\Gamma_n=\Gamma(R_{n-1})$ if $\Gamma(R_{n-1})\neq\emptyset$ $R_n=BR(\Gamma_n,m)$

$=\Gamma_{n-1}$ if $\Gamma(R_{n-1})=\emptyset$

where $BR(\Gamma_n,m)$ is the set of best responses to m when the beliefs are in Γ_n

$$\Gamma^*(m)=\bigcap_{n=0}^{\infty}\Gamma_n \text{ and } R^*(m)=\bigcap_{n=0}^{\infty}R_n$$

(2) If there exists an out of equilibrium message m such that E's beliefs after observing m do not belong to $\Gamma^*(m)$, then E does not resist the D criterion.

The definition may look complicated but the idea is quite simple. For any out of equilibrium message m Banks and Sobel define, for each type of player 1, a propensity µ to deviate to m for any response r after m. Then they require, on the one hand, that the beliefs γ respect this propensity, on the other hand that the responses after m respect these beliefs γ. And they repeat the process with these new responses. For the equilibrium E to satisfy the D criterion, it is necessary that the beliefs after m in E belong to the set of beliefs obtained at the end of the above process.

> **STUDENTS' INSERT**
>
> By contrast to Kohlberg and Mertens, Banks and Sobel do not dismiss the equilibrium E in Figure 6.7 for the following reason. For all responses after the out of equilibrium message m_2, either only t_1 is best off deviating, either t_1 and t_2 are best off deviating, or nobody is best off deviating. It follows that in Γ_1 player 2 assigns m_2 to t_1 with at least the probability ½. By doing so, he can still play r_1 after m_2, because r_1 is his best reply as soon as he assigns at most probability 2/3 to t_1. So, at the end of the first step of reasoning, R_1 is still the set of initial possible responses after m_2 (player 2 can play r_1 and or r_2 after m_2 with any probability). It follows that $\Gamma_2=\Gamma_1$, so that player 2 will again, in the second step, assign m_2 to t_1 with at least probability ½ and so on, to infinity. And $\mu(t_1/m_2)=0.6$, a system of beliefs that sustains E, is in Γ_∞. In other words, the iterative process that leads to assign m_2 more to t_1 than to t_2, is not sufficiently strong to reject E, because it still allows beliefs that sustain the equilibrium.
>
> This is clearly due to the fact that, given his own payoffs after m_2, player 2 can still play r_1 after m_2, even if he assigns this message to t_1 with a probability higher than the prior one.

So Banks and Sobel as well as Kohlberg agree on the fact that m_2 is played more by t_1 than by t_2. Yet, given that t_2 can also get more with m_2 than with m_1, they do not conclude that m_2 is *inferior* for t_2, so they do not exclusively attribute m_2 to t_1; consequently, player 2 is not compelled to exclusively play r_2 after m_2 and the equilibrium is not destroyed.

Yet both Banks and Sobel and Kohlberg (and Kohlberg and Mertens) go in the same direction (m_2 is more often played by t_1 than by t_2) and there is a link between both criteria: any stable outcome corresponds to a PBE path that resists the D criterion.

SOLVING A GAME DIFFERENTLY

We now switch to completely different criteria (only developed for signalling games), that take into account a remark attributed to Stiglitz, who wondered *how we should interpret equilibrium planned actions when we change the interpretation of out of equilibrium actions*. As a matter of fact, if an unexpected action is able to convey new signals, and hence to induce new actions after it, then an equilibrium action may also have to be reinterpreted. In other words, the new interpretations of *all* the actions have to be *consistent*; i.e. they *have to belong to an equilibrium*. The Consistent Forward Induction Equilibrium Path (CFIEP) (Umbhauer 1991)[27] belongs to these criteria. The idea is to test a PBE by seeing if there exists another PBE that contains a deviating action, such that the deviating players are better off even if recognized.

Definition 12: Umbhauer's CFIEP (1991)
The PBE E is a CFIEP if:

∀ m out of E's equilibrium path, ∄ PBE E' /:

1. m is played in E'

2. For D={t∈T(the set of types)/$\pi_{1E'}$(m/t)>0 and $u_{1E'}$(t, m)>u_{1E}(t)} and S={r∈R(m) (set of responses after m),

$$r \in \arg\max_{r\in R(m)} \sum_{t\in D} u_2(t,m,r)\rho(t)\pi_{1E'}(m/t) / \left(\sum_{t\in D} \rho(t)\pi_{1E'}(m/t)\right)\}$$

iii) D≠∅

iv) ∀t∈T\D, $u_{1E'}$(t,m)≤u_{1E}(t)

v) ∃r∈S/∀t∈D, u_1(t,m,r)≥u_{1E}(t)

where u_{1E}(t) is t's utility in equilibrium E and $u_{1E'}$(t,m) is t's utility when she plays m in equilibrium E'.

So a PBE E can only be dismissed by another PBE E' such that: an out of equilibrium E message m is played in E', and there are some types of player 1 who are happy to play this new message (they get more than in E), i.e. D≠∅ (condition (i)). Given that only these types are happy to induce the change, the criterion requires that they are still better off deviating even if recognized (condition iii). Condition (ii) avoids that types not playing m in E' may be induced to switch to m (even if they do not play m in E').

> **STUDENTS' INSERT**

The CFIEP does not dismiss the PBE E in Figure 6.7 because this game has no other PBE leading to a higher payoff for t_1 or t_2. Notably, a separating PBE where only t_1 would play m_2 doesn't exist.
    The motivation for this non elimination is easy to understand: if one assigns m_2 to t_1, as it is done by Kohlberg, and Kohlberg and Mertens, then t_1 deviates to m_2, but t_2 deviates also, not only because 5>4, *but because* 5>1: as a matter of fact, if t_1 plays m_2, then m_1, if played, can only be assigned to t_2 and followed by r_2! So assigning m_2 only to t_1 is not consistent.

The difference between consistent criteria (see also Mailath et al. 1993),[28] which require a global equilibrium interpretation of all the messages, and the other criteria is even more obvious in some other games *(see Exercise 12)*.

We now show how forward induction works in concrete examples, like plea bargaining and repetition of the battle of the sexes.

4.3 The power of forward induction through applications

4.3.1 French plea-bargaining and forward induction

We return to the plea bargaining game introduced in chapter 1 (section 3.2), given in Figure 6.8.

Remember *(see chapter 5, section 2.2.2)* that this game, regardless of the value of ρ, has a PBE such that both types of player 1 plead guilty, because the jury convicts at trial with probability 1, in that he can assign the action "don't plead guilty" to the guilty defendant.

If $\rho \leq 1/2$ there exists a second "go to trial" PBE where both types of defendant go to trial and are released.

And if $\rho \geq 1/2$ there exists a semi-separating PBE such that the guilty defendant goes to trial with probability $(1-\rho)/\rho$, pleads guilty with probability $(2\rho-1)/\rho$, the innocent defendant only goes to trial, the jury releases with probability u and convicts with probability 1–u.

When $\rho \leq 1/2$, only the equilibrium where both types of defendant go to trial and are released is a forward induction equilibrium, regardless of the chosen criterion.

To simplify, we apply only the contrasted criteria of Kohlberg (1990) and Umbhauer (1991).

According to Kohlberg, the equilibrium where both plead guilty requires that: q≤u and q≤v, where q is the probability of releasing at trial. Given that q≤v<u, going to trial is an inferior strategy for the guilty defendant, so this action has to be assigned to the innocent defendant; it is therefore followed by the action "release" which destroys the plead guilty equilibrium.

According to Umbhauer, the plead guilty PBE is destroyed by the go to trial PBE because, in this new PBE, both types of defendants play the out of (the first) equilibrium action "go to trial" and both are happy to do so when the jury takes this fact into account.

When $\rho > 1/2$, the semi separating PBE is the only forward induction equilibrium, but only according to some criteria.

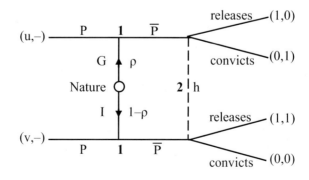

Figure 6.8

SOLVING A GAME DIFFERENTLY

For example Banks and Sobel do not eliminate the plead guilty equilibrium for the following reason: given that q≥u ⇒ q>v, player 2 should, in the first step of the process, assign the out of equilibrium action "go to trial" to the innocent defendant with at least probability 1−ρ. The problem is that, despite this minimal probability, the jury can still convict at trial. The set of best responses after the deviation "go to trial", at this first step, is still the initial set of responses. It derives that the beliefs and actions at the second step are the ones of the first step and so on, to infinity. So it is possible to assign the action "go to trial" to the guilty defendant with probability ρ, which sustains the tested equilibrium.

By contrast, both Kohlberg and Umbhauer eliminate the plead guilty equilibrium but not for the same reasons:

Kohlberg proceeds exactly as in the previous case. Umbhauer proceeds differently. The plead guilty PBE is this time destroyed by the semi separating PBE because:

A new message "going to trial" is played in this semi separating equilibrium. D, the set of types who play this message and are better off by doing so than in the original equilibrium is the singleton {innocent defendant} (condition i). The innocent defendant is still better off when the jury focuses his beliefs on her (condition iii). And the guilty defendant does not get more in this new equilibrium than in the old one (condition ii). So Umbhauer eliminates the plead guilty equilibrium because the innocent defendant is happy to switch to the semi separating equilibrium even if she is recognized.

In fact, in this study, we exploit two properties: first the innocent defendant is always more induced to go to trial than the guilty one, and, second, when she is recognized, she is even happier to go to trial. In other words, it is easier for a "good" type, i.e. a type that does not fear to be recognized, to initiate a switch. So forward induction equilibria will often be in favour of 'good' types, which is a rather pleasing result. It follows, in this game, that forward induction (at least some forward induction criteria) is in favour of the "ethical" equilibrium where the innocent type does not plead guilty.

4.3.2 The repeated battle of the sexes or another version of burning money

We now propose a funny application of forward induction to the repeated battle of the sexes game, given in Matrix 6.10:

		Player 2	
		A_2	B_2
Player 1	A_1	(4,1)	(0,0)
	B_1	(0,0)	(1,4)

Matrix 6.10

Suppose that the game is played twice (with observation of the actions between two rounds) and that the planned SPNE consists of playing (B_1,B_2) in the first period, and (B_1,B_2) in the second period regardless of the played actions in period 1.

This SPNE does not resist forward induction.

As proposed by Ponssard (1989),[29] suppose that player 1 deviates to A_1 in the first period. By doing so, she deliberately loses 1. The only way to rationalize this action is to play A_1 in the second period, so that she gets 0+4=4 which is better than 1+1, the payoff she gets at equilibrium. So player 2, when he interprets player 1's deviation as a rational act, has to anticipate that player 1 will play again A_1 in period 2, so he plays A_2, which indeed incites player 1 to deviate in the first period (and in the second), so as to get a payoff 4 higher than the equilibrium payoff 2.

Yet, to my mind, forward induction can even lead us farther!

Player 2 is a clever player too. He is able to foresee that player 1 will deviate to A_1 in the first period. So he is induced to play A_2 *in the first period*. By doing so, he gives the payoff 4 to player 1 in the first period. So player 1 has no longer any reason to deviate to A_1 in the second period (A_1 is no longer the only possible rational action), so, in the second period, both players can only (rationally) play (B_1,B_2). By doing so, both players get a total payoff 5 on the two periods. Player 1 is much better off than in the tested SPNE, and player 2 is better off than in the first deviating scenario (where he got 0+1=1).

This game has links with the burning money game *(see Exercise 7 in chapter 2)* in that deviating to A_1 in the first period amounts to burning 1.

Well, that's surely asking a bit too much from forward induction, isn't it? In real life, it might be quite dangerous to play A_2 in period 1: just imagine that player 1 is a very modest player, and happy to be sure to get 1 in each period. In that case you can only regret to have deviated to A_2 in period 1!

CONCLUSION

Nothing is obvious.

The alternative approaches we chose to develop in this chapter all have good features. Forward induction, in contrast to backward induction, tries to cope rationally with unexpected actions. And intersection between forward induction and Nash equilibria is in general not empty. Level-*k* reasoning copes with the heterogeneity of the players' ability to run a reasoning. Regret minimization copes with the idea that important regrets (losses of payoffs) are difficult to accept. Best reply matching proposes a new way to cope with probabilities: a probability associated with an action reflects how many times it is a best reply. Yet, none of these alternative approaches is exempt from flaws. Forward induction is too multiform and often ad hoc. Level-*k* reasoning is too restrictive with regard to the behaviour of opponents. Regret minimization is asymmetric in that it only focuses on the regrets. Best reply matching is also asymmetric, in that it only takes into account how often a strategy is a best reply, but ignores how often it is a worst reply.

And the same is true for other criteria not developed in this book. Game theory is still an open, very stimulating field: you surely have some good ideas to develop!

NOTES

1. Kosfeld, M., Droste, E. and Voorneveld, M., 2002. A myopic adjustment process leading to best reply matching, *Games and Economic Behaviour*, 40, 270–298.
2. Bernheim, D., 1984. Rationalizable strategic behavior, *Econometrica*, 52, 1007–1028.
3. Pearce, D., 1984. Rationalizable strategic behaviour and the problem of perfection, *Econometrica*, 52, 1029–1050.
4. Umbhauer, G., 2007. Best-reply matching in an experience good model, Social Science Research Network, SSRN-id 983708.pdf, April 2007.
5. We later write the probabilities next to the actions they are assigned to, like in Matrix 6.3b, so we no longer have to define them.

6   Basu, K., 1994. The traveler's dilemma: paradoxes of rationality in game theory, *American Economic Review*, 84 (2), 391–395.
7   Halpern, J.Y. and Pass, R., 2012. Iterated regret minimization: a new solution concept, *Games and Economic Behavior*, 74, 184–207.
8   Becker, T., Carter, M. and Naeve, J., 2005. Experts playing the traveller's dilemma. Discussion paper 252/2005, Universität Hohenheim.
9   Capra, M., Goere, J.K., Gomez, R. and Holt, C.A., 1999. Anomalous behaviour in a traveller's dilemma, *American Economic Review*, 89 (3), 678–690.
10   Savage, L.J., 1951. The theory of statistical decision, *Journal of American Statistical Association*, 46, 55–67.
11   Niehans, J., 1948. Zur Preisbildung bei ungewissen Erwartungen. *Schweizerische Zeitschrift für Volkswirtschaft und Statistik*, 84 (5), 433–456.
12   Observe that the same reasoning doesn't hold for $x_1=2$ and $x_1=3$. If $x_1=2$, then the opponent cannot play less so the maximal regret is 99–2=97. And if $x_1=3$, then the opponent, if he plays less, can only play 2 which yields the regret P. So the maximal regret is max (P, 96). It derives that if P<97 and 96<2P–1, i.e. P>97/2, then 3 minimizes the regret. Observe also that if 100–2P<2, i.e. P>49, then 3 again minimizes the regret. And if P>97, then 2 minimizes the regret.
13   You observe that in that game, where Nature also intervenes, we combine Halpern's criterion and the min max regret decision rule in decision theory, because we consider the worst scenario, with regard to the opponent's strategy and to Nature's move. For example, when a player does not exchange, we suppose that the opponent exchanges his value K, and that he has this value in the envelope (Nature's move).
14   Stahl, D.O. and Wilson P., 1994. Experimental evidence on players' models of other players, *Journal of Economic Behavior and Organization*, 25 (3), 309–327.
15   Nagel, R., 1995. Unravelling in guessing games: an experimental study, *American Economic Review*, 85 (5), 1313–1326.
16   The behavior of the level-0 players is not obvious. We come back to it later.
17   Camerer, C.F., Ho, T.H. and Chong, J.K., 2004. A cognitive hierarchy model of games, *Quarterly Journal of Economics*, 119 (3), 861–898.
18   Moulin, H. 1986. *Game theory for social sciences*, New York, New York University Press.
19   Agranov, M., Potamites, E., Schotter, A. and Tergiman, C., 2012. Beliefs and endogenous cognitive levels: an experimental study, *Games and Economic Behavior*, 75, 449–463.
20   Heap, S.H., Arjona, D.R. and Sudgen, R., 2014. How portable is level-0 behavior? A test of level-*k* theory in games with non neutral frames, *Econometrica*, 82 (3), 1133–1151.
21   Breitmoser, Y., 2012. Strategic reasoning in *p*-beauty contests, *Games and Economic Behavior*, 75, 555–569.
22   As we know from section 3.2, this is not necessarily true but the remarks on the decreasing payoff are still true.
23   Cognitive hierarchy theory resists this flaw better.
24   Kohlberg, E. and Mertens, J.F., 1986. On the strategic stability of equilibria, *Econometrica*, 54, 1003–1037.
25   Kohlberg, E., 1990. Refinement of Nash equilibrium: the main ideas, in Ichiishi, T., Leyman, A., and Tauman, Y. (eds), 1993, *Game Theory and Applications*, , San Diego, Academic Press, 1–45.
26   Banks, J. and Sobel, J., 1987. Equilibrium selection in signaling games, *Econometrica*, 55, 647–661.
27   Umbhauer, G., 1991. Information asymétrique et principe d'induction, *Revue Economique*, 42 (6), 1089–1110.
28   Mailath, G.J., Okuno-Fujywara, M. and Postlewaite, A., 1993. Belief based refinements in signalling games, *Journal of Economic Theory*, 60, 241–276.
29   Ponssard, J.P., 1989. Concurrence imparfaite et rendements croissants, une approche en termes de fair-play, Working paper 314, Laboratoire d'Econométrie, Ecole Polytechnique, France.

Chapter 6

Exercises

♣ EXERCISE 1 BEST REPLY MATCHING IN A NORMAL FORM GAME

Consider the game studied in chapters 2 and 3, given in Matrix E6.1:

		Player 2 A_2	Player 2 B_2	Player 2 C_2
Player 1	A_1	(7,8)	(3,3)	(0,8)
	B_1	(0,0)	(4,4)	(1,4)
	C_1	(7,1)	(3,0)	(1,1)

Matrix E6.1

Questions:
1. Find the BRM equilibrium. Comment.
2. Show that the GBRM concept yields couples of strategies surviving iterative elimination of weakly dominated strategies, but also other couples of strategies. Comment. What happens if you try to take the payoffs better into account?

♣ ♣ EXERCISE 2 BEST REPLY MATCHING IN THE TRAVELLER'S DILEMMA

In the traveller's dilemma *(see section 2.1 in Chapter 6, the exercises 11, 12 in chapter 2 and the exercises 15, 16, 17 in chapter 3)* each traveller i=1,2 asks for a given amount x_i, an integer in [2,M] (in the games studied up to now we fixed M=100). The company observes the two demands and proposes to reimburse the travellers in a given way. In this exercise, we choose the two following ways:

First way (close to Basu's way):
- If $x_1=x_2$, each traveller gets x_1

If $x_i>x_j$ i≠j=1,2, the traveller j gets x_j+P, and the traveller i only gets max (0, x_j–P), where P is a fixed amount higher than 1.

Second way (Student's way):
- If $x_1=x_2$, each traveller gets x_1
- If $x_i>x_j$ i≠j=1,2, the traveller j gets x_j+P, and the traveller i only gets x_i–P.

EXERCISES

Questions

1. Work with the first reimbursement rule. Give the matrix of best responses when x is an integer in [2, M], P=2 and M=8. Find the unique BRM equilibrium. Generalize for any M and P>1. Comment.
2. Consider the second reimbursement rule. Give the matrix of best responses when x is an integer in [2,M], P=2 and M=8. Find the unique BRM equilibrium. Show that the mixed Nash equilibrium can be obtained by switching to the GBRM concept.
3. Generalize the result in question 2 for any M and $1<P<(M-2)/2$. Comment.

♣♣ EXERCISE 3 BEST REPLY MATCHING IN THE PRE-EMPTION GAME

Consider the pre-emption game reproduced in Figure E6.1 (see the exercises 5, 8, 12 and 3 in chapters 1, 2, 3 and 4). The normal form and the reduced normal forms of this game are given in Matrices E6.2a and E6.2b.

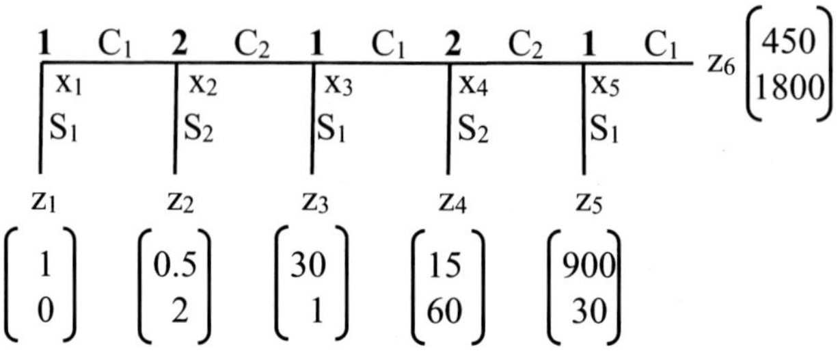

$$\underset{z_1}{\overset{1\quad C_1}{\begin{bmatrix}x_1\\S_1\end{bmatrix}}} \underset{z_2}{\overset{2\quad C_2}{\begin{bmatrix}x_2\\S_2\end{bmatrix}}} \underset{z_3}{\overset{1\quad C_1}{\begin{bmatrix}x_3\\S_1\end{bmatrix}}} \underset{z_4}{\overset{2\quad C_2}{\begin{bmatrix}x_4\\S_2\end{bmatrix}}} \underset{z_5}{\overset{1\quad C_1}{\begin{bmatrix}x_5\\S_1\end{bmatrix}}} \underset{z_6}{\begin{bmatrix}450\\1800\end{bmatrix}}$$

$$\begin{bmatrix}1\\0\end{bmatrix}\quad \begin{bmatrix}0.5\\2\end{bmatrix}\quad \begin{bmatrix}30\\1\end{bmatrix}\quad \begin{bmatrix}15\\60\end{bmatrix}\quad \begin{bmatrix}900\\30\end{bmatrix}$$

■ **Figure E6.1**

		Player 2		
	$(S_2/x_2 S_2/x_4)$	$(S_2/x_2 C_2/x_4)$	$(C_2/x_2 S_2/x_4)$	$(C_2/x_2 C_2/x_4)$
$(S_1/x_1 S_1/x_3 S_1/x_5)$	(1, 0)	(1, 0)	(1, 0)	(1, 0)
$(S_1/x_1 S_1/x_3 C_1/x_5)$	(1, 0)	(1, 0)	(1, 0)	(1, 0)
$(S_1/x_1 C_1/x_3 S_1/x_5)$	(1, 0)	(1, 0)	(1, 0)	(1, 0)
$(S_1/x_1 C_1/x_3 C_1/x_5)$	(1, 0)	(1, 0)	(1, 0)	(1, 0)
$(C_1/x_1 S_1/x_3 S_1/x_5)$	(0.5, 2)	(0.5, 2)	(30, 1)	(30, 1)
$(C_1/x_1 S_1/x_3 C_1/x_5)$	(0.5, 2)	(0.5, 2)	(30, 1)	(30, 1)
$(C_1/x_1 C_1/x_3 S_1/x_5)$	(0.5, 2)	(0.5, 2)	(15, 60)	(900, 30)
$(C_1/x_1 C_1/x_3 C_1/x_5)$	(0.5, 2)	(0.5, 2)	(15, 60)	(450,1800)

Player 1

Matrix E6.2a

		Player 2	
	S_2/x_2	$(C_2/x_2 S_2/x_4)$	$(C_2/x_2 C_2/x_4)$
S_1/x_1	(1,0)	(1,0)	(1,0)
$(C_1/x_1 S_1/x_3)$	(0.5,2)	(30,1)	(30,1)
$(C_1/x_1 C_1/x_3 S_1/x_5)$	(0.5,2)	(15,60)	(900,30)
$(C_1/x_1 C_1/x_3 C_1/x_5)$	(0.5,2)	(15,60)	(450,1800)

Player 1

Matrix E6.2b

Questions
1. Find the BRM equilibrium in the normal form game. Give the equivalent behavioural strategies.
2. Find the BRM equilibrium in the reduced normal form game. Comment. Give the equivalent behavioural strategies.
3. Find the two extreme GBRM equilibria in the reduced normal form game. Comment.

♣ EXERCISE 4 MINIMIZING REGRET IN A NORMAL FORM GAME

We come back to the game in Matrix E6.1.

Question
Find the strategy profile that minimizes regret. Comment.

♣ ♣ EXERCISE 5 MINIMIZING REGRET IN BERTRAND'S DUOPOLY

We apply Halpern and Pass's concept to Bertrand's duopoly price competition *(see the exercises 1, 10, and 13 in chapters 1, 2 and 3)*. We recall that firm 1 and firm 2's utility (profit) functions, $u_1(p_1,p_2)$ and $u_2(p_1,p_2)$, are given by:

- if $p_1=p_2$, $u_1(p_1,p_2)=\max(0,K-ap_1)(p_1-c)/2=u_2(p_1,p_2)$
- if $p_1>p_2$, $u_1(p_1,p_2)=0$ and $u_2(p_1,p_2)=\max(0,K-ap_2)(p_2-c)$
- if $p_1<p_2$, $u_2(p_1,p_2)=0$ and $u_1(p_1,p_2)=\max(0,K-ap_1)(p_1-c)$

Question
Find the strategy that minimizes regret in a continuous setting. Comment.

♣ ♣ EXERCISE 6 MINIMIZING REGRET IN THE PRE-EMPTION GAME

Consider the variant of the centipede game given in Figure E6.2 I proposed to my L3 and M1 students *(see Exercise 3 in chapter 4)*.

$$\begin{array}{cccccccccccccccccccc}
\mathbf{1} & C_1 & \mathbf{2} & C_2 & \mathbf{1} & C_1 & \mathbf{2} & C_2 & \mathbf{1} & C_1 & \mathbf{2} & C_2 & \mathbf{1} & C_1 & \mathbf{2} & C_2 & \mathbf{1} & C_1 & \mathbf{2} & C_2 \\
\end{array} \begin{pmatrix} 400 \\ 40 \end{pmatrix}$$

$$\begin{vmatrix} x_1 \\ S_1 \end{vmatrix} \begin{vmatrix} x_2 \\ S_2 \end{vmatrix} \begin{vmatrix} x_3 \\ S_1 \end{vmatrix} \begin{vmatrix} x_4 \\ S_2 \end{vmatrix} \begin{vmatrix} x_5 \\ S_1 \end{vmatrix} \begin{vmatrix} x_6 \\ S_2 \end{vmatrix} \begin{vmatrix} x_7 \\ S_1 \end{vmatrix} \begin{vmatrix} x_8 \\ S_2 \end{vmatrix} \begin{vmatrix} x_9 \\ S_1 \end{vmatrix} \begin{vmatrix} x_{10} \\ S_2 \end{vmatrix}$$

$$\begin{pmatrix} 5 \\ 0 \end{pmatrix} \begin{pmatrix} 0 \\ 5 \end{pmatrix} \begin{pmatrix} 50 \\ 1 \end{pmatrix} \begin{pmatrix} 1 \\ 50 \end{pmatrix} \begin{pmatrix} 100 \\ 10 \end{pmatrix} \begin{pmatrix} 10 \\ 100 \end{pmatrix} \begin{pmatrix} 150 \\ 15 \end{pmatrix} \begin{pmatrix} 15 \\ 150 \end{pmatrix} \begin{pmatrix} 250 \\ 25 \end{pmatrix} \begin{pmatrix} 25 \\ 250 \end{pmatrix}$$

▬ **Figure E6.2**

I also recall in Table E6.1 their answers to my two questions: how far would you go if you were player 1 and player 2? Clearly my students go on in the game by contrast to the backward induction stopping behaviour *(see Exercise 3 in chapter 4 for a contrasted comment)*.

EXERCISES

■ Table E6.1

Player 1	L3 Students	M1 Students	Player 2	L3 Students	M1 Students
Stops at x_1	11.5%	7.7%	Stops at x_2	13.1%	7.7%
Stops at x_3	3.3%	3.8%	Stops at x_4	6.6%	11.5%
Stops at x_5	9.8%	19.2%	Stops at x_6	11.5%	19.2%
Stops at x_7	10.7%	19.2%	Stops at x_8	13.1%	15.4%
Stops at x_9	29.5%	23.1%	Stops at x_{10}	39.3%	30.8%
Never stops	35.2%	27%	Never stops	16.4%	15.4%

The reduced normal form of this game is given in Matrix E6.3.

                                   Player 2
           S_2/x_2 $C_2/x_2S_2/x_4$ $C_2/x_2C_2/x_4$ $C_2/x_2C_2/x_4$ $C_2/x_2C_2/x_4C_2/$ $C_2/x_2C_2/x_4C_2/$
                                        S_2/x_6 $C_2/x_6S_2/x_8$ $x_6\,C_2/x_8S_2/x_{10}$ $x_6\,C_2/x_8C_2/x_{10}$

	S_1/x_1	a	(5,0)	(5,0)	(5,0)	(5,0)	(5,0)	(5,0) a
P1	$C_1/x_1S_1/x_3$		(0,5)	(50,1)	(50,1)	(50,1)	(50,1)	(50,1)
	$C_1/x_1C_1/x_3S_1/x_5$		(0,5)	(1,50)	(100,10)	(100,10)	(100,10)	(100,10) b
	$C/x_1C/x_3C/x_5S/x_7$		(0,5)	(1,50)	(10,100)	(150,15)	(150,15)	(150,15)
	$C/x_1C/x_3C/x_5C/x_7S/x_9$		(0,5)	(1,50)	(10,100)	(15,150)	(250,25)	(250,25)
	$C/x_1C/x_3C/x_5C/x_7C/x_9$		(0,5)	(1,50)	(10,100)	(15,150)	(25,250)	(400,40)

Matrix E6.3

Questions
1. Find the strategies that minimize regret in the reduced normal form game.
2. Adapt Halpern and Pass's concept to the extensive form game. Do you get the same conclusion than in the normal form game? Comment.
3. Replace the payoff 400 by the payoff 500. What happens? Study the reduced game starting at node x_8 with these new payoffs. What's the problem with regret minimization?

♣ ♣ ♣ EXERCISE 7 MINIMIZING REGRET IN THE TRAVELLER'S DILEMMA

Consider the traveller's dilemma I proposed to my students, inspired from but different from Basu's, where: if $x_i=x_j$, both players get x_i and if $x_i>x_j$, player j get x_j+P and player i gets max (0, x_j-P). We suppose that the travellers can ask for an integer in [2,M].

Question
Find the minimizing regret strategy when the players ask for an integer in [2,M], 1<P<M−3, and M and P are two integers. What happens for M=100 and P=4? What happens for M=100 and P=40?

EXERCISES

♣ EXERCISE 8 LEVEL-K REASONING IN AN ASYMMETRIC NORMAL FORM GAME

Question

Level-k reasoning is usually applied to symmetric games, with one population of players, but can be applied to asymmetric games. So apply it to the asymmetric game in Matrix E6.1. Comment.

♣♣ EXERCISE 9 LEVEL-1 AND LEVEL-K REASONING IN BASU'S TRAVELLER'S DILEMMA

Consider the traveller's dilemma studied in Exercise 7, close to Basu's version (if $x_i=x_j$, both players get x_i and if $x_i>x_j$, player j gets x_j+P and player i gets max $(0, x_j-P)$).

Questions:
1. Consider the two simplified games P=5, M=12 and P=2, M=12. Find the level-1 and level-k behaviour in these games. Comment
2. Find the level-1 behaviour for any integers P and M, with P<M. Comment by comparing to the minimizing regret strategy (Exercise 7).
3. What is the consequence of the response to question 2 in another game?

♣♣ EXERCISE 10 LEVEL-1 AND LEVEL-K REASONING IN THE STUDENTS' TRAVELLER'S DILEMMA

Consider the students' version of the traveller's dilemma, where, if $x_i=x_j$, both players get x_i and if $x_i>x_j$, player j gets x_j+P and player i gets x_i-P. We suppose that the travellers can ask for an integer in [2,M].

Question:
1. Find the level-1 behaviour for M=8 and P=2. Generalize for any M and P with P<<M/2. Comment. What about the level-k behaviour for k large, for M=8 and P=2?
2. Find the level-1 behaviour for M=15 and P=8. Generalize for any M and P with P>>M/2. Comment. What about the level-k behaviour for k large?

♣♣ EXERCISE 11 STABLE EQUILIBRIUM, PERFECT EQUILIBRIUM AND PERTURBATIONS

Consider the game studied in Exercise 7 of chapter 5, whose normal and extensive forms are recalled in Matrix E6.4 and Figure E6.3.

		Player 2			
		A_2a_2	A_2b_2	B_2a_2	B_2b_2
Player 1	A_1a_1	(5,2)	(5,2)	(5,2)	(5,2)
	A_1b_1	(5,2)	(5,2)	(5,2)	(5,2)
	B_1a_1	(0,1)	(0,1)	(4,4)	(0,0)
	B_1b_1	(0,1)	(0,1)	(0,0)	(1,1)

Matrix E6.4

EXERCISES

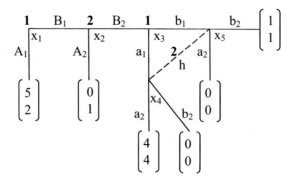

Figure E6.3

Questions
1. Show that (A_1a_1, A_2a_2) is not a stable equilibrium.
2. Show that (A_1a_1, B_2a_2) is a perfect equilibrium regardless of the minimal probabilities introduced but that it is not a stable equilibrium. Comment.
3. Show that the set $(A_1a_1, (A_2a_2, B_2a_2))$, corresponding to the unique outcome (5,2), is stable.

♣♣ EXERCISE 12 FOUR DIFFERENT FORWARD INDUCTION CRITERIA AND SOME DYNAMICS

Consider the signalling game in Figure E6.4, out of Umbhauer (1991).

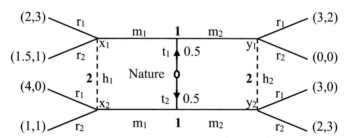

Figure E6.4

Questions
1. Consider the PBE E in which player 1 plays m_1 regardless of type, followed by r_1, m_2 being followed by r_2, being for example assigned to t_2. Show that E is not stable, not self-enforcing, does not satisfy the D criterion but that it is a CFIEP. Comment.
2. Interchange player 2's payoffs 2 and 3 after m_2. What does it change?

♣♣ EXERCISE 13 FORWARD INDUCTION AND THE EXPERIENCE GOOD MARKET

We come back to the experience good market with two qualities (Figure E6.5), studied in exercises 7, 27 and 10 in chapters 1, 3 and 5. We have:
$p_1H_1+p_2H_2<h_2$, $H_2>h_2>H_1>h_1$, $h_1=40$, $H_1=60$, $h_2=100$, $H_2=120$ $p_1=p_2=0.5$. We know (see Exercise 10 in chapter 5) that there are semi separating PBE – we call them E – such that t_1 plays

60 and P and t_2 only plays P with 100<P<120. And there exists a separating PBE, E*, where t_1 only plays 60 and t_2 only plays 120.

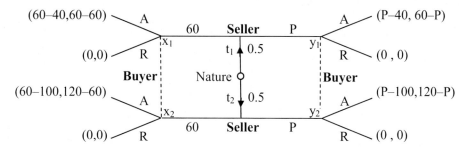

Figure E6.5

Questions
1. Show that all the equilibria E are eliminated by E* with the CFIEP criterion, but that E* can't be eliminated by an equilibrium E.
2. Show that all the equilibria E are eliminated by Kohlberg's criterion but that E* can't be eliminated. Comment.

♣ EXERCISE 14 FORWARD INDUCTION AND ALTERNATE NEGOTIATION

In Exercise 12 in chapter 5 we studied all the PBE of the alternate negotiation problem with incomplete information on the number of periods of negotiation. We reproduce the game in Figure E6.6 and recall the equilibria in Tables E6.2 and E6.3.

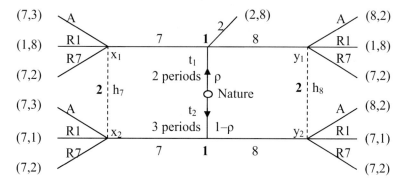

Figure E6.6

Table E6.2

ρ	0	1/7	2/7	1
equilibria	E_8	E_{82}	E_{82}	
	E_7	E_7	E_{72}	

EXERCISES

Table E6.3

equilibria	E_8	E_7	E_{82}	E_{72}
payoffs	$t_1=8$	$t_1=7$	$t_1=2$	$t_1=2$
	$t_2=8$	$t_2=7$	$t_2=50/7$	$t_2=7$
	player 2=2	player 2=3	player 2=from 2 to 8 (when $\rho \to 1$)	player 2=from 3 to 8 (when $\rho \to 1$)

Questions

1. Show that only E_8 and E_{82} are self-enforcing.
2. Show that only E_8 and E_{82} are CFIEP. Comment.

Answers to Exercises

Chapter 1

Answers to Exercises

♣ EXERCISE 1 EASY STRATEGY SETS

1. $S_1 = S_2 = \{2,3,....,99,100\}$.
2. $S_1 = S_2 = [0, +\infty]$. Yet a price can lead to a positive or null payoff only if it is higher or equal to c and lower or equal to K/a. So we will later reduce these sets to [c, K/a].
3. $S_i = [0,100]$, i from 1 to N.

♣ EXERCISE 2 CHILDREN GAME, HOW TO SHARE A PIE EQUALLY

The mother can propose the following game. She asks one of the children to cut the pie in half, knowing that the other child will choose first her piece. This game is very efficient: the daughter who cuts the pie is best off cutting two equal pieces, given that her sister will always choose the largest piece.

Observe that the same game works with N children, the child who cuts the pie being the last to choose her piece.

♣♣ EXERCISE 3 CHILDREN GAME, THE BAZOOKA GAME, AN ENDLESS GAME

1. The extensive form game is given in Figure A1.1a.

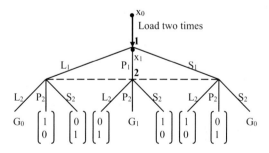

Figure A1.1a G_0 means the whole game starting at x_0 and G_1 means the game starting at x_1. The arrow means that this action is chosen by both players.

The game is not so easy because it is *an endless game* (it may last for ever if both children always choose the same action!). But it can be simplified. Loading two times at the beginning is just fun

ANSWERS

for the children. The strategic actions begin at node x_1, so the extensive form game becomes the tree in Figure A1.1b.

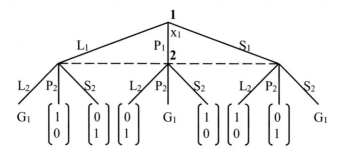

Figure A1.1b G_1 means the game starting at x_1.

2. It follows that the normal form game is given by Matrix A1.1:

		child 2 L_2	child 2 P_2	child 2 S_2
child 1	L_1	(g_1,g_2)	$(1,0)$	$(0,1)$
	P_1	$(0,1)$	(g_1,g_2)	$(1,0)$
	S_1	$(1,0)$	$(0,1)$	(g_1,g_2)

Matrix A1.1

We have here a special game, in that the game is embedded in itself. We will see how to solve such a game in the exercises of chapter 3.

♣ EXERCISE 4 SYNDICATE GAME: WHO WILL BE THE FREE RIDER?

1. There are three players. The pure strategy set of each player i is $S_i=\{A_i,B_i\}$, where A_i, respectively B_i, means that the worker i (player i) joins the syndicate, respectively does not join the syndicate, i=1,2,3.

   Player i's preferences are given by: $(B_i,A_j,A_k) \succ (A_i,A_j,A_k) \succ (A_i,A_j,B_k) \sim (A_i,B_j,A_k) \succ (B_i,B_j,A_k) \sim (B_i,A_j,B_k) \sim (A_i,B_j,B_k) \sim (B_i,B_j,B_k)$ with $j \neq i \neq k$ and i,j,k from 1 to 3.

2. We translate the preferences in utilities as follows:

   $U_i(B_i,A_j,A_k)=1>U_i(A_i,A_j,A_k)=u_i>U_i(A_i,B_j,A_k)=U_i(A_i,A_j,B_k)=v_i>U_i(B_i,B_j,A_k)=U_i(B_i,A_j,B_k)=U_i(A_i,B_j,B_k)=U_i(B_i,B_j,B_k)=0$, with $j \neq i \neq k$ and i,j,k from 1 to 3.

The normal form of the game is given in Matrices A1.2.

ANSWERS

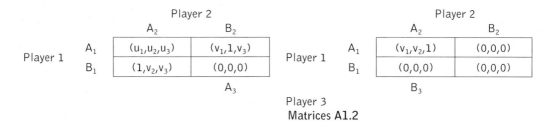

Matrices A1.2

Player 1 plays the rows, player 2 plays the columns and player 3 the matrices.

3. Given that two players are enough to create the syndicate, we may conjecture that there will not be more than two players in the syndicate; these two members would prefer the third player joining them ($u_i > v_i$, i=1,2,3) but the latter is better off not doing so and just benefiting from the activities of the syndicate ($1 > u_i$, i=1,2,3). And it is better for all that the syndicate exists, i.e. that at least two players join it ($v_i > 0$, i=1,2,3). So we can deduce that a stable situation is such that two players join the syndicate: the player out of the syndicate is most happy (he gets 1), and no member in the syndicate will go out, because this would induce the collapse of the syndicate, which lowers his payoff ($v_i > 0$, i=1,2,3). So the problem is: who are the two players who sacrifice themselves?

♣ EXERCISE 5 ROSENTHAL'S CENTIPEDE (PRE-EMPTION) GAME, REDUCED NORMAL FORM GAME

1. Rosenthal's centipede games, more generally the *pre-emption games*, are the companion games of war of attrition games.

    Two players play in a sequential way, with only two actions at each information set, stopping the game or going on. By contrast to the war of attrition game, payoffs are increasing when the players go on (in our game, payoffs grow exponentially), yet there is a problem: it is better to go on if you think that your opponent goes on at the next step, but it is always better to stop before him, so it is better to stop if you think that he stops at the next round. And the last player is always better off stopping the game (here player 1 gets 900 instead of 450).

    We may conjecture that this last stopping action may induce player 2 to stop if he reaches the node before the last one (because 60>30), which may induce player 1 to stop if she reaches the node x_3 (30>15) and so on, yet it seems silly to stop at the beginning of the game. So it is really difficult to conjecture what will happen. You can see how my L3 students play a longer centipede game in Exercise 3 in chapter 4.

    Player 1 has three information sets $\{x_1\}, \{x_3\}$ and $\{x_5\}$, two possible actions at each information set, S_1 and C_1. So she has $2^3=8$ pure strategies; $(S_1/x_1, C_1/x_3, C_1/x_5)$ is one of these strategies. Player 2 has two information sets $\{x_2\}$ and $\{x_4\}$ and two possible actions at each information set, S_2 and C_2. So he has $2^2=4$ pure strategies: $(C_2/x_2, C_2/x_4)$ is one of these strategies.

    The normal form game is given in Matrix A1.3a.

303

ANSWERS

Player 1

	Player 2			
	$(S_2/x_2 S_2/x_4)$	$(S_2/x_2 C_2/x_4)$	$(C_2/x_2 S_2/x_4)$	$(C_2/x_2 C_2/x_4)$
$(S_1/x_1 S_1/x_3 S_1/x_5)$	(1,0)	(1,0)	(1,0)	(1,0)
$(S_1/x_1 S_1/x_3 C_1/x_5)$	(1,0)	(1,0)	(1,0)	(1,0)
$(S_1/x_1 C_1/x_3 S_1/x_5)$	(1,0)	(1,0)	(1,0)	(1,0)
$(S_1/x_1 C_1/x_3 C_1/x_5)$	(1,0)	(1,0)	(1,0)	(1,0)
$(C_1/x_1 S_1/x_3 S_1/x_5)$	(0.5,2)	(0.5,2)	(30,1)	(30,1)
$(C_1/x_1 S_1/x_3 C_1/x_5)$	(0.5,2)	(0.5,2)	(30,1)	(30,1)
$(C_1/x_1 C_1/x_3 S_1/x_5)$	(0.5,2)	(0.5,2)	(15,60)	(900,30)
$(C_1/x_1 C_1/x_3 C_1/x_5)$	(0.5,2)	(0.5,2)	(15,60)	(450,1800)

Matrix A1.3a

2. $(S_1/x_1, S_1/x_3, S_1/x_5), (S_1/x_1, S_1/x_3, C_1/x_5), (S_1/x_1, C_1/x_3, S_1/x_5)$ and $(S_1/x_1, C_1/x_3, C_1/x_5)$ are equivalent in that, whatever player 2 plays, both players get the same payoffs (1,0) with these four strategies.

To get the reduced normal form game, we replace them by the unique action S_1/x_1 which characterizes the four strategies. $(S_2/x_2, S_2/x_4)$ and $(S_2/x_2, C_2/x_4)$ are also equivalent in that, whatever player 1 plays, both players get the same payoffs with these two strategies, either (1, 0) or (0.5,2). We replace them by the unique action S_2/x_2 which characterizes the two strategies. $(C_1/x_1, S_1/x_3, S_1/x_5)$ and $(C_1/x_1, S_1/x_3, C_1/x_5)$ are also equivalent because, whatever player 2 plays, both players get the same payoffs with these two strategies, either (0.5, 2) or (30, 1). We replace them by the unique couple of actions $(C_1/x_1 S_1/x_3)$ which characterizes the two strategies and we get the reduced normal form game in Matrix A1.3b:

Player 1

	Player 2		
	S_2/x_2	$(C_2/x_2 S_2/x_4)$	$(C_2/x_2 C_2/x_4)$
S_1/x_1	(1,0)	(1,0)	(1,0)
$(C_1/x_1 S_1/x_3)$	(0.5,2)	(30,1)	(30,1)
$(C_1/x_1 C_1/x_3 S_1/x_5)$	(0.5,2)	(15,60)	(900,30)
$(C_1/x_1 C_1/x_3 C_1/x_5)$	(0.5,2)	(15,60)	(450,1800)

Matrix A1.3b

♣ ♣ EXERCISE 6 DUEL GUESSING GAME, A ZERO SUM GAME

1. The game lasts at most 10 periods, given that, with 10 clues, player 2 is sure to give the right answer, so he answers. The extensive form game is described in Figure A1.2a. A and P respectively mean Answers and Passes.

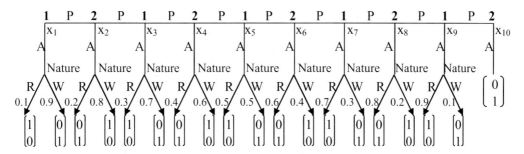

Figure A1.2a

We can eliminate the actions played by Nature by including them directly in the payoffs. So, for example, when player 1 answers at x_1 his payoff is $0.1 \times 1 + 0.9 \times 0 = 0.1$, and player 2's payoff is $0.1 \times 0 + 0.9 \times 1 = 0.9$. So we get the (friendlier) new extensive form game in Figure A1.2b:

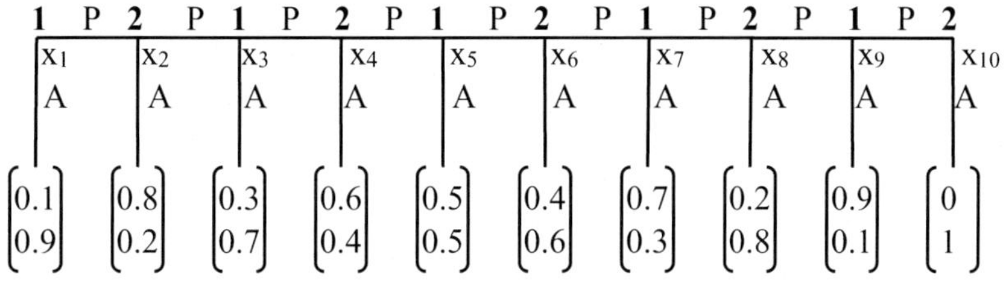

Figure A1.2b

This game is a zero sum game; this would be more obvious if we rescaled the payoffs by adding -0.5 to each of them. For example, instead of the payoffs $(0.3, 0.7)$, we would get the payoffs $(-0.2, 0.2)$: by doing so, what is got by one player is lost by the other, and the sum of the payoffs is always null (without the rescaling the sum of the payoffs is always equal to 1).

Player 1 has five information sets, $\{x_1\},\{x_3\},\{x_5\},\{x_7\}$ and $\{x_9\}$ and two possible actions at each information set, so she has $2^5=32$ pure strategies, among them, for example $(A/x_1,A/x_3,P/x_5,P/x_7,A/x_9)$. Player 2 has five information sets, $\{x_2\},\{x_4\},\{x_6\},\{x_8\}$ and $\{x_{10}\}$ and two possible actions at each information set, except at $\{x_{10}\}$, where he can only answer. So he has $2^4 \times 1=16$ pure strategies, among them, for example $(A/x_2, P/x_4, P/x_6, A/x_8, A/x_{10})$.

2. Many of these strategies are equivalent. In the sequel * means any available action, A or P. If player 1 answers at x_1, the game stops (either she wins or she loses), so what is done at further nodes has no impact. In other words all the strategies $(A/x_1,*/x_3,*/x_5,*/x_7,*/x_9)$ lead to the same couple of payoffs $(0.1, 0.9)$ for each strategy played by player 2. So they are equivalent and we replace them by the strategy A/x_1 (the action that characterizes them).

ANSWERS

If player 2 answers at x_2, the game stops latest at x_2. So the actions chosen by player 2 at further nodes have no impact, neither on his nor on player 1's payoffs: all the strategies $(A/x_2,*/x_4,*/x_6,*/x_8, A/x_{10})$ are equivalent (they lead either to (0.1, 0.9) if player 1 answers at x_1, or to (0.8, 0.2) if she does not). We replace them by the strategy A/x_2 (the action that characterizes them).

All the strategies $(P/x_1,A/x_3,*/x_5,*/x_7,*/x_9)$ are equivalent, the actions played at x_5, x_7 and x_9 having no impact on the payoffs because the game stops latest at x_3. We replace them by the strategy $(P/x_1\ A/x_3)$ (the actions that characterize them). And so on. All the strategies $(P/x_2,A/x_4,*/x_6,*/x_8, A/x_{10})$ are equivalent and replaced by the strategy $(P/x_2A/x_4)$. All the strategies $(P/x_1,P/x_3,A/x_5,*/x_7,*/x_9)$ are equivalent and replaced by the strategy $(P/x_1P/x_3A/x_5)$. All the strategies $(P/x_2,P/_4,A/x_6,*/x_8,A/x_{10})$ are equivalent and replaced by the strategy $(P/x_2 P/x_4A/x_6)$. All the strategies $(P/x_1,P/x_3,P/x_5,A/x_7,*/x_9)$ are equivalent and replaced by the strategy $(P/x_1P/x_3P/x_5A/x_7)$.

So we get the reduced normal form game in Matrix A1.4:

		Player 2				
		A/x_2	$P/x_2A/x_4$	$P/x_2P/x_4A/x_6$	$P/x_2P/x_4P/x_6A/x_8$	$P/x_2P/x_4P/x_6P/x_8A/x_{10}$
Player 1	A/x_1	(0.1,0.9)	(0.1,0.9)	(0.1,0.9)	(0.1,0.9)	(0.1,0.9)
	$P/x_1A/x_3$	(0.8,0.2)	(0.3,0.7)	(0.3,0.7)	(0.3,0.7)	(0.3,0.7)
	$P/x_1P/x_3A/x_5$	(0.8,0.2)	(0.6,0.4)	(0.5,0.5)	(0.5,0.5)	(0.5,0.5)
	$P/x_1P/x_3P/x_5A/x_7$	(0.8,0.2)	(0.6,0.4)	(0.4,0.6)	(0.7,0.3)	(0.7,0.3)
	$P/x_1P/x_3P/x_5P/x_7A/x_9$	(0.8,0.2)	(0.6,0.4)	(0.4,0.6)	(0.2,0.8)	(0.9,0.1)
	$P/x_1P/x_3P/x_5P/x_7P/x_9$	(0.8,0.2)	(0.6,0.4)	(0.4,0.6)	(0.2,0.8)	(0,1)

Matrix A1.4

3. This time the game ends earlier, latest at period 7, because player 1 gives the right answer with seven clues with probability 1. So the new extensive form game is given in Figure A1.2c.

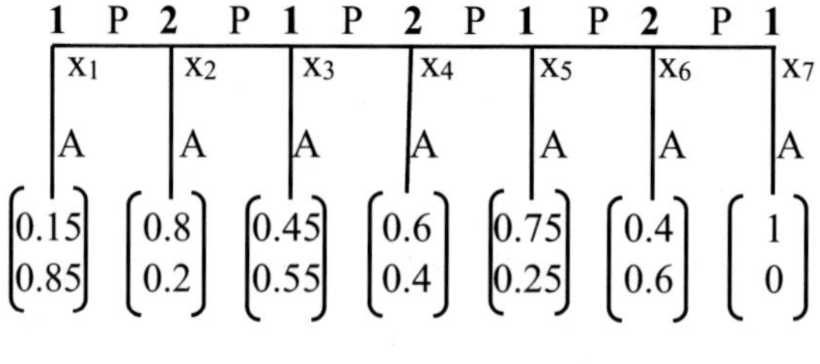

```
  1    P   2    P   1    P   2    P   1    P   2    P   1
  x1       x2       x3       x4       x5       x6       x7
  |A       |A       |A       |A       |A       |A       |A
 [0.15]  [0.8 ]  [0.45]  [0.6]  [0.75]  [0.4]  [1]
 [0.85]  [0.2 ]  [0.55]  [0.4]  [0.25]  [0.6]  [0]
```

Figure A1.2c

ANSWERS

♣♣ EXERCISE 7 AKERLOF'S LEMON CAR, EXPERIENCE GOOD MODEL, SWITCHING FROM AN EXTENSIVE FORM SIGNALLING GAME TO ITS NORMAL FORM

1. The extensive form game is given in Figure A1.3a. This game tree is often replaced by the one given in Figure A1.3b (the starting node is of course still Nature's decision node x_0).

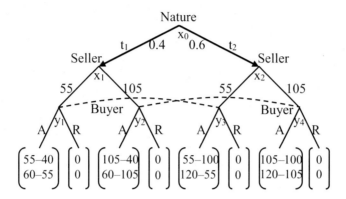

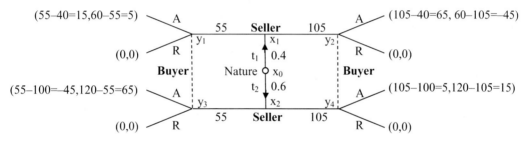

Figure A1.3a and Figure A1.3b the first (respectively the second) coordinate of each payoff vector is the seller's (respectively the buyer's) payoff.

2. The seller has two information sets $\{x_1\}$, the node after Nature's choice t_1, and $\{x_2\}$, the node after Nature's choice t_2.

   Given that she can choose two prices at each node, 55 and 105, she has 2×2=4 strategies: $(55/\{x_1\},55/\{x_2\})$, $(55/\{x_1\},105/\{x_2\})$, $(105/\{x_1\},55/\{x_2\})$ and $(105/\{x_1\},105/\{x_2\})$, we write as follows: $(55/t_1,55/t_2)$, $(55/t_1,105/t_2)$, $(105/t_1,55/t_2)$ and $(105/t_1,105/t_2)$. The buyer has two information sets $\{y_1,y_3\}$, the set after the seller's price 55, and $\{y_2,y_4\}$, the set after the seller's price 105. Given that he can accept or refuse the transaction at each information set, he has 2×2=4 strategies: $(A/\{y_1,y_3\},A/\{y_2,y_4\})$, $(A/\{y_1,y_3\},R/\{y_2,y_4\})$, $(R/\{y_1,y_3\},A/\{y_2,y_4\})$ and $(R/\{y_1,y_3\},R/\{y_2,y_4\})$, we write as follows : (A/55, A/105), (A/55, R/105), (R/55, A/105) and (R/55, R/105).

   Nature has two moves, t_1 and t_2.

   So we first get an (intermediate) normal form game (Matrices A1.5a), where the seller plays the rows, the buyer plays the columns and Nature plays the matrices.

ANSWERS

Seller

Buyer

	A/55 A/105	A/55 R/105	R/55 A/105	R/55 R/105
55/t_1 55/t_2	(15,5)	(15,5)	(0,0)	(0,0)
55/t_1 105/t_2	**(15,5)**	(15,5)	(0,0)	(0,0)
105/t_1 55/t_2	(65,−45)	(0,0)	(65,−45)	(0,0)
105/t_1 105/t_2	(65,−45)	(0,0)	(65,−45)	(0,0)

t_1
Nature

Seller

Buyer

	A/55 A/105	A/55 R/105	R/55 A/105	R/55 R/105
55/t_1 55/t_2	(−45,65)	(−45,65)	(0,0)	(0,0)
55/t_1 105/t_2	(5,15)	(0,0)	(5,15)	(0,0)
105/t_1 55/t_2	(−45,65)	(−45,65)	(0,0)	(0,0)
105/t_1 105/t_2	(5,15)	(0,0)	**(5,15)**	(0,0)

t_2
Nature

Matrices A1.5a

In the first, respectively in the second matrix, only the action chosen by the seller of type t_1, respectively type t_2, matters. So, for example, the bold underlined couple (15,5) is obtained as follows: Nature chooses t_1, the seller of type t_1 chooses 55, and the buyer accepts 55. The bold underlined couple (5, 15) is obtained as follows: Nature chooses t_2, the seller of type t_2 chooses 105, and the buyer accepts 105.

Yet Nature is not a real player. Her choice is given. She chooses t_1 with probability 0.4 and t_2 with probability 0.6. So the Matrix (A1.5b), the true normal form game, is obtained by summing the two Matrices A1.5a, the payoffs in the first Matrix A1.5a being weighted by 0.4, the payoffs in the second Matrix A1.5a being weighted by 0.6. So we get:

Seller

Buyer

	A/55 A/105	A/55 R/105	R/55 A/105	R/55 R/105
55/t_1 55/t_2	(−21,41)	(−21,41)	(0,0)	(0,0)
55/t_1 105/t_2	(9,11)	(6,2)	(3,9)	(0,0)
105/t_1 55/t_2	(−1,21)	(−27,39)	(26,−18)	(0,0)
105/t_1 105/t_2	**(29,−9)**	(0,0)	(29,−9)	(0,0)

Matrix A1.5b

For example, the bold underlined payoffs are obtained as follows: 29=65×0.4+5×0.6 and −9=−45×0.4+15×0.6.

Of course, you can write Matrix A1.5b directly, without using the Matrices A1.5a. For example, if the seller plays (105/t_1, 105/t_2) and the buyer plays (A/55, A/105), given that Nature chooses t_1 with probability 0.4, and t_2 with probability 0.6, we get the payoffs (65, −45)

ANSWERS

– linked to the sequence (t_1, 105, A) – with probability 0.4, and we get the payoffs (5, 15) – linked to the sequence (t_2, 105, A) – with probability 0.6. So we get immediately (29, –9), with 29=0.4×65+0.6×5, and –9=–0.4×45+0.6×15.

♣ ♣ ♣ EXERCISE 8 BEHAVIOURAL STRATEGIES AND MIXED STRATEGIES, HOW TO SWITCH FROM THE FIRST TO THE SECOND AND VICE VERSA?

1. $S_1=\{A_1, B_1\}$ and $S_2=\{(A_2/x_1,a_2/x_2), (A_2/x_1,b_2/x_2), (B_2/x_1,a_2/x_2), (B_2/x_1,b_2/x_2)\}$.

   Call ($p_1(A_1)$, $p_1(B_1)$) a mixed strategy for player 1 and ($p_2(A_2/x_1,a_2/x_2)$, $p_2(A_2/x_1,b_2/x_2)$, $p_2(B_2/x_1,a_2/x_2)$, $p_2(B_2/x_1,b_2/x_2)$) a mixed strategy for player 2. Player 1's and player 2's payoffs are linked to the probability to reach the end decision nodes.

   Call $p(z_i)$, i from 1 to 4, the probability to reach the end node z_i. In the sequel * means any available action.

   For z_1 to be reached, player 1 has to play A_1 and player 2 has to play A_2. We get:
   $$p(z_1)=p_1(A_1)(p_2(A_2/x_1,a_2/x_2)+p_2(A_2/x_1,b_2/x_2))=p_1(A_1) p_2(A_2/x_1,*/x_2). \qquad (1)$$

In the same way we get:

$$p(z_2)=p_1(A_1)(p_2(B_2/x_1,a_2/x_2)+p_2(B_2/x_1,b_2/x_2))=p_1(A_1)p_2(B_2/x_1,*/x_2) \qquad (2)$$
$$p(z_3)=p_1(B_1)(p_2(A_2/x_1,a_2/x_2)+p_2(B_2/x_1,a_2/x_2))=p_1(B_1)p_2(*/x_1,a_2/x_2) \qquad (3)$$
$$p(z_4)=p_1(B_1)(p_2(A_2/x_1,b_2/x_2)+p_2(B_2/x_1,b_2/x_2))=p_1(B_1)p_2(*/x_1,b_2/x_2) \qquad (4)$$

A profile of behavioural strategies leads to the same payoffs than the profile of mixed strategies if it leads to the end nodes with the same probabilities.

We call ($\pi_1(.)$, $\pi_2(.)$) a profile of behavioural strategies.

Observe that $\pi_{2x1}(A_2)$ is the probability to reach z_1 when you are at x_1. We write $p(z_i/x_j)$, the probability to reach z_i knowing that the player is at x_j.

So we get: $\pi_{2x1}(A_2)=p(z_1/x_1)=p(z_1 \cap x_1)/p(x_1)$ (by Bayes's rule)
=$p(z_1)/[p(z_1)+p(z_2)]$ (because you can't reach z_1 without reaching x_1 and because if you reach x_1, you reach either z_1 or z_2, which are two incompatible events).

So $\pi_{2x1}(A_2)=p_1(A_1)p_2(A_2/x_1,*/x_2)/((p_1(A_1)p_2(A_2/x_1,*/x_2)+p_1(A_1)p_2(B_2/x_1,*/x_2))=p_2(A_2/x_1,*/x_2)/((p_2(A_2/x_1,*/x_2)+p_2(B_2/x_1,*/x_2))=p_2(A_2/x_1,*/x_2)=p_2(A_2/x_1,a_2/x_2)+p_2(A_2/x_1,b_2/x_2)$

In the same way, $\pi_{2x2}(a_2)=p(z_3/x_2)=p(z_3)/(p(z_3)+p(z_4))$

=$p_1(B_1)p_2(*/x_1,a_2/x_2)/(p_1(B_1) p_2(*/x_1,a_2/x_2)+p_1(B_1) p_2(*/x_1,b_2/x_2))$
=$p_2(*/x_1,a_2/x_2)/(p_2(*/x_1,a_2/x_2)+p_2(*/x_1,b_2/x_2))=p_2(*/x_1,a_2/x_2)=p_2(A_2/x_1,a_2/x_2)+p_2(B_2/x_1,a_2/x_2)$

Finally $\pi_1(A_1)=p(x_1/x_0)=(p(z_1)+p(z_2))/(p(z_1)+p(z_2)+p(z_3)+p(z_4))=(p(z_1)+p(z_2))/1=p_1(A_1)p_2(A_2/x_1,*/x_2)+p_1(A_1)p_2(B_2/x_1,*/x_2)=p_1(A_1)$

So we have found the equivalent profile of behavioural strategies. Let us however observe that, to get $\pi_{2x1}(A_2)$, we divided the fraction by $p_1(A_1)$ and to get $\pi_{2x2}(a_2)$, we divided the fraction by $p_1(B_1)$. This is only possible if $p_1(A_1)$ and $p_1(B_1)$ are not equal to 0.

ANSWERS

Let us give a numerical example: suppose that $p_1(A_1)=0.2$, $p_1(B_1)=0.8$, $p_2(A_2/x_1,a_2/x_2)=0.2$, $p_2(A_2/x_1,b_2/x_2)=0.3$, $p_2(B_2/x_1,a_2/x_2)=0.1$ and $p_2(B_2/x_1,b_2/x_2)=0.4$

The equivalent profile of behavioural strategies is: $\pi_1(A_1)=0.2$, $\pi_1(B_1)=1-\pi_1(A_1)=0.8$, $\pi_{2x1}(A_2)=0.2+0.3=0.5$, $\pi_{2x1}(B_2)=1-\pi_{2x1}(A_2)=0.5$, $\pi_{2x2}(a_2)=0.2+0.1=0.3$, $\pi_{2x2}(b_2)=1-\pi_{2x2}(a_2)=0.7$

You can check that each end node is reached with the same probability with both the mixed and the behavioural profiles of strategies. You get: $p(z_1)=0.1$, $p(z_2)=0.1$, $p(z_3)=0.24$ and $p(z_4)=0.56$

More generally, by proceeding as above, you find a unique equivalent profile of behavioural strategies as soon as the profile of mixed strategies is completely mixed.

2. We start now with a profile of behavioural strategies and we look for an equivalent profile of mixed strategies. So we look for $p_1(.)$ and $p_2(.)$ such that:

$$p(z_1)=\pi_1(A_1)\pi_{2x1}(A_2)=p_1(A_1)p_2(A_2/x_1,*/x_2)=p_1(A_1)(p_2(A_2/x_1,a_2/x_2)+p_2(A_2/x_1,b_2/x_2)) \quad (5)$$

$$p(z_2)=\pi_1(A_1)\pi_{2x1}(B_2)=p_1(A_1)p_2(B_2/x_1,*/x_2)=p_1(A_1)(p_2(B_2/x_1,a_2/x_2)+p_2(B_2/x_1,b_2/x_2)) \quad (6)$$

$$p(z_3)=\pi_1(B_1)\pi_{2x2}(a_2)=p_1(B_1)p_2(*/x_1,a_2/x_2)=p_1(B_1)(p_2(A_2/x_1,a_2/x_2)+p_2(B_2/x_1,a_2/x_2)) \quad (7)$$

$$p(z_4)=\pi_1(B_1)\pi_{2x2}(b_2)=p_1(B_1)p_2(*/x_1,b_2/x_2)=p_1(B_1)(p_2(A_2/x_1,b_2/x_2)+p_2(B_2/x_1,b_2/x_2)) \quad (8)$$

Adding (5) and (6) leads to $\pi_1(A_1)=p_1(A_1)$, adding (7) and (8) leads to $\pi_1(B_1)=p_1(B_1)$ Hence the equations (5), (6), (7) and (8) become :

$$\pi_{2x1}(A_2)=p_2(A_2/x_1,a_2/x_2)+p_2(A_2/x_1,b_2/x_2) \quad (5')$$

$$\pi_{2x1}(B_2)=p_2(B_2/x_1,a_2/x_2)+p_2(B_2/x_1,b_2/x_2) \quad (6')$$

$$\pi_{2x2}(a_2)=p_2(A_2/x_1,a_2/x_2)+p_2(B_2/x_1,a_2/x_2) \quad (7')$$

$$\pi_{2x2}(b_2)=p_2(A_2/x_1,b_2/x_2)+p_2(B_2/x_1,b_2/x_2) \quad (8')$$

The equations (5') and (6') are duplicates, as well as the equations (7') and (8').
So the system of equations reduces to:

$$\pi_{2x1}(A_2)=p_2(A_2/x_1,a_2/x_2)+p_2(A_2/x_1,b_2/x_2) \quad (5')$$

$$\pi_{2x2}(a_2)=p_2(A_2/x_1,a_2/x_2)+p_2(B_2/x_1,a_2/x_2) \quad (7')$$

To which we add:

$$p_2(A_2/x_1,a_2/x_2)+p_2(A_2/x_1,b_2/x_2)+p_2(B_2/x_1,a_2/x_2)+p_2(B_2/x_1,b_2/x_2)=1 \quad (9)$$

$$p_2(A_2/x_1,a_2/x_2)\geq 0 \; p_2(A_2/x_1,b_2/x_2)\geq 0 \; p_2(B_2/x_1,a_2/x_2)\geq 0, \; p_2(B_2/x_1,b_2/x_2)\geq 0 \quad (10)$$

So we get a system of three equations with four unknowns, and there can exist an infinite number of solutions.

ANSWERS

For example, let us suppose that $\pi_{2x1}(A_2) > \pi_{2x2}(a_2)$ and that $\pi_{2x1}(A_2) + \pi_{2x2}(a_2) < 1$. We propose three solutions (among an infinity):

$p_1(A_1) = \pi_1(A_1)$, $p_1(B_1) = 1 - \pi_1(A_1)$, $p_2(A_2/x_1, a_2/x_2) = \pi_{2x2}(a_2)$, $p_2(A_2/x_1, b_2/x_2) = \pi_{2x1}(A_2) - \pi_{2x2}(a_2)$, $p_2(B_2/x_1, a_2/x_2) = 0$ and $p_2(B_2/x_1, b_2/x_2) = 1 - \pi_{2x1}(A_2)$.

$p_1(A_1) = \pi_1(A_1)$, $p_1(B_1) = 1 - \pi_1(A_1)$, $p_2(A_2/x_1, a_2/x_2) = 0$, $p_2(A_2/x_1, b_2/x_2) = \pi_{2x1}(A_2)$, $p_2(B_2/x_1, a_2/x_2) = \pi_{2x2}(a_2)$ and $p_2(B_2/x_1, b_2/x_2) = 1 - \pi_{2x1}(A_2) - \pi_{2x2}(a_2)$.

$p_1(A_1) = \pi_1(A_1)$, $p_1(B_1) = 1 - \pi_1(A_1)$, $p_2(A_2/x_1, a_2/x_2) = \pi_{2x1}(A_2)\pi_{2x2}(a_2)$, $p_2(A_2/x_1, b_2/x_2) = \pi_{2x1}(A_2)\pi_{2x2}(b_2)$, $p_2(B_2/x_1, a_2/x_2) = \pi_{2x1}(B_2)\pi_{2x2}(a_2)$, $p_2(B_2/x_1, b_2/x_2) = \pi_{2x1}(B_2)\pi_{2x2}(b_2)$

The last solution is perhaps the most intuitive one. *But it is not the only one.*

Let us give a numerical example. We take the behavioural strategies of question 1: $\pi_1(A_1) = 0.2$, $\pi_1(B_1) = 0.8$ $\pi_{2x1}(A_2) = 0.5$, $\pi_{2x2}(a_2) = 0.3$. The three proposed solutions are:

$p_1(A_1) = 0.2$, $p_1(B_1) = 0.8$ $p_2(A_2/x_1, a_2/x_2) = 0.3$, $p_2(A_2/x_1, b_2/x_2) = 0.2$, $p_2(B_2/x_1, a_2/x_2) = 0$ et $p_2(B_2/x_1, b_2/x_2) = 0.5$

$p_1(A_1) = 0.2$, $p_1(B_1) = 0.8$, $p_2(A_2/x_1, a_2/x_2) = 0$, $p_2(A_2/x_1, b_2/x_2) = 0.5$, $p_2(B_2/x_1, a_2/x_2) = 0.3$ et $p_2(B_2/x_1, b_2/x_2) = 0.2$

$p_1(A_1) = 0.2$, $p_1(B_1) = 0.8$, $p_2(A_2/x_1, a_2/x_2) = 0.15$, $p_2(A_2/x_1, b_2/x_2) = 0.35$, $p_2(B_2/x_1, a_2/x_2) = 0.15$, $p_2(B_2/x_1, b_2/x_2) = 0.35$

None of these three solutions is the profile of mixed strategies in question 1. But we can check that the profile $p_1(A_1) = 0.2$, $p_1(B_1) = 0.8$, $p_2(A_2/x_1, a_2/x_2) = 0.2$, $p_2(A_2/x_1, b_2/x_2) = 0.3$, $p_2(B_2/x_1, a_2/x_2) = 0.1$ and $p_2(B_2/x_1, b_2/x_2) = 0.4$ also satisfies the equations (5') (7') and (9).

3. We recall in Figure A1.4 the extensive form of the pre-emption game by replacing the end payoffs by the end nodes.

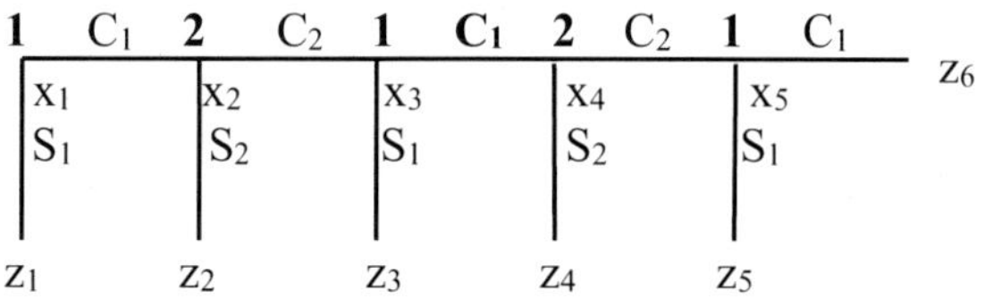

■ **Figure A1.4**

If $p_1(S_1/x_1, S_1/x_3, S_1/x_5) = p_1(S_1/x_1, S_1/x_3, C_1/x_5) = p_1(S_1/x_1, C_1/x_3, S_1/x_5) = p_1(S_1/x_1, C_1/x_3, C_1/x_5) = p_1(C_1/x_1, S_1/x_3, S_1/x_5) = p_1(C_1/x_1, S_1/x_3, C_1/x_5) = p_1(C_1/x_1, C_1/x_3, S_1/x_5) = 1/7$, $p_1(C_1/x_1, C_1/x_3, C_1/x_5) = 0$ and $p_2(S_2/x_2, S_2/x_4) = p_2(S_2/x_2, C_2/x_4) = p_2(C_2/x_2, S_2/x_4) = 2/7$, $p_2(C_2/x_2, C_2/x_4) = 1/7$.

We get the equivalent behavioural strategies:

$\pi_{1x5}(S_1) = p(z_5)/(p(z_5) + p(z_6)) = p_1(C_1/x_1, C_1/x_3, S_1/x_5)/(p_1(C_1/x_1, C_1/x_3, S_1/x_5) + p_1(C_1/x_1, C_1/x_3, C_1/x_5)) = (1/7)/(1/7 + 0) = 1$

311

ANSWERS

$\pi_{2\times4}(S_2)=p(z_4)/(p(z_4)+p(z_5)+p(z_6))=p_1(C_1/x_1,C_1/x_3,*/x_5)p_2(C_2/x_2,S_2/x_4)/[p_1(C_1/x_1,C_1/x_3,*/x_5)\,p_2(C_2/x_2,S_2/x_4)+p_2(C_2/x_2,C_2/x_4)\,p_1(C_1/x_1,C_1/x_3,S_1/x_5)+p_2(C_2/x_2,C_2/x_4)\,p_1(C_1/x_1,C_1/x_3,C_1/x_5)]=p_1(C_1/x_1,C_1/x_3,*/x_5)p_2(C_2/x_2,S_2/x_4)/[p_1(C_1/x_1,C_1/x_3,*/x_5)\,p_2(C_2/x_2,S_2/x_4)+p_2(C_2/x_2,C_2/x_4)\,p_1(C_1/x_1,C_1/x_3,*/x_5)]$
$=p_2(C_2/x_2,S_2/x_4)/[(p_2(C_2/x_2,S_2/x_4)+p_2(C_2/x_2,C_2/x_4)]=(2/7)/(2/7+1/7)=2/3$

In the same way we get: $\pi_{1\times3}(S_1)=p(z_3)/(p(z_3)+p(z_4)+p(z_5)+p(z_6))=p_1(C_1/x_1,S_1/x_3,*/x_5)/(p_1(C_1/x_1,S_1/x_3,*/x_5)+p_1(C_1/x_1,C_1/x_3,*/x_5))=(p_1(C_1/x_1,S_1/x_3,S_1/x_5)+p_1(C_1/x_1,S_1/x_3,C_1/x_5))/[p_1(C_1/x_1,S_1/x_3,S_1/x_5)+p_1(C_1/x_1,S_1/x_3,C_1/x_5)+p_1(C_1/x_1,C_1/x_3,S_1/x_5)+p_1(C_1/x_1,C_1/x_3,C_1/x_5)]$
$=(1/7+1/7)/(1/7+1/7+1/7+0)=2/3$

$\pi_{2\times2}(S_2)=p(z_2)/(p(z_2)+p(z_3)+p(z_4)+p(z_5)+p(z_6))=p_2(S_2/x_2,*/x_4)/(p_2(S_2/x_2,*/x_4)+p_2(C_2/x_2,*/x_4))=p_2(S_2/x_2,*/x_4)/1=2/7+2/7=4/7$ (we can also write $\pi_{2\times2}(S_2)=p(z_2)/(1-p(z_1))=p_1(C_1/x_1,*/x_3,*/x_5)p_2(S_2/x_2,*/x_4)/p_1(C_1/x_1,*/x_3,*/x_5)=p_2(S_2/x_2,*/x_4)=p_2(S_2/x_2,S_2/x_4)+p_2(S_2/x_2,C_2/x_4)=2/7+2/7=4/7$

$\pi_{1\times1}(S_1)=p(z_1)/(p(z_1)+p(z_2)+p(z_3)+p(z_4)+p(z_5)+p(z_6))=p_1(S_1/x_1,*/x_3,*/x_5)/1=p_1(S_1/x_1,S_1/x_3,S_1/x_5)+p_1(S_1/x_1,S_1/x_3,C_1/x_5)+p_1(S_1/x_1,C_1/x_3,S_1/x_5)+p_1(S_1/x_1,C_1/x_3,C_1/x_5)=1/7+1/7+1/7+1/7=4/7$

4. If $p_1(S_1/x_1)=½$, $p_1(C_1/x_1\,S_1/x_3)=1/3$, $p_1(C_1/x_1,C_1/x_3,S_1/x_5)=1/6$, $p_1(C_1/x_1,C_1/x_3,C_1/x_5)=0$,

$p_2(S_2/x_2)=½$, $p_2(C_2/x_2,S_2/x_4)=1/3$ and $p_2(C_2/x_2,C_2/x_4)=1/6$, we get the equivalent behavioural strategies:

$\pi_{1\times5}(S_1)=p(z_5)/(p(z_5)+p(z_6))=p_1(C_1/x_1,C_1/x_3,S_1/x_5)/(p_1(C_1/x_1,C_1/x_3,S_1/x_5)+p_1(C_1/x_1,C_1/x_3,C_1/x_5)=(1/6)/(1/6+0)=1$

$\pi_{2\times4}(S_2)=p(z_4)/(p(z_4)+p(z_5)+p(z_6))=p_2(C_2/x_2,S_2/x_4)/(p_2(C_2/x_2,S_2/x_4)+p_2(C_2/x_2,C_2/x_4))=(1/3)/(1/3+1/6)=2/3$

$\pi_{1\times3}(S_1)=p(z_3)/(p(z_3)+p(z_4)+p(z_5)+p(z_6))=p_1(C_1/x_1\,S_1/x_3)/\,(p_1(C_1/x_1\,S_1/x_3)+p_1(C_1/x_1,C_1/x_3,S_1/x_5)+p_1(C_1/x_1,C_1/x_3,C_1/x_5))=(1/3)/(1/3+1/6)=2/3$

$\pi_{2\times2}(S_2)=p(z_2)/(p(z_2)+p(z_3)+p(z_4)+p(z_5)+p(z_6))=p_2(S_2/x_2)=1/2$

$\pi_{1\times1}(S_1)=p(z_1)/(\,p(z_1)+p(z_2)+p(z_3)+p(z_4)+p(z_5)+p(z_6))=p_1(S_1/x_1)/1=1/2$

♣ ♣ ♣ EXERCISE 9 DUTCH AUCTION AND FIRST PRICE SEALED BID AUCTION, STRATEGIC EQUIVALENCE OF DIFFERENT GAMES

1. The FPSBA in extensive form is given (without the payoffs) in Figure A1.5a, and the Dutch auction is given (without the payoffs) in Figure A1.5b.
2. In the FPSBA, player 1 has three information sets, according to the value she assigns to the object. Given that she bids 0 if she values the object 0, 0 or 1 if she values the object 1, and 0, 1 or 2 if she values the object 2, the cardinal of her pure strategy set is 1×2×3=6. The same is true for player 2. An example of strategy for player 1 is (bid 0/value 0, bid 1/value 1, bid 1/value 2).

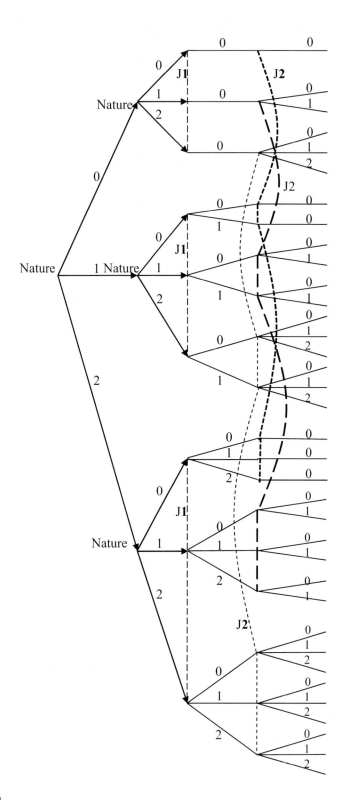

Figure A1.5a

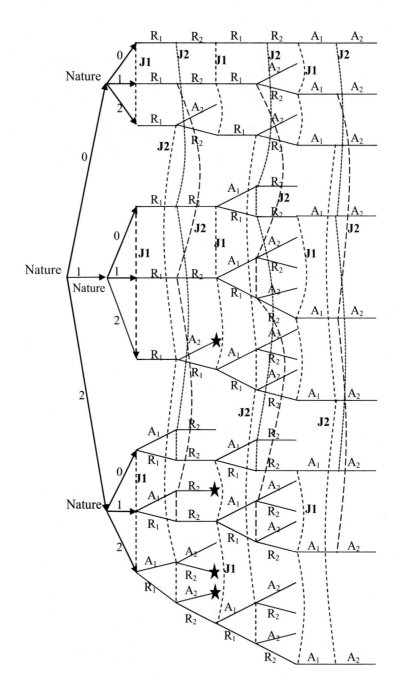

Figure A1.5b

ANSWERS

The Dutch Auction game is more complicated. Player 1 has three information sets (according to the value she assigns to the object) in the first period, where the auctioneer proposes the price 2. She can only stay in if she values the object 0 or 1 and she stays in or accepts the price if she values the object 2.

In the second period (where the auctioneer proposes the price 1, because the price 2 has been refused by both bidders), she has again three information sets: she stays in if she values the object 0 and she can stay in or accept the price if she values the object 1 or 2.

In the third period (where the auctioneer proposes the price 0 because the price 1 has been refused by both bidders), she has again three information sets and she accepts the price in each set.

So player 1 has nine information sets but the cardinal of player 1's pure strategy set is only: $1\times1\times2\times1\times2\times2\times1\times1\times1=8$. And the same is true for player 2. By noting R and A for stay in (refuse the price) and accept the price, an example of strategy for player 1 is:

(R/price2Value0, R/price2Value1, R/price2Value2, R/price1Value0, A/price1Value1, A/price1Value2, A/price0Value0, A/price0Value1, A/price0Value2). If we omit saying the action each time there is only one action available, this strategy becomes: (R/price2Value2, A/price1Value1, A/price1Value2).

At first view, the strategies are rather different in a FPSBA and in a Dutch auction. In a FPSBA the bidders just have to propose a price for the object, not higher than its value, without knowing the value nor what is proposed by the opponent. We are in a *simultaneous game*. The Dutch auction is a sequential game, in that the bidders potentially play in several periods: if nobody wants to pay 2, the game begins a second round, where each bidder is called on to accept or refuse the price 1 (and if nobody accepts the price 1, the game turns to a third round, where each bidder accepts the price 0). Observe that, in the second round, each bidder knows that the price 2 has not been accepted by the opponent. This may help to infer some information about the opponent's type. For example, if the opponent usually accepts the highest price he can accept, the refusal of price 2 means that he values the object 0 or 1. So past actions may convey information on the nature of the opponent and perhaps on the way he will play in future periods.

3. There are 36 payoffs vectors (terminal nodes) in the FPSBA, and only 32 payoff vectors (terminal nodes) in the Dutch auction. This is due to the following fact:

In a Dutch Auction, if a player, who values the object 2, accepts the price 2, whereas the opponent does not accept it, we will not know if the opponent accepts or not the price 1 (if he values the object 1 or 2), because the game stops as soon as a price is accepted. This happens at the nodes with a star in Figure A1.5b.

In a FPSBA, we always know what each player proposes because the two offers are compared.

4. At first view, we might be astonished by the fact that equilibrium concepts lead to a same behaviour in both games, yet the incentives of both players in the two games are the same. We will see later (in Exercise 28 of chapter 3) that if a player proposes a price p in a FPSBA, then he accepts all prices lower or equal to p in a Dutch auction. And this is not surprising. Especially, if he gets the object at price p in the FPSBA, this means that he is happy to get the object at price p and that he can't bid less without losing the auction. So he is also happy to accept p in the Dutch auction, because waiting for a lower price would lead him to lose the auction. So this game illustrates that two games of different structure may have the same strategic content.

Chapter 2

Answers to Exercises

♣ EXERCISE 1 DOMINANCE IN A GAME IN NORMAL FORM

B_2 is strictly dominated by D_2 because: if player 1 plays A_1, player 2 gets 1 with B_2 and the higher amount 2 with D_2, if player 1 plays B_1, player 2 gets 2 with B_2 and the higher amount 3 with D_2, if player 1 plays C_1, player 2 gets 0 with B_2 and the higher amount 4 with D_2, and if player 1 plays D_1, player 2 gets 0 with B_2 and the higher amount 1 with D_2 (we later simply write 1<2, 2<3, 0<4 and 0<1).

Once B_2 is eliminated, C_1 becomes strictly dominated by B_1 because: if player 2 plays A_2, player 1 gets 2 with C_1 and the higher amount 3 with B_1, if player 2 plays C_2, player 1 gets 1 with C_1 and the higher amount 2 with B_1 and if player 2 plays D_2, player 1 gets 0 with C_1 and the higher amount 1 with B_1 (we later simply write 2<3, 1<2 and 0<1).

Once C_1 is eliminated, D_2 becomes weakly dominated by C_2 because 2=2, 3<5 and 1=1.

Once D_2 is eliminated, D_1 becomes strictly dominated by A_1 (0<5, 0<1) and by B_1 (0<3 and 0<2).

There are no additional dominated strategies in the game.

♣ EXERCISE 2 DOMINANCE BY A MIXED STRATEGY

1. B_2 is for example strictly dominated by the mixed strategy p_2 that assigns the probability 0.4 to A_2 and the probability 0.6 to D_2 (but B_2 is also strictly dominated by other mixed strategies, for example the one which assigns the probability 0.6 to C_2 and the probability 0.4 to D_2).

B_2 is strictly dominated by p_2 because:

if player 1 plays A_1, player 2 gets 1 with B_2 and the higher amount 1.2=0×0.4+2×0.6 with p_2,
if player 1 plays B_1, player 2 gets 4 with B_2 and the higher amount 4.2=6×0.4+3×0.6 with p_2,
if player 1 plays C_1, player 2 gets 0 with B_2 and the higher amount 2.4=0×0.4+4×0.6 with p_2,
and if player 1 plays D_1, player 2 gets 0 with B_2 and the higher amount 0.4=1×0.4+0×0.6 with p_2.

Once B_2 is eliminated, C_1 becomes strictly dominated by, for example, the mixed strategy p_1 that assigns the probability 0.3 to A_1 and the probability 0.7 to B_1 because:

■ 316

ANSWERS

if player 2 plays A_2, player 1 gets 2 with C_1 and the higher amount $2.2=5\times0.3+1\times0.7$ with p_1, if player 2 plays C_2, player 1 gets 1 with C_1 and the higher amount $1.4=0\times0.3+2\times0.7$ with p_1, and if player 2 plays D_2, player 1 gets 0 with C_1 and the higher amount $0.7=0\times0.3+1\times0.7$ with p_1.

2. Once C_1 is eliminated, D_2 becomes weakly dominated by C_2 because 2=2, 3<5 and 0<1. Once D_2 is eliminated, D_1 becomes strictly dominated by B_1 because 0<1 and 0<2.

There are no additional dominated strategies.

♣ EXERCISE 3 ITERATED DOMINANCE, THE ORDER MATTERS

We start by eliminating the strictly dominated strategies because the order in which strictly dominated strategies are eliminated does not matter.

A_2 is strictly dominated by D_2 (1<2, 2<5, 0<1 and 0<4). Once A_2 is eliminated, D_1 becomes strictly dominated by B_1 (2<3, 1<2 and 0<1).

Now we can proceed in different ways:

First way:
A_1 is weakly dominated by B_1 (3=3, 2=2 and 0<1). Once A_1 is eliminated, B_2 becomes strictly dominated by C_2 (0<5, 0<2).

Once B_2 is eliminated, we can proceed in two ways:

- We can eliminate D_2 because it is weakly dominated by C_2 (5=5, 1<2), in which case there remain the two couples (B_1,C_2) and (C_1,C_2) (which yield the two couples of payoffs (2,5) and (2,2)).
- But we can also eliminate B_1 which is weakly dominated by C_1 (2=2, 1<4), after which D_2 becomes strictly dominated by C_2 (1<2), and there only remains the couple (C_1,C_2), so the couple of payoffs (2,2).

Second way
D_2 is weakly dominated by C_2 (2=2, 5=5 and 1<2).

Once D_2 is eliminated, C_1 becomes weakly dominated by A_1 (0<3, 2=2) and there remain the four couples (A_1,B_2), (A_1,C_2), (B_1,B_2) and (B_1,C_2), so the four couples of payoffs (3,3), (2,2), (3,0) and (2,5).

Clearly the order matters: for example, only the second way can lead to the payoffs (3,3).

♣♣ EXERCISE 4 ITERATED DOMINANCE IN ASYMMETRIC ALL PAY AUCTIONS

1. Player 1, by bidding 4.5 or 5, whether winning or losing the auction, loses money, because she values the object only 4. So the bids 4.5 and 5 are strictly dominated by bid 0 and eliminated.

   Similarly, player 2, by bidding 3.5, 4, 4.5 or 5, loses money because he values the object only 3. So the bids 3.5, 4, 4.5 and 5 are strictly dominated by bid 0 and eliminated.

   After these first eliminations, player 1's bid 4 becomes strictly dominated by bid 3.5, because she wins the auction with both bids, but pays more with 4 than with 3.5. By eliminating bid 4, we get Matrix E2.4a.

ANSWERS

2. We successively eliminate:
   - *Player 1's bid 0 and player 2's bid 3*. Player 1's bid 0 is weakly dominated by the mixed strategy that assigns probability 0.5 to bid 0.5 and probability 0.5 to bid 3.5 (7=8.5½+5.5½, 5<6.5½+5.5½, and 5=4.5½+5.5½). Player 2's bid 3 is weakly dominated by bid 0 (5<6.5, 5=5, 3.5<5, 2<5)
   - *Player 1's bid 3.5 and player 2's bid 0.5*. Player 1's bid 3.5 is strictly dominated by bid 3 (5.5<6) and player 2's bid 0.5 is weakly dominated by the strategy that assigns probability 0.5 to bid 0 and probability 0.5 to bid 1 (6=5½+7½, 4.5<5½+5.5½, 4.5=5½+4½)
   - *Player 1's bid 1*. It is weakly dominated by the strategy that assigns probability 0.5 to bid 0.5 and probability 0.5 to bid 1.5 (8=8.5½+7.5½, 6=4.5½+7.5½, 4<4.5½+5.5½, 4=4.5½+3.5½)
   - *Player 2's bid 1.5*. It is weakly dominated by the strategy that assigns probability 0.5 to bid 1 and probability 0.5 to bid 2 (6.5=7½+6½, 5=4½+6½, 3.5<4½+4.5½, 3.5=4½+3½)
   - *Player 1's bid 2*. It is weakly dominated by the strategy that assigns probability 0.5 to bid 1.5 and probability 0.5 to bid 2.5 (7=7.5½+6.5½, 5=3.5½+6.5½, 3<3.5½+4.5½).
   - *Player 2's bid 2.5*. It is weakly dominated by the strategy that assigns probability 0.5 to bid 0 and probability 0.5 to bid 2 (5.5=5½+6½, 4=5½+3½, 2.5<5½+3½).
   - *Player 1's bid 3*. It is strictly dominated by bid 2.5 (6<6.5).

   We get the game in Matrix E2.4b.

3. Let us first point out a strange phenomenon: *a strategy is eliminated at step t because of a strategy eliminated at step t+1*.

   For example, player 1's bid 0 is weakly dominated by the mixed strategy that puts probability ½ on bids 0.5 and 3.5: yet both strategies yield the same payoff except if player 2 bids 0.5. But player 2's bid 0.5 is eliminated in the second round, because it is weakly dominated by the mixed strategy that puts probability ½ on the bids 0 and 1. Yet both strategies lead to the same payoff except if player 1 bids 1, but player 1's bid 1 is eliminated in the third round, because it is weakly dominated by the mixed strategy that puts probability ½ on the bids 0.5 and 1.5. And so on.

   So we systematically eliminate a strategy thanks to a strategy eliminated at the next round. It is as if you refuse a person A in a committee because of the presence of another person B, who anyway will not be present if A is not present. This may seem strange.

   It explains why at (the further studied) equilibrium some eliminated strategies also lead to the equilibrium payoff (because they yield a lower payoff only in presence of strategies that are not played), and that the "holes" in the players' strategies seem somewhat artificial.

   Let us make another observation. All happens as if player 2 is more pushed to the left than player 1 (player 2 plays 0, 1, 2, whereas player 1 plays 0.5, 1.5 and 2.5). At first sight, this seems natural: given that player 2 values the object less than player 1, it seems logical that he bids less than player 1. Yet this asymmetry between the players only depends on *the increment used in the game: it does not depend on the difference between V_1 and V_2*. In fact *the asymmetry goes to 0 when the increment goes to 0*, so it is quite artificial.

ANSWERS

In fact, we will see in Exercise 19 in chapter 3, by looking for the Nash equilibria of the continuous version of the game, that the bids played by both players are the same: only the probability distribution on the bids differs among the two players. So, in fact, the shift and the holes completely disappear: the asymmetry between the players only appears in the equilibrium probabilities.

♣ EXERCISE 5 DOMINANCE AND VALUE OF INFORMATION

1. In Figure E2.1a, x_2 and x_3 belong to the same information set, so player 2 plays in the same way at x_2 and at x_3. It follows that B_1 is strictly dominated by A_1 because:

   if player 2 plays A_2, player 1 gets 4 with B_1 and 6 with A_1;

   if player 2 plays B_2, player 1 gets 1 with B_1 and 2 with A_1.

   In the game in Figure E2.1b, player 2 can behave differently at x_2 and at x_3. B_1 is no longer dominated by A_1 because B_1 yields the payoff 4 and A_1 only yields the payoff 2 when player 2 plays the strategy $(B_2/x_2 A_2/x_3)$.

2. *Game in Figure E2.1a*

   B_1 is strictly dominated by A_1. Once B_1 is eliminated, A_2 becomes weakly dominated by B_2 (because A_2 and B_2 yield the payoff 5 if player 1 plays C_1 but A_2 yields the payoff 1 whereas B_2 yields the payoff 2 when player 1 plays A_1). Once A_2 is eliminated, A_1 becomes strictly dominated by C_1 because 2<3.
   What remains is the action C_1 and the couple of payoffs (3, 5).

   Game in Figure E2.1b

   $(A_2/x_2,*/x_3)$ is weakly dominated by $(B_2/x_2,*/x_3)$ (where * is any available action, the same in both strategies), because both strategies lead to the same payoff when player 1 plays B_1 or C_1, but the first strategy leads to the payoff 1 whereas the second leads to the payoff 2 when player 1 plays A_1.

   $(*/x_2,B_2/x_3)$ is weakly dominated by $(*/x_2,A_2/x_3)$ (where * is any available action, the same in both strategies) because both strategies lead to the same payoff when player 1 plays A_1 or C_1, but the first strategy leads to the payoff 1 whereas the second leads to the payoff 2 when player 1 plays B_1.

   Once these dominated strategies are eliminated, A_1 and C_1 become strictly dominated by B_1 because A_1 is followed by B_2 and yields the payoff 2, B_1 is followed by A_2 and yields the payoff 4 and C_1 yields the payoff 3.
   It remains the path (B_1,A_2) and the couple of payoffs (4, 2).

3. Player 2 is better off in the game in Figure E2.1a (where he gets 5 according to the iterative elimination of dominated strategies) than in the game in Figure E2.1b, where he only gets 2. In other words, he is better off when he doesn't know if he is at x_2 or x_3, than when he knows at which node he is.

   So information has a negative value for player 2. This is due to the fact that player 1 knows the level of information of player 2, and he knows and exploits the fact that player 2 behaves differently at x_2 and at x_3 when he knows at which node he is. *The fact that more information*

319

ANSWERS

can be negative for a player i is often observed in game theory as soon as the other players know that player i has this additional information.

♣ EXERCISE 6 STACKELBERG FIRST PRICE ALL PAY AUCTION

Let us first eliminate player 2's weakly dominated strategies.
The strategy $(1/x_0, a/x_1, b/x_2, c/x_3, d/x_4, e/x_5)$ weakly dominates the strategies $(x/x_0, a/x_1, b/x_2, c/x_3, d/x_4, e/x_5)$, with $x=0,2,3,4,5$, and a, b, c, d, e any available bid, because all strategies lead to the same payoff if player 1 bids 1,2,3,4 or 5, and the first strategy leads to the payoff 7 whereas the second ones lead to the lower payoffs 6.5, 6, 5, 4 or 3, depending on the value of x, when player 1 bids 0. For similar reasons the strategy $(a/x_0, 2/x_1, b/x_2, c/x_3, d/x_4, e/x_5)$ weakly dominates the strategies $(a/x_0, x/x_1, b/x_2, c/x_3, d/x_4, e/x_5)$, with $x=0,1,3,4,5$, and a, b, c, d, e any available bid. The strategies $(a/x_0, b/x_1, 0/x_2, c/x_3, d/x_4, e/x_5)$ and $(a/x_0, b/x_1, 3/x_2, c/x_3, d/x_4, e/x_5)$ weakly dominate the strategies $(a/x_0, b/x_1, x/x_2, c/x_3, d/x_4, e/x_5)$, with $x=1,2,4,5$, and a, b, c, d, e any available bid. The strategy $(a/x_0, b/x_1, c/x_2, 0/x_3, d/x_4, e/x_5)$, respectively the strategy $(a/x_0, b/x_1, c/x_2, d/x_3, 0/x_4, e/x_5)$ and the strategy $(a/x_0, b/x_1, c/x_2, d/x_3, e/x_4, 0/x_5)$ weakly dominate the strategies $(a/x_0, b/x_1, c/x_2, x/x_3, d/x_4, e/x_5)$, respectively the strategies $(a/x_0, b/x_1, c/x_2, d/x_3, x/x_4, e/x_5)$ and the strategies $(a/x_0, b/x_1, c/x_2, d/x_3, e/x_4, x/x_5)$, with $x=1,2,3,4,5$, and a, b, c, d, e any available bid.

Once the dominated strategies are eliminated, it remains the game in Figure A2.1a.

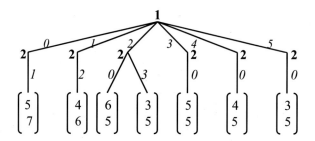

■ **Figure A2.1a**

It immediately follows that player 1's bids 1, 4 and 5 are strictly dominated by bid 0. So there remains the game in Figure A2.1b:

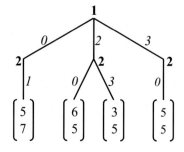

■ **Figure A2.1b**

ANSWERS

Let us now start by eliminating player 1's strictly dominated bids 4 and 5 (by bid 0), and her weakly dominated bid 3 (by bid 0). Then one eliminates player 2's weakly dominated strategies as follows:

The strategy $(1/x_0, a/x_1, b/x_2)$ weakly dominates the strategies $(x/x_0, a/x_1, b/x_2)$, with $x=0,2,3,4,5$, and a, b any available bid. The strategy $(a/x_0, 2/x_1, b/x_2)$ weakly dominates the strategies $(a/x_0, x/x_1, b/x_2)$, with $x=0,1,3,4,5$, and a, b any available bid. The strategies $(a/x_0, b/x_1, 0/x_2)$ and $(a/x_0, b/x_1, 3/x_2)$ weakly dominate the strategies $(a/x_0, b/x_1, x/x_2)$, with $x=1,2,4,5$ and a, b any available bid. Once these strategies are eliminated, player 1's bid 1 becomes strictly dominated by bid 0. And it remains the game tree in Figure A2.1c.

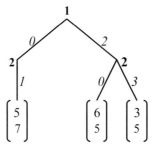

Figure A2.1c

So, according to the chosen order of elimination, we do not get the same remaining game trees. So the order matters.

♣♣ EXERCISE 7 BURNING MONEY

*means any available action.

The strategies $(b, */\bar{b}, B_1/b)$ are strictly dominated by the strategies $(\bar{b}, A_1/\bar{b}, */b)$ because, regardless of the strategy played by player 2, the first strategies lead at best to the payoff −1, whereas the second lead at worst to the payoff 0.

Once these dominated strategies are eliminated, the strategies $(*/\bar{b}, B_2/b)$ become weakly dominated by the strategies $(*/\bar{b}, A_2/b)$ (* is the same action in both strategies), because all strategies lead to the same payoff when player 1 does not burn money, but the first strategies lead to the payoff 0, whereas the second strategies lead to the payoff 1 if player 1 burns money (and necessarily plays A_1).

Once these dominated strategies are eliminated, the strategies $(\bar{b}, B_1/\bar{b}, */b)$ become strictly dominated by the strategies $(b, */\bar{b}, A_1/b)$ because, regardless of player 2's action after $\bar{b}$, the first strategies lead at best to the payoff 1, whereas the second ones lead to the payoff 2.

Once $(\bar{b}, B_1/\bar{b}, */b)$ are eliminated, $(B_2/\bar{b}, A_2/b)$ becomes weakly dominated by $(A_2/\bar{b}, A_2/b)$ because both strategies lead to the same payoff if player 1 burns money but the first strategy leads to the payoff 0, whereas the second leads to the payoff 1 if player 1 doesn't burn money.

Once $(B_2/\bar{b}, A_2/b)$ is eliminated, the strategies $(b, */\bar{b}, A_1/b)$ become strictly dominated by the strategies $(\bar{b}, A_1/\bar{b}, */b)$ because the first strategies lead to the payoff 2 whereas the second ones lead to the payoff 4.

So there only remain the strategies $((\bar{b}, A_1/\bar{b}, */b), (A_2/\bar{b}, A_2/b))$ and the played actions are $\bar{b}$, A_1 and A_2. Thanks to the possibility of burning money, player 1 gets the high payoff 4, whereas player

ANSWERS

2 only gets 1. Player 1 *never burns money*, but *the possibility* of burning money, and *the message she sends by doing so* – I can't play B_1 after burning money; it would be silly because I can always get more by not burning money – *ensures that she gets 2* by burning money and so that she *wants more* if she does not burn money. Player 2 is constrained to understand her messages, so to expect that player 1 plays A_1 after $\bar{b}$ (the only way to get more than 2), and so to play A_2 after $\bar{b}$.

The possibility of losing money may be attractive for a player, in that it gives rise to a sequence of messages that finally benefit the player who can lose money. The reasoning is very close to a forward induction reasoning (see chapter 6, section 4.3.2).

What is interesting is that this observation is still true even if people do not make all the iterations. With only two iterations (only $(b, */\bar{b}, B_1/b)$ and $(*/\bar{b}, B_2/b)$ are eliminated) player 1 is sure to get 2 (by burning money), which is a nice payoff. The possibility of burning 2 dollars ensures that she gets the maximal payoff minus 2 dollars, even if the players only make two iterations; i.e. even if the degree of crossed rationality is only 1.

♣♣ EXERCISE 8 PRE-EMPTION GAME IN EXTENSIVE FORM AND NORMAL FORM, AND CROSSED RATIONALITY

1. * means any available action.

   $(C_1/x_1, C_1/x_3, C_1/x_5)$ is weakly dominated by $(C_1/x_1, C_1/x_3, S_1/x_5)$ (0.5=0.5, 15=15 and 450<900). Once $(C_1/x_1, C_1/x_3, C_1/x_5)$ is eliminated, $(C_2/x_2, C_2/x_4)$ becomes weakly dominated by $(C_2/x_2, S_2/x_4)$ (0=0, 1=1 and 30<60). Once $(C_2/x_2, C_2/x_4)$ is eliminated, $(C_1/x_1, C_1/x_3, S_1/x_5)$ becomes weakly dominated by $(C_1/x_1, S_1/x_3, */x_5)$ (0.5=0.5 and 15<30). Once $(C_1/x_1, C_1/x_3, S_1/x_5)$ is eliminated, $(C_2/x_2, S_2/x_4)$ becomes weakly dominated by $(S_2/x_2, */x_4)$ (0=0 and 1<2). Once $(C_2/x_2, S_2/x_4)$ is eliminated, $(C_1/x_1, S_1/x_3, C_1/x_5)$ and $(C_1/x_1, S_1/x_3, S_1/x_5)$ become strictly dominated by $(S_1/x_1, */x_3, */x_5)$ because 0.5<1. So there remain the eight couples of strategies where player 1 stops at node x_1 and player 2 stops at node x_2: both players stop as soon as possible.

2. $(C_1/x_1, C_1/x_3, C_1/x_5)$ is weakly dominated by $(C_1/x_1, C_1/x_3, S_1/x_5)$: both strategies lead to the same payoffs when player 1 does not reach x_5, i.e. each time player 2 stops at x_2 or x_4, and $(C_1/x_1, C_1/x_3, C_1/x_5)$ yields the payoff 450 whereas $(C_1/x_1, C_1/x_3, S_1/x_5)$ yields the higher payoff 900 if player 1 reaches x_5, which happens when player 2 goes on each time he is called on to play.

   Eliminating $(C_1/x_1, C_1/x_3, C_1/x_5)$ induces that player 1 stops if she reaches x_5. So $(C_2/x_2, C_2/x_4)$ becomes weakly dominated by $(C_2/x_2, S_2/x_4)$ (both strategies lead to the same payoff if player 1 stops at x_1 or x_3, but, if not, the first strategy leads to the payoff 30, whereas the second leads to 60).

   Eliminating $(C_2/x_2, C_2/x_4)$ induces that player 2 stops if he reaches x_4. It follows that $(C_1/x_1, C_1/x_3, S_1/x_5)$ becomes weakly dominated by $(C_1/x_1, S_1/x_3, */x_5)$ (all these strategies lead to the same payoff if player 2 stops at x_2, but, if not, the first strategy leads to the payoff 15, whereas the second ones lead to the payoff 30).

   Eliminating $(C_1/x_1, C_1/x_3, S_1/x_5)$ induces that player 1 stops if she reaches x_3. So $(C_2/x_2, S_2/x_4)$ becomes weakly dominated by $(S_2/x_2, */x_4)$ (all these strategies lead to the same payoff if player 1 stops at x_1, but, if not, $(S_2/x_2, */x_4)$ leads to the payoff 2 whereas $(C_2/x_2, S_2/x_4)$ leads to the payoff 1).

ANSWERS

Eliminating (C_2/x_2, S_2/x_4) induces that player 2 stops if he reaches x_2. This induces that player 1 gets 0.5 if she goes on at x_1, whereas she gets 1 by stopping immediately. So the strategies (C_1/x_1, S_1/x_3, */x_5) become strictly dominated by the strategies (S_1/x_1, */x_3, */x_5).

Again, both players stop as soon as possible.

Clearly, the two eliminating processes follow the same steps in both forms.

3. Consider the extensive form game and suppose that player 2 is called on to play at x_2. What is his problem?

Iterative elimination of dominated strategies asks to stop at x_2, *but also at x_1*, yet player 1 didn't stop at x_1. So it is logically inconsistent to require that player 2 plays as planned by the iterative elimination process, given that player 1 proved that she does not conform to this process at x_1.

Let us count the levels of crossed rationality necessary to stop at x_1 (by iterative elimination of dominated strategies).

- To suppress (C_1/x_1, C_1/x_3, C_1/x_5), i.e. to stop at x_5, player 1 has to be rational.
- To suppress (C_2/x_2, C_2/x_4), i.e. to stop at x_4, player 2 has to be rational and to know that player 1 is rational.
- To suppress (C_1/x_1, C_1/x_3, S_1/x_5), i.e. to stop at x_3, player 1 needs to be rational and to know that player 2 is rational and that he knows that player 1 is rational.
- To suppress (C_2/x_2, S_2/x_4), i.e. to stop at x_2, player 2 has to be rational and to know that player 1 is rational and that she knows that player 2 is rational and that he knows that player 1 is rational.
- To suppress (C_1/x_1, S_1/x_3, */x_5), i.e. to stop at x_1, player 1 needs to be rational and to know that player 2 is rational and that he knows that player 1 is rational and that she knows that player 2 is rational and that he knows that player 1 is rational. So, to stop at x_1, player 1 has to <u>know</u> that player 2 <u>knows</u> that player 1 <u>knows</u> that player 2 <u>knows</u> that player 1 is rational, so she needs four levels of crossed rationality.

Well, it may be that player 1 does not believe in so much rationality, so does not stop at x_1. But, if she goes on, what does she play next? Perhaps she isn't rational at all and always goes on, in which case player 2 is better off going on than stopping at x_2. But perhaps she only believes in 2 levels of crossed rationality, so she stops at x_3 (because she is rational and <u>knows</u> that player 2 is rational and <u>knows</u> that player 1 is rational). If so, player 2 is better off stopping at x_2.

Be that as it may, being at x_2 is really problematic for player 2. Let us observe that this problem doesn't arise in the normal form game, because, in the normal form, you do not observe anything before playing.

♣ EXERCISE 9 GUESSING GAME AND CROSSED RATIONALITY

Even if everybody plays 100, 100(2/3)=200/3, so 200/3 weakly dominates any higher price. Why is that?

Consider y such that 200/3<y≤100. Call X the mean proposed value. Necessarily (2/3)X ≥(2/3)100=200/3. Let us first suppose that playing y leads to win the game. If so, 200/3 also leads to win because 200/3 is closer to (2/3)X than y: (2/3)X≤200/3<y. And there are values for X such that 200/3 leads to win the game whereas y leads to lose it. For example, if all the other players play (200/3), then playing 200/3 leads to win the game (and sharing the win), whereas playing y leads

323

ANSWERS

to lose it because (2/3)X is very close to 400/9 which is closer to 200/3 than y. So 200/3 weakly dominates y.

It follows that we suppress, at a first step, any real number larger than 200/3. So everybody plays at most 200/3, and 2/3(200/3)=400/9 weakly dominates any number higher than 400/9. The proof is similar to the one above. And so on. After eliminating, at a second step, any real number larger than 400/9, 2/3(400/9)=800/27 weakly dominates any value higher than 800/27, and so on. The t's step of iterative elimination leads to not play values higher than $100(2/3)^t$. And of course this value converges to 0 when t goes to ∞.

An infinite number of steps of eliminations are needed to reach 0, but few steps are enough to reach low values: for example only 4 steps of elimination are enough to play a number lower than 20, 12 steps are enough to play a number lower than 1. Given that each additional step needs an additional level of crossed rationality, 4 steps of iterative elimination (which require 3 levels of crossed rationality) seem to be a maximum, and we may expect that not too many players propose numbers lower than 20.

For information, I proposed this game to my M1 students and the mean played value was 37.6 ($29.63=100(2/3)^3<37.6<44.44=100(2/3)^2$: so the mean value is between the value obtained after 2 and 3 steps of elimination; see chapter 6 for more details, section 3.1.1).

♣♣ EXERCISE 10 BERTRAND DUOPOLY

Any price p'≥K/a is weakly dominated by any price p in {c+0.01,..., K/a–0.01} because p' leads to a null profit (given that the demand shrinks to 0), whereas p can lead to a null or positive profit, whether the opponent's price is lower or higher (or equal) to p.

p in {c+0.01,...,K/a–0.01} weakly dominates any price lower than or equal to c, because a price lower than c leads to a null or negative payoff (whether the opponent's price is lower or higher), c leads to a null payoff, whereas p leads to a null or positive payoff.

Observe also that (K+ac)/2a weakly dominates any price p in {(K+ac)/2a+0.01,...,K/a–0.01}: if the opponent plays p'<(K+ac)/2a, p and (K+ac)/2a lead to a null payoff, if (K+ac)/2a≤p'<p, then (K+ac)/2a leads to a positive payoff whereas p leads to a null payoff. And if p'≥p, then (K+ac)/2a leads to the payoff $(K-ac)^2/4a$, whereas p leads to the payoff (K–ap)(p–c) which is lower than $(K-ac)^2/4a$ because (K+ac)/2a maximizes (K–ax)(x–c).

So, at this stage, both firms only play prices in {c+0.01,..., (K+ac)/2a}.

But now observe the property: (K–ap)(p–c)/2<(K–a(p–0.01))(p–0.01–c) if p≥c+0.02 (because (K–a(p–0.01))(p–0.01–c)–(K–ap)(p–c)/2=(K–ap)(p–c–0.02)/2+0.01a(p–c–0.01)). So it is better to get the full demand at price p–0.01 than the half demand at price p, for any price p≥c+0.02.

It follows that (K+ac)/2a–0.01 weakly dominates (K+ac)/2a. As a matter of fact, both prices lead to a null payoff when the opponent plays p<(K+ac)/2a–0.01. If p=(K+ac)/2a–0.01, then (K+ac)/2a –0.01 leads to a positive payoff, whereas (K+ac)/2a leads to a null payoff. And if p=(K+ac)/2a, (K+ac)/2a–0.01 leads to a higher payoff than (K+ac)/2a, because of the above property.

Once (K+ac)/2a is eliminated in both strategy sets, (K+ac)/2a–0.02 weakly dominates (K+ac)/2a –0.01 (the proof is similar to the one above). And so on, till there only remain c+0.02 and c+0.01 in the strategy sets. If the opponent plays c+0.01, c+0.02 leads to a null payoff whereas c+0.01 leads to a positive payoff. And, if the opponent plays c+0.02, c+0.01 leads to a higher payoff than c+0.02, because of the above property. So c+0.01 strictly dominates c+0.02.

ANSWERS

So, a very long iterative elimination of weakly and strictly dominated strategies leads to the unique very low price c+0.01. If the increment is ε, this price goes to c: this result is known as Bertrand's very strong competition result.

But many iterations, and hence many levels of crossed rationality are necessary to get this result.

♣ EXERCISE 11 TRAVELLER'S DILEMMA (BASU'S VERSION)

1. Clearly, the strategy "asking for x=100" is weakly dominated by the strategy "asking for x=99".

    We further simply write "100 is weakly dominated by 99". As a matter of fact, both strategies lead to the same payoff max (0, y–P) if the other traveller asks for y<99. But, if he asks for 99, then 99 leads to the payoff 99, whereas 100 leads to the lower payoff max (0, 99–P). And if he asks for 100, 99 leads to the payoff 99+P, whereas 100 leads to the payoff 100, which is lower than 99+P.

    Yet, once 100 is eliminated by both players, 99 becomes weakly dominated by 98 (both strategies lead to the same payoff max (0,y–P) if the other traveller asks for y<98, 98 leads to the payoff 98, whereas 99 leads to the lower payoff max (0,98–P) if he asks for 98, and 98 leads to the payoff 98+P, whereas 99 leads to the lower payoff 99 if he asks for 99).

    And so on. Once 99 is suppressed by both players, 98 becomes weakly dominated by 97. So we successively eliminate all the integers until 3, which is strictly dominated by 2 because, if the other player asks for 3, 3 leads to the payoff 3 whereas 2 leads to the higher payoff 2+P, and if he asks for 2, 3 leads to the payoff max (0, 2–P) whereas 2 leads to the higher payoff 2. So both travellers end up asking for 2 and the company is very happy given that she has almost nothing to pay.

2. Nothing changes when a player j gets x_j–P instead of max (0, x_j–P). Just reproduce the above reasoning by replacing max (0, a–P) by (a–P). *I chose to study the first version of the game with my students (the one in question 1), because a traveller, in the first version, never gets a negative payoff: as a matter of fact, my students simply could not accept that a traveller gets a negative payoff (which is possible in Basu's version because a–P may be negative), which amounts to saying that the traveller has to pay the company that damaged his luggage! You surely agree with them.*

3. Clearly, the fact that both travellers ask for 2 is not convincing. On the one hand, most of the strategies are only weakly dominated, on the other hand, the number of necessary steps to get the result is 98. So we need 97 levels of crossed rationality which is clearly unrealistic.

    For example, let us just illustrate the levels of crossed rationality needed to eliminate x=97.

    Let us suppose that a rational player does not play a weakly dominated strategy. If so, we can say that:

    Player 1 eliminates x=100 if she is rational. *Player 2 eliminates x=100 if he is rational.*

    Given that 99 is weakly dominated only once 100 is eliminated, player 1 eliminates 99 if she is rational and if she knows that player 2 has eliminated 100, that is to say if she is rational and if she knows that player 2 is rational. And *player 2 eliminates 99 if he is rational and if he knows that player 1 is rational.*

    Given that 98 is weakly dominated only once 99 and 100 are eliminated, player 1 eliminates 98 if she is rational and if she knows that player 2 has eliminated 99 and 100. So she eliminates

ANSWERS

98 if she is rational and if she knows that player 2 is rational and knows that player 1 is rational. And player 2 *eliminates 98 if he is rational and <u>knows</u> that player 1 is rational and <u>knows</u> that player 2 is rational.*

Given that 97 is weakly dominated only once 98, 99 and 100 are eliminated, player 1 eliminates 97 if she is rational and if she knows that player 2 has eliminated 98, 99 and 100. Hence she eliminates 97 if she is rational and if she <u>knows</u> that player 2 is rational and <u>knows</u> that player 1 is rational and <u>knows</u> that player 2 is rational.

And player 2 eliminates 97 if he is rational and if he <u>knows</u> that player 1 is rational and <u>knows</u> that player 2 is rational and <u>knows</u> that player 1 is rational.

So each player i has to <u>know</u> that player j <u>knows</u> that player i <u>knows</u> that player j is rational: three levels of crossed rationality are required to eliminate x=97.

Many of my students played this game, and their behaviour was never motivated by iterative elimination of dominated strategies. For example, for low values of P, P=4 for example, 31.3% of my (112) L3 students played 100: so many students did not even make the first step of elimination. See Exercise 15 in chapter 3 and chapter 6, section 2.1 for more details on the students' behaviour.

♣ EXERCISE 12 TRAVELLER'S DILEMMA, THE STUDENTS' VERSION

1. We work with P<49, so 100–2P>2. Observe that x<100–2P is strictly dominated by 100, because 100 leads to the payoff 100 or 100–P (depending on whether the other traveller asks for 100 or less), whereas x leads at most to the payoff x+P<100–P.

   x=100–2P is weakly dominated by 100 (because 100–2P leads to the payoffs 100–3P, 100–2P, or 100–P). To put it more precisely 100 and 100–2P lead to the same payoff 100–P when the other traveller asks for y, with 100–2P<y<100. In any other case 100 yields a higher payoff.

   There are no other dominated strategies. No strategy in]100–2P,100] is dominated by another one. For example 100 is not dominated by 99, because you get more with 100 than with 99 when the other traveller asks for less than 99 (99–P<100–P). More generally, consider x and y in]100–2P,100[, with x>y>100–2P. x does better than y when the other player asks for more than x or less than y, but it does worse when the other traveller asks for an amount between x and y, because y+P>x–P (because y>100–2P>x–2P); if x=y+1, x does worse than y when the other traveller asks for y, because y=x–1>x–P.

   It follows that, for low values of P, for example P=5, the students' reimbursement rule is not good for the airline company: no player asks less than 100–2P+1=91!

2. Here dominance is clearly very natural in that there is no iteration. The players do not need high levels of crossed rationality. It is enough to be rational to eliminate all the strategies lower than 100–2P. So, when P is low, we can be quite sure that most of the travellers will play high amounts.

And this is indeed the case for my students. For P=4, 100–2P=92. About 4/5 of my (137) L3 students played 92 or more and even 65% asked for 100. See Exercise 16 in chapter 3 for more details on the students' behaviour.

ANSWERS

♣♣♣ EXERCISE 13 DUEL GUESSING GAME, THE COWBOY STORY

1. * means any available action.
   We propose two iterative processes starting from the beginning of the game.

   First process
   All the strategies (A/x_1,*/x_3,*/x_5,*/x_7,*/x_9) are strictly dominated by the strategies (P/x_1,A/x_3, */x_5,*/x_7,*/x_9) because the first strategies lead to the payoff 0.1, whereas the second ones lead to the payoff 0.8 if player 2 answers at x_2 and to the payoff 0.3 if player 2 passes at x_2.

   Once the strategies (A/x_1, */x_3, */x_5,*/x_7,*/x_9) are eliminated, the strategies (A/x_2,*/x_4,*/x_6, */x_8,A/x_{10}) become strictly dominated by the strategies (P/x_2,A/x_4,*/x_6,*/x_8,A/x_{10}) because the first strategies lead to the payoff 0.2 whereas the second ones lead to the payoffs 0.7 or 0.4.

   Once the strategies (A/x_2,*/x_4,*/x_6,*/x_8,A/x_{10}) are eliminated, the strategies (P/x_1,A/x_3, */x_5,*/x_7,*/x_9) become strictly dominated by the strategies (P/x_1,P/x_3,A/x_5,*/x_7,*/x_9), given that the first strategies lead to the payoff 0.3 whereas the second ones yield the payoffs 0.6 or 0.5.

   Once the strategies (P/x_1,A/x_3,*/x_5,*/x_7,*/x_9) are eliminated, the strategies (P/x_2,A/x_4,*/x_6, */x_8,A/x_{10}) become strictly dominated by the strategies (P/x_2,P/x_4,A/x_6,*/x_8,A/x_{10}), given that the first strategies lead to the payoff 0.4 whereas the second ones lead to the payoffs 0.5 or 0.6.

   It derives that iterative elimination of strictly dominated strategies leads player 1 and player 2 to pass till x_5.

   Now observe that (P/x_1,P/x_3,P/x_5,P/x_7,P/x_9) is weakly dominated by (P/x_1,P/x_3,P/x_5,P/x_7, A/x_9) given that both strategies lead to the same payoffs if player 2 answers at x_6 or x_8, but the first yields the payoff 0 and the second the payoff 0.9 if player 2 goes on at x_6 and x_8.

   Once (P/x_1, P/x_3, P/x_5,P/x_7,P/x_9) is eliminated, player 1 answers if she reaches x_9 and (P/x_2, P/x_4,P/x_6,P/x_8,A/x_{10}) becomes weakly dominated by (P/x_2,P/x_4,P/x_6,A/x_8,A/x_{10}), because both strategies lead to the same payoffs if player 1 answers at x_5 or x_7, but the first yields the payoff 0.1 whereas the second yields the payoff 0.8, if player 1 goes on at x_5 and x_7.

   Once (P/x_2, P/x_4, P/x_6,P/x_8,A/x_{10}) is eliminated, player 2 answers if he reaches x_8 and (P/x_1, P/x_3,P/x_5,P/x_7,A/x_9) becomes weakly dominated by (P/x_1,P/x_3, P/x_5,A/x_7,A/x_9) (both strategies lead to the same payoff if player 2 answers at x_6, but the first yields the payoff 0.2 and the second the payoff 0.7 if player 2 goes on at x_6).

   Once (P/x_1,P/x_3,P/x_5,P/x_7,A/x_9) is eliminated, player 1 answers if she reaches x_7 and (P/x_2, P/x_4, P/x_6,A/x_8,A/x_{10}) becomes weakly dominated by (P/x_2, P/x_4, A/x_6,A/x_8,A/x_{10}) (both strategies lead to the same payoff if player 1 answers at x_5, but the first yields the payoff 0.3 and the second the payoff 0.6 if player 1 goes on at x_5).

   Once (P/x_2, P/x_4, P/x_6,A/x_8,A/x_{10}) is eliminated, player 2 answers if she reaches x_6, and (P/x_1, P/x_3, P/x_5,A/x_7,A/x_9) becomes strictly dominated by (P/x_1,P/x_3,A/x_5,A/x_7,A/x_9) given that the first strategy yields the payoff 0.4 whereas the second yields the payoff 0.5.

   So we get a unique profile of strategies such as player 1 passes till x_5 and answers at x_5 and thereafter, and player 2 passes till x_6 and answers at x_6 and thereafter. So the game stops at x_5 and the players get the couple of payoffs (0.5,0.5).

ANSWERS

Second process

It starts without iterations. We first observe that the strategies (A/x_1,*/x_3,*/x_5,*/x_7,*/x_9) are strictly dominated by the strategies (P/x_1, A/x_3,*/x_5,*/x_7,*/x_9). We also observe that the strategies (P/x_1,A/x_3,*/x_5, */x_7,*/x_9) are weakly dominated by the strategies (P/x_1,P/x_3,A/x_5, */x_7,*/x_9), given that all lead to the same payoff if player 2 answers at x_2, but, if he passes, the first strategies yield the payoff 0.3 whereas the second ones yield the payoffs 0.6 or 0.5. The strategies (A/x_2,*/x_4,*/x_6,*/x_8,A/x_{10}) are weakly dominated by the strategies (P/x_2,A/x_4,*/x_6, */x_8,A/x_9), given that all lead to the same payoff if player 1 answers at x_1, but, if she passes, the first strategies yield the payoff 0.2 whereas the second ones yield the payoffs 0.7 or 0.4. And the strategies (P/x_2,A/x_4,*/x_6,*/x_8,A/x_{10}) are weakly dominated by the strategies (P/x_2, P/x_4,A/x_6,*/x_8,A/x_{10}), given that all lead to the same payoffs if player 1 answers at x_1 or x_3 but, if she passes, the first strategies yield the payoff 0.4 whereas the second ones yield the payoffs 0.5 or 0.6.

And the strategy (P/x_1,P/x_3,P/x_5,P/x_7,P/x_9) is weakly dominated by the strategy (P/x_1,P/x_3, P/x_5,P/x_7,A/x_9) given that both lead to the same payoffs if player 2 answers at x_2, x_4, x_6 or x_8, but the first yields the payoff 0 and the second the payoff 0.9 if player 2 goes on at these nodes.

So, player 1 and player 2 again pass until x_5 and answer before x_{10} without iterative elimination.

Then the process goes on as the previous one. It is worth noting that this process only needs four iterations to obtain the stop at x_5.

In these two processes, we first eliminated actions at the beginning of the game tree, then we switched to the actions at the end of the game tree and worked in a recursive way.

We could also start at the end of the game and again obtain the same result.

2. The reduced normal form is given in Matrix A2.1.

		Player 2				
		A/x_2	P/x_2A/x_4	P/x_2P/x_4A/x_6	P/x_2P/x_4P/x_6A/x_8	P/x_2P/x_4P/x_6P/x_8A/x_{10}
Pl1	A/x_1	(0.1,0.9)	(0.1,0.9)	(0.1,0.9)	(0.1,0.9)	(0.1,0.9)
	P/x_1A/x_3	(0.8,0.2)	(0.3,0.7)	(0.3,0.7)	(0.3,0.7)	(0.3,0.7)
	P/x_1P/x_3A/x_5	(0.8,0.2)	(0.6,0.4)	**(0.5,0.5)**	(0.5,0.5)	(0.5,0.5)
	P/x_1P/x_3P/x_5A/x_7	(0.8,0.2)	(0.6,0.4)	(0.4,0.6)	(0.7,0.3)	(0.7,0.3)
	P/x_1P/x_3P/x_5P/x_7A/x_9	(0.8,0.2)	(0.6,0.4)	(0.4,0.6)	(0.2,0.8)	(0.9,0.1)
	P/x_1P/x_3P/x_5P/x_7P/x_9	(0.8,0.2))	(0.6,0.4)	(0.4,0.6)	(0.2,0.8)	(0,1)

Matrix A2.1

We get again different iterative processes.

First process:

(A/x_1) is strictly dominated by (P/x_1A/x_3) (0.1<0.8 and 0.1<0.3). Once (A/x_1) is eliminated, (A/x_2) becomes strictly dominated by (P/x_2A/x_4) (0.2<0.7 and 0.2<0.4). Once (A/x_2) is eliminated, (P/x_1A/x_3) becomes strictly dominated by (P/x_1P/x_3A/x_5) (0.3<0.6 and 0.3<0.5). Once (P/x_1A/x_3) is eliminated, (P/x_2A/x_4) becomes strictly dominated by (P/x_2P/x_4A/x_6) (0.4<0.5 and 0.4<0.6).

At this level, (P/x_1, P/x_3,P/x_5,P/x_7,P/x_9) is weakly dominated by (P/x_1,P/x_3,P/x_5,P/x_7,A/x_9) (0.4=0.4, 0.2=0.2, 0<0.9). Once (P/x_1,P/x_3, P/x_5,P/x_7,P/x_9) is eliminated, (P/x_2,P/x_4,P/x_6, P/x_8,A/x_{10}) becomes weakly dominated by (P/x_2P/x_4P/x_6A/x_8) (0.5=0.5, 0.3=0.3 and 0.1<0.8).

328

Once $(P/x_2,P/x_4,P/x_6,P/x_8,A/x_{10})$ is eliminated, $(P/x_1,P/x_3,P/x_5,P/x_7,A/x_9)$ becomes weakly dominated by $(P/x_1P/x_3P/x_5A/x_7)$ (0.4=0.4 and 0.2<0.7). Once $(P/x_1,P/x_3,P/x_5,P/x_7,A/x_9)$ is eliminated, $(P/x_2P/x_4P/x_6A/x_8)$ becomes weakly dominated by $(P/x_2P/x_4A/x_6)$ (0.5=0.5 and 0.3<0.6). Finally, once $(P/x_2P/x_4P/x_6A/x_8)$ is eliminated, $(P/x_1P/x_3P/x_5A/x_7)$ becomes strictly dominated by $(P/x_1P/x_3A/x_5)$ (0.4<0.5).

And so we end with the state where player 1 and player 2 pass till x_5 and x_6 and answer at these nodes and thereafter.

Second process

We start by observing that (A/x_1) is strictly dominated by $(P/x_1A/x_3)$, that (A/x_2) is weakly dominated by $(P/x_2A/x_4)$ (0.9=0.9, 0.2<0.7 and 0.2<0.4), that $(P/x_1A/x_3)$ is weakly dominated by $(P/x_1P/x_3A/x_5)$ (0.8=0.8, 0.3<0.6, 0.3<0.5), that $(P/x_2A/x_4)$ is weakly dominated by $(P/x_2P/x_4A/x_6)$ (0.9=0.9, 0.7=0.7, 0.4<0.5, 0.4<0.6) and that $(P/x_1, P/x_3, P/x_5,P/x_7,P/x_9)$ is weakly dominated by $(P/x_1,P/x_3,P/x_5,P/x_7,A/x_9)$ (0.8=0.8, 0.6=0.6, 0.4=0.4, 0.2=0.2, 0<0.9). Then we proceed as above.

Clearly, the elimination processes are the same in the extensive form and in the reduced normal form.

3. Given that the reduced normal form and the extensive form lead to the same results we work on the reduced normal form given in Matrix A2.2:

		Player 2			
		A/x_2	$P/x_2A/x_4$	$P/x_2P/x_4A/x_6$	$P/x_2P/x_4P/x_6$
	A/x_1	(0.15,0.85)	(0.15,0.85)	(0.15,0.85)	(0.15,0.85)
Player1	$P/x_1A/x_3$	(0.8,0.2)	(0.45,0.55)	(0.45,0.55)	(0.45,0.55)
	$P/x_1P/x_3A/x_5$	(0.8,0.2)	**(0.6,0.4)**	(0.75,0.25)	(0.75,0.25)
	$P/x_1P/x_3P/x_5A/x_7$	(0.8,0.2)	(0.6,0.4)	(0.4,0.6)	(1,0)

Matrix A2.2

We use the second elimination process:

(A/x_1) is strictly dominated by $(P/x_1A/x_3)$ (0.15<0.8, 0.15<0.45), $(P/x_1A/x_3)$ is weakly dominated by $(P/x_1P/x_3A/x_5)$ (0.8=0.8, 0.45<0.6, 0.45<0.75), (A/x_2) is weakly dominated by $(P/x_2A/x_4)$ (0.85=0.85, 0.2<0.55 and 0.2<0.4), and $(P/x_2,P/x_4,P/x_6)$ is weakly dominated by $(P/x_2,P/x_4,A/x_6)$ (0.85=0.85, 0.55=0.55, 0.25=0.25, 0<0.6).

Once these strategies are eliminated, $(P/x_1,P/x_3,P/x_5,A/x_7)$ becomes weakly dominated by $(P/x_1P/x_3A/x_5)$ (0.6=0.6, 0.4<0.75). And once $(P/x_1,P/x_3,P/x_5,A/x_7)$ is eliminated, $(P/x_2P/x_4A/x_6)$ becomes strictly dominated by $(P/x_2A/x_4)$ (0.25<0.4).

And so we end up in the state where player 2 passes till x_4 and answers at this node (and at the further nodes). And player 1 passes till x_5 and answers at x_5 (and at all further nodes).

Observe that player 2, the less gifted player, answers before player 1! This may seem strange but it is in fact quite logical. Given that player 1 guesses better than player 2, if player 2 waits too long, player 1 may answer and give the right answer with a very high probability. So player 2 is compelled to answer fast in order to not give the opportunity to player 1 to give the right answer.

4. According to the studied examples, it seems that iterative elimination of weakly and strictly dominated strategies leads to a unique state, the one we get after the first round of elimination

ANSWERS

of strictly dominated strategies when starting from the beginning of the game (in the first game, these eliminations lead players 1 and 2 to pursue at least till x_5 and x_6, in the second, they lead players 1 and 2 to pursue at least till x_4 and x_5). Is this always true?

Let us focus on the first game where both players pass till x_5 and x_6. The (by strict dominance) suppressed strategies are such that if player i answers at x_t, he gets less than if he awaits the next decision round, i.e. $\boxed{p_i(x_t)<1-p_j(x_{t+1})}$ and $p_i(x_t)<p_i(x_{t+2})$, where $p_i(x)$, respectively $p_j(x)$, is player i's, respectively player j's probability to give the right answer at node x; given that the last inequality is always satisfied, player i doesn't answer at x_t because $p_i(x_t)<1-p_j(x_{t+1})$. And we indeed observe in the first game that $p_1(x_1)<1-p_2(x_2)$, $p_1(x_3)<1-p_2(x_4)$, $p_2(x_2)<1-p_1(x_3)$ and $p_2(x_4)<1-p_1(x_5)$.

By contrast $p_1(x_5)>1-p_2(x_6)$ and $p_2(x_6)>1-p_1(x_7)$. So all happens as if a player i answers as soon as the inequation $p_i(x_t)<1-p_j(x_{t+1})$ is no more fulfilled, i.e. if $p_i(x_t)\geq 1-p_j(x_{t+1})$. And this is indeed the case. Why is that?

Remember the elimination process that starts at the end of the game. Let us suppose that there are T decision nodes. The last player answers with probability 1 at x_T so, for the other player, say player i, always passing is dominated by always passing except at node x_{T-1} (because both strategies lead to the same payoffs except if the opponent goes on till x_T, where he answers, in which case always passing yields a null payoff whereas answering at x_{T-1} yields the positive payoff $p_i(x_{T-1})$). Yet, if player i answers at x_{T-1}, passing before x_{T-2} and answering at x_{T-2} and thereafter weakly dominates passing before x_T and answering at x_T, if and only if $p_j(x_{T-2})>1-p_i(x_{T-1})$. Continuing the argument ensures that, for player i, the strategy "pass before x_t, answer at x_t and thereafter" weakly dominates the strategy "pass before x_{t+2}, answer at x_{t+2} and thereafter" each time $\boxed{p_i(x_t)>1-p_j(x_{t+1})}$ i≠j=1,2.

By putting together the two conditions (in the two boxes), we see that the first time t a player i answers is such as $p_j(x_{t-1})\leq 1-p_i(x_t)$ and $p_i(x_t)\geq 1-p_j(x_{t+1})$ (player i answers at t and player j does not answer before). So we need $1-p_j(x_{t-1})\geq p_i(x_t)\geq 1-p_j(x_{t+1})$. If the game is continuous in time, t−1 and t+1 go to t and the condition becomes $\boxed{p_i(x_t)+p_j(x_t)=1}$, *a well-known result in duel games*.

Let us tell the story of the duel game: two cowboys walk towards one another; each has a gun with only one bullet and each has to choose the good moment to shoot in order to kill the opponent. And of course the probability of killing the opponent is decreasing in the distance between the two cowboys. If one cowboy shoots and does not kill the other, the other kills him, because he can wait to be next to him before shooting.

So, three things characterize a duel game: the game is a zero sum game, the more a player waits to act, the higher is the probability of winning, but the higher is also the probability that the other plays before him and wins the game.

The guessing game is a duel game: it is a zero sum game, and both players give the right answer with a probability that grows in time. Duel games are numerous in economic sciences. For example, if you come too early on to a market with a new product, you may not be able to sell it, but if you wait too long, another firm may have invested in the market and you no longer have the possibility of doing it.

So the problem in a duel game is to find the good time to do an action and this time is determined by the equality: $p_i(t)+p_j(t)=1$, where $p_i(t)$ and $p_j(t)$ are player i's and player j's probabilities of winning at time t.

♣ ♣ ♣ EXERCISE 14 SECOND PRICE SEALED BID AUCTION

1. Suppose that player 1 values the object x_1. Call b her bid.

   We first show that bidding more than x_1 (b>x_1) is weakly dominated by bidding x_1.

   Call b' player 2's bid. If b'>b, both b and x_1 yield a null payoff (player 1 doesn't get the object). If b'<x_1, player 1 gets the object at price b' with both bids b and x_1. If b'=x_1, x_1 and b lead to the payoff 0. If x_1<b'<b, x_1 leads to a null payoff (because player 1 doesn't get the object), but b leads to the negative payoff x_1–b'. Finally, if x_1<b'=b, player 1 gets a null payoff with x_1 and she gets the negative payoff (x_1–b)/2 with b. So b (>x_1) is weakly dominated by x_1.

   Symmetrically, bidding less than x_1 (b<x_1) is also weakly dominated by bidding x_1.

   Call again b' player 2's bid. If b'>x_1, both b and x_1 lead to a null payoff. If b'=x_1, x_1 leads to the payoff (x_1–b')/2=0 and b leads to a null payoff too. If b<b'<x_1, x_1 yields the payoff (x_1–b')>0, whereas b yields a null payoff. If b'=b<x_1, x_1 leads to the payoff x_1–b'>0 and b leads to the payoff (x_1–b')/2<(x_1–b'). Finally, if b'<b<x_1, x_1 and b yield the same positive payoff x_1–b'. So b (<x_1) is weakly dominated by x_1.

   Comment: this is a nice result for many reasons. It doesn't require *iterative* eliminations, so no player has to believe that the opponent is rational: regardless of the degree of rationality of the other, bidding the value assigned to the object is a (weakly) dominant strategy. And this result is independent of the distribution functions of the values. The players do not even need to know these distribution functions. So the result is very robust. Last but not least, from a revelation point of view, the players are spontaneously led to reveal their private information.

2. Unfortunately, contrary to a widespread opinion, this result does not automatically generalize to N players with N>2 if we opt for the first payoff rule.

   Suppose that player i values the object x_i. Call b her bid.

   We first show that there is just one case that impedes from concluding that bidding more than x_i (b>x_i) is weakly dominated by bidding x_i.

   Call b' the highest bid proposed by the opponents and b" the second highest bid offered by the opponents (b"<b'). If b'>b, both b and x_i yield a null payoff. If b'<x_i, both x_i and b yield the same positive payoff x_i–b'. If b'=x_i, x_i leads to the payoff (x_i–b")/K>0, where K is the number of players (including player i) playing b', whereas b leads to the null payoff x_i–b'. If x_i<b'<b, x_i leads to a null payoff but b leads to the negative payoff x_i–b'. If b'=b, we have to distinguish three subcases:

   x_i<b"<b'=b. In that case, x_i leads to a null payoff and b leads to the negative payoff (x_i–b")/K (where K is the number of players playing b' including player i).

   x_i=b"<b'=b. In that case, x_i yields a null payoff and b also yields a null payoff, (x_i–b")/K=0

   b"<x_i<b'=b. This is the problematic case. If player i bids b, she gets (x_i–b")/K>0. But if she plays x_i she gets a null payoff because she doesn't get the object. It is this last configuration that impedes us from concluding that bidding b (>x_i) is weakly dominated by bidding x_i.

   Symmetrically, bidding less than x_i (b<x_i) is not weakly dominated by bidding x_i because of a symmetric problematic case.

   Call again b' the highest proposed bid by the opponents and b" the second highest bid offered by them (b"<b').

ANSWERS

If b'>x_i, both b and x_i yield a null payoff. If b<b'=x_i, x_i leads to the payoff $(x_i-b")/K>0$ (where K is the number of players playing b' including player i), and b leads to a null payoff. If b<b'<x_i, x_i yields the payoff $x_i-b'>0$ and b yields a null payoff. If b'<b<x_i, both x_i and b lead to the positive payoff (x_i-b').

b'=b<x_i is the problematic case, because b"<b'=b<x_i. So x_i leads to the payoff $x_i-b'=x_i-b$ >0, but b leads to the payoff $(x_i-b")/K$ (where K is the number of players playing b' including player i). Yet, if K is low (K=2 for example), and b" is low (<<b), it may be that $(x_i-b")/K$ is higher than (x_i-b). This configuration impedes from concluding that b (<x_i) is weakly dominated by x_i.

So, because of two special configurations, we cannot conclude that, regardless of N, bidding the object's value is the weakly dominant strategy. Yet we can observe that the first problematic case appears when b"<x_i<b'=b, the second when b"<b'=b<x_i. In both cases, b leads (or can lead) to a higher payoff than x_i because b leads to the payoff $(x_i-b")/K$. So there is a problem only if several players play b (=b'), leading them to only pay b". *If we switch to the second payoff rule; i.e. if we suppose that, if several players play the same highest price, they pay this price,* then our two problematic cases disappear. In the first case, x_i leads to the payoff 0 and b leads to the negative payoff $(x_i-b)/K$, in the second case, x_i leads to the payoff x_i-b and b leads to the lower payoff $(x_i-b)/K$.

The proposed change induces some changes in the previous comparisons of payoffs, but you can check that bidding the object's value now becomes a weakly dominant strategy.

So switching to the new payoff rule (which only differs from the first one when K>1 players propose the highest price, in which case each of them gets the object and pays this price with probability 1/K) allows to keep the dominance of the value revealing strategy.

This second payoff rule is not shocking at least for three reasons:
- First we already applied it when the N players play the same price.
- Second, when K players play the highest price p, the list of decreasing prices played in the game is p p p...p p'<p...and the second price in the list is p.
- Third, this change is not very important from an equilibrium point of view. Later we will look for equilibrium strategies and we will show that the bids are growing in x_i. So, saying that K players play a same price amounts to saying that they assign the same value to the object. Given that the values distribution functions are defined on intervals, the probability of such an event goes to 0. It follows that the change has in fact almost no impact at equilibrium.

So, given that our change has no impact at equilibrium, but ensures that bidding the value is a (weakly) dominant strategy, without iteration, regardless of N, regardless of the values distribution functions, regardless of the knowledge of these distributions, regardless of the rationality of the opponents, we just think that it is worth working with the second rule!

ANSWERS

♣ ♣ ♣ EXERCISE 15 ALMOST COMMON VALUE AUCTION, BIKHCHANDANI AND KLEMPERER'S[1] RESULT

We use the approach given in Avery and Kagel (1997).[2] The elimination goes as follows.

■ Step 1

First, player 1's bid necessarily belongs to $[x_1+K, x_1+1+K]$ in that a bid b_1 higher than x_1+1+K is weakly dominated by $b_{1max}=x_1+1+K$. As a matter of fact, b_{1max} and b_1 ($>b_{1max}$) have a different impact on player 1's payoff only if player 2's bid b' checks $b_{1max} \leq b' \leq b_1$, in which case player 1 is better off playing b_{1max} than b_1 (if $b_{1max}=b'<b_1$, player 1 gets x_2-1 with b_1 and the higher or equal payoff $(x_2-1)/2$ with b_{1max}, if $b_{1max}<b' \leq b_1$, b_1 leads to losing money (because the paid price is too high), whereas b_{1max} leads to a null payoff).

In a similar way, a bid lower than x_1+K is weakly dominated by the bid x_1+K.

For similar reasons player 2's bid belongs to $[x_2, x_2+1]$, after elimination of the weakly dominated strategies.

■ Step 2

Knowing that player 1 bids at least x_1+K, player 2, if his type is lower than K, prefers losing the auction because, if winning, he gets less than x_1+K and pays at least this amount. So, for $x_2<K$, a weakly dominant way to play consists of playing $b_2(x_2)<K$, in order to lose the auction (bidding x_2 is a good way to play).

Similarly, knowing that player 2' bid is at most x_2+1, player 1, if his type is higher than 1–K, prefers winning the auction because he gets x_1+x_2+K and pays at most x_2+1; so, for $x_1>1-K$, a weakly dominant way to play consists to bid $b_1(x_1)>2$, in order to always win the auction, (playing x_1+1+K is a good way to play).

It follows that the non weakly dominated strategies become:
player 1:
If $x_1>1-K$, $b_1(x_1)>2$;   if $x_1 \leq 1-K$, $b_1(x_1) \in [x_1+K, x_1+1+K]$

player 2:
If $x_2<K$, $b_2(x_2)<K$;   if $x_2 \geq K$, $b_2(x_2) \in [x_2, x_2+1]$

■ Step 3

Knowing that player 2 plays as in step 2, player 1 knows that she can lose the auction only against a player 2's type higher or equal to K. So she can focus on types higher than K and she should at least play x_1+2K. $b_{1min}=x_1+2K$ weakly dominates a lower bid b_1 (b_1 between x_1+K and x_1+2K) (for similar reasons than the ones given in step 1).

Knowing that player 1 bids as in step 2, player 2 knows that, each time he wins the auction, he faces a type x_1 lower or equal to 1–K; so he should never bid more than $b_{2max}=x_2+1-K$; $b_{2max}=x_2+1-K$ weakly dominates any higher bid (for similar reasons than the ones given in step 1).

It follows that the non weakly dominated strategies become:
player 1:
If $x_1>1-K$, $b_1(x_1)>2$;   if $x_1 \leq 1-K$, $b_1(x_1) \in [x_1+2K, x_1+1+K]$

333

player 2:
If $x_2<K$, $b_2(x_2)<K$; if $x_2 \geq K$, $b_2(x_2) \in [x_2, x_2+1-K]$

■ **Step 4**

Given player 1's behaviour in step 3, player 2 observes that, if his type is lower than 2K, it is better to lose the auction because, by winning, he gets x_1+x_2 but pays at least x_1+2K. So, for $x_2<2K$, a weakly dominant way to play is to bid $b_2(x_2)<2K$, in order to lose the auction (bidding x_2 is a good way to play).

Similarly, given player 2's behaviour in step 3, player 1 observes that, if $x_1>1-2K$, she prefers winning the auction because she gets x_1+x_2+K and pays at most x_2+1-K; so, for $x_1>1-2K$, a weakly dominant way to play is to bid $b_1(x_1)>2-K$ in order to win the auction (because player 2 does not bid more than $2-K$). Bidding x_1+1+K is a good way to play.

It follows that the non weakly dominated strategies become:
player 1:
If $x_1>1-K$, $b_1(x_1)>2$; if $1-K \geq x_1 > 1-2K$, $b_1(x_1)>2-K$, if $x_1 \geq 1-2K$, $b_1(x_1) \in [x_1+2K, x_1+1+K]$

player 2:
If $x_2<K$, $b_2(x_2)<K$; if $K \leq x_2 < 2K$, $b_2(x_2)<2K$, if $x_2 \geq 2K$, $b_2(x_2) \in [x_2, x_2+1-K]$

■ **Step 5**

Following a similar reasoning than above, in step 5, knowing that he can win against player 1 only if $x_1 \leq 1-2K$, player 2, for x_2 higher or equal to 2K, will play $b_2(x_2)$ in $[x_2, x_2+1-2K]$. For similar opposite reasons, player 1, for x_1 lower or equal to $1-2K$, bids $b_1(x_1)$ in $[x_1+3K, x_1+1+K]$ and so on, so that, at the end, the behaviours become:

Player 1:
If $x_1>1-K$, $b_1(x_1)>2$; if $1-K \geq x_1>1-2K$, $b_1(x_1)>2-K$,... if $1-iK \geq x_1>1-(i+1)K$, $b_1(x_1)>2-iK$... if $x_1 \geq 1-NK$ $b_1(x_1) \in [x_1+NK, x_1+1+K]$ where $1-NK>0>1-(N+1)K$.

Player 2
If $x_2<K$, $b_2(x_2)<K$; if $K \geq x_2<2K$, $b_2(x_2)<2K$... if $iK \geq x_2<(i+1)K$ $b_2(x_2)<(i+1)K$ ... if $x_2 \geq NK$ $b_2 \in [x_2, x_2+1-(N-1)K]$

We observe that the often tested continuous behaviours, such that player 1 of type x_1 plays x_1+1+K and player 2 of type x_2 plays x_2, belong to this family of behaviours that resist the iterative elimination of weakly dominated strategies.

So we get very asymmetric behaviours, in which player 2 never wins ($x_2<x_1+1+K$), whereas player 1 always wins, pays the price x_2, much lower than the value of the good (x_1+x_2+K), hence gets the high payoff $x_1+x_2+K-x_2=x_1+K>0$.

This very strong asymmetry between player 1 and player 2 contrasts with the very small asymmetry in the players' valuation of the good (K is small and goes to 0). But, once again, the number of levels of crossed rationality necessary to come to this result is very high, in that N=1/K depends on K and goes to ∞ when K goes to 0.

It is here worth noting that we again only suppressed weakly dominated strategies (this way of doing it may suppress equilibria (see Umbhauer 2015)).[3]

NOTES

1   Klemperer, P.D.,1998. Auctions with almost common values: the wallet game and its applications. *European Economic Review*, May 42 (3–5), 757–769.
2   Avery, C. and Kagel, J.H., 1997. Second-price auction with asymmetric payoffs: an experimental investigation, *Journal of Economics and Management Strategy*, 6, 573–603.
3   Umbhauer, G., 2015. Almost common value auctions and discontinuous equilibria, *Annals of Operations Research*, 225 (1), 125–140.

Chapter 3

Answers to Exercises

♣ **EXERCISE 1 NASH EQUILIBRIA IN A NORMAL FORM GAME**

C_2 is strictly dominated by A_2 (3<4, 0<2, 3<10). After its elimination, C_1 becomes strictly dominated by B_1 (5<7, 0<1).

Nash equilibria resist the iterative elimination of strictly dominated strategies. So the Nash equilibria of the game in Matrix E3.1 are the Nash equilibria of the game in Matrix A3.1:

		Player 2 A_2	B_2
Player1	A_1	(1,4)	(2,5)
	B_1	(7,2)	(1,0)

Matrix A3.1

(A_1, B_2) is a pure strategy Nash equilibrium of this game because:

- A_1 is player 1's best response to B_2 (she gets 2 by playing A_1 and only 1 by playing B_1)
- and B_2 is player 2's best response to A_1 (he gets 5 by playing B_2 and only 4 by playing A_2).

(B_1, A_2) is also a pure strategy Nash equilibrium because B_1 is player 1's best response to A_2 (she gets 7 by playing B_1 and 1 by playing A_1) and A_2 is player 2's best response to B_1 (he gets 2 by playing A_2 and 0 by playing B_2).

Let us look for the mixed Nash equilibria. Call p and (1–p), respectively q and (1–q), the probabilities assigned to A_1 and B_1 by player 1, respectively to A_2 and B_2 by player 2. Player 1 plays A_1 and B_1 only if both actions yield the same payoff. With A_1 player 1 gets: q+2(1–q), with B_1 she gets 7q+(1–q). So we need q+2(1–q)=7q+(1–q), i.e. q=1/7. Player 2 accepts playing A_2 with probability 1/7 and B_2 with probability 6/7 only if both actions yield the same payoff. With A_2 player 2 gets 4p+2(1–p), with B_2 he gets 5p+0(1–p), so we need 4p+2(1–p)=5p, i.e. p=2/3. So the mixed strategy Nash equilibrium of the game in Matrix E3.1 is defined by: ((2/3,1/3,0), (1/7,6/7,0)), where the six probabilities are the probabilities respectively assigned to A_1, B_1, C_1, A_2, B_2 and C_2. Player 1's expected payoff is 1×1/7+2×6/7=13/7 and player 2's expected payoff is 4×2/3+2×1/3=10/3.

ANSWERS

♣ EXERCISE 2 STORY NORMAL FORM GAMES

1. The syndicate game with three players is given in Matrices A3.2:

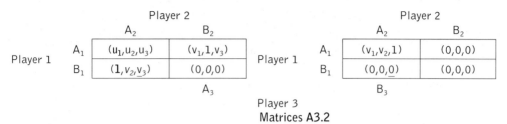

Player 3
Matrices A3.2

with $1>u_i>v_i>0$, i=1 to 3.

There are three pure strategy Nash equilibria, (B_1,A_2,A_3), (A_1,B_2,A_3) and (A_1,A_2,B_3). We prove the first equilibrium (the two other proofs are similar).

B_1 is player 1's best response to A_2 and A_3, because she gets 1 with B_1 and only u_1 with A_1 (bold payoffs). A_2 is player 2's best response to B_1 and A_3, because he gets v_2 with A_2 and only 0 with B_2 (payoffs in italics). A_3 is player 3's best response to B_1 and A_2, because he gets v_3 with A_3 and only 0 with B_3 (underlined payoffs).

The Nash equilibria are the expected ones *(see Exercise 4 in chapter 1)*.

2. The chicken game is given in Matrix A3.3:

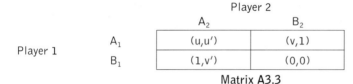

Matrix A3.3

with $1>u>v>0$ an $1>u'>v'>0$

(A_1,B_2) is a Nash equilibrium (because A_1 is the best response to B_2 ($v>0$), and B_2 is the best response to A_1($1>u'$). (B_1,A_2) is a Nash equilibrium for the symmetric reasons.

The game has a mixed strategy Nash equilibrium. Call p, respectively q, the probability assigned by player 1 to A_1, respectively by player 2 to A_2. Player 1 plays A_1 and B_1 with positive probabilities if they yield the same payoff, i.e. if uq+v(1−q)=1q+0(1−q), i.e. q=v/(1−u+v). And player 2 plays A_2 and B_2 with positive probability if u'p+v'(1−p)=1p+0(1−p), i.e. p=v'/(1−u'+v'). So we get a mixed strategy Nash equilibrium where player 1 swerves with probability v'/(1−u'+v') and player 2 swerves with probability v/(1−u+v). Player 1's and player 2's mean expected payoffs are respectively uq+v(1−q)=q=v/(1−u+v) and v'/(1−u'+v').

We now consider the matching pennies game (in Matrix A3.4):

	Player 2	
	A_2	B_2
A_1	(−1,1)	(1,−1)
B_1	(1,−1)	(−1,1)

Player 1

Matrix A3.4

337

ANSWERS

There is no pure strategy Nash equilibrium. The game is finite, so the existence theorem ensures the existence of at least one mixed strategy Nash equilibrium.

Call p, respectively q, the probability assigned by player 1 to A_1, respectively by player 2 to A_2. Player 1 accepts playing A_1 and B_1 with positive probabilities if $-q+(1-q)=q+(-1)(1-q)$, i.e. $q=\frac{1}{2}$. And player 2 accepts playing A_2 and B_2 with positive probabilities if $p+(-1)(1-p)=-p+(1-p)$, i.e. $p=\frac{1}{2}$. So the matching pennies game has a unique mixed strategy Nash equilibrium, where both players play both their actions with equal probability. The mean expected payoff of each player is 0.

♣ EXERCISE 3 BURNING MONEY

The normal form is given in Matrix A3.5:

		Player 2			
		$A_2/\bar{b}\,A_2/b$	$A_2/\bar{b}\,B_2/b$	$B_2/\bar{b}\,A_2/b$	$B_2/\bar{b}\,B_2/b$
Player 1	$\bar{b}\,A_1/\bar{b}\,A_1/b$	(4,1)	(4,1)	(0,0)	(0,0)
	$\bar{b}\,A_1/\bar{b}\,B_1/b$	(4,1)	(4,1)	(0,0)	(0,0)
	$\bar{b}\,B_1/\bar{b}\,A_1/b$	(0,0)	(0,0)	(1,4)	(1,4)
	$\bar{b}\,B_1/\bar{b}\,B_1/b$	(0,0)	(0,0)	(1,4)	(1,4)
	$b\,A_1/\bar{b}\,A_1/b$	(2,1)	(-2,0)	(2,1)	(-2,0)
	$b\,A_1/\bar{b}\,B_1/b$	(-2,0)	(-1,4)	(-2,0)	(-1,4)
	$b\,B_1/\bar{b}\,A_1/b$	(2,1)	(-2,0)	(2,1)	(-2,0)
	$b\,B_1/\bar{b}\,B_1/b$	(-2,0)	(-1,4)	(-2,0)	(-1,4)

Matrix A3.5

There are eight pure strategy Nash equilibria (that lead to the payoffs in the circles) and three Nash equilibrium paths.

On the first path player 1 does not burn money and players 1 and 2 respectively play A_1 and A_2. This first path, that leads to the payoffs (4,1), corresponds to the profiles of strategies that resist the iterative elimination of strictly and weakly dominated strategies, $((\bar{b},A_1/\bar{b},*/b), (A_2/\bar{b},A_2/b))$ (but there are two additional Nash equilibria with the payoffs (4,1)).

On the second path, player 1 does not burn money and players 1 and 2 respectively play B_1 and B_2; it leads to the payoffs (1,4). On the third path, player 1 burns money and players 1 and 2 respectively play A_1 and A_2; it leads to the payoffs (2,1).

It follows that eliminating weakly dominated strategies leads to eliminating six Nash equilibria and two equilibrium paths. We observe that the third equilibrium path corresponds to the partial elimination of dominated strategies we talked about in Exercise 7 in chapter 2: player 2 understands that player 1 does not play B_1 after burning money, because burning money and playing B_1 is strictly dominated by not burning money. So player 1 expects that player 2 plays A_2 after her choice of burning money, and she chooses to burn money and then play A_1 in order to be sure to get 2.

In the last Nash equilibria that lead to the payoffs (1,4), all happens as if level 1 of crossed rationality is not fulfilled, given that player 2 plays B_2 after b, so is not sure that player 1 is rational (because a rational player 1 does not play B_1 after b, leading player 2 to not play B_2 after b). But remember that player 2, according to the Nash concept, has not to justify his action after b, given that b is not played (so all player 2's actions after b are automatically optimal).

♣ EXERCISE 4 MIXED NASH EQUILIBRIA AND WEAK DOMINANCE

D_2 is strictly dominated by B_2 (1<2,0<2,3<4). After its elimination, A_1 becomes strictly dominated by C_1 (3<4,1<2,0<4). It follows that the Nash equilibria of the game in Matrix E3.2 are the Nash equilibria of the game in Matrix A3.6:

		Player 2		
		A_2	B_2	C_2
Player 1	B_1	(1,2)	(5,2)	(3,0)
	C_1	(4,3)	(2,4)	(4,6)

Matrix A3.6

A_2 is weakly dominated by B_2 (2=2,3<4): given that it leads to a lower payoff as soon as player 1 doesn't exclusively play B_1, we first study the case where player 1 only plays B_1 before turning to the complementary case.

Case 1: player 1 only plays B_1
Player 2's best responses are A_2 and B_2 (2=2>0) and so he can play A_2 and B_2 with any probabilities q and 1−q. Yet B_1 is player 1's best response only if 1q+5(1−q) ≥4q+2(1−q), i.e. q≤1/2.

So we get the family of Nash equilibria (for the game in Matrix E3.2) defined by: {(B_1,(q,1−q,0,0)) with q≤1/2}. This family contains the pure strategy Nash equilibrium (B_1,B_2).

Case 2: player 1 plays C_1 with positive probability
If player 1 only plays C_1, player 2 only plays C_2 (6>4>3), and C_1 is a best response to C_2 (4>3). So we get the pure strategy Nash equilibrium (C_1,C_2).

If player 1 plays B_1 and C_1 with positive probabilities, player 2 does not play A_2. B_1 and C_1 have to yield the same payoff, so we need 5q+3(1−q)=2q+4(1−q), i.e. q=¼, where q and 1−q are the probabilities assigned to B_2 and C_2. Yet player 2 plays B_2 and C_2 with positive probabilities only if 2p+4(1−p)=0p+6(1−p), i.e. p=½, where p and 1−p are the probabilities assigned to B_1 and C_1.

So we get for the game in Matrix E3.2, the mixed Nash equilibrium defined by ((0,½,1/2), (0, ¼,3/4,0)).

♣ EXERCISE 5 A UNIQUE MIXED NASH EQUILIBRIUM

1. We proceed by contradiction, so suppose that player 2 only plays A_2 and B_2 with positive probabilities. If so, player 1 doesn't play A_1 because B_1 yields higher payoffs (3<4,0<1). It follows that player 2 doesn't play B_2 (a contradiction to our assumption) because C_2 yields higher payoffs (2>1,−1>−2).

2. We again proceed by contradiction, so suppose that player 2 only plays A_2 and C_2 with positive probabilities. If so, player 1 doesn't play C_1 because A_1 yields higher payoffs (2<3,−1<0). It follows that player 2 doesn't play A_2 (a contradiction to our assumption) because B_2 yields higher payoffs (4>3,1>0).

339

ANSWERS

3. We again proceed by contradiction, so suppose that player 2 only plays B_2 and C_2 with positive probabilities. Then player 1 doesn't play B_1 because C_1 yields higher payoffs (1<2, –2<–1). It follows that player 2 doesn't play C_2 (a contradiction to our assumption) because A_2 yields higher payoffs (3>2, 0>–1).

4. It follows that player 2, and player 1 by symmetry, necessarily play the three actions at equilibrium. Call p_1, p_2, p_3 the probabilities assigned to A_1, B_1, C_1 and q_1, q_2, q_3 the probabilities assigned to A_2, B_2 and C_2. A_1, B_1 and C_1 have to yield the same payoff, so we need $3q_1=4q_1+q_2-2q_3=2q_1+2q_2-q_3$ and $q_1+q_2+q_3=1$. The unique solution is $q_1=q_2=q_3=1/3$. And $p_1=p_2=p_3=1/3$ for symmetric reasons. So, at equilibrium, each player plays each action with the same probability 1/3.

♣ EXERCISE 6 FRENCH VARIANT OF THE ROCK PAPER SCISSORS GAME

1. Rock is weakly dominated by well (0<1, 1=1, –1=–1, –1<0). This doesn't mean that it can't be played at equilibrium (a Nash equilibrium may contain weakly dominated strategies).
2. a) If player 2 plays rock or well with positive probability, then player 1 is better off with well than with rock (because 0<1, –1<0).

   It derives that player 1 can play rock with positive probability only if player 2 only plays scissors and/or paper, i.e. $q_2+q_3=1$, $q_1=q_4=0$. It follows that player 1 can't play paper because scissors yield higher payoffs (0>–1, 1>0), so $p_3=0$.

   b) $p_3=0$ induces player 2 to not play scissors because well yields higher payoffs (1>–1, 1>0, 0>–1), so $q_2=0$.

   It follows that player 2 only plays paper, leading player 1 to only play scissors. So she cannot play rock with positive probability.

   Given the symmetry of the game, player 2 can't play rock at equilibrium either.
3. It derives that the equilibria are the equilibria of the game in Matrix A3.7.

		player 2 scissors	paper	well
player 1	scissors	(0,0)	(1,–1)	(–1,1)
	paper	(–1,1)	(0,0)	(1,–1)
	well	(1,–1)	(–1,1)	(0,0)

Matrix A3.7

The structure of this game is the same as the structure of the traditional rock paper scissors game, so the unique Nash equilibrium consists in playing scissors, paper and well, each with the probability 1/3.

From a theoretical point of view, *adding well just excludes rock*. So this variant is not very exciting. But don't forget that it is a children's game. Adding a fourth action yields an additional difficulty *and more fun, because there are more things to mime*!

ANSWERS

♣ ♣ EXERCISE 7 BAZOOKA GAME

1. Given that g_1 and g_2 are lower than 1 (because a child does not necessarily win when the game starts again), the game has no pure strategy Nash equilibrium. Let us find the mixed strategy Nash equilibrium. You can check, by proceeding as in Exercise 5, that the three actions are played at equilibrium.

   Call p_1, p_2 and p_3, respectively q_1, q_2 and q_3, the probabilities assigned to L, P and S by player 1, respectively by player 2.

   Player 1 plays L_1, P_1 and S_1 if they all lead to the same payoff, i.e. $q_1g_1+q_21+q_30=q_10+q_2g_1+q_31=q_11+q_20+q_3g_1$ (and $q_1+q_2+q_3=1$)

   Moreover, given that the game played in the future is the same than the one played at the studied period, player 1 gets the same payoff in both games, hence g_1 is player 1's payoff in the played game. So we get: $g_1=q_1g_1+q_21+q_30=q_10+q_2g_1+q_31=q_11+q_20+q_3g_1$.

   The unique solution is $g_1=1/2$ and $q_1=q_2=q_3=1/3$. By symmetry we also get $g_2=1/2$ and $p_1=p_2=p_3=1/3$. So, as in the rock paper scissors game, each action has to be chosen with the same probability and the expected payoff is ½, which amounts to saying that each child wins half of the time.

   Both this game and the rock paper scissors game have perfectly symmetric equilibria (same strategy for both players, same weight on each action). No action is silly. This is very nice: any way you play, you play well; you are never stupid! Any child can play and win, whether she/he is logical or not: she never gets the impression of playing wrong or being less skilled than her/his friends. There is no frustration; there is only fun! That's perhaps why these games are so much played in schoolyards.

2. What is more, patient children play like impatient ones. To see why, introduce a discount factor δ_i, i=1,2 on the payoffs obtained in the future, so that Matrix E3.5 becomes Matrix A3.8:

		Player 2		
		L_2	P_2	S_2
	L_1	$(\delta_1 g_1, \delta_2 g_2)$	$(1,0)$	$(0,1)$
Player 1	P_1	$(0,1)$	$(\delta_1 g_1, \delta_2 g_2)$	$(1,0)$
	S_1	$(1,0)$	$(0,1)$	$(\delta_1 g_1, \delta_2 g_2)$

Matrix A3.8

The mixed strategy Nash equilibrium requires: $g_1=q_1\delta_1g_1+q_21+q_30=q_10+q_2\delta_1g_1+q_31=q_11+q_20+q_3\delta_1g_1$ and $q_1+q_2+q_3=1$. The Nash equilibrium still consists in playing L, P and S with probability 1/3 each: the symmetry of the game is not affected by the introduction of discount factors (each action leads to pursue the game in front of one of the three opponents' possible actions). Only the payoffs decrease with impatience: g_1 becomes $1/(3-\delta_1)$ and g_2 becomes $1/(3-\delta_2)$.

It is funny that, whether you are patient or impatient, there is no reason to play one action more than another. *Once again, whatever you do, it's OK. There are only heroes in this game!*

ANSWERS

♣ EXERCISE 8 PURE STRATEGY NASH EQUILIBRIA IN AN EXTENSIVE FORM GAME

1. If player 1 plays A_1 and player 2 plays A_2, player 3 is at x_4, so he is induced to play C_3 at his information set (2>1>0; see the bold numbers).
    But, if player 1 and player 3 play A_1 and C_3, A_2 is not player 2's best answer, because he gets 0 with A_2 and 3 with B_2 (bold numbers). So we cannot construct a Nash equilibrium in which player 1 and player 2 play A_1 and A_2.
2. Let us now turn to the three other possibilities.

Case 1: player 1 plays A_1 and player 2 plays B_2
If so, player 3 is at x_5 and his best response is A_3 (3>1>0, see the numbers in italic). If player 1 and player 3 play A_1 and A_3, B_2 is a player 2's best response, because he gets 2 with B_2 and A_2 (numbers in italics). And if player 2 and player 3 play B_2 and A_3, A_1 is player 1's best response, because she gets 2 with A_1 and 1 with B_1 (numbers in italics). So each player's strategy is a best response to the strategies of the two other players and (A_1, B_2, A_3) is a Nash equilibrium.

Case 2: player 1 plays B_1 and player 2 plays A_2
If so, player 3 is at x_6 and his best response is B_3 (5>1>0, see the underlined numbers). But if player 2 and player 3 play A_2 and B_3, B_1 is not player 1's best response: she gets 3 with B_1 and 7 with A_1 (underlined numbers). So there does not exist a Nash equilibrium in which player 1 and player 2 play B_1 and A_2.

Case 3: player 1 plays B_1 and player 2 plays B_2
If so, player 3 is at x_7 and his best response is C_3 (4>2>0, see the double underlined numbers). If player 1 and player 3 play B_1 and C_3, B_2 is a player 2's best response, because he gets 1 with B_2 and A_2 (double underlined numbers). And if player 2 and player 3 play B_2 and C_3, B_1 is player 1's best response, because she gets 9 with B_1 and only 1 with A_1 (double underlined numbers). So each player best responds to the two other players and (B_1, B_2, C_3) is a Nash equilibrium.

♣ EXERCISE 9 GIFT EXCHANGE GAME

If the employer offers w_2 with positive probability, the employee is at x_2 with positive probability, so offers e_1 at x_2(1>u'). It follows that the employer gets 0 by playing w_2, and that she deviates to (or only plays) w_1 (she gets at least v>0).

If the employer only plays w_1, player 1 is at x_1 and plays e_1 at x_1 (v'>0). He can play any action at x_2 because x_2 is not reached. Call q the probability he assigns to e_1. The employer is best off playing w_1 as soon as v≥(1–q)u, i.e. q≥(u–v)/u.

It derives the family of behavioural Nash equilibria defined by:

$\{(w_1, (\pi_{employee/x1}(e_1)=1, \pi_{employee/x2}(e_1)=q), (u-v)/u \leq q \leq 1\}$. This family corresponds to a unique Nash equilibrium path: the employer offers the lowest salary and the employee offers the lowest effort for this salary.

ANSWERS

This negative result generalizes as soon as we work with continuous and derivable utility functions $U_{ER}(e,w)$ and $U_{EE}(e,w)$, with $U_{ER}'_e(e,w)>0$, $U_{ER}'_w(e,w)<0$, $U_{EE}'_e(e,w)<0$, $U_{EE}'_w(e,w)>0$, where $U_{ER}(e,w)$ and $U_{EE}(e,w)$ are respectively the employer's and the employee's utility functions, $e \in [e_0, E]$ and $w \in [w_0, W]$, where e_0 and w_0, respectively E and W, are the minimal, respectively the maximal levels of effort and salary.

As a matter of fact, if the employer offers, with a positive probability, a salary w^* higher than w_0, the employee systematically chooses e_0 after w^*, given that $U_{EE}(e, w^*)$ is decreasing in effort. It follows that the employer is better off switching to (only playing) w_0, given that, by doing so, she gets at least $U_{ER}(e_0,w_0)$ and $U_{ER}(e_0,w_0) > U_{ER}(e_0,w^*)$ (because $U_{ER}(e,w)$ is decreasing in w). So, at equilibrium, the employer can only offer w_0. And the employee offers e_0 after observing w_0. After each non equilibrium salary w, he can play any distribution π_w on efforts that checks that the employer gets a higher payoff with (e_0, w_0) than with (π_w, w); such a distribution always exists, given that $U_{ER}(e_0,w) < U_{ER}(e_0,w_0)$. But the Nash equilibrium path is always the same: the employer offers w_0 and the employee offers e_0 for this salary. This situation is particularly harmful when there exist couples (e,w) such as $U_{ER}(e,w) > U_{ER}(e_0,w_0)$ and $U_{EE}(e,w) > U_{EE}(e_0,w_0)$, an assumption usually introduced in this game.

♣ ♣ EXERCISE 10 BEHAVIOURAL NASH EQUILIBRIA IN AN EXTENSIVE FORM GAME

Case 1: player 1 plays B_1

Subcase 11: player 2 plays B_2

Player 3 is not called on to play, so he can play any action. Yet, if he plays A_3 with positive probability, player 2 will deviate to A_2, in that he gets 0 with B_2 and more with A_2 as soon as A_2 is not exclusively followed by B_3. If player 3 plays B_3, B_2 is a player 2's best response (he gets 0 with B_2 and A_2) and B_1 is player 1's best response (she gets 2 with B_1 and 0 with A_1). We have a first Nash equilibrium, **(B_1, B_2, B_3)**.

Subcase 12: player 2 plays A_2

Player 3 is at x_4 and plays B_3 (1>0). A_2 is a best response for player 2 (he gets 0 with A_2 and B_2), and B_1 is player 1's best response (she gets 1 with B_1 and 0 with A_1).

We have a second Nash equilibrium, **(B_1, A_2, B_3)**.

Subcase 13: player 2 plays A_2 with probability q and B_2 with probability 1–q, 0<q<1.

Player 3 is at x_4 with positive probability and plays B_3 (1>0). So player 2 is indifferent between A_2 and B_2 and can play both with positive probabilities. Player 1 gets 2(1–q)+q=2–q>0 with B_1 and 0 with A_1, so B_1 is her best response.

By adding the two previous equilibria we get the family of Nash equilibria: **{(B_1, $\pi_2(A_2)$, B_3), $0 \leq \pi_2(A_2) \leq 1$}**.

Case 2: player 1 plays A_1.

Player 3 is at x_3 and plays A_3 (1>0). Player 2 is not called on to play, so can play any action. Call q and 1–q the probabilities he assigns to A_2 and B_2. A_1 is player 1's best response if it leads to a higher (or equal) payoff than B_1, i.e. if $1 \geq 2(1-q)$, i.e. $q \geq 1/2$.

343

ANSWERS

We get the family of Nash equilibria: $\{(A_1, \pi_2(A_2), A_3), 1/2 \leq \pi_2(A_2) \leq 1\}$. This family includes the pure strategy Nash equilibrium (A_1, A_2, A_3).

Case 3: player 1 plays A_1 with probability p and B_1 with probability 1–p, 0<p<1

Subcase 31: player 2 only plays B_2.
Player 3 is at x_3 and plays A_3 (1>0). Player 2 is called on to play, and B_2 is not his best response (he gets 0 with B_2 and 1 with A_2). So there is no equilibrium in this case.

Subcase 32: player 2 only plays A_2.
Player 3 gets p with A_3 and 1–p with B_3. If he only plays A_3, player 1 only plays A_1 (1>0). If he only plays B_3 she only plays B_1 (1>0). So we need that player 3 plays A_3 and B_3, which is only possible if A_3 and B_3 lead to the same payoff, i.e. p=1–p, i.e. p=½. Call r and 1–r the probabilities assigned to A_3 and B_3. Player 1 is indifferent between B_1 and A_1 only if r=(1–r), i.e. r=1/2. And A_2 is player 2's best response (he gets ½ with A_2 and 0 with B_2).
We have a new Nash equilibrium: $(\pi_1(A_1)=1/2, A_2, \pi_3(A_3)=½)$.

Subcase 33: player 2 plays A_2 with probability q, B_2 with probability 1–q, 0<q<1.
This is only possible if player 3 exclusively plays B_3 (if not, A_2 yields a higher payoff). But if player 3 only plays B_3, then player 1 only plays B_1 because B_1 yields q+2(1–q)>0 and A_1 yields 0. There is no equilibrium in this case.

♣ ♣ EXERCISE 11 DUEL GUESSING GAME

The normal reduced form game is recalled in Matrix A3.9:

			Player 2			
		A/x_2	$P/x_2 A/x_4$	$P/x_2 P/x_4 A/x_6$	$P/x_2 P/x_4 P/x_6 A/x_8$	$P/x_2 P/x_4 P/x_6 P/x_8 A/x_{10}$
Pl1	A/x_1	(0.1,0.9)	(0.1,0.9)	(0.1,0.9)	(0.1,0.9)	(0.1,0.9)
	$P/x_1 A/x_3$	(0.8,0.2)	(0.3,0.7)	(0.3,0.7)	(0.3,0.7)	(0.3,0.7)
	$P/x_1 P/x_3 A/x_5$	(0.8,0.2)	(0.6,0.4)	**(0.5,0.5)**	(0.5,0.5)	(0.5,0.5)
	$P/x_1 P/x_3 P/x_5 A/x_7$	(0.8,0.2)	(0.6,0.4)	(0.4,0.6)	(0.7,0.3)	(0.7,0.3)
	$P/x_1 P/x_3 P/x_5 P/x_7 A/x_9$	(0.8,0.2)	(0.6,0.4)	(0.4,0.6)	(0.2,0.8)	(0.9,0.1)
	$P/x_1 P/x_3 P/x_5 P/x_7 P/x_9$	(0.8,0.2)	(0.6,0.4)	(0.4,0.6)	(0.2,0.8)	(0,1)

Matrix A3.9

The only pure strategy Nash equilibrium leads player 1, respectively player 2, to pass till x_5 and answer at x_5, respectively to pass till x_6 and answer at x_6. Given that the normal form and the reduced normal form lead to the same Nash equilibrium outcomes, (0.5, 0.5) is the only pure strategy Nash equilibrium outcome of this game. And it is the unique outcome after iterative elimination of strictly and weakly dominated strategies.

We know *(see Exercise 13, chapter 2)* that this game is a duel game. If time were continuous, both players would answer at the time t checking $p_1(t)+p_2(t)=1$, where $p_1(t)$ and $p_2(t)$ are player 1's and player 2's probability to give the right answer at time t. Given that our game is discrete, player 1 answers at the first time t (with t=1, 3, 5, 7 or 9) checking $p_1(t)>1-p_2(t+1)$, and player 2 answers at the first time t (with t=2, 4, 6, 8 or 10) checking $p_2(t)>1-p_1(t+1)$.

ANSWERS

Why is that? Why couldn't we have a mixed strategy Nash equilibrium? Let us work on a generalized discrete guessing game.

If player 2 answers at time t, the only best response for player 1 is, either not to answer before him (so she gets $1-p_2(t)$), or to answer just one period before (so she gets $p_1(t-1)$), because the probability of giving the right answer grows in time (so it would be silly to answer at period t–3 or earlier). Now consider the sequence of inequations:

$$p_1(1)<1-p_2(2),\ p_2(2)<1-p_1(3)\ldots\ldots p_1(t)<1-p_2(t+1),\ p_2(t+1)<1-p_1(t+2)$$

The first inequation holds because $p_1(1)=0$. This sequence says that answering at period 1 (t=1) is strictly dominated for player 1, because, even if player 2 answers at period 2, it is better to wait. And, after the elimination of this strategy, answering at t=2 becomes strictly dominated for player 2, because, even if player 1 answers at period 3, it is better to wait, and so on. It follows that answering before t+2 is strictly dominated for player 1, and answering before t+3 is strictly dominated for player 2. So no Nash equilibrium puts a positive weight on these strategies.

The above sequence stops at a time d, because the probabilities grow in time. Let us suppose, w.l.o.g, that $p_2(d-1)<1-p_1(d)$ but $p_1(d)>1-p_2(d+1)$ (it derives that $p_2(d+1)>1-p_1(d+2)$ because $p_1(d)+p_2(d+1)>1 \Rightarrow p_1(d+2)+p_2(d+1)>1$). It follows that, for any time t>d, it is always better for a player to answer just before the opponent.

We already know that at equilibrium, the players answer at time d or later. Let us suppose, w.l.o.g., that they play a mixed strategy so that answering at time D_2 (>d+1) is the last action on the equilibrium path. W.l.o.g. suppose that player 2 is the player playing at time D_2. Now consider player 1 at date D_2-1 (>d by assumption). She is called on to play with positive probability at this date and supposed to pass with positive probability given that the game reaches date D_2 with positive probability. She also knows, when she is called on to play at date D_2-1, that player 2 has not answered before but will answer at date D_2 (because it is the last date on the equilibrium path). Yet $p_1(D_2-1)>1-p_2(D_2)$, given that $p_1(t)>1-p_2(t+1)$ as soon as t>d. So she has to answer with probability 1 at D_2-1, a contradiction to the fact that date D_2 is on the equilibrium path. So necessarily player 2 latest answers at d+1 and player 1 latest answers at d, and we get the unique Nash equilibrium where player 1 answers at date d and player 2 at date d+1.

♣ ♣ EXERCISE 12 PRE-EMPTION GAME (IN EXTENSIVE AND NORMAL FORMS)

1. * means any available action.

   We first establish that all the profiles $((S_1/x_1,*/x_3,*/x_5)(S_2/x_2,*/x_4))$ are pure strategy Nash equilibria of the game.

   If player 2 stops at x_2, player 1 is best off stopping at x_1 because 1>0.5, whatever player 2 does at decision nodes after x_2. And her actions at decision nodes after x_1 are without impact given that player 2 stops at x_2. So any strategy leading her to stop at x_1 is a best response to player 2's strategy. And given that player 1 stops at x_1, any strategy of player 2 is a best response because he is not called on to play. So all the strategies $(S_2/x_2,*/x_4)$ are best responses.

   We now establish that each Nash equilibrium path (in pure or behavioural strategies) leads player 1 to stop at x_1 in any pre-emption game. We proceed by contradiction.

   Suppose that player 1, on the equilibrium path, plays C_1 at x_1 with positive probability. Call x_i the highest node (i.e. the node with the largest i) on the equilibrium path, i.e. the highest node reached with a positive probability. The game is finite, so x_i exists, except if both players

ANSWERS

go on with positive probability at each node, even the last: but this last profile is not a Nash equilibrium, the last player being, by assumption in a pre-emption game, better off stopping the game with probability 1. So x_i exists, and the player at x_i stops the game. In a pre-emption game, it is better for the player at x_{i-1} to stop if the opponent stops at x_i. So the player at x_{i-1} stops with probability 1 (because x_{i-1} is reached with positive probability and the game stops at x_i). So we get a contradiction, except if $x_i = x_1$. Given that this result does not depend on the number of nodes of the game, each Nash equilibrium path, in any pre-emption game, is such that player 1 immediately stops the game with probability 1.

Player 2 may go on with a positive probability, but, given that player 1 stops at x_1 at equilibrium, he goes on with a probability small enough to prevent player 1 from going on at x_1.

2. The game in normal form is given in Matrix A3.10:

	Player 2			
Player 1	$(S_2/x_2 S_2/x_4)$	$(S_2/x_2 C_2/x_4)$	$(C_2/x_2 S_2/x_4)$	$(C_2/x_2 C_2/x_4)$
$(S_1/x_1 S_1/x_3 S_1/x_5)$	(1,0)	(1,0)	(1,0)	(1,0)
$(S_1/x_1 S_1/x_3 C_1/x_5)$	(1,0)	(1,0)	(1,0)	(1,0)
$(S_1/x_1 C_1/x_3 S_1/x_5)$	(1,0)	(1,0)	(1,0)	(1,0)
$(S_1/x_1 C_1/x_3 C_1/x_5)$	(1,0)	(1,0)	(1,0)	(1,0)
$(C_1/x_1 S_1/x_3 S_1/x_5)$	(0.5,2)	(0.5,2)	(30,1)	(30,1)
$(C_1/x_1 S_1/x_3 C_1/x_5)$	(0.5,2)	(0.5,2)	(30,1)	(30,1)
$(C_1/x_1 C_1/x_3 S_1/x_5)$	(0.5,2)	(0.5,2)	(15,60)	(900,30)
$(C_1/x_1 C_1/x_3 C_1/x_5)$	(0.5,2)	(0.5,2)	(15,60)	(450,1800)

Matrix A3.10

The only pure strategy Nash equilibria are such that player 1 and player 2 stop at their first decision nodes. They are circled in the matrix.

We can add some mixed strategy Nash equilibria, in which player 2 goes on at x_2 and x_4 with positive probability. Player 1 sticks on $(S_1/x_2, */x_3, */x_5)$ as long as $1 \leq 0.5(q_1+q_2)+30(q_3+q_4)$ and $1 \geq 0.5(q_1+q_2)+15q_3+900q_4$ where q_1, q_2, q_3 and q_4 are the probabilities assigned to $(S_2/x_2, S_2/x_4)$, $(S_2/x_2,C_2/x_4)$, $(C_2/x_2,S_2/x_4)$ and $(C_2/x_2,C_2/x_4)$. Yet these probabilities do not change the nature of the equilibrium path: player 1 stops immediately with probability 1.

3. There is a difference between the pre-emption game and the ascending all pay auction/war of attrition game.

In the pre-emption game, there is only one pure strategy Nash equilibrium path (player 1 stops at x_1). In the ascending all pay auction game (studied in chapter 3), there are two Nash equilibrium paths: in the first, player 1 goes on and player 2 stops immediately after, in the second, player 1 stops immediately because player 2 goes on after (with a sufficiently high probability). This difference is due to the fact that, in the pre-emption game, we expect the same behaviour from both players: a player goes on only if the other goes on, a behaviour which is impossible because it leads both players to the end decision node, where the player does not go on! In the ascending all pay auction, a player expects the opposite behaviour from the opponent: when you go on, you hope that the opponent stops after you and vice versa. So player 1 can go on because she hopes that player 2 stops at the next round (which he does because he fears that player 1 goes on at the next round). And player 1 can also stop immediately because she fears that player 2 goes on at the next round. None of these opposite expectations lead to the end of the game and to a contradictory behaviour at this node.

♣ EXERCISE 13 BERTRAND DUOPOLY

Consider p the lowest price paid at equilibrium. If p is lower than c, the firm who plays p gets necessarily a negative payoff and is better off switching to c. So no player plays a price p<c at equilibrium.

We first study the pure strategy Nash equilibria of the continuous version.

We can't have a Nash equilibrium (p_1^*, p_2^*) with $p_1^* \neq p_2^*$. Suppose w.l.o.g. that $p_1^* > p_2^*$. If $p_1^* > p_2^* > c$, player 1 deviates to p, with $p_2^* > p > c$, to get a positive instead of a null payoff. If $p_1^* > p_2^* = c$, player 2 deviates to p with $p_1^* > p > c$, to get a positive instead of a null payoff.

So both players play the same price p* at equilibrium. If p*>(K+ac)/2a, then player 1 is better off switching to (K+ac)/2a, because she gets the whole market at the price that maximizes her monopoly payoff. If c<p*≤(K+ac)/2a, then player 1 is better off switching to p*–ε, with ε positive and going to 0, to get the whole market at price p*–ε, instead of the half market at price p*. So p* can only be equal to c. And (c, c) is a Nash equilibrium. Both firms get a null payoff but can't get more by unilaterally deviating to a higher price (null demand).

We now turn to the discrete version of the model. There are two Nash equilibria (c,c) and (c+0.01, c+0.01). The proof is similar to the previous one, with some small changes.

We can't have a Nash equilibrium (p_1^*, p_2^*) with $p_1^* \neq p_2^*$, for example $p_1^* > p_2^*$. If $p_1^* > p_2^* > c$, player 1 can deviate to p_2^* to get a positive instead of a null payoff. If $p_1^* > p_2^* = c$, player 2 can deviate to p_1^* to get a positive instead of a null payoff.

So both players play the same price p* at equilibrium. If p*>(K+ac)/2a, player 1 is better off switching to (K+ac)/2a for the same reasons as above. If c+0.01<p*≤(K+ac)/2a, player 1 is better off switching to p*–0.01, because, for p>c+0.01, it is better to get the whole market at price p–0.01, rather than the half market at price p (see Exercise 10 in chapter 2).

If p*=c+0.01, then switching to p*–0.01, i.e. to c, yields a null payoff instead of the positive one obtained with c+0.01. Switching to a higher price also leads to a null payoff (null demand). It follows that (c+0.01, c+0.01) is a Nash equilibrium.

(c, c) is also a Nash equilibrium for the same reasons as above.

So we get two (very close) equilibria (c, c) and (c+0.01, c+0.01). The second corresponds to the only state that resists the iterative elimination of weakly and strictly dominated strategies.

ANSWERS

♣♣ EXERCISE 14 GUESSING GAME

Consider a profile of guesses $(G_1, G_2, \ldots G_N)$. Either the N players win, or at least one of them does not win.

Case 1: at least one player does not win
Suppose, without loss of generality, that player 1 does not win. Then she is best off deviating and guessing G, with $G=(2/3)(G+G_2+\ldots+G_N)/N$, because, by doing so, she wins the game and gets a positive payoff. We check that G is a possible guess, i.e. that $0 \leq G \leq 100$.

$G=(2/3)(G+G_2+\ldots+G_N)/N \Leftrightarrow G=(2G_2+\ldots+2G_N)/(3N-2)$

$0 \leq (2G_2+\ldots+2G_N)/(3N-2) \leq (2(N-1)100)/(3N-2) < 100$

Case 2: everybody wins, hence shares the amount of the win.
There are only two possibilities: either everybody plays the same guess G*, either k persons guess G_1 and (N–k) guess G_2 ($>G_1$) with

$\bar{G}-G_1=G_2-\bar{G}$ and $\bar{G}=(2/3)(kG_1+(N-k)G_2)/N$

Subcase 1: everybody makes the same guess G.*
If so, let player 1 deviate to G, with $G=(2/3)(G+(N-1)G^*)/N$, i.e. $G=(2(N-1)G^*)/(3N-2)$;

If G*>0, then $0 \leq G < G^*$ and player 1 is the only player who wins the game (so gets a higher payoff by deviating because she doesn't have to share the win).
Observe also that a simple switch to G*–ε leads to winning the game.
If G*=0, then G=G*=0 and it is not possible to get more by deviating. **The profile such that everybody guesses 0 is a Nash equilibrium**.

Subcase 2: k players guess G_1 and (N–k) players guess G_2 ($>G_1$) with $\bar{G}-G_1=G_2-\bar{G}$ and $\bar{G}=(2/3)(kG_1+(N-k)G_2)/N$
Each player shares the win with the other N–1 players.
Suppose, w.l.o.g. that player 1 plays G_2. If so, she will be better off switching to G to be the only winner with $G=(2/3)(kG_1+G+(N-k-1)G_2)/N$, i.e. $G=(2kG_1+2(N-k-1)G_2)/(3N-2)$.

To conclude, there is only one Nash equilibrium: everybody guesses 0.

♣ EXERCISE 15 TRAVELLER'S DILEMMA (BASU)

1. We first show that (x_1, x_2), with $x_1 \neq x_2$, cannot be a Nash equilibrium.
   Suppose that $x_1 > x_2$. Player 1 does not play her best response, because she gets max $(0, x_2-P)$, whereas she can get x_2 by switching from x_1 to x_2.
   For similar reasons, (x_1, x_2), with $x_1=x_2>2$, can't be a Nash equilibrium. Player 1 does not play her best response, because she gets $x_1=x_2$, whereas she can get $x_2-1+P>x_2$ by switching from $x_1=x_2$ to $x_1=x_2-1$.
   So the only possible equilibrium is (2,2) and (2, 2) is a Nash equilibrium. Nobody is better off switching to a higher price (he gets max (0, 2–P) instead of 2) and nobody can switch to a lower price.

Nothing changes if we switch to Basu's reimbursement rule (just replace max (0, x_2–P) by x_2–P and max (0,2–P) by (2–P) in the reasoning); (2,2) is still the only Nash equilibrium.

2. Well, will real players really play (2,2)?

I propose here the histograms that give the amounts asked by my L3 students, for P=4 and P=40.

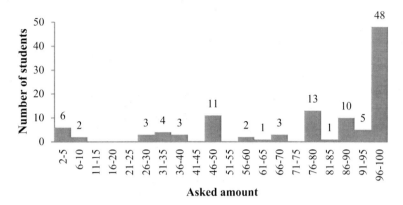

Figure A3.1a Traveller's dilemma, 'Basu's' version, asked amounts for P=4

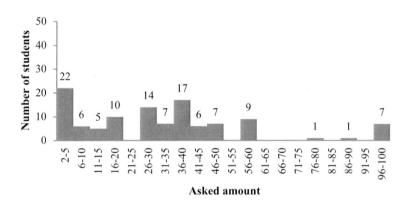

Figure A3.1b Traveller's dilemma, 'Basu's' version, asked amounts for P=40

Players may ask for low amounts when P is very high, because they may fear to lose P. And real players indeed play in this way. For example, 18.8% of my (112) L3 students played 2 for P=40, and 50.9% asked for 30 or less (see Figure A3.1b and chapter 6, section 2.1 for more details). But the reason that leads them to ask for a low amount is not the Nash equilibrium logic, but just *a comparison between potential loss and potential win*. When P is high, a player *both fears* to get y–P (y being the lower amount of the opponent) and *appreciates* to get x+P (x being her amount), even if x is low. For example, if P=70, 2+P=72 is an interesting amount whereas 90–P=20 is a bad one. *But observe that the players play 2 in order to get 2+P, and not to get 2 as in the Nash equilibrium.* More generally, they play a low amount x because they expect the payoff x+P, something that does not happen in the Nash equilibrium. *We have here an illustration of incompatible beliefs in real life. Clearly, if both players play 2 because they*

ANSWERS

hope to get 2+P, both hope to be the player playing the lowest amount, which is of course not possible. But wrong beliefs are possible in reality, whereas they are impossible in a Nash equilibrium.

When P is low, for example P=4, people do not play 2 in reality. So, for P=4, 31.3% of my (112) L3 students played 100 and 56.3% played 90 or more (see Figure A3.1a and chapter 6, section 2.1 for more details). When P is low, a player may *both appreciate* getting y–P (y being the lower but still high amount of the opponent) and *not appreciate* getting x+P (x being her low amount). For example, for P=4, he may prefer playing 95 than playing 40, because, with 40, he gets at most 44, whereas, with 95, if the opponent plays for example 90, he gets 86, a much higher amount (see chapter 6 for additional comments). And real players are satisfied with a high amount of money, *even if it is not the highest possible*. In the example above, the player is satisfied with the amount 86, albeit he could have obtained 89+P=93 by switching from 95 to 89. And the opponent is satisfied with 90+P=94, albeit he could have obtained 94+P=98 by switching to 94. So the players may be aware that they do not play the optimal strategies, but play 95 and 90 all the same, just because both hope that the opponent plays a high amount y so as to get at least y–P, which is still a high amount. And, this time, *their beliefs are right and compatible! The reason that explains the gap between real behaviour and Nash behaviour lies not in the existence of wrong beliefs in the reality, but in the ability to be satisfied with an amount of money that is not the maximal one.*

♣ ♣ EXERCISE 16 TRAVELLER'S DILEMMA, STUDENTS' VERSION, P<49

1. Each strategy x<100–2P is strictly dominated by 100 (see Exercise 12 in chapter 2), so cannot be played at equilibrium.

   So consider a couple (x_1, x_2), with x_1 and x_2 belonging to [100–2P, 100].

   Suppose first that $x_1 \neq x_2$, w.l.o.g. $x_1 > x_2$. If $x_2 < x_1 - 1$, player 2 is better off deviating from x_2 to $x_1 - 1$ ($x_1 - 1 + P > x_2 + P$). If $x_2 = x_1 - 1$, player 1 is better off switching to x_2, because $x_1 - 1 > x_1 - P$. Suppose now that $x_1 = x_2$. Player 1 is better off switching to $x_1 - 1$ ($x_1 - 1 + P > x_1$).

   So there does not exist a Nash equilibrium in pure strategies.

2. In the discrete version, each x, with 100>100–2P, is the best response to x+1 (because x+P>x+1, x+P>z+P with z<x, and x+P>100–P); so we may conjecture that any strategy in]100–2P,100[is played with positive probability.

   It follows that in the continuous version of the game, given that 100–2P is not strictly dominated and that 100 is a best response to 100–2P, we can conjecture that any price in [100–2P, 100] is played at equilibrium.

   Let us prove that this conjecture is right. If player 1 plays x, she gets x–P if the opponent plays y<x and she gets x+P if the opponent plays y>x. The probability that the opponent plays x goes to 0 given that we do not expect an atom on x.

   So player 1 gets:

$$A(x) = \int_{100-2P}^{x} (x-P)g(y)dy + \int_{x}^{100} (x+P)g(y)dy =$$

ANSWERS

$$\int_{100-2P}^{100} xg(y)dy + \int_{100-2P}^{x} -Pg(y)dy + \int_{x}^{100} Pg(y)dy$$

where g(y) is the equilibrium density function of player 2.

Given that player 1 plays each x in [100–2P,100] with positive probability, she has to get the same payoff with each x. So A(x) has to be constant, which implies A'(x)=0 for any x in [100–2P, 100].

It derives:

$$\int_{100-2P}^{100} g(y)dy - Pg(x) - Pg(x) = 0, \text{ i.e. } 1-2Pg(x)=0, \text{ i.e. } g(x)=1/2P \text{ regardless of x.}$$

So player 2's equilibrium strategy is the uniform distribution on [100–2P, 100]. By symmetry player 1's equilibrium strategy is also the uniform distribution on [100–2P, 100]. The unique Nash equilibrium of the continuous version of the students' game consists in asking for each amount in [100–2P, 100] with the same probability.

This equilibrium is rather intuitive, given that each strategy x, with 100–2P≤x<100, is the best response to x+1, and 100 is the best response to 100–2P.

3. Figures E3.5a and E3.5b show the amounts asked by my students for P=4 and P=40. 79.6% of my (137) L3 students played 92 (100–2P) or more for P=4, but they do not play a uniform distribution (you can observe an atom on 100, given that 65% of them played 100). For P=40, 95% played 20 or more, but again they didn't play in a uniform way (38% of them played 100 and 45.3% played 90 or more). So theory only partially fits with real behaviour.

♣ EXERCISE 17 TRAVELLER'S DILEMMA, STUDENTS' VERSION, P>49, AND CAUTIOUS BEHAVIOUR

Clearly, two groups of students emerge, one group playing 2, the other 100: 37.2% of my students play 2 and 21.2% play 100. So there are two atoms, 2, as in the Nash equilibrium, but also 100, which is not an atom in the Nash equilibrium.

I should also mention that many of the students playing 2 were convinced of getting 72, because they were convinced of playing less than the opponent by playing 2. This belief was partly right, but also partly wrong given that they were numerous to play 2!

Well, let us turn to the players playing 100. I asked them why they played 100, and they systematically answered that by doing so, they were sure to get at least 30 (and perhaps 100 if the opponent also plays 100).

Well, playing 100 is the cautious strategy of this game. If you ask for x, the opponent minimizes your payoff by asking for x–1 (or 2 if x=2). So it is cautious to play 100.

So it seems that the students' behaviour is between the Nash equilibrium and the cautious strategy.

ANSWERS

♣ EXERCISE 18 FOCAL POINT IN THE TRAVELLER'S DILEMMA

Well, when I asked my students to play the game, either the version in Exercise 15, or the version in exercises 16 and 17 (or even other versions), some of my students asked me *what was the value of the destroyed luggage*? Up to now, I gave no answer, but I really wonder in which way such an answer would change the way of playing in the different versions of this game. Especially in Basu's version, where the players are reluctant to play the equilibrium value 2, it may be that many students would ask for an amount around the luggage's value, namely because *it is fair* asking for such an amount. It may be that if the luggage's value is high, students focus their demands around this value, even for high values of P. And it may be that if the luggage's value is low, students focus their demands around this value, even for low values of P. Be that as it may, the information on the luggage's value will become a focal point that will surely affect the students' way of playing. Rest assured I will check this conjecture with my future students!

♣ ♣ ♣ EXERCISE 19 ASYMMETRIC ALL PAY AUCTIONS

1. At equilibrium, each bid is played with a positive probability. We prove it *in a cascade way*:
    0.5 can be played at equilibrium only if 0 is played at equilibrium (if not, player 1 prefers bidding 2.5). But 0 can be played only if 2.5 is played (if not, player 2 prefers bidding 2). But 2.5 can be played only if 2 is played (if not, player 1 prefers bidding 1.5). But 2 can be played only if 1.5 is played (if not, player 2 prefers bidding 1). But 1.5 can be played only if 1 is played (if not, player 1 prefers bidding 0.5). But 1 can be played only if 0.5 is played (if not, player 2 prefers bidding 0).
    So all the bids are played at equilibrium. Call $p_{0.5}$, $p_{1.5}$, $p_{2.5}$, q_0, q_1 and q_2 the probabilities assigned to the bids 0.5, 1.5, 2.5, 0, 1 and 2.
    For player 2, we need:

    $5p_{0.5}+5p_{1.5}+5p_{2.5}=7p_{0.5}+4p_{1.5}+4p_{2.5}$

    $5p_{0.5}+5p_{1.5}+5p_{2.5}=6p_{0.5}+6p_{1.5}+3p_{2.5}$

    And $p_{0.5}+p_{1.5}+p_{2.5}=1$. It follows that $p_{0.5}=p_{1.5}=p_{2.5}=1/3$. Player 2's payoff is 5.
    For player 1 we need:

    $6.5q_0+6.5q_1+6.5q_2=7.5q_0+7.5q_1+3.5q_2$

    $6.5q_0+6.5q_1+6.5q_2=8.5q_0+4.5q_1+4.5q_2$

    and $q_0+q_1+q_2=1$. It follows that $q_0=1/2$, $q_1=0.25$ and $q_2=0.25$. Player 1's payoff is 6.5.
    We already observe that q_0 is larger than q_1, which is equal to q_2.

2. For an increment of 0.1, we get:

    $6.1q_0+6.1(1-q_0)=8.9q_0+4.9(1-q_0)$ which leads to $q_0=1.2/4$. In fact, for an increment going to 0, this probability goes to $¼=(V_1-V_2)/V_1=1-V_2/V_1$

    Let us look for the other probabilities. The last equality becomes:

    $6.3(1-q_{2.8})+2.3\ q_{2.8}=6.1(1-q_{2.8})+6.1\ q_{2.8}$

Hence $q_{2.8}=0.2/4$. In a recursive way, by comparing two adjacent lines, we get that all the (14) bids different from 0 played by player 2 are played with the same probability $0.2/4$. When the increment goes to 0, all the strategies in $]0,V_2]$ are played with the same probability $[1-(1-V_2/V_1)]/V_2=1/V_1$ multiplied by dv.

So we have an atom on 0 (and only on 0).

For player 2, we get $5(1-p_{0.1})+5p_{0.1}=7.8p_{0.1}+4.8(1-p_{0.1})$, i.e. $p_{0.1}=0.2/3=1/15$. By comparing two adjacent columns, going from the first to the last one, we show that each player 1's bid is played with the same probability. When the increment goes to 0, this probability becomes $(1/V_2)$ dv. Here dv=0.2 and $dv/V_2=1/15$.

3. Given the previous results, we conjecture that in a continuous setting, at the Nash equilibrium, player 1 plays the uniform distribution on $[0,V_2]$ and player 2 bids each price in $]0,V_2[$ with probability dv/V_1 and bids 0 with probability $1-V_2/V_1$.

We prove that this conjecture is right.

Given that player 2 doesn't play more than V_2, the highest bid for player 1 is also V_2. Consider now player 2. We look for an equilibrium where he plays all the prices in $[0,V_2]$. So he gets the same payoff with 0, x and V_2, where x is any real number in $]0,V_2[$.

The payoff with 0 is equal to $M-0+(V_2$ multiplied by the probability that player 1 plays 0)/2=M (if there is no atom on 0 for player 1)

The payoff with V_2 is equal to $M-V_2+(V_2$ multiplied by the probability that player 1 plays less than $V_2)=M-V_2+V_2F_1(V_2)=M$ (if there is no atom on V_2), where $F_1(x)$ is player 1's cumulative distribution function.

The payoff with x is equal to $M-x+(V_2$ multiplied by the probability that player 1 plays less than $x)=M-x+V_2F_1(x)$ (if there is no atom on x)

It derives: $M=M-x+V_2F_1(x)$, hence $F_1(x)=x/V_2$.

So player 1's equilibrium strategy is the uniform distribution on $[0,V_2]$ (there are no atoms on the bids).

Let us now consider player 1. We look for an equilibrium where she plays the uniform distribution on $[0,V_2]$. To do so, she has to get the same payoff with ε, x and V_2, where x is any real number in $]0,V_2[$ and ε is positive and close to 0.

The payoff with ε is equal to $M-ε+(V_1$ multiplied by the probability that player 2 plays less than $ε)=M-ε+V_1F_2(ε)$ (if there is no atom on $ε)=M+V_1F_2(0)$ when ε goes to 0, where $F_2(x)$ is player 2's cumulative distribution function).

The payoff with V_2 is equal to $M-V_2+(V_1$ multiplied by the probability that player 2 plays less than $V_2)=M-V_2+V_1F_2(V_2)=M-V_2+V_1$ (if there is no atom on V_2)

The payoff with x is equal to $M-x+(V_1$ multiplied by the probability that player 2 plays less than $x)=M-x+V_1F_2(x)$ (if there is no atom on x).

It derives $M-V_2+V_1=M+V_1F_2(0)$, i.e. $F_2(0)=f_2(0)=1-V_2/V_1$, so we have an atom on 0. $f_2(x)$ is player 2's density function.

And we get $M-V_2+V_1=M-x+V_1F_2(x)$, i.e. $F_2(x)=x/V_1+1-V_2/V_1=x/V_1+f_2(0)$.

It follows $F_2'(x)=f_2(x)=1/V_1$. So player 2 plays each action in $]0,V_2[$ with the probability dx/V_1, and he bids 0 with the probability $1-V_2/V_1$.

ANSWERS

♣♣ EXERCISE 20 WALLET GAME, FIRST PRICE AUCTION, WINNER'S CURSE AND A ROBUST EQUILIBRIUM

1. Given that a player has no idea about the private signal of the opponent, a first reaction may be to add ½, the mean value of a signal, to his own value, in order to try to get the wallet.

    But $b_1(x_1)=x_1+1/2$ and $b_2(x_2)=x_2+1/2$, where $b_i(x_i)$ is player i's bid given his private information x_i, cannot constitute a Nash equilibrium: the bids are too high.

    To see why, imagine that both players bid in this way. So player 1 wins the wallet if $x_1+1/2 \leq x_2+1/2$, i.e. if $x_1 \leq x_2$. So she wins if player 2's private signal x_2 is lower than x_1. It derives that the mean value of x_2 is not 1/2, but $x_1/2$; so player 1 pays $x_1+1/2$ for a wallet whose expected value is only $x_1+x_1/2$. Given that $x_1 \in [0,1]$, she loses money, except in the very special case where $x_1=1$. *This result is called the winner's curse: the winner of the auction, instead of being happy, is unhappy because he paid too much!*

    Let us establish the best response $b_1(x_1)$ to player 2's bid function $b_2(x_2)=x_2+1/2$. Call b player 1's bid. Player 1 wins the auction if $b \leq x_2+1/2$. So b is optimal if it maximizes

    $$\int_0^{b-1/2} (x_1+x_2-b) f_2(x_2) dx_2$$

    Deriving on b leads to:

    $$-\int_0^{b-1/2} f_2(x_2) dx_2 + (x_1+b-1/2-b)f_2(b-1/2)=0 \text{ at equilibrium, i.e. for } b=b_1(x_1)$$

    So $-b_1(x_1)+1/2+x_1+b_1(x_1)-1/2-b_1(x_1)=0$, i.e. $b_1(x_1)=x_1$. This result is only compatible with $x_1 \geq 1/2$ (otherwise $b-1/2<0$). But you can check that, for $x_1<1/2$, a best answer is also to play x_1 (by doing so, player 1 gets 0, whereas she would get $\left[(x_1-b)x_2+\frac{x_2^2}{2}\right]_0^{b-1/2} = (b-1/2)$ $(x_1-b/2-1/4)<0$ for any value $b>1/2$). So the best response to player 2's bid function $b_2(x_2)=x_2+1/2$ is the bid function $b_1(x_1)=x_1$. You can check that this best response is *robust*, in that *it is also the best response to player 2's strategy $b_2(x_2)=x_2+a$, where a is any constant from 0 to 1*. This result is interesting in real life because it says that, provided you expect that your opponent plays in an additive way (he adds a constant to his private value), your best answer is to bid your private signal regardless of the opponent's added constant. Useful, isn't it?

    Let me observe that I often played this game with postdoctoral students and that they *never suffered* from the winner's curse. On the contrary, they generally offered low bids, and many of them even bid less than their private signal! As I asked them why they offered an amount necessarily lower than the value of the wallet, they answered that they hoped to get a lot of money if winning. Well is that the Internet generation impact (surfers on the Internet are often used to getting things without paying) or is that the expression of a very extreme cautious behaviour?

2. In fact the only symmetric Nash equilibrium consists for both players in offering their private signal.

    We look for $b_1(x_1)$. Call b player 1's bid. She wins each time $x_2 \leq b_2^{-1}(b)$ (player 2 is supposed to bid $b_2(x_2)$). So player 1's programme is

ANSWERS

$$\max_b \int_0^{b_2^{-1}(b)} ((x_1+x_2)-b) f(x_2) dx_2$$

We get $-\int_0^{b_2^{-1}(b)} f(x_2) dx_2 + [(x_1+b_2^{-1}(b)-b)]f(b_2^{-1}(b))/b_2'(b_2^{-1}(b))=0$ for $b=b_1(x_1)$

i.e. $-b_2^{-1}(b_1(x_1))+[x_1+b_2^{-1}(b_1(x_1))-b_1(x_1)]/b_2'(b_2^{-1}(b_1(x_1)))=0$

We look for a symmetric equilibrium, so $b_1(x)=b_2(x)=b(x)$, and the equation becomes:

$-x_1+[x_1+x_1-b(x_1)]/b'(x_1)=0$

$b(x_1)=x_1$ is the solution of this differential equation.

At first sight, this equilibrium contrasts with intuition: a player only plays the maximum he could pay if the wallet only contained the amount he observes, even if he knows that the wallet surely contains a higher amount. But the equilibrium is more intuitive than it seems at first view. We should not forget that, in a first price sealed bid auction, when a player values an object x, he only bids x/2 (when the number of players is 2). So bidding x when observing x, just means bidding two times the amount bid for an object of value x in the first price sealed-bid auction.

♣ ♣ EXERCISE 21 WALLET GAME, FIRST PRICE AUCTION, NASH EQUILIBRIUM AND NEW STORIES

1. At the symmetric Nash equilibrium of the wallet game (see Exercise 20), x_1 and x_2 being uniformly distributed on [0,1], player 1 bids $b_1(x_1)=x_1=x_1/2+x_1/2$ and player 2 bids $b_2(x_2)=x_2=x_2/2+x_2/2$.

   Each player bids two times what he bids in a first price sealed bid auction. So in the wallet game, *all happens as if a player just thinks that there are two first price sealed bid auctions on two identical objects.*

2. We now suppose that x_1 is uniformly distributed on [0,1] and x_2 is uniformly distributed on [0,½]. We check that $b_1(x_1)=x_1/2+x_1/4=3x_1/4$ and $b_2(x_2)=x_2/2+2x_2/2=3x_2/2$ is a Nash equilibrium.

   We first establish $b_1(x_1)$. Suppose that player 1 bids b. Given that $b_2(x_2)=3x_2/2$, she wins as long as $x_2 \leq 2b/3$. 2b/3 is in [0,1/2] if $b \leq 3/4$, which will be the case.

   Player 1's programme is:

$$\max_b \int_0^{2b/3} (x_1+x_2-b) f_2(x_2) dx_2$$

We get $-\int_0^{2b/3} f(x_2) dx_2 + (x_1+2b/3-b)f(2b/3)2/3=0$ for $b=b_1(x_1)$

Given that $x_2 \in [0,1/2]$, $f(2b/3)=2$ and we get:

355

ANSWERS

$-4b/3+(x_1-b/3)4/3=0$ i.e. $b=b_1(x_1)=3x_1/4$ ($\leq 3/4$).

We now check that $b(x_2)=3x_2/2$ is a best response to $b_1(x_1)=3x_1/4$. If player 2 bids b, he wins as long as $x_1 \leq 4b/3$. This value is in [0,1] if $b \leq 3/4$, which will be the case.

Player 2's programme is

$$\max_b \int_0^{4b/3} (x_1 + x_2 - b) f_1(x_1) dx_1$$

We get $-\int_0^{4b/3} f(x_1) dx_1 + (x_2 + 4b/3 - b) f(4b/3) 4/3 = 0$ for $b = b_2(x_2)$

Given that $x_1 \in [0,1]$, $f(x_1)=1$, and we get:

$-4b/3+(x_2+b/3)4/3=0$ hence $b=b_2(x_2)=3x_2/2$ ($\leq 3/4$).

So all happens as if player 1 just plays as if there were two first price sealed bid auctions on two objects, the second one having the half the value $-x_1/2$ – of the first one $-x_1$. And player 2 just plays as if there were two first price sealed bid auctions on two objects, the second one having double the value $-2x_2$ – of the first one $-x_2$.

♣♣ EXERCISE 22 TWO PLAYER WALLET GAME, SECOND PRICE AUCTION, A ROBUST SYMMETRIC EQUILIBRIUM

1. We look for $b_1(x_1)$. Call $b_2(x_2)$ player 2's bid function. By playing b, player 1 wins each time $x_2 \leq b_2^{-1}(b)$, provided $b_2(1) \leq b \geq b_2(0)$ (which will be the case).

Player 1's programme is: $\max_b \int_0^{b_2^{-1}(b)} ((x_1 + x_2) - b_2(x_2)) f(x_2) dx_2$

Deriving on b leads to $[x_1+b_2^{-1}(b)-b_2(b_2^{-1}(b))] f(b_2^{-1}(b))/b_2'(b_2^{-1}(b))=0$, i.e. $x_1+b_2^{-1}(b)-b=0$ for $b=b_1(x_1)$. So we have $x_1+b_2^{-1}(b_1(x_1))-b_1(x_1)=0$

Given that we look for a symmetric equilibrium, we get:

$x_1+x_1-b(x_1)=0$, i.e. $b_1(x_1)=2x_1$ and, by symmetry, $b_2(x_2)=2x_2$.

The bids at equilibrium do not depend on the density functions. This is quite interesting from a practical point of view, given that players in reality often ignore the density functions of the other players.

We observe a link with the second price sealed bid auction. Recall that in the latter, when the value x_i, $i=1,2$ is distributed on [0,1], each player i bids x_i, $i=1,2$. Hence, all happens as if a player just thinks that there are two second price sealed bid auctions on two identical objects.

2. The Nash equilibrium checks the *no- ex post regret property*, that is to say:

   ■ *First, when a player wins, he gets a positive payoff when playing against any losing opponent.*

■ 356

Given that $b_1(x_1)=2x_1$ and $b_2(x_2)=2x_2$, player 1 wins the auction if $b_1(x_1)>b_2(x_2)$, i.e. if $2x_1>2x_2$, i.e. $x_1>x_2$. So she gets, regardless of the value of x_2, $x_1+x_2-2x_2=x_1-x_2>0$

■ *Second, when a player loses the auction, he would get a negative payoff by winning it, against any (actually) winning opponent.*

As a matter of fact, player 1 loses the auction when $b_1(x_1)<b_2(x_2)$, i.e. when $x_1<x_2$. To win, she should at least bid $b_1(x_1)=b_2(x_2)$. So she would get: $x_1+x_2-b_2(x_2)=x_1-x_2<0$.

We should add the case "player 1 gets the object half of the time" but this would not change the conclusions.

♣ ♣ ♣ EXERCISE 23 TWO PLAYER WALLET GAME, SECOND PRICE AUCTION, ASYMMETRIC EQUILIBRIA

1. We know (see exercise 22), that at equilibrium we have:

$$x_1+b_2^{-1}(b_1(x_1))-b_1(x_1)=0 \tag{1}$$

and $b_1^{-1}(b_2(x_2))+x_2-b_2(x_2)=0.$ (2)

Replacing $b_1(x_1)$ by $x_1+h(x_1)$ in (1) leads to

$x_1+b_2^{-1}(x_1+h(x_1))=x_1+h(x_1)$, so $b_2^{-1}(x_1+h(x_1))=h(x_1)$, so $x_1+h(x_1)=b_2(h(x_1))$, which is true, by definition of $b_2(x)$.

Replacing $b_2(x_2)$ by $x_2+h^{-1}(x_2)$ in (2) leads to

$x_2+b_1^{-1}(x_2+h^{-1}(x_2))=x_2+h^{-1}(x_2)$ hence $b_1^{-1}(x_2+h^{-1}(x_2))=h^{-1}(x_2)$, hence $x_2+h^{-1}(x_2)=b_1(h^{-1}(x_2))$ which is true by definition of $b_1(x)$.

So $b_1(x_1)=x_1+h(x_1)$ and $b_2(x_2)=x_2+h^{-1}(x_2)$ satisfy equations (1) and (2), so they constitute a Nash equilibrium.

$h(x)=100x$ leads to:

$b_1(x_1)=101x_1$ and $b_2(x_2)=x_2+x_2/100=101x_2/100$

This strongly asymmetric equilibrium favours player 1: as soon as $x_2<100x_1$, player 2 loses the auction. So, as soon as $x_1>0.01$, player 1 always wins the auction.

This strange equilibrium still satisfies the no ex-post regret property. Consider the most frequent case, $x_2<100x_1$. Player 2 is happy to lose the auction because, to win the auction, he would have to play $b>101x_1$ and so he would get $x_2-100x_1<0$. And player 1 is happy to win because, despite the fact that she bids a lot, she pays a low amount: she gets $x_2+x_1-101x_2/100=x_1-x_2/100>0$.

Observe that there exist other asymmetric equilibria, for example $b_1(x_1)=x_1+10$ and $b_2(x_2)=x_2$. Player 1 always wins (because she plays at least 10), whereas player 2 always loses (because he bids at most 1). These strategies constitute a Nash equilibrium. Player 1, by winning, gets $x_1+x_2-b_2(x_2)=x_1\geq0$, so she gets more by winning than by losing the auction; and she cannot get more with another winning bid, in that her payoff only depends on player 2's bid. Player 2 is happy to lose, because, to win, he has to bid $b>x_1+10$, so he gets $x_1+x_2-x_1-10=x_2-10<0$. So nobody wants to unilaterally deviate.

ANSWERS

Second price auctions allow the coexistence of cautious behaviour and hothead behaviour. When a player is able to convince that he bids a lot, whereas the other player is known to be cautious, then the players can converge to an equilibrium where the cautious player bids a low amount and mainly loses the auction, whereas the other bids a high amount and mainly wins the auction. Nobody is incited to deviate: the hothead player is happy because he only pays the low amount of the cautious player (so he can afford to bid very high amounts and to win), and the cautious player is happy to lose the auction, because to win he would have to overbid (and pay) the high bids of the opponent.

2. The proposed strategy is *discontinuous*[1]: the bidding function is discontinuous in c for player 1, in d for player 2.

   In some way, a player is rather cautious when getting a low signal, and more courageous when getting a high signal. *I mention here that my post doctoral students, who played this game, often adopted a discontinuous behaviour in common value auctions: when their signal was low, they rarely added a high amount to their signal, whereas they dared adding higher amounts to high signals. For example adding 0.05 to low signals and 0.25 to high ones was not uncommon.*

   Let us prove that the strategy profile is a Nash equilibrium:

   We here only show that $b_1(x_1)$ is a best response:

   If $x_1 > c$ player 1 best replies: she wins against any type of player 2, because $x_1 + 1 > c + 1$ and player 2 never bids more than $c+1$, and she gets a positive payoff by winning: she gets either $x_1 + x_2 - x_2 - c > 0$ or $x_1 + x_2 - x_2 > 0$, depending on player 2's type. And the payoff she gets cannot be higher, because she pays player 2's bid.

   If $x_1 \leq c$, player 1 also best replies:

   - She loses against any type of player 2 higher than d, because $x_1 + d \leq c + d < x_2 + c$. And losing gives her the best payoff, 0, because winning would lead to the payoff $x_1 + x_2 - x_2 - c \leq 0$.
   - She wins against any type of player 2 lower than d, because $x_2 < d \leq x_1 + d$, and she gets a positive or null payoff by winning, because $x_1 + x_2 - x_2 \geq 0$; and this payoff cannot be higher, given that she pays player 2's bid.
   - If $x_1 < c$, then she loses against player 2 of type d because $x_1 + d < c + d$, and losing gives her the best payoff 0, because winning would lead to the payoff $x_1 + x_2 - x_2 - c < 0$.
   - If $x_1 = c$, she makes the same bid as player 2 of type d; proposition 1 does not give the sharing rule in case of a tie. This is of no importance because player 1 gets a null payoff from getting the wallet, and she cannot get more by bidding in a different way.

   One establishes that $b_2(x_2)$ is a best response in a similar way (see Umbhauer 2015).

   Observe that the no ex-post regret property is satisfied:

3. We observe that the distribution functions play no role in the above equilibria.

   There are no dominated strategies in the discontinuous Nash equilibrium but there are weakly dominated strategies in the equilibria of question 1. As a matter of fact, as soon as $h(x_1) > 1$ for a given x_1, $b(x_1) = x_1 + h(x_1)$ is weakly dominated by the strategy $b(x_1) = x_1 + 1$. To see why, call b the strategy played by player 2. If $b < x_1 + 1$, both $x_1 + 1$ and $x_1 + h(x_1)$ lead to the same payoff $x_1 + x_2 - b$. If $b > x_1 + h(x_1)$, both $x_1 + 1$ and $x_1 + h(x_1)$ lead to the same payoff 0. But if $x_1 + 1 < b < x_1 + h(x_1)$, then $x_1 + 1$ leads to the payoff 0, whereas $x_1 + h(x_1)$ leads to the payoff $x_1 + x_2 - b < x_1 + x_2 - x_1 - 1 < 0$, when $x_2 \neq 1$. If $b = x_1 + 1$ or $b = x_1 + h(x_1)$, you can check that $x_1 + 1$ leads to a higher payoff than $x_1 + h(x_1)$.

The weak dominance highlights the risk linked to a hothead strategy. If you are not sure that your opponent is bidding low amounts, you may win and pay a high amount you do not want to pay.

♣♣ EXERCISE 24 N PLAYER WALLET GAME, SECOND PRICE AUCTION, MARGINAL APPROACH

1. Suppose that player 1 plays b at the Nash equilibrium when her signal is x_1.

    The only change induced by a switch to b+ε (ε positive and going to 0), is that she also wins against player 2 when he plays b. The payoff she gets with a player 2 playing less than b is not changed (she still wins and still pays player 2's bid), and the payoff she gets with a player 2 playing more than b is still null. Moreover, given that we look for a symmetric Nash equilibrium, player 2 plays b for the same signal as player 1, i.e. when $x_2=x_1$. So the only payoff change for player 1, when she switches from b to b+ε, is that she gets an additional payoff when facing a player 2 of type x_1, who plays b by assumption of a symmetric equilibrium. This additional payoff is x_1+x_1-b, multiplied by the probability that player 2 has the signal x_1. So, given that player 1 should not be induced to deviate from b (b is supposed to be her equilibrium message), this additional payoff has to be null, i.e. $b=b(x_1)=2x_1$, which is the bid function we got in a more standard way.

    This marginal payoff approach is only possible because we work with a second price auction. It is because what we pay is not what we bid that the switch from b to b+ε has no impact on the payoff player 1 gets with all types of player 2 who do not bid b. It follows that the marginal payoff approach cannot be applied in a first price auction.

2. Let us now look for an N–player wallet game, when the distribution functions are uniform.

    We again suppose that player 1 plays b at the Nash equilibrium when her signal is x_1.

    The only change induced by a switch from b to b+ε (ε positive and going to 0) is that player 1 also wins against a player j, j from 2 to N, who plays b, the other players bidding less than b. Given the symmetry of the equilibrium, a player j bids b only if he is of type x_1. And the other players bid less than b only if their signal is lower than x_1. So their mean signal is $x_1/2$ (because of the uniform distributions) and player 1's additional payoff is:

    $$\sum_{j=2}^{N}(x_1+x_1+(N-2)x_1/2-b)f_j(x_1)dx_j \prod_{l\neq j, l=2}^{N} F_l(x_1),$$ where $f_k(x)$ and $F_k(x)$, k=1...N are player k's density function and cumulative distribution function. This additional payoff has to be null, so it follows that $b=b(x_1)=(N+2)x_1/2$, when the signals are uniformly distributed.

    Observe that, for N=3, we get $b(x_i)=5x_i/2$, i=1 to 3: this is a lower bid than the one a player would make if there were three second price sealed bid auctions on three identical objects $(3x_i)$; so the nice story we told for the 2 player wallet game does not hold for N>2.

ANSWERS

♣ ♣ ♣ EXERCISE 25 SECOND PRICE ALL PAY AUCTIONS

1. With bid 0, player 1 gets 6.5×83/293+5 (1–83/293)=1589.5 /293
   With bid 1, player 1 gets 8×83/293+5.5×57/293+4(1–83/293–57/293)=1589.5/293
   With bid 2, player 1 gets 8×83/293+7×57/293+4.5×45/293+3×27/293+3×81/293=1589.5/293
   With bid 3, player 1 gets 8×83/293+7×57/293+6×45/293+3.5×27/293+2×81/293=1589.5/293
   With bid 4, player 1 gets 8×83/293+7×57/293+6×45/293+5×27/293+1×81/293=1549/293
   With bid 5, player 1 gets 8×83/293+7×57/293+6×45/293+5×27/293+1.5×81/293=1589.5/293

   Given that 1589.5/293>1549/293, player 1 and, by symmetry, player 2 get the same payoff with the bids 0,1,2,3 and 5 and get more with these bids than with the bid 4. So they play a mixed strategy Nash equilibrium. We observe that the bids 0 and 5 are played with the highest probabilities and that the probabilities decrease in the bids from 0 to 4 (with a probability equal to 0 for bid 4). Bid 4 is weakly dominated by bid 5.

2a. In the previous game, M–V/2=5–1.5=3.5. And 4, the only bid strictly between M–V/2 and 5 is not played and weakly dominated by 5.

   In the general case, let us compare the payoffs obtained by player 1 with a bid b and with bid M, with M–V/2≤b<M. Assume that player 2 plays x.

   If player 2 plays x<b, player 1 wins the auction and gets M+V–x with b and M. If player 2 plays x=b, player 1 gets M+V–b if she plays M and only M+V/2 –b if she plays b. If player 2 plays x, with b<x<M, player 1 loses the auction and gets M–b with b, and she gets M+V–x with M. Given that M–V/2≤b and x<M, we have M–b<M+V–x, i.e. x<V+b, because x<M≤b+V/2. So player 1 is better off with M than with b. Finally, if player 2 plays M, player 1 gets M–b with b, and M+V/2–M with M. M–b≤V/2, because M–V/2≤b.

   It follows that b, with M–V/2≤b<M, is weakly dominated by M.

2b. Given that all bids between M–V/2 and M are weakly dominated by M, and given the probabilities we got in the discrete case studied in question 1, we conjecture that the Nash equilibrium strategy is a density function f(.) that decreases from 0 to M–V/2 and has an atom on M.

   So call $f_2(.)$ player 2's equilibrium strategy. Suppose that player 1 plays b. She wins the auction each time player 2 bids less than b. So she gets:

$$G(b)=M+\int_0^b (V-x)f_2(x)dx - b\left(\int_b^{M-V/2} f_2(x)dx + f_2(M)\right)$$

   It follows, as expected, that G(b)<G(M–V/2) for b>M–V/2, because

$$M+\int_0^{M-V/2}(V-x)f_2(x)dx - \left(M-\frac{V}{2}\right)f_2(M) > M+\int_0^{M-V/2}(V-x)f_2(x)dx - bf_2(M)$$

   G(b) has to be constant for each b in [0, M–V/2] ∪ {M}. So G'(b)=0 for b in [0, M–V/2].

   We get (V–b)f₂(b)–F₂(M–V/2)+F₂(b)–f₂(M)+bf₂(b)=0

   where $F_2(.)$ is the cumulative distribution of the density function $f_2(.)$.

   By construction f₂(M)=1–F₂(M–V/2), so we get the differential equation: Vf₂(b)–1+F₂(b)=0

   whose solution is: $F_2(b)=1+Ke^{-b/V}$ where K is a constant determined as follows:

$F_2(0)=0$ because there is no atom on 0, so $1+K=0$ and $K=-1$.

It follows $F_2(b)=1-e^{-b/V}$ for b in $[0, M-V/2]$, $f_2(M)=1-F_{21}(M-V/2)=e^{1/2-M/V}$ (<1), $f_2(b)=e^{-b/V}/V$ for b in $[0, M-V/2]$ (and $f_2(b)=0$ for b in $]M-V/2,M[$)

By symmetry, we get $f_1(b)=e^{-b/V}/V$ for b in $[0, M-V/2]$, $f_1(M)=1-F_1(M-V/2)=e^{1/2-M/V}$ (and $f_1(b)=0$ for b in $]M-V/2,M[$).

We write $f_1(b)=f_2(b)=f(b)$. Let us compare $f(b)$ for b in $[0,M-V/2]$ and $f(M)$.

$f(b)$ only depends on V, but $f(M)$ is growing in V and decreasing in M. This last point is logical; for a given value V, the interval of bids in $[0, M-V/2]$ is growing in M and, given that the probability of a bid in $[0,M-V/2]$ only depends on V, this immediately implies that, if M grows, there is less weight remaining for $f(M)$.

For M=5 and V=3, we get: $f(0)=1/3$, $f(1)=0.239$, $f(2)=0.171$, $f(3)=0.123$ and $f(5)=0.311$. Except for $f(5)$, these probabilities have to be multiplied by db, but we can observe that by dividing them respectively by p_0, p_1, p_2, p_3 and p_5 we get:

$f(0)/p_0=1.18$, $f(1)/p_1=1.23$ $f(2)/p_2=1.11$ $f(3)/p_3=1.33$ $f(5)/p_5=1.13$. We are still far from proportionality but this is due to the fact that there are too few values in the discrete model in comparison to the continuous model. We could establish that the probabilities come close when M and V grow.

♣ EXERCISE 26 SINGLE CROSSING IN A FIRST PRICE SEALED BID AUCTION

We say that a player is of type t when he values the object t.

Consider two values, t and t' with t'>t, and two bids for player 1, b and b' with b'>b. Call S the set of types of player 2 who bid less than b, and S' the set of types of player 2 who bid less than b'. By construction $S \subseteq S'$, and even $S \subset S'$, because, if S=S', no type of player 2 bids in [b,b'], and it would be suboptimal for player 1 to bid b'.

Suppose that player 1 of type t weakly prefers b' to b. So she gets a higher or equal payoff with b' than with b, i.e. $\int_{S'} (t-b')f(x_2)\,dx_2 \leq \int_S (t-b)f(x_2)\,dx_2$

where $f(x_2)$ is player 2's probability of signal x_2.

This induces:

$$\int_{S'}(t'-b')f(x_2)\,dx_2 = \int_{S'}(t'-t)f(x_2)\,dx_2 + \int_{S'}(t-b')f(x_2)\,dx_2 \geq$$

$$\int_{S'}(t'-t)f(x_2)\,dx_2 + \int_S(t-b)f(x_2)\,dx_2 > \int_S(t'-t)f(x_2)\,dx_2 + \int_S(t-b)f(x_2)\,dx_2$$

(because S⊂S' and t'>t)

$$> \int_S(t'-b)f(x_2)\,dx_2.$$ So player 1 of type t' prefers b' to b.

ANSWERS

♣ ♣ EXERCISE 27 SINGLE CROSSING AND AKERLOF'S LEMON CAR

1. We call type t_h the seller who has the reservation price h (a higher h corresponds to a higher quality). Consider two prices p and p' with p'>p, p'>h' and p>h, and two types t_h and $t_{h'}$, with $t_{h'}>t_h$ (h'>h). Suppose that p, respectively p', is accepted with probability q(p), respectively q(p').

   Suppose that a seller of type t_h weakly prefers p' to p. We have to show that the seller of type $t_{h'}$ prefers p' to p. So we show that:

   (p'–h)q(p')≤(p–h)q(p) ⇔ (p'–h')q(p')>(p–h')q(p)

   If q(p')≤q(p) or p<h' this is automatically true

   If q(p')<q(p)

   (p'–h)q(p')≤(p–h)q(p) ⇔ (p'–h'+h'–h)q(p')≥(p–h'+h'–h)q(p)

   ⇔ (p'–h')q(p')≥(p–h')q(p)+(h'–h)(q(p)–q(p'))   (1)

   ⇒ (p'–h')q(p')>(p–h')q(p) (because (h'–h)(q(p)–q(p'))>0)

2. Suppose that the played price is p, with $h_j \leq p < h_{j+1}$, j higher or equal to 2.

   Only qualities lower or equal to t_j can be sold at price p. So the expected quality of the sold car is $\dfrac{\sum_{i=1}^{j}\rho_i t_i}{\sum_{i=1}^{j}\rho_i}$ and the highest price the consumer accepts paying is $\dfrac{\sum_{i=1}^{j}\rho_i H_i}{\sum_{i=1}^{j}\rho_i}$. Yet this price, by assumption (a), is lower than h_j and therefore lower than p. So trade will not occur at price p. As a consequence, trade can only occur at a price p lower than h_2. This price is necessarily assigned to the quality t_1 and will be accepted, provided it is lower or equal to H_1. So only the worst quality can be sold. Akerlof says that the worst quality throws all the other qualities out of the market.

3. Akerlof's reasoning is a pure strategy reasoning. Each seller is supposed to choose one price with probability 1.

   And, given that it is always more interesting to sell (with probability 1) at a higher than at a lower price, each type of seller necessarily plays the same price. Akerlof's reasoning follows from this fact. When we switch to mixed strategies, trade does not necessarily occur at a unique price, because higher prices may be accepted with a lower probability (so a player is not necessarily better off with a higher than with a lower price). So we will find mixed Nash equilibria such that many prices (that do not lead to the same probabilities of buying) coexist on the market and this coexistence allows all qualities to be sold.

ANSWERS

♣ ♣ ♣ EXERCISE 28 DUTCH AUCTION AND FIRST PRICE SEALED BID AUCTION

1. Let us show that player 1 plays her best response regardless of her type.

   Type 0: she can only bid 0

   Type 1: if she bids 1 she gets 0, if she bids 0, she gets 0.5(1–0)2/6=1/6, because she wins (the auction) half of the time when facing a type 0 or 1. Hence bidding 0 is her best response.

   Type 2: if she bids 2, she gets 0. If she bids 1, she gets:
   (2–1)2/6+0.5(2–1)2/6=3/6 because she wins when she faces a type 0 or 1, she wins half of the time when facing a type 2 or 3, and she loses (the auction) when facing a type 4 or 5.

   If she bids 0, she gets 0.5 (2–0). 2/6=2/6 because she wins half of the time when facing a type 0 or 1. Given that 3/6>2/6, bidding 1 is her best response.

   Type 3: if she bids 3, she gets 0. If she bids 2, she gets (3–2)4/6+0.5(3–2) 2/6=5/6 because she wins when facing the types 0, 1, 2 and 3 and she wins half of the time when facing a type 4 or 5. If she bids 1, she gets (3–1)2/6+0.5 (3–1)2/6=6/6 because she wins when facing the types 0 and 1 and wins half of the time when facing the types 2 and 3.

   She can't bid 0 because of the single crossing property (given that type 2 bids 1, she can't bid less than 1, see Exercise 26) (for future comparisons, let us observe that she wins 0.5 (3–0)2/6=3/6<6/6 by bidding 0). So, given that 6/6>5/6, she is best off bidding 1.

   Type 4: if she bids 4, she gets 0. If she bids 3, she gets (4–3). 6/6=6/6 given that the opponent bids less regardless of type. If she bids 2, she gets: (4–2). 4/6+0.5(4–2)2/6=10/6, because she wins when confronted with the types 0,1,2 and 3 and wins half of the time when facing a type 4 or 5. If she bids 1, she gets: (4–1)2/6+0.5(4–1)2/6=9/6 because she wins when facing the types 0 and 1, and wins half of the time when facing a type 2 or 3. She can't bid less than 1 because of the single crossing property (for future comparisons, let us observe that she wins 0.5(4–0)2/6=4/6<10/6 by bidding 0).

   So, given that 10/6>9/6, her best response is bidding 2.

   Type 5: Given that no opponent bids more than 2, she always prefers bidding 3 than bidding 4 or 5. If she bids 3, she gets (5–3). 6/6=12/6, because she wins all the time. If she bids 2, she gets: (5–2)4/6+0.5(5–2)2/6=15/6>12/6, because she wins when facing the types 0,1,2 and 3 and wins half of the time when facing a type 4 or 5. She can't bid less than 2 because of the single crossing property (for future comparisons, let us observe that she wins (5–1)2/6+0.5 (5–1)2/6=12/6 when bidding 1 and 0.5(5–0)2/6=5/6 by bidding 0). So bidding 2 is her best response.

   So the proposed profile is a Nash equilibrium.
   Let us summarize the calculi in Table A3.1.

363

ANSWERS

■ Table A3.1

	Payoffs	
Type 0	Bid 0: 0	
Type 1	Bid 1: 0	(m)
	Bid 0: 0.5(1–0)2/6=1/6	(n)
Type 2	Bid 2: 0	
	Bid 1: (2–1)2/6+0.5(2–1)2/6=3/6	(k)
	Bid 0: 0.5(2–0)2/6=2/6	(l)
Type 3	Bid 3: 0	
	Bid 2: (3–2)4/6+0.5(3–2)2/6=5/6	(h)
	Bid 1: (3–1)2/6+0.5(3–1)2/6=6/6	(i)
	Bid 0: 0.5(3–0)2/6=3/6	(j)
Type 4	Bid 4: 0	
	Bid 3: (4–3)6/6=6/6	
	Bid 2: (4–2)4/6+0.5(4–2)2/6=10/6	(e)
	Bid 1: (4–1)2/6+0.5(4–1)2/6=9/6	(f)
	Bid 0: 0.5(4–0)2/6=4/6	(g)
Type 5	Bid 5: 0	
	Bid 3: (5–3)6/6=12/6	(a)
	Bid 2: (5–2)4/6+0.5(5–2)2/6=15/6	(b)
	Bid 1: (5–1)2/6+0.5(5–1)2/6=12/6	(c)
	Bid 0: 0.5(5–0)2/6=5/6	(d)

2. We now switch to the Dutch auction.

   Knowing that the opponent doesn't accept a price higher than 2, the first price somebody may accept is the price 3. Given the single crossing property, if a type t is better off accepting this price, than a type t'>t is also better off accepting this price.

   So let us study if player 1 of type 5 is better off accepting 3, or if she is better off accepting 2. If she accepts 3, she gets (5–3)6/6=12/6, because she is the only player who accepts this price. If she waits for the auctioneer to propose 2, she gets (5–2)4/6+0.5(5–2)2/6=15/6. So she prefers waiting (15/6>12/6). Observe that we just compare (a) and (b) from Table A3.1, as we did in the FPSBA.

   Given the single crossing property, each type lower than 5 also prefers waiting.

   Let us now turn to the next period, where the auctioneer proposes 2.

   Player 1 of type 5 compares the payoff she gets by accepting 2, i.e. 15/6=(5–2)4/6+0.5 (5–2)2/6 (because she gets the object at this price when facing the types 0,1,2,3 who only accept lower prices, and gets the object half of the time when facing a type 4 or 5, who accept the price 2) to the payoff she gets by waiting for lower prices. If she waits for the next period, where the auctioneer proposes 1, she gets (0)2/6+(5–1)2/6+0.5(5–1)2/6=12/6 (because the auction stops before this period (hence she gets 0), when she faces a type 4 or 5, she gets the object at price 1 when facing the types 0,1 who only accept the lower price 0, and she gets the object half of the time at price 1 when facing a type 2 or 3). So she compares (b) and (c) in Table A3.1 and prefers accepting the price 2. If she waits two additional periods, i.e. the period where the auctioneer proposes 0, she gets 0.5(5–0)2/6=5/6, so she compares (b) and (d) in Table A3.1 and prefers accepting the price 2.

ANSWERS

Player 1 of type 4 compares the payoff she gets by accepting 2, i.e. (4–2)4/6+0.5(4–2)2/6=10/6, to the payoff she gets by waiting for lower prices. If she waits for the next period, where the auctioneer proposes 1, she gets (0)2/6+(4–1)2/6+0.5(4–1)2/6=9/6; if she waits two additional periods, i.e. the period where the auctioneer proposes 0, she gets 0.5(4–0)2/6=4/6. So she compares (e) to (f) and to (g) in Table A3.1 and prefers accepting the price 2.

Player 1 of type 3 compares the payoff she gets by accepting 2, i.e. (3–2)4/6+0.5 (3–2)2/6=5/6, to the payoff she gets by waiting for the next period where the auctioneer proposes 1, i.e. (0)2/6+(3–1)2/6+0.5(3–1)2/6=6/6>5/6. So she compares (h) and (i) in Table A3.1 and prefers waiting.

Player 1 of types 2, 1 and 0, by single crossing, also prefer waiting.

We now switch to the period where the auctioneer proposes 1. If the game reaches this stage, there is no opponent of types 4 and 5, because these types would have accepted the price 2. So, if the game reaches this stage, the probability of being confronted with a type 4 or 5 is equal to 0, which means that the probability of being confronted with each type 0, 1, 2 or 3 is ¼ (and no longer 1/6).

Player 1 of type 3 compares the payoff she gets by accepting 1, i.e. (3–1)2/4+0.5 (3–1)2/4=6/4, to the payoff she gets by waiting for the next period where the auctioneer proposes 0, i.e. 0.5(3–0)2/4=3/4. So she compares (i) and (j) in Table A3.1, *but with the denominator 4 instead of 6, which does not change anything.* And so she accepts the price 1.

Player 1 of type 2 compares the payoff she gets by accepting 1, i.e. (2–1)2/4+0.5 (2–1)2/4=3/4, to the payoff she gets by waiting for the next period where the auctioneer proposes 0, i.e. 0.5(2–0)2/4=2/4. So she compares (k) to (l) in Table A3.1, but with the new denominator 4 instead of 6, which does not change anything. And so she accepts the price 1.

Player 1 of types 1 and 0 prefer waiting (player 1 of type 1 gets 0 by accepting 1 whereas she gets 0.5(1–0)2/4=1/4 by waiting for the next period where the auctioneer proposes 0 (so she compares (m) and (n) with a change – without impact – of the denominator).

Now let us switch to the period where the auctioneer proposes 0. If the game reaches this stage, there is no opponent of types 2, 3, 4 and 5, because these types would have accepted the higher prices 1 or 2. And player 1 of types 0 and 1 accept the price 0 by assumption.

So we have established, by using the same comparisons as in the FPSBA, that a player of types 4 and 5 prefers waiting for price 2 rather than accepting a higher price, and prefers accepting 2 than waiting for lower prices. It derives that refusing a price higher than 2 and accepting each price lower or equal to 2 is a best response for these types of players.

Similarly, we have established that a best response for a player of types 2 and 3 consists of refusing a price higher than 1 and accepting each price lower or equal to 1, and that a best response for a player of types 0 and 1 consists of refusing any price higher or equal to 1, and to accept the price 0.

So we get the same equilibrium prices in both the FPSBA and in the Dutch auction for exactly the same reasons, despite these two games seeming strongly different at first view.

NOTE

1 For more on discontinuous equilibria see Umbhauer, G., 2015. Almost common value auctions and discontinuous equilibria, *Annals of Operations Research*, 225 (1), 125–140.

Chapter 4

Answers to Exercises

♣ EXERCISE 1 STACKELBERG ALL PAY AUCTION, BACKWARD INDUCTION AND DOMINANCE

There are six subgames, one at each node x_i, i from 0 to 5. Bid 1 is player 2's best response at the subgame starting at x_0, bid 2 is player 2's best response at the subgame starting at x_1, bid 0 is player 2's best response at the subgames starting at x_3, x_4 and x_5, and bids 0 and 3 are player 2's best responses at the subgame starting at x_2 (5=5>4.5>4=4>3).

So there are two possible reduced games at y, depending on player 2's choice at x_2.

Case 1: player 2 bids 0 at x_2.
The reduced game is given in Figure A4.1a. Player 1's best response is bid 2, which leads to the SPNE E_2=(2, (1/x_0,2/x_1,0/x_2,0/x_3,0/x_4,0/x_5)).

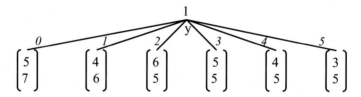

■ Figure A4.1a

Case 2: player 2 bids 3 at x_2.
The reduced game is given in Figure A4.1b. Player 1 is indifferent between bidding 0 and bidding 3 (5=5>4=4>3=3), so there are two additional SPNE,

E_0=(0, (1/x_0,2/x_1,3/x_2,0/x_3,0/x_4,0/x_5)) and E_3=(3, (1/x_0,2/x_1,3/x_2,0/x_3,0/x_4,0/x_5)

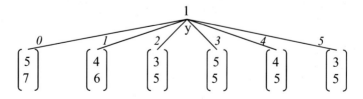

■ Figure A4.1b

ANSWERS

In Exercise 6 in chapter 2, we showed that, by first eliminating player 2's weakly dominating strategies and then player 1's strictly dominated strategies, we also get these three SPNE.

By contrast, by first eliminating player 1's strictly and weakly dominated strategies, we eliminate the possibility for player 1 to bid 3, which leads to only two SPNE, E_0 and E_2. So backward induction and iterative elimination of dominated strategies do not necessarily lead to the same results.

♣ EXERCISE 2 DUEL GUESSING GAME

1. This game has as many subgames as decision nodes.

    In the last subgame, at x_{10}, player 2 plays her only possible action A. It follows that in the reduced game starting at x_9, player 1 compares 0.9 to 0, so plays A (Figure A4.2a). And so on. In the reduced game starting at x_8, player 2 plays A (0.8>0.1, Figure A4.2b). In the reduced game starting at x_7, player 1 plays A (0.7>0.2, Figure A4.2c). In the reduced game starting at x_6, player 2 plays A (0.6>0.3, Figure A4.2d). In the reduced game starting at x_5, player 1 plays A (0.5>0.4, Figure A4.2e). By contrast, in the reduced game starting at x_4, player 2 plays P (0.5>0.4, Figure A4.2f), in the reduced game starting at x_3 player 1 plays P (0.5>0.3, Figure A4.2g), in the reduced game starting at x_2, player 2 plays P(0.5>0.2, Figure A4.2h). At last, in the reduced game starting at x_1, player 1 plays P (0.5>0.1, Figure A4.2i). The actions are indicated by a ● on Figure E4.2.

$$\begin{array}{c} 1 \\ x_9 \end{array} \xrightarrow{P} (0,1) \quad \begin{array}{c} 2 \\ x_8 \end{array} \xrightarrow{P} (0.9, 0.1) \quad \begin{array}{c} 1 \\ x_7 \end{array} \xrightarrow{P} (0.2, 0.8) \quad \begin{array}{c} 2 \\ x_6 \end{array} \xrightarrow{P} (0.7, 0.3) \quad \begin{array}{c} 1 \\ x_5 \end{array} \xrightarrow{P} (0.4, 0.6)$$

$$A \begin{pmatrix} 0.9 \\ 0.1 \end{pmatrix} \qquad A \begin{pmatrix} 0.2 \\ 0.8 \end{pmatrix} \qquad A \begin{pmatrix} 0.7 \\ 0.3 \end{pmatrix} \qquad A \begin{pmatrix} 0.4 \\ 0.6 \end{pmatrix} \qquad A \begin{pmatrix} 0.5 \\ 0.5 \end{pmatrix}$$

$$\begin{array}{c} 2 \\ x_4 \end{array} \xrightarrow{P} (0.5, 0.5) \quad \begin{array}{c} 1 \\ x_3 \end{array} \xrightarrow{P} (0.5, 0.5) \quad \begin{array}{c} 2 \\ x_2 \end{array} \xrightarrow{P} (0.5, 0.5) \quad \begin{array}{c} 1 \\ x_1 \end{array} \xrightarrow{P} (0.5, 0.5)$$

$$A \begin{pmatrix} 0.6 \\ 0.4 \end{pmatrix} \qquad A \begin{pmatrix} 0.3 \\ 0.7 \end{pmatrix} \qquad A \begin{pmatrix} 0.8 \\ 0.2 \end{pmatrix} \qquad A \begin{pmatrix} 0.1 \\ 0.9 \end{pmatrix}$$

▪ Figures A4.2a, A4.2b, A4.2c, A4.2d, A4.2e, A4.2f, A4.2g, A4.2h, A4.2i

So we get a unique SPNE (($P/x_1, P/x_3, A/x_5, A/x_7, A/x_9$), ($P/x_2, P/x_4, A/x_6, A/x_8, A/x_{10}$)) which corresponds to the unique pure strategy Nash equilibrium.

2. The backward induction procedure corresponds to one possible way of eliminating the dominated strategies, starting at the end of the game.

    Yet, as observed in chapter 2, this way of eliminating dominated strategies is not the most intuitive one. It is more intuitive to first iteratively eliminate strictly dominated strategies like ($A/x_1,./x_3,./x_5,./x_7,./x_9$), ($A/x_2,./x_4,./x_6,./x_8, A/x_{10}$), ($P/x_1, A/x_3,./x_5,./x_7,./x_9$) and ($P/x_2, A/x_4,./x_6,./x_8, A/x_{10}$). Backward induction doesn't first eliminate these strategies, just because they focus on the beginning of the game and not on the end! Clearly iterative elimination of dominated strategies doesn't follow an arrow of time, by contrast to backward induction.

367

ANSWERS

The Nash equilibrium concept also doesn't follow any arrow of time. It only focuses on unilateral stability. So, for example, it highlights that answering at x_i can be a best answer only if the opponent answers at x_{i+1}, regardless of i (at the beginning, in the middle or at the end of the game (i<10)). So the way to get the Nash equilibrium is quite different from a backward induction reasoning.

♣ EXERCISE 3 CENTIPEDE PRE-EMPTION GAME, BACKWARD INDUCTION AND THE STUDENTS' WAY OF PLAYING

1. By backward induction, player 1 plays S_1 at x_5 (Figure A4.3a), because 900>450. So player 2 plays S_2 (60>30) in the reduced game starting at x_4 (Figure A4.3b), then player 1 plays S_1 (30>15) in the reduced game starting at x_3 (Figure A4.3c), player 2 plays S_2 (2>1) in the reduced game starting at x_2 (Figure A4.3d), and player 1 plays S_1 (1>0.5) in the reduced game starting at x_1 (Figure A4.3e). So the game has a unique SPNE: ((S_1/x_1, S_1/x_3, S_1/x_5),(S_2/x_2, S_2/x_4)).

$$\underset{S_1}{\overset{1}{\underset{x_5}{|}}}\overset{C_1}{\longrightarrow}(450,1800) \quad \underset{S_2}{\overset{2}{\underset{x_4}{|}}}\overset{C_2}{\longrightarrow}(900,30) \quad \underset{S_1}{\overset{1}{\underset{x_3}{|}}}\overset{C_1}{\longrightarrow}(15,60) \quad \underset{S_2}{\overset{2}{\underset{x_2}{|}}}\overset{C_2}{\longrightarrow}(30,1) \quad \underset{S_1}{\overset{1}{\underset{x_1}{|}}}\overset{C_1}{\longrightarrow}(0.5,2)$$

$$\begin{pmatrix}900\\30\end{pmatrix} \quad \begin{pmatrix}15\\60\end{pmatrix} \quad \begin{pmatrix}30\\1\end{pmatrix} \quad \begin{pmatrix}0.5\\2\end{pmatrix} \quad \begin{pmatrix}1\\0\end{pmatrix}$$

■ Figures A4.3a, A4.3b, A4.3c, A4.3d, A4.3e

The backward induction reasoning here follows the same steps as the elimination of dominated strategies. And it leads to the same equilibrium path as the Nash equilibrium concept. But there are no mixed SPNE (by contrast to the mixed Nash equilibria).

2. First observe that the only SPNE of the game in Figure E4.3b is ((S_1/x_1, S_1/x_3, S_1/x_5, S_1/x_7, S_1/x_9), (S_2/x_2, S_2/x_4, S_2/x_6, S_2/x_8, S_2/x_{10})). Well, it is a pity to stop a game and get 0 or 5, when everybody can at least get 40. In reality, in a centipede game, everybody wishes to go on, in order to get a maximal amount of money: *if both players could sign a contract, they would decide to go to the end of the game and share equally the amount of money*. This is quite different from what happens in the ascending all pay auction/war of attrition game where *nobody is happy to reach the end of the game*. And this surely explains why my students play very differently in both games, despite backward induction leads to stopping both games at the first nodes.

My (122) L3 students are not an exception. I add in Table A4.1 the results I got with my (26) M1 students.

■ Table A4.1

Player 1		M1 students	Player 2		(M1 students)
Stops at x_1	11.5%	7.7%	Stops at x_2	13.1%	7.7%
Stops at x_3	3.3%	3.8%	Stops at x_4	6.6%	11.5%
Stops at x_5	9.8%	19.2%	Stops at x_6	11.5%	19.2%
Stops at x_7	10.7%	19.2%	Stops at x_8	13.1%	15.4%
Stops at x_9	29.5%	23.1%	Stops at x_{10}	39.3%	30.8%
Never stops	35.2%	27%	Never stops	16.4%	15.4%

We know that by backward induction, both players should always stop, so they should stop at x_1 and x_2. We may observe some backward induction reasoning in that much more students advise player 2 to stop at x_{10} than to go on (39.3% against 16.4% and 30.8% against 15.4%), and perhaps in that 29.5% of the L3 players advise player 1 to stop at x_9. Moreover, with regard to the L3 students, observe that if they hold a backward induction reasoning, they run it to the end: so, for example, 11.5% advise player 1 to stop immediately (at x_1) whereas only 3.3% advise him to stop at x_3. The reason for this difference seems in the same spirit as backward induction: if it is rational for player 1 to stop at x_3, then it is rational for player 2 to stop at x_2 and so it is better for player 1 to stop at x_1!

Let me also add a piece of information which is difficult to interpret: in fact, 82% of my L3 students and 96.2% of my M1 students advise player 1 and player 2 to play in a similar way, that is to say, if they advise player 1 to stop at x_i, they advise player 2 to stop at x_{i+1} or x_{i-1}. To put it more precisely, Table A4.2 gives the links between the behaviour for player 1 and player 2.

Table A4.2

Player 1	Player 2	L3 students	M1 students
Stops at x_i	Stops at x_{i+1}	42.6%	57.7%
Never stops	Never stops	9.85%	15.4%
Stops at x_i	Stops at x_{i-1}	9.85%	11.55%
Never stops	Stops at x_{10}	19.7%	11.55%
Different behaviour		18%	3.8%

If we suppose that the students thought at player 1 and player 2's behaviour simultaneously, we may say that Table A4.2 shows some backward induction behaviour: the fact that a player stops just before the other is in accordance with a partial backward induction reasoning, and the large percentages 42.6% and 57.7% show that the students think that player 1 has to stop before player 2, perhaps because player 2 may be the last player to play. Yet, it is possible that the students *didn't think of player 1 and player 2's behaviour simultaneously.* It may be that they just looked at *the payoffs*, which are *symmetric at adjacent nodes*, and this may explain similar advice for player 1 and player 2. So it is difficult to deduce something from Table A4.2, with regard to backward induction.

Anyway, even if there is some backward induction in their results, *backward induction is clearly not what drives the students' behaviour*: so, even if my M1 students go less far than the L3 students (a higher percentage stops at the nodes x_5 to x_8), it is obvious that most of the students do not stop at x_1, and more than half of them advise reaching end nodes (x_9, x_{10}, never stop). Almost 2/3 of the L3 students (1/2 of the M1 students) advise player 1 to only stop at x_9 or to never stop, and more than 45% of the L3 and M1 students advise player 2 to stop at x_{10} or never. So clearly, the students try to grab the high amounts of money and so go on, which is in complete opposition with the backward induction behaviour. Moreover, there is some altruism in the students' behaviour. As a matter of fact, player 2 is much worse off by going on at x_{10} than by stopping (40<<250), yet about 1/6 of the students advise him to always go on. It is as if player 2 should understand that the only motivation for player 1 to go on, especially at x_9, is that player 2 goes on at the next node, x_{10}, and so they compel him to go on, so that nobody is induced to start a harmful backward induction reasoning.

ANSWERS

This is in sharp contrast with the ascending all pay auction (which is a war of attrition game), where backward induction is not harmful, in that it leads everyone to not lose money, and where always going on is harmful for the players: this explains why the students advise the players more to stop at early nodes. So the nature of the game (the possibility for both to get a lot of money in the centipede game, the shrinking of the payoffs in the ascending all pay auction) influences the students' behaviour more than backward induction arguments.

♣♣ EXERCISE 4 HOW TO SHARE A SHRINKING PIE

1. The scheduled extensive form of the game is given in Figure A4.4.

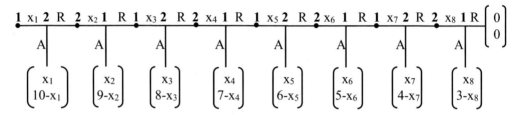

Figure A4.4

2. By backward induction, player 1, in the last subgames in period 8, accepts any offer x_8 (x_8 is either 1 or 2). So player 2, in the previous reduced games, chooses $x_8=1$ and gets 2. It follows that he refuses any offer $x_7>2$ in the previous reduced games in period 7, given that he gets $4-x_7$ by accepting, and 2 by waiting for period 8. It follows that player 1 offers $x_7=2$ in these reduced games (it is accepted by player 2 and player 1 prefers getting 2 than waiting for period 8 where she gets only 1).

So, by waiting for period 7, player 1 is able to get 2. It follows that she accepts $x_6 \geq 2$ but refuses $x_6=1$ in the previous reduced games in period 6. So player 2 offers $x_6=2$ in these reduced games (it is accepted by player 1 and player 2 prefers getting $5-x_6=3$ than waiting for period 7 where he gets 2).

So, by waiting for period 6, player 2 can get 3. It derives that he accepts $x_5 \leq 3$ ($6-x_5 \geq 3$) but refuses $x_5>3$ in the reduced games in period 5. So player 1 offers $x_5=3$ in these reduced games (it is accepted by player 2 and player 1 prefers getting 3 than waiting for period 6 where she gets 2).

So, by waiting for period 5, player 1 can get 3. It follows that she accepts $x_4 \geq 3$ but refuses $x_4<3$ in the previous reduced games in period 4. So player 2 offers $x_4=3$ in these reduced games (it is accepted by player 1 and player 2 prefers getting $7-x_4=4$ than waiting for period 5 where he gets 3).

So, by waiting for period 4, player 2 can get 4. It follows that he accepts $x_3 \leq 4$ ($8-x_3 \geq 4$) but refuses $x_3>4$ in the previous reduced games in period 3. So player 1 offers $x_3=4$ in these reduced games (it is accepted by player 2 and player 1 prefers getting 4 than waiting for period 4 where she gets 3).

So, by waiting for period 3, player 1 can get 4. It follows that she accepts $x_2 \geq 4$ but refuses $x_2<4$ in the previous reduced games in period 2. So player 2 offers $x_2=4$ in these reduced games (it is accepted by player 1 and player 2 prefers getting $9-x_4=5$ than waiting for period 3 where he gets 4).

So, by waiting for period 2, player 2 can get 5. It follows that he accepts $x_1 \leq 5$ ($10-x_1 \geq 5$) but refuses $x_1 > 5$ in the previous reduced games in period 1. So player 1 offers $x_1 = 5$ in these reduced games (it is accepted by player 2 and player 1 prefers getting 5 than waiting for period 2 where she gets 4).

It derives that the only SPNE path leads player 1 to share in a fair way the sum 10 in the first period, player 2 accepting this fair offer. Observe that, as expected, the negotiation ends immediately in period 1 (because the sum to share decreases over time). Observe also that the power of the last offering player is very loose, because the sum to share in period 8 is very low. *So the threat to wait for the last period is not convincing* – nobody wants to reach period 8 – which explains why player 2 gets exactly the same amount as player 1.

♣♣ EXERCISE 5 ENGLISH ALL PAY AUCTION

1. We prove that $E_1=((\pi_{1x1}(S_1)=1/3, \pi_{1x2}(S_1)=1/2, 3/x_3), (\pi_{2h1}(S_2)=1/3, \pi_{2h2}(S_2)=1/2, 3/h_3))$, where both players always overbid with positive probability, is a SPNE by backward induction. We know (see chapter 4, section 1), that $(3/x_3, 3/h_3)$ is the Nash equilibrium of the last subgame given in Figure A4.5a.

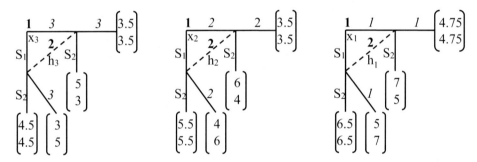

■ Figures A4.5a, A4.5b, A4.5c

In the reduced game starting at x_2 (Figure A4.5b) player 2 gets 4.75 (=5.5/2+4/2) by stopping and 4.75(=6/2+3.5/2) by bidding 2, so he can play both actions with probability ½. A same reasoning holds for player 1, so $(\pi_{1x2}(S_1)=1/2, \pi_{2h2}(S_2)=½)$ is a Nash equilibrium of the reduced game in Figure A4.5b.

In the reduced game starting at x_1 (Figure A4.5c), player 2 gets 5.5 (=6.5/3+5×2/3) by stopping and 5.5 (=7/3+4.75×2/3) by bidding 1. So he can stop and bid 1 with probability 1/3 and 2/3. A same reasoning holds for player 1, so $(\pi_{1x1}(S_1)=1/3, \pi_{2h1}(S_2)=1/3)$ is a Nash equilibrium of the reduced game in Figure A4.5c.

Hence E_1 is a SPNE.

To prove that $E_2=((S_1/x_1, \pi_{1x2}(S_1)=1/2, 3/x_3), (1/h_1, \pi_{2h2}(S_2)=1/2, 3/h_3))$ is a SPNE, it is enough to show that $(S_1/x_1, 1/h_1)$ is a Nash equilibrium of the reduced game in Figure A4.5c (we already know that $(3/x_3, 3/h_3)$ is a Nash equilibrium of the subgame in Figure A4.5a and that $(\pi_{1x2}(S_1)=1/2, \pi_{2h2}(S_2)=½)$ is a Nash equilibrium of the reduced game in Figure A4.5b): player 1 is best off stopping (5>4.75) and player 2 is best off bidding 1 (7>6.5). So E_2 is a SPNE. For symmetric reasons, $E_3=((1/x_1, \pi_{1x2}(S_1)=1/2, 3/x_3), (S_2/h_1, \pi_{2h2}(S_2)=1/2, 3/h_3))$ is also a SPNE.

ANSWERS

2. We now turn to the game in Figure A4.6a, where both players can bid 4 and 5, and we construct a SPNE in which both players always overbid with positive probability.

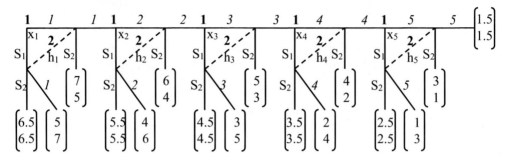

■ Figure A4.6a

In the last subgame (Figure A4.6b), the only Nash equilibrium leads both players to bid 5.

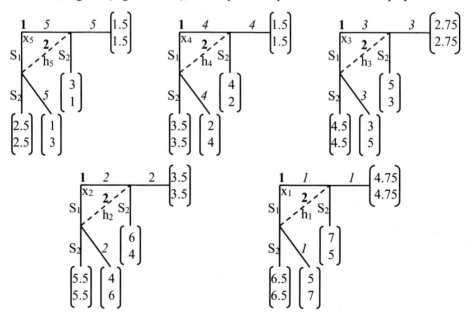

■ Figures A4.6b, A4.6c, A4.6d, A4.6e, A4.6f

In the reduced game in Figure A4.6c, we call p and 1–p, respectively q and 1–q, the probabilities assigned by player 1, respectively by player 2, to stopping and bidding 4. Player 2 can stop and bid 4 if $3.5p+2(1-p)=4p+1.5(1-p)$, i.e. $p=0.5$. By symmetry, player 1 can stop and bid 4 if $q=0.5$. It follows that $((\pi_{1x4}(S_1)=1/2, \pi_{2h4}(S_2)=½)$ is a Nash equilibrium of the reduced game in Figure A4.6c. Player 1 and player 2's expected payoff is $(3.5+2)/2=2.75$.

So we get the reduced game in Figure A4.6d. Player 1 stops and bids 3 with probability p and 1–p, player 2 stops and bids 3 with probability q and 1–q. Player 2 can stop and bid 3 if $4.5p+3(1-p)=5p+2.75(1-p)$, i.e. $p=1/3$. By symmetry, player 1 can stop and bid 3 if $q=1/3$. It follows that $((\pi_{1x3}(S_1)=1/3, \pi_{2h3}(S_2)=1/3)$ is a Nash equilibrium of the reduced game in Figure A4.6d. Player 1's and player 2's expected payoff is $(4.5+3\times2)/3=3.5$.

ANSWERS

It derives the reduced game in Figure A4.6e. Player 1 stops and bids 2 with probability p and 1–p, player 2 stops and bids 2 with probability q and 1–q. Player 2 can stop and bid 2 if 5.5p+4(1–p)=6p+3.5(1–p), i.e. p=1/2. By symmetry, player 1 can stop and bid 2 if q=0.5. So $((\pi_{1x2}(S_1)=1/2, \pi_{2h2}(S_2)=1/2)$ is a Nash equilibrium of the reduced game in Figure A4.6e, and player 1's and player 2's expected payoff is (5.5+4)/2=4.75.

It follows the reduced game in Figure A4.6f. Player 1 stops and bids 1 with probability p and 1–p, player 2 stops and bids 1 with probability q and 1–q. Player 2 can stop and bid 1 if 6.5p+5(1–p)=7p+4.75(1–p), i.e. p=1/3. By symmetry, player 1 can stop and bid 1 if q=1/3. So $((\pi_{1x1}(S_1)=1/3, \pi_{2h1}(S_2)=1/3)$ is a Nash equilibrium of the reduced game in Figure A4.6f, and player 1's and player 2's expected payoff is (6.5+5×2)/3=5.5.

So there exists a SPNE such that both players always go on with positive probability. It is defined by: $((\pi_{1x1}(S_1)=1/3, \pi_{1x2}(S_1)=1/2, \pi_{1x3}(S_1)=1/3, \pi_{1x4}(S_1)=½, 5/x_5), (\pi_{2h1}(S_2)=1/3, \pi_{2h2}(S_2)=1/2, \pi_{2h3}(S_2)=1/3, \pi_{2h4}(S_2)=½, 5/h_5))$.

Let us comment on the probabilities of this equilibrium.

First, strangely enough, the probabilities and payoffs in Figures A4.6e and A4.6f are the same as in Figures A4.5b and A4.5c. And the mean payoff of the SPNE, 5.5, is the same for the game in Figure E4.4 and the game in Figure A4.6a. *Being able to bid up to 3 or up to 5 does not change anything.*

Second, *we have no nice story to tell, like, the players overbid with decreasing (or increasing) probability over time.* As a matter of fact, both players overbid in period 1 with probability 2/3, with lower probability (½) in period 2, with higher probability (2/3) in period 3, with lower probability (1/2) in period 4, and finally with probability 1 in period 5. Clearly, the probabilities have no economic meaning: they do not express that the players have more or less difficulty to overbid when the games goes on. The probabilities are just here to stabilize the behaviour of the opponent, which is not very intuitive.

♣♣ EXERCISE 6 GENERAL SEQUENTIAL ALL PAY AUCTION, INVEST A MAX IF YOU CAN

1. Each non terminal node initializes a subgame, and so a reduced game.

   We first observe, regardless of the assumptions in questions 1, 2 and 3, that player 1 bids 5 at x_8 (3>2), player 2 stops at y_4(5>3), bids 5 at y_8(3>2), and bids 5 at y_7 (3>2) leading player 1 to stop at x_3(4>3>1). Player 1 stops at x_4 (4>3) and bids 5 at x_7(3>2), leading player 2 to stop at y_3(5>3>1). So we get Figure A4.7a (without the points). We clearly observe, at y_7, x_3 x_7 and y_3, that being able to bid 5 compels the opponent to stop at the previous node.

   Player 2 is indifferent between S_2 and 5 at y_5 (3=3>1) and at y_6 (3=3), so is supposed to bid 5 at y_5 and y_6, which compels player 1 to stop at x_2 (4>3>2>1), and encourages player 2 to overbid (by bidding 2) at y_1 (6>5=5>4>3). Player 1 is indifferent between stopping and bidding 5 at x_5 (3=3>1) and x_6 (3=3), so is supposed to bid 5, which compels player 2 to stop at y_2 (5>3>2>1). We observe, in accordance with our reasoning in chapter 4, that bidding 5 when being indifferent between stopping and bidding 5 compels the opponent to stop at the previous node (at x_2 and y_2). Finally player 1 bids 2 (M–V) at x_1 in accordance with our reasoning in chapter 4 (6>5=5>4=4>3), to avoid being trapped in the stopping action at x_2 (if she only bids 1 at x_1). The strategies are marked with points on Figure A4.7a.

ANSWERS

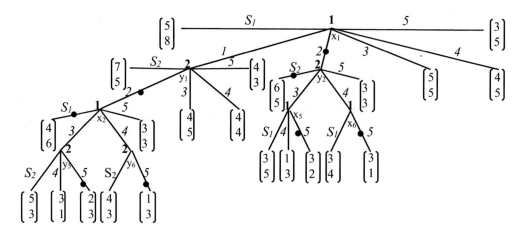

■ **Figure A4.7a**

2. We can start with Figure A4.7a (without the points). Player 2 is indifferent between S_2 and 5 at y_5 and at y_6. He is now supposed to stop, which encourages player 1 to overbid at x_2 (by bidding 3) (5>4=4>3), which dissuades player 2 from bidding 2 at y_1; he is indifferent between stopping and bidding 3 (5=5>4>3), so is supposed to stop at y_1. We immediately deduce that this encourages player 1 to bid 1 at x_1 because this low bid is enough to get the object (player 1 gets her highest possible payoff, 7, in the game). We also observe that player 1 is indifferent between stopping and bidding 5 at x_5 and x_6, is supposed to stop, so player 2 is indifferent between stopping and overbidding (by bidding 3) at y_2 (5=5>4>3), and stops by assumption. The strategies are marked with points on Figure A4.7b.

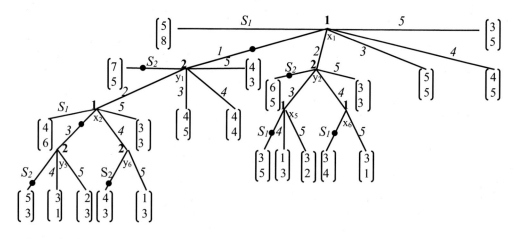

■ **Figure A4.7b**

Clearly a stopping player (in case of indifference) generally encourages the opponent to overbid in the previous round, which may in turn encourage the stopping player to stop before this round. So a stopping behaviour generally benefits the other player.

3. The above remarks are well illustrated in the case under study. Player 2 bids 5 at y_5 and y_6, given that he is indifferent between S_2 and 5, which compels player 1 to stop at x_2, and

encourages player 2 to overbid (by bidding 2) at y_1. Player 1 stops at x_5 and x_6 because she is indifferent between stopping and bidding 5, which encourages player 2 to overbid (by bidding 3) at y_2, because he is indifferent between stopping and bidding 3. So, given that, due to player 1's stopping behaviour, player 2 is encouraged to overbid at y_1 and at y_2, player 1, at x_1, stops the game, because she is indifferent between stopping and bidding 3. The strategies are marked with points on Figure A4.7c.

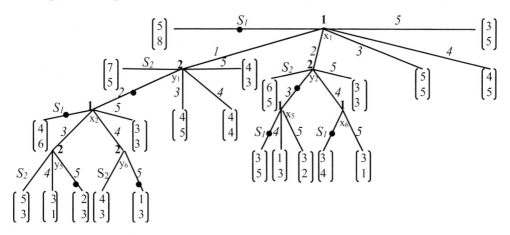

Figure A4.7c

4. We have found 3 SPNE paths. In the first, player 1 stops the game immediately and player 2 is best off. In the second, player 1 bids 1 and player 2 stops the game, in the third player 1 bids 2 and player 2 stops the game. There is one additional SPNE path, the one in which player 1 bids 3 and player 2 stops the game. There is no other SPNE path, given that player 1 can't bid 4 or 5 (she would lose money) and given that the losing player necessarily stops at his first decision node.

♣♣ EXERCISE 7 GIFT EXCHANGE GAME

1. Each wage leads to a subgame, in which the employee chooses his effort. Obviously the best effort is the lowest one in each of these subgames (i.e. after each wage), because 2>1>0, 4>3>2, and 5.5>4.5>3.5.

    By backward induction, given the employee's behaviour, the employer's reduced game is such that w_1 yields the payoff 2, w_2 the payoff 1 and w_3 the payoff 0, so she chooses the lowest wage w_1. So, as expected, the unique SPNE of this game fits with the unique Nash equilibrium path: w_1 followed by E_1 (see Exercise 9 of chapter 3).

2. The game is now more complicated, given that the employee plays first, promising an effort for each wage: given that he can announce three levels of effort for three levels of wage, his set of pure strategies contains 3×3×3=27 strategies; an example of strategy is: (E_3/w_1, E_1/w_2, E_2/w_3).

    Each strategy leads to a subgame for the employer (27 subgames), where she has to choose among three pure strategies w_1, w_2 or w_3. So she has 3^{27} strategies.

ANSWERS

To solve this game, first observe that the employee can never get 5.5 at equilibrium because the employer will never offer a wage w_3 for the lowest effort. But he can get the second highest payoff 4.5. Only one strategy allows to get it: $(E_1/w_1, E_1/w_2, E_2/w_3)$. As a matter of fact, on the one hand, the employer's best answer to this strategy is the wage w_3 as she gets 2.5 with w_3 and only 1 and 2 with the wages w_2 and w_1. On the other hand, she will never play w_3 (when w_3 is linked to E_2) if the employee offers more than the lowest effort for w_1 or w_2. It derives that the SPNE path is such as the employee promises $(E_1/w_1, E_1/w_2, E_2/w_3)$ and the employer offers w_3.

3. In this new game, the employer plays first, and has to promise a wage for each effort (hence she has 27 strategies). In front of these 27 strategies, the employee provides an effort and gets the associated wage (3^{27} strategies).

    The employer cannot get the payoff 6 because the employee will never offer the effort E_3 for the lowest wage. But the employer can get her second highest payoff 5. This is possible only if she plays the strategy $(w_1/E_1, w_1/E_2, w_2/E_3)$ (if she proposes more than w_1 for the efforts E_1 or E_2, the employee will not offer the effort E_3), and if the employer, who is indifferent between E_1 and E_3 in front of this strategy (because he gets 2 with E_1 and E_3 and 1 with E_2) plays E_3. So we have a first SPNE path where the employer plays $(w_1/E_1, w_1/E_2, w_2/E_3)$ followed by the effort E_3.

    But there are two other SPNE paths, in that the employee can offer the effort E_1 when faced with the strategy $(w_1/E_1, w_1/E_2, w_2/E_3)$. If so, the employer can only get her third highest payoff, 4, at equilibrium. And there are two ways to get it. She can play either $(w_1/E_1, w_1/E_2, w_3/E_3)$ or $(w_1/E_1, w_2/E_2, w_3/E_3)$: when faced with the first strategy, the employee compares the payoffs 2, 1, 3.5 and chooses E_3, faced with the second he compares the payoffs 2, 3 and 3.5 and again chooses E_3. So we get two additional SPNE paths. In the first, the employer offers $(w_1/E_1, w_1/E_2, w_3/E_3)$ and the employee offers the effort E_3. In the second, the employer offers $(w_1/E_1, w_2/E_2, w_3/E_3)$ and the employee again offers the effort E_3. Both paths rest on the fact that the employee offers the effort E_1 when faced with the strategy $(w_1/E_1, w_1/E_2, w_2/E_3)$.

    Let us observe, that, whether the employee is able to commit himself to an effort for a wage, or whether the employer is able to commit herself to a wage for an effort, the commitment can benefit both players. In the first case, the players get the payoffs (4.5, 2.5), which most benefit the employee; in the second they can get the payoffs (3.5, 4) which most benefit the employer. So the ability to commit oneself most benefits the player who is able to do so, but it can also benefit the other player: both (4.5, 2.5) and (3.5, 4) are better than (2, 2).

♣ EXERCISE 8 REPEATED GAMES AND STRICT DOMINANCE

1. The only pure strategy Nash equilibrium is (C_1, A_2). Given that A_1 is strictly dominated by C_1, the mixed strategy Nash equilibria are the same than the ones obtained in Matrix A4.1.

		Player 2 A_2	B_2
Player 1	B_1	(6,6)	(17,2)
	C_1	(8,1)	(16,1)

Matrix A4.1

So we get a family of mixed Nash equilibria $\{(C_1,q)$, with $q \geq 1/3$, q being the probability to play $A_2\}$.

2. The only way for player 2 to get 16 is to get 15 in period 1 and 1 in period 2. So the players should play (A_1,B_2) in the first period, and a Nash equilibrium in period 2.

The problem is that player 1 may deviate in period 1, in order to get 17 instead of 15. To avoid this deviation, we reward her in period 2 if she didn't deviate in period 1 and we punish her if she did. Given that the worst equilibrium for player 1 is (C_1,A_2) and the best is $(C_1,q=1/3)$, we punish her with (C_1,A_2) and reward her with $(C_1,q=1/3)$. So player 1 will not deviate from A_1 in period 1, because she gets $15+(8/3+32/3)=15+40/3>25$ if she does not deviate and only, at most, $17+8=25$ if she deviates.

A possible SPNE is the following one:

Period 1: the players play (A_1,B_2)

Period 2: if (A_1, B_2) has been played in period 1, then the players play $(C_1, q=1/3)$, otherwise they play (C_1, A_2).

Proof:
- All the subgames in period 2 are strategically equivalent to the stage game and $(C_1,q=1/3)$ and (C_1,A_2) are Nash equilibria of the stage game (hence of the subgames in period 2).
- The reduced game in period 1 is given in Figure A4.8.

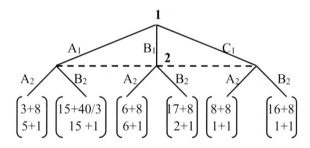

Figure A4.8

(A_1,B_2) is a Nash equilibrium of this reduced game (A_1 is the best answer to B_2 because $15+40/3>17+8>16+8$, and B_2 is the best answer to A_1 because $15+1>5+1$).

So the proposed profile is a SPNE.

3. A_1 is strictly dominated by both B_1 and C_1 in the stage game but it can be played in the repeated game in period 1. This is due to the fact that A_1 will not be followed, in period 2, by the same Nash equilibrium as B_1 and C_1; so A_1 can lead to a higher payoff than B_1 and C_1 in the first period reduced game. Of course, this remark generalizes to any game: *strictly dominated strategies in a stage game may be played in a repeated game context.*

This game has another particularity: the stage game has several equilibria, some of them rewarding player 1, some of them punishing her. But player 2 gets the same payoff 1 in all the equilibria. So there is no possibility of punishing or rewarding player 2. Yet this is not a

ANSWERS

problem in this game, given that player 2 has no interest in deviating from B_2 in period 1 if player 1 plays A_1 (15>5). If he deviates he punishes himself in period 1! There is no need to punish him again in period 2.

♣ EXERCISE 9 THREE REPETITIONS ARE BETTER THAN TWO

1. To get a SPNE in which both players get 10 in the first period, hence play (B_1,B_2), it is necessary to reward and to punish player 1 and player 2 because both can get more in period 1 by deviating from the action B.

    We can prevent player 1 from deviating to C_1 (to get 15 instead of 10), by rewarding her with the Nash equilibrium (A_1,A_2) in period 2 if she plays B_1 in period 1, and by punishing her with the equilibrium (C_1,C_2) if she doesn't. So she gets 10+7=17 by playing B_1 and only at best 15+1=16 by deviating.

    But we can't prevent player 2 from deviating to C_2 (to get 14 instead of 10). One can reward him with the best Nash equilibrium (A_1,A_2) in period 2 if he plays B_2 in period 1, and punish him with the worst Nash equilibrium, the mixed one, if he deviates. But this is not sufficient to avoid the deviation: by deviating to C_2, player 2 gets (at least) 14+0.8, whereas, by playing B_2, he only gets, at best, 10+4=14. The punishment is not strong enough – or the reward is not attractive enough – to avoid the deviation.

    To get a SPNE in which both players play B in period 1 we can replace the couple (7,4) by the couple (7,5). In that case, player 2 is no more induced to deviate because 10+5=15>14.8. We give a possible SPNE:

    > Period 1: the players play (B_1,B_2)
    >
    > Period 2: if (B_1, B_2) has been played in period 1, then the players play (A_1, A_2), otherwise they play the mixed Nash equilibrium with payoffs (1, 0.8)

    Proof
    - All the subgames in period 2 are strategically equivalent to the stage game, so Nash equilibria of the stage game are Nash equilibria of these subgames.
    - (B_1,B_2) is a Nash equilibrium of the reduced game in period 1. B_1 is the best answer to B_2 because 10+7>15+1>9+1 (because (B_1,B_2) is followed by (A_1,A_2) whereas (≠B_1,B_2) is followed by the mixed equilibrium), and B_2 is the best answer to B_1 because 10+5>14+0.8>0+0.8 (because (B_1,B_2) is followed by (A_1,A_2) whereas (B_1,≠B_2) is followed by the mixed equilibrium).
    - So the proposed profile is a SPNE.

2. When the game is played three times, we do not have to change the payoffs to induce both players to play B in the first period, because we can reward two times with the payoffs (7,4) and punish two times with the payoffs (1,0.8).

    We propose the following SPNE:

ANSWERS

> SPNE:
>
> Period 1: the players play (B_1, B_2)
>
> Period 2 and period 3: if (B_1, B_2) has been played in period 1, then the players play (A_1, A_2), otherwise they play the mixed Nash equilibrium with payoffs $(1, 0.8)$

Proof:
- Period 3: All the subgames in period 3 are strategically equivalent to the stage game, so Nash equilibria of the stage game are Nash equilibria of these subgames.
- Period 2: *Given that the behaviour in the subgames in period 3 is independent of the behaviour in period 2* (it only depends on the strategies in period 1), all the reduced games in period 2 are strategically equivalent to the stage game. So any Nash equilibrium of the stage game is a Nash equilibrium of any reduced game in period 2.
- Period 1: (B_1, B_2) is a Nash equilibrium of the reduced game in period 1: B_1 is the best answer to B_2 because $10+7\times2=24>15+1\times2=17>9+1\times2=11$ (because (B_1, B_2) is followed by (A_1, A_2) played twice, whereas $(\neq B_1, B_2)$ is followed by the mixed equilibrium played twice) and B_2 is the best answer to B_1 because $10+4\times2=18>14+0.8\times2=15.6>0+0.8\times2=1.6$ (because (B_1, B_2) is followed by (A_1, A_2) played twice, whereas $(B_1, \neq B_2)$ is followed by the mixed equilibrium played twice).

  In other words, by adding two times the Nash payoff 4 (reward), instead of two times the Nash payoff 0.8 (punishment), we more than outweigh the additional payoff due to the deviation $(14-10)+2\times0.8<2\times4$.

♣ EXERCISE 10 ALTERNATE REWARDS IN REPEATED GAMES

1. Both players are a priori incited to deviate from (B_1, A_2): player 1 prefers A_1 to B_1 (7>5) and player 2 prefers B_2 to A_2 (9>7). It is possible to reward and punish each player sufficiently to avoid his deviation, but the problem is that it is not possible to reward both players simultaneously.

   So, to avoid player 1's deviation, we can reward her in the second period with the Nash equilibrium (C_1, C_2) and punish her with the Nash equilibrium (A_1, B_2) (she gets $5+4=9$ by playing B_1 and at most $7+1=8$ by deviating). To avoid player 2's deviation, we can reward him with the Nash equilibrium (A_1, B_2) and punish him with the Nash equilibrium (C_1, C_2) (he gets $7+5=12$ by playing A_2 and at most $9+2=11$ by deviating).

   Yet what equilibrium is played in period 2 if nobody deviates in period 1?
   - If it is (A_1, B_2), player 1 does not play B_1 in period 1, because she gets $5+1=6$ by playing B_1, and at least $7+1$ by deviating to A_1 (because she can't get less than 1 in any Nash equilibrium of the stage game).
   - If it is (C_1, C_2), player 2 deviates to B_2 because he gets at least $9+5/3$ by doing so (he can't get less than $5/3$ at any Nash equilibrium of the stage game) and he gets only $7+2=9$ by not deviating.
   - If it is the mixed equilibrium, player 2 deviates because he gets $7+5/3$ by not deviating, and he gets at least $9+5/3$ by deviating to B_2 (player 1 is also incited to deviate).

ANSWERS

Hence it is not possible to play (B_1,A_2) in the first period because there does not exist a Nash equilibrium in the stage game that rewards everybody.

2. When a game is played three or more times, the inexistence of a Nash equilibrium in the stage game that rewards everybody may not be problematic, because it becomes possible to reward the players alternately.

   We proceed as follows: if player 1 deviates in period 1, we punish her two times with the equilibrium (A_1,B_2). If she doesn't deviate, we play (C_1,C_2) in period 2 and (A_1,B_2) in period 3. So, by not deviating she gets 5+4+1=10, and by deviating she only gets, at most, 7+1+1=9.

   If player 2 deviates in period 1, we punish him two times with the equilibrium (C_1,C_2). If he doesn't deviate, we play, as above, (C_1,C_2) in period 2 and (A_1,B_2) in period 3. So, by not deviating he gets 7+2+5=14, and by deviating he only gets, at most, 9+2+2=13.

   We propose the following SPNE:

   Period 1: the players play (B_1,A_2).

   Period 2 and period 3:

   ■ If the players played (B_1,A_2) in period 1 or if both deviated from this couple of actions, they play (C_1,C_2) in period 2 and (A_1,B_2) in period 3.

   ■ If only player 1 deviated from (B_1,A_2) in period 1, they play (A_1,B_2) in period 2 and period 3.

   ■ If only player 2 deviated from (B_1,A_2) in period 1, they play (C_1,C_2) in period 2 and period 3.

   Proof:
   ■ Period 3: All the subgames in period 3 are strategically equivalent to the stage game, so Nash equilibria of the stage game are Nash equilibria of these subgames.
   ■ Period 2: Given that the behaviour in the subgames in period 3 is independent of the behaviour in period 2, all the reduced games in period 2 are strategically equivalent to the stage game. Hence a Nash equilibrium of the stage game is a Nash equilibrium of any reduced game in period 2.
   ■ Period 1: (B_1,A_2) is a Nash equilibrium of the reduced game in period 1 (B_1 is the best answer to A_2 because 5+4+1=10>7+2×1=9>0+2×1=2 (because (B_1,A_2) is followed by (C_1,C_2) and (A_1,B_2), whereas $(\neq B_1,A_2)$ is followed by (A_1,B_2) played twice) and A_2 is the best answer to B_1 because 7+2+5=14>9+2×2=13>4+2×2=8 (because (B_1,A_2) is followed by (C_1,C_2) and (A_1,B_2) whereas $(B_1,\neq A_2)$ is followed by (C_1,C_2) played twice)).

♣♣ EXERCISE 11 SUBGAME PERFECTION IN RUBINSTEIN'S[1] FINITE BARGAINING GAME

For T=1, the game is an ultimatum game: player 1 gets the whole pie in the unique SPNE (because player 2 accepts any offer even if it goes to 0).

To solve the T period game, we exploit the structure of the game:

ANSWERS

First, in a T period game, a subgame that starts in period 2 is a T−1 period game, except that the players' roles are reversed (player 2 starts the game). Second, a SPNE in the T period game has to induce a Nash equilibrium in the T−1 period subgames, but also in the T−2 subgames and so on: so it has to induce a SPNE in the T−1 period subgames.

We first illustrate these remarks on a 2 period and on a 3 period game.

Case T=2

In a 2 period game, the second period subgames are the ultimatum game, player 2 being the player making the offer: he gets 1 and player 1 gets 0. So, if player 1 and player 2 fail to agree in period 1, they respectively get the discounted payoffs 0 and δ_2. It follows that, in period 1, they play the reduced game in Figure A4.9a, whose unique equilibrium is such that player 2 is indifferent between accepting and refusing player 1's offer, so $1-x=\delta_2$. So we get a unique SPNE path where player 1 offers (and gets) $1-\delta_2$ in period 1 and player 2 accepts this offer (and gets δ_2).

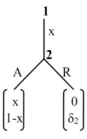

Figure A4.9a

Case T=3

The second period subgames are 2 period games, player 2 starting the games. Given our previous result, he offers δ_1 to player 1 who immediately accepts the offer. It follows that if player 1 and player 2 fail to agree in period 1, they get the (discounted) payoffs δ_1^2 and $(1-\delta_1)\delta_2$. So we get the period 1 reduced game in Figure A4.9b, whose unique equilibrium is such that player 2 is indifferent between accepting and refusing player 1's offer, so $1-x=\delta_2(1-\delta_1)$. We get a unique SPNE path where player 1 offers (and gets) $1-\delta_2(1-\delta_1)$ in period 1 and player 2 accepts this offer (and gets $\delta_2(1-\delta_1)$).

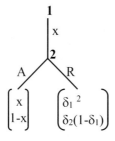

Figure A4.9b

ANSWERS

In other words, switching from a 2 period game to a 3 period game just amounts to solving an additional equality. The same procedure holds when switching from a T–1 period game to a T period game.

Case T, where T is any positive integer (higher than 1)

> In the T period game, the only SPNE leads player 1 to offer x_T at period 1, this offer being immediately accepted by player 2.
>
> $x_T = (1-\delta_2)((\delta_1\delta_2)^0 + (\delta_1\delta_2)^1 + (\delta_1\delta_2)^2 + \ldots + (\delta_1\delta_2)^{(T-1)/2 - 1}) + (\delta_1\delta_2)^{(T-1)/2}$
> for T odd, $T \geq 3$ (1)
>
> and $x_T = (1-\delta_2)((\delta_1\delta_2)^0 + (\delta_1\delta_2)^1 + (\delta_1\delta_2)^2 + \ldots + (\delta_1\delta_2)^{(T/2)-1})$
> for T even, $T \geq 2$ (2)
>
> $\lim_{T \to \infty} x_T = (1-\delta_2)/(1-\delta_1\delta_2)$
>
> Player 1 gets x_T and player 2 gets $1-x_T$.

Proof by recurrence
(2) holds for T=2 ($x_2 = 1-\delta_2$).
(1) holds for T=3 ($x_3 = 1-\delta_2+\delta_1\delta_2$)
We replace $(\delta_1\delta_2)^0$ by 1 in the proof. We suppose that the proof holds for T, and we show that it holds for T+1.

Case 1 T is even
By assumption, $x_T = (1-\delta_2)(1 + (\delta_1\delta_2)^1 + (\delta_1\delta_2)^2 + \ldots + (\delta_1\delta_2)^{(T/2)-1})$. Player 1 gets x_T and player 2 gets $1-x_T$ in period 1 in the T period game.

Consider the T+1 period game. The subgames that start in the second period are T period games, except that player 2 starts the game. So player 2 gets $y_T = (1-\delta_1)(1+(\delta_1\delta_2)^1 + (\delta_1\delta_2)^2 + \ldots + (\delta_1\delta_2)^{(T/2)-1})$ in these subgames. It follows that if player 1 and player 2 fail to agree in period 1, player 2 gets the discounted payoff $\delta_2 y_T$

$$= \delta_2(1-\delta_1)(1+(\delta_1\delta_2)^1 + (\delta_1\delta_2)^2 + \ldots + (\delta_1\delta_2)^{(T/2)-1})$$

It derives that player 1, in period 1, offers x_{T+1} such that

$$1-x_{T+1} = \delta_2(1-\delta_1)(1+(\delta_1\delta_2)^1 + (\delta_1\delta_2)^2 + \ldots + (\delta_1\delta_2)^{(T/2)-1})$$

So $x_{T+1} = (1-\delta_2)(1+(\delta_1\delta_2)^1 + (\delta_1\delta_2)^2 + \ldots + (\delta_1\delta_2)^{(T/2)-1}) + (\delta_1\delta_2)^{(T/2)}$ (which conforms with (1) because T+1 is odd)

Case 2 T is odd
By assumption, $x_T = (1-\delta_2)(1+(\delta_1\delta_2)^1 + (\delta_1\delta_2)^2 + \ldots + (\delta_1\delta_2)^{(T-1)/2 - 1}) + (\delta_1\delta_2)^{(T-1)/2}$. Player 1 gets x_T and player 2 gets $1-x_T$ in period 1 in the T period game.

Consider the T+1 period game. The subgames that start in the second period are T period games, except that player 2 starts the game. So player 2 gets $y_T = (1-\delta_1)(1+(\delta_1\delta_2)^1 + (\delta_1\delta_2)^2 + \ldots + (\delta_1\delta_2)^{(T-1)/2 - 1}) + (\delta_1\delta_2)^{(T-1)/2}$ in these subgames. It follows that if player 1 and player 2 fail to agree in period 1, player 2 gets the discounted payoff $\delta_2 y_T = \delta_2(1-\delta_1)((1+(\delta_1\delta_2)^1+(\delta_1\delta_2)^2+\ldots+(\delta_1\delta_2)^{(T-1)/2-1}) + \delta_2(\delta_1\delta_2)^{(T-1)/2}$

It derives that player 1, in period 1, offer x_{T+1} such that

$$1 - x_{T+1} = \delta_2(1-\delta_1)(1+(\delta_1\delta_2)^1+(\delta_1\delta_2)^2+\ldots+(\delta_1\delta_2)^{(T-1)/2-1}) + \delta_2(\delta_1\delta_2)^{(T-1)/2}$$

So $x_{T+1} = (1-\delta_2)((1+(\delta_1\delta_2)^1+(\delta_1\delta_2)^2+\ldots+(\delta_1\delta_2)^{(T+1)/2-1})$ (which conforms with (2) because T+1 is even)

♣♣ EXERCISE 12 SUBGAME PERFECTION IN RUBINSTEIN'S INFINITE BARGAINING GAME

> When T is infinite, player 1, in the unique SPNE of this game, offers in period 1 the share $x^* = (1-\delta_2)/(1-\delta_1\delta_2)$. Player 2 accepts the offer; hence player 1 gets x^* and player 2 gets $1-x^*$.

Observe that this SPNE is the limit of the T period game SPNE, when $T \to +\infty$. So, there is no discontinuity when switching from the finite to the infinite game.

The way to prove the result is worthy of interest. It rests on the fact that, in an infinite alternating bargaining game, a subgame that starts in period t+2 is exactly the same as a subgame starting in period t.

The proof also takes into account that both players will agree in period 1 because the pie is shrinking in size (given the discount factors).

So let us call **M the maximal payoff** player 1 can get in a subgame starting in period 3. This offer is immediately accepted by player 2, given that each SPNE leads to a first period offer which is immediately accepted. It derives that in period 2, player 2 offers at most $M\delta_1$ to player 1, the maximal payoff player 1 can expect by waiting for period 3. So player 2 gets at least the payoff $1-M\delta_1$ in period 2 and player 1 offers him at least this discounted payoff, $\delta_2(1-M\delta_1)$, in period 1 (and player 2 accepts it). So player 1 gets at best $1-\delta_2(1-M\delta_1)$ in period 1.

Yet, given that the game starting in period 3 is equivalent to the game starting in period 1, the maximal payoff M player 1 can achieve in a game starting in period 3, is also the maximal payoff she can achieve in a game starting in period 1. It follows that $M=1-\delta_2(1-M\delta_1)$, so $M=(1-\delta_2)/(1-\delta_1\delta_2)$.

We can hold the same reasoning with **m, the minimal payoff** player 1 can achieve in a game starting in period 3. This offer is immediately accepted by player 2, given that each SPNE leads to an immediately accepted first period offer. It derives that in period 2, player 2 offers at least $m\delta_1$ to player 1, the minimal payoff player 1 can expect by waiting for period 3. So player 2 gets at most the payoff $1-m\delta_1$ in period 2 and player 1 offers him at best this discounted payoff $\delta_2(1-m\delta_1)$ in period 1 (and player 2 accepts it). So player 1 gets at least $1-\delta_2(1-m\delta_1)$ in period 1. Given that the game starting in period 3 is equivalent to the game starting in period 1, the minimal payoff m player 1 can achieve in a game starting in period 3, is the same than the one she can achieve in a game starting in period 1, so $m=1-\delta_2(1-m\delta_1)$ i.e. $m=(1-\delta_2)/(1-\delta_1\delta_2)$.

ANSWERS

To conclude, given that M=m, player 1's share of the pie is necessarily $x^* = (1-\delta_2)/(1-\delta_1\delta_2)$, and player 2's share is $1-x^* = (\delta_2 - \delta_1\delta_2)/(1-\delta_1\delta_2)$.

♣♣♣ EXERCISE 13 GRADUALISM AND ENDOGENOUS OFFERS IN A BARGAINING GAME, LI'S INSIGHTS

1. We fix, w.l.o.g. i=1, j=2. Player 2 necessarily accepts player 1's clinching offer at period t because, if he accepts, both players get $z_{1t} + \max(\delta - z_{1t}, z_{2t}) \geq \delta$. If he refuses, both players are not allowed to get less in period t+1 than in period t, but this is not possible because the whole pie in period t+1 is at maximum worth δ in period t. So refusing leads to an impasse, which is not possible at equilibrium.

   And player 2 will not accept less than the clinching offer.

   If $\delta < z_{1t} + z_{2t} < 1$, then the clinching offer is z_{2t}, and player 2 cannot accept less by definition.

   If $z_{1t} + z_{2t} \leq \delta$, and $z_{2t} \leq c_1 \leq \delta - z_{1t}$, player 2 is better off refusing c_1 and proposing the clinching offer max $(\delta - c_1/\delta, z_{1t}/\delta)$ in the next period (we divide by δ, because the payoff x/δ in period t+1 is worth $\delta x/\delta = x$ in period t). As a matter of fact, by doing so, if $z_{1t}/\delta > \delta - c_1/\delta$, player 2 gets the (at period t discounted) payoff $\delta(1 - z_{1t}/\delta) = \delta - z_{1t} \geq c_1$. If $z_{1t}/\delta < \delta - c_1/\delta$, player 2 gets the (at period t discounted) payoff $\delta(1 - (\delta - c_1/\delta)) = \delta - \delta^2 + c_1 > c_1$.

2. Given than an impasse can't be rational, we now know that the game ends at a period t with a clinching offer.

   At period 1, $z_{11} = z_{21} = 0$ because nobody made any offer up to now. So, making the clinching offer at period 1 leads player 1 to offer δ and she gets $1-\delta$. If player 1 waits, hence makes no offer, and if player 2 makes the clinching offer at period 2, player 2 offers δ; if player 1 accepts, she gets δ^2 (because the payoff has to be discounted). Yet, for $\delta = 0.7$, we have $\delta^2 > 1-\delta$, so player 1 prefers waiting for player 2 making the clinching offer at period 2.

   More generally, when starting with z_{1t} and z_{2t}, with $z_{1t} + z_{2t} \leq \delta^2$, making the clinching offer at period t leads player 1 to offer $\delta - z_{1t}$. So she gets $1-(\delta-z_{1t})$. If player 1 waits, makes no offer, and player 2 makes the clinching offer at period t+1, player 2 offers max $(\delta - z_{2t}/\delta, z_{1t}/\delta) = \delta - z_{2t}/\delta$ (because $z_{1t} + z_{2t} \leq \delta^2$); if player 1 accepts, she gets $\delta(\delta - z_{2t}/\delta)$ (payoff discounted in period t). So player 1 refuses to make the clinching offer at period t if $1-(\delta - z_{1t}) \leq \delta(\delta - z_{2t}/\delta)$, i.e. if $z_{1t} + z_{2t} \leq \delta + \delta^2 - 1$, and accepts in the other case.

   This is true for any period and any player. For $\delta = 0.7$, $\delta + \delta^2 - 1 > 0$, so, given that $z_{11} = z_{21} = 0$, player 1 refuses to make the clinching offer in period 1, player 2 refuses to make the clinching offer in period 2, and so on if nobody makes an offer. Yet players have to come to an agreement, which suggests that they make offers which are not necessarily clinching offers.

   So suppose that player 1 makes an offer $c_1 < \delta$ in period 1, that player 2 refuses this offer, so that, at period 2, $z_{12} = 0$ and $z_{22} = c_1/\delta$. We know that player 2 accepts making the clinching offer, i.e. $\delta - c_1/\delta$, in period 2, only if $z_{12} + z_{22} = c_1/\delta \geq \delta + \delta^2 - 1$. In that case, player 1 gets $\delta - c_1/\delta$ in period 2 (and her highest payoff is obtained for $c_{1*} = \delta^2 + \delta^3 - \delta$, and it is equal to $1-\delta^2$ (in period 2)).

   Now suppose that player 1 waits in period 1, player 2 makes in period 2 a rejected offer c_2 ($\leq \delta^2$) and player 1 makes the accepted clinching offer $(\delta - c_2/\delta)$ in period 3. So she gets $\delta(1-(\delta-c_2/\delta))$ (payoff discounted in period 2). And we know that she accepts making this offer if $c_2/\delta \geq \delta + \delta^2 - 1$. Hence $c_{2*} = \delta^2 + \delta^3 - \delta$ (because it is better for a player to make the lowest offer), and player 1 gets $\delta(1-(\delta - c_{2*}/\delta)) = \delta^3$.

It follows that player 1 prefers making the refused offer c_{1*} in period 1 (followed by the clinching offer by player 2 in period 2), rather than waiting for and refusing player 2's offer c_{2*} in period 2 (followed by her clinching offer in period 3) if $\delta-c_{1*}/\delta \geq \delta(1-(\delta-c_{2*}/\delta))$, i.e. $\delta^3+\delta^2 \leq 1$. This is indeed the case for $\delta=0.7$

So player 1, by making the refused offer c_1^* ($=\delta^2+\delta^3-\delta=0.133$) in period 1, followed by the clinching offer, $\delta-c_{1*}/\delta=1-\delta^2=0.51$ in period 2, gets $1-\delta^2$ (in period 2), i.e. $\delta-\delta^3=0.357$ in period 1. And player 2 gets $1-(\delta-c_{1*}/\delta)=\delta^2$ in period 2, i.e. $\delta^3=0.343$ in period 1.

Finally, for $\delta=0.7$, both players are better off agreeing before period 4, because the whole pie in period 4 is worth $1\delta^3=0.343$ in period 1.

Let us summarize.

Player 1 doesn't make the clinching offer in period 1 because $0<\delta+\delta^2-1$. But she makes the refused offer $c_1=\delta^2+\delta^3-\delta=0.133$ in period 1, so that player 2 makes the clinching offer $\delta-c_1/\delta=0.51$ in period 2, this offer being accepted by player 1. These are the SPNE contributions. The transaction happens in period 2, player 1 gets the payoff 0.357 and player 2 gets the payoff 0.343 (0.357 and 0.343 are the payoffs discounted in period 1).

We clearly observe a gradualism, in that player 1 starts by offering 0.133 in period 1, this offer being refused, and then, in period 2, player 2 offers 0.51, this offer being accepted. It follows that the transaction happens only in period 2. We also observe a loss of efficiency, given that the sum of the (in period 1 discounted) payoffs is only 0.7.

Let us stress the crucial role of the offer c_{1*} by player 1 in period 1: this offer allows player 2 to make a lower clinching offer $\delta-c_{1*}/\delta$ (instead of δ) in period 2, so he can keep a larger share of the pie: as a consequence, it encourages player 2 to make the clinching offer.

♣ ♣ EXERCISE 14 INFINITE REPETITION OF THE TRAVELLER'S DILEMMA GAME

STUDY 1: tough or moderate reprisals

Easy state transition diagrams that lead to the expected behaviour are given in Figure 4A.10a and A4.10b.

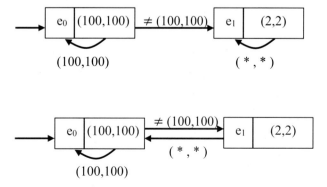

Figure A4.10a and A4.10b * means any action. This convention holds for all the Figures A4.10.

ANSWERS

According to the strategy profiles in Figures A4.10a and A4.10b, the travellers start by playing (100,100) and then play (100,100) as long as this couple has been played in the past. So, if nobody deviates, they will always ask for and get 100.

But, if somebody deviates, they switch to the Nash equilibrium (2,2) forever in Figure A4.10a. In Figure A4.10b, they switch to the Nash equilibrium but then come back to the state e_0, where they stay as long as everybody plays 100.

Clearly, the strategy profile in Figure A4.10a leads to tough reprisals if somebody doesn't play 100, given that this deviation implies a payoff 2 for each player in any further period. By contrast, the reprisals in Figure A4.10b are very moderate. A deviation leads both players to only play one time the Nash equilibrium. Then they come back to the couple (100,100). So, logically, the conditions on the discount factors will be stronger in Figure A4.10b than in Figure A4.10a, because a deviation is less punished in Figure A4.10b than in Figure A4.10a.

So consider player 1 in a state e_0 in Figure A4.10a. If she plays 100, both players stay in e_0 forever and she gets $100+100\delta_1/(1-\delta_1)$. If she asks for another amount, the players, in the next period, switch to e_1 and stay there forever, so she gets at best $99+P+2\delta_1/(1-\delta_1)$

So she is better off not deviating if $P-1 \leq 98\delta_1/(1-\delta_1)$, i.e. $\delta_1 \geq (P-1)/(97+P)$

By symmetry, the same condition holds for player 2's discount factor.

Nobody is induced to deviate in e_1, given that the future behaviour does not depend on the present behaviour in e_1 and given that (2,2) is a Nash equilibrium of the stage game.

So the profile in Figure A4.10a is a SPNE if $\delta_i \geq (P-1)/(97+P)$, i=1, 2.

This condition is very weak: even for high values of P, which a priori incite more deviating, we get very weak conditions on the discount factors (for P=50, $\delta \geq 49/147 = 1/3$, for P=96, $\delta \geq 95/193 = 0.492$).

Now consider player 1 in a state e_0 in Figure A4.10b. If she plays 100, both players stay in e_0 forever and she gets $100+100\delta_1+100\delta_1^2/(1-\delta_1)$. If she asks for another amount,, the players, next time, switch to e_1, but then come back to e_0 forever. So she gets at best $99+P+2\delta_1+100\delta_1^2/(1-\delta_1)$.

She is better off not deviating if $100+100\delta_1 \geq 99+P+2\delta_1$, i.e. $\delta_1 \geq (P-1)/98$

By symmetry the same condition holds for player 2's discount factor. For the same reasons than in Figure A4.10a, nobody is induced to deviate in e_1.

So the profile in Figure A4.10b is a SPNE if $\delta_i \geq (P-1)/98$, i=1,2.

Clearly, the condition $\delta \geq (P-1)/98$ is stronger than the condition $\delta \geq (P-1)/(97+P)$. But it is still rather soft for not too large values of P ($\delta \geq 49/98 = 0.5$ for P=50).

STUDY 2: the folk theorem strategy profile

ANSWERS

The minmax value is 2, each player getting at least 2 by playing 2, and being able to limit to 2 the opponent's payoff by playing 2. So playing the Nash equilibrium amounts to giving the minmax values to the players. We introduce the couples U(1)=(99–P,99+P) and U(2)=(99+P,99–P) which both belong to the set of strictly individually rational payoffs, and check: $2<U_1(1)=99-P<100$, $2<U_2(2)=99-P<100$, $U_2(1)=99+P>U_2(2)=99-P$ and $U_1(2)=99+P>U_1(1)=99-P$.

In the spirit of the folk theorem, after player 1's (player 2's) deviation, we switch to the couple of minmax values (2,2) for a while, before playing (100, 99) ((99,100)) forever. The easiest state transition diagram of this kind is given in Figure A4.10c.

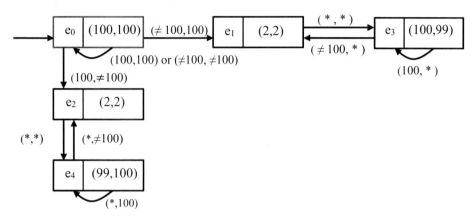

Figure A4.10c

But there is a problem with the state e_3: to see why, look at the evolution Table A4.3a for player 1:

Table A4.3a

	t	t+1	t+2
Action 100	e_3	e_3	e_3
	(100,99)	(100,99)	(100,99)
payoff	99–P	$(99-P)\delta_1$	$(99-P)\delta_1^2$
Action≠100	e_3	e_1	e_3
	(≠100,99)	(2,2)	(100,99)
Payoff	at best 98+P	$2\delta_1$	$(99-P)\delta_1^2$

Player 1 doesn't deviate if $99-P+(99-P)\delta_1 \geq 98+P+2\delta_1$ i.e. if $\delta_1 \geq (2P-1)/(97-P)$.

Yet his condition can't be fulfilled if P>98/3.

And this is quite logical: 99–P is a very bad payoff for high values of P. When P is large, a player prefers getting one time 98+P, even if he gets 2 the next time, rather than getting two times 99–P.

A way to bypass this problem is to introduce more states, say K states, in which the players play (2,2), so that player 1, if she deviates in e_3, suffers more from the difference (99–P) –2. Yet you immediately observe that this difference is small when P is large. Moreover, if P is large, the initial payoff induced by deviating, 98+P, is very large in comparison to the equilibrium payoff 99–P. In other words, for P=95 for example, by deviating to 98 in e_3, player 1 gets 193 instead of 99–95=4 by not deviating. And the payoff 193 will be followed by K payoffs 2, whereas the non deviating

ANSWERS

payoff 4 will be followed by K payoffs 4. So, to avoid the deviation, even for a discount factor equal to 1, we need: 193+K2≤4+4K, i.e. K≥95!

More generally, with a discount factor δ and k states e_{1i}, i from 1 to k, we get the new evolution Table A4.3b for player 1:

■ Table A4.3b

	t	t+1	...	t+k	t+k+1
Action 100	e_3 (100,99)	e_3 (100,99)	...	e_3 (100,99)	e_3 (100,99)
Payoff	99–P	(99–P)δ_1		(99–P)δ_1^k	(99–P)δ_1^{k+1}
Action≠100	e_3 (≠100,99)	e_{11} (2,2)	...	e_{1k} (2,2)	e_3 (100,99)
Payoff	at best 98+P	2δ_1		2δ_1^k	(99–P)δ_1^{k+1}

Player 1 doesn't deviate if

$$99-P+(99-P)(\delta_1+\delta_1^2+\ldots+\delta_1^k) \geq 98+P+2(\delta_1+\delta_1^2+\ldots \delta_1^k)$$

i.e. if $\delta_1+\delta_1^2+\ldots+\delta_1^k \geq (2P-1)/(97-P)$, which can only be fulfilled if $k \geq (2P-1)/(97-P)$. For P=95, we get k≥95, and for P=96 we get k≥191!

So it is possible to construct a folk theorem state transition diagram for any value of P, but with very many states leading to the play of the couple (2,2) when P is large.

Let us restrict P to lower values, P≤50 for example. For P≤50, 3 states e_1 are sufficient (we need k≥(2P–1)/(97–P) (i.e. higher than 99/47)) to prevent player 1 from deviating in e_3. More exactly, player 1 doesn't deviate if $\delta_1+\delta_1^2+\delta_1^3 \geq (2P-1)/(97-P)$. For example, for P=10, we need $\delta \geq 0.19$, for P=50 we need $\delta \geq 0.84$.

We now check, for P≤50, that the strategy profile in the state transition diagram in Figure A4.10d is a SPNE for sufficiently large discount factors.

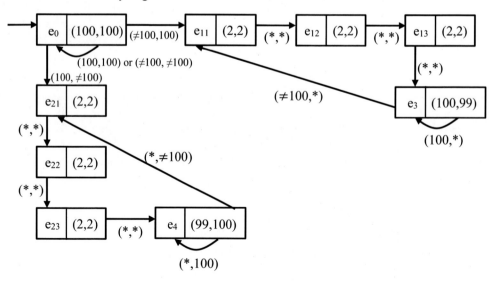

■ Figure A4.10d

ANSWERS

In state e_0, we get for player 1 the evolution Table A4.3c:

Table A4.3c

	t	t+1	t+2	t+3	t+4
Action 100	e_0 (100,100)	e_0 (100,100)	e_0 (100,100)	e_0 (100,100)	e_0 (100,100)
Payoff	100	$100\delta_1$	$100\delta_1^2$	$100\delta_1^3$	$100\delta_1^4$
Action≠100	e_0 (≠100,100)	e_{11} (2,2)	e_{12} (2,2)	e_{13} (2,2)	e_3 (100,99)
Payoff	at best 99+P	$2\delta_1$	$2\delta_1^2$	$2\delta_1^3$	$(99-P)\delta_1^4$

Player 1 doesn't deviate if

$$100+100\delta_1+100\delta_1^2+100\delta_1^3+100\delta_1^4/(1-\delta_1) \geq 99+P+2\delta_1+2\delta_1^2+2\delta_1^3+(99-P)\delta_1^4/(1-\delta_1)$$

i.e. $(\delta_1+\delta_1^2+\delta_1^3) \geq (P-1)/98-(1+P)\delta_1^4/(1-\delta_1)98$

This condition is easier to fulfil than the previous condition $\delta_1+\delta_1^2+\delta_1^3 \geq (2P-1)/(97-P)$ because $(2P-1)/(97-P)>(P-1)/98>(P-1)/98-(1+P)\delta_1^4/98(1-\delta_1)$ for any P and any δ_1.

By symmetry, the same condition holds for player 2's discount factor.

In $e_{11}, e_{12}, e_{13}, e_{21}, e_{22}$ and e_{23} nobody is better off deviating because the future behaviour does not depend on the present behaviour, and because (2,2) is a Nash equilibrium of the stage game. It remains to check that player 2 does not deviate in e_3. This is obvious for any discount factor, given that the future behaviour does not depend on player 2's present behaviour and given that 99 is the best response to 100 in the stage game.

The state e_4 is symmetric to the state e_3.

> It derives that the strategy profile in Figure A4.10d is a SPNE, provided that $\delta_i+\delta_i^2+\delta_i^3 \geq (2P-1)/(97-P)$, i=1,2.

Observe that we do not in Figure A4.10d systematically punish a player when he deviates from the expected behaviour, by contrast to the folk theorem. This is due to the fact that in the game under study, many deviations are not fruitful for the deviator (so he punishes himself by deviating).

> STUDY 3: No punishment for the player who does not deviate

The couple (2,2) punishes both the deviator and the player who doesn't deviate from the couple (100,100). The idea is now to find strategies that do not punish the latter, and that allow him to recover at least part of the loss due to the deviation.

The construction of the profile will be progressive in order to highlight the difficulties it gives rise to. We first propose the easy state transition diagram in Figure A4.10e:

ANSWERS

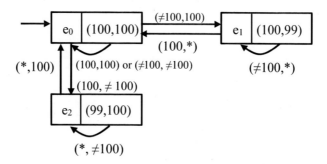

■ **Figure A4.10e**

The proposed strategy profile has nice properties. If each player plays what he is expected to, both players get 100 forever. If one player tries to get more, by switching to 99 for example, in order to get 99+P, he is punished in the next state where he only gets 99–P. What is more, the other player, who got 99–P during the deviation, gets 99+P in the punishing state. So both players are even, the deviator gets 99+P and then 99–P, the non deviating one gets 99–P and then 99+P. Moreover, once even, both players come back to (100,100).

Yet there is a problem: player 1 deviates in e_1. By deviating to 98, she first gets 98+P, then she is punished with a payoff 99–P, before turning back to the payoff 100. If she does not deviate, she gets 99–P and turns back to the payoff 100. So she doesn't deviate if: $98+P+(99-P)\delta_1 \leq 99-P+100\delta_1$, i.e. $\delta_1 \geq (2P-1)/(1+P)$, which is impossible as soon as P is higher than 2.

Adding a second state e_1 would not help (the condition would become stronger). A good idea is simply to add a Nash state, e_3, as in the state transition diagram in Figure A4.10f.

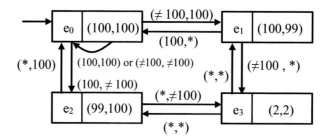

■ **Figure A4.10f**

Let us show that the strategy profile in this diagram is a SPNE, but requires strong conditions on the discount factors.

We first study the evolution Table A4.3d for player 1 in e_0.

ANSWERS

Table A4.3d

	t	t+1	t+2
Action 100	e_0 (100,100)	e_0 (100,100)	e_0 (100,100)
Payoff	100	$100\,\delta_1$	$100\delta_1^2$
Action≠100	e_0 (≠100,100)	e_1 (100,99)	e_0 (100,100)
Payoff	At best 99+P	$(99-P)\delta_1$	$100\,\delta_1^2$

Player 1 doesn't deviate if $100+100\delta_1 \geq 99+P+(99-P)\delta_1$ i.e. $\delta_1 \geq (P-1)/(P+1)$.

By symmetry, the same condition holds for player 2's discount factor.

We now study for player 1 the evolution Table A4.3e in e_1:

Table A4.3e

	t	t+1	t+2	t+3
Action 100	e_1 (100,99)	e_0 (100,100)	e_0 (100,100)	e_0 (100,100)
Payoff	99–P	$100\,\delta_1$	$100\delta_1^2$	$100\delta_1^3$
Action≠100	e_1 (≠100,99)	e_2 (2,2)	e_1 (100,99)	e_0 (100,100)
Payoff	At best 98+P	$2\delta_1$	$(99-P)\,\delta_1^2$	$100\,\delta_1^3$

Player 1 doesn't deviate in e_1 if $99-P+100\delta_1+100\delta_1^2 \geq 98+P+2\delta_1+(99-P)\delta_1^2$, i.e.

$$(1+P)\delta_1^2 + 98\delta_1 - (2P-1) \geq 0, \text{ i.e. } \delta_1 \geq (-49+(49^2+(2P-1)(1+P))^{0.5})/(1+P)$$

This condition can always be fulfilled (for P=50, $\delta_1 \geq 0.74$, for P=90, $\delta_1 \geq 0.97$).

Player 2 is not induced to deviate in e_1 because the future behaviour does not depend on his present behaviour and because 99 is the best response to 100 in the stage game.

e_2 is symmetric to e_1. In e_3, nobody deviates because the future behaviour does not depend on the present behaviour in e_3 and because (2,2) is a Nash equilibrium of the stage game.

> It derives that the strategy profile in Figure A4.10f is a SPNE provided that $\delta_i \geq (P-1)/(P+1)$, and $\delta_i \geq (-49+(49^2+(2P-1)(1+P))^{0.5})/(1+P)$, i=1,2.

Yet $\delta \geq (P-1)/(P+1)$ is a very strong condition, even for low values of P; this is mainly due to the fact that the deviator is not much punished. So it may be interesting to add a second punishing state, as in the state diagram in Figure A4.10g.

ANSWERS

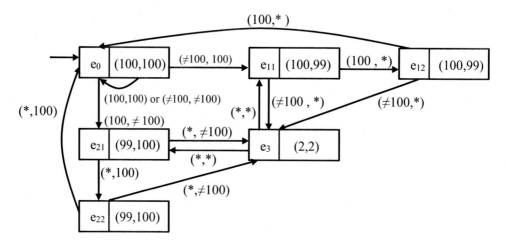

Figure A4.10g

In Figure A4.10g, player 1 will not deviate in e_0 if : $100+100\delta_1+100\delta_1^2 \geq 99+P+(99-P)\delta_1+(99-P)\delta_1^2$ i.e. $(1+P)(\delta_1+\delta_1^2) \geq P-1$, i.e. $\delta_1 \geq [-1+((5P-3)/(P+1))^{0.5}]/2$.

This condition is much easier to fulfil than the condition $\delta_1 \geq (P-1)/(P+1)$: we now get, for P=10, $\delta_1 \geq 0.54$, for P=50, $\delta_1 \geq 0.61$, for P=96, $\delta_1 \geq 0.61$.

By symmetry, the same condition holds for player 2's discount factor.

Player 1 will not deviate in e_{11} if $99-P+(99-P)\delta_1+100\delta_1^2+100\delta_1^3 \geq 98+P+2\delta_1+(99-P)\delta_1^2+(99-P)\delta_1^3$, i.e. $(1+P)(\delta_1^2+\delta_1^3) \geq (2P-1)-(97-P)\delta_1$

This condition can always be fulfilled but, unfortunately, it is easier to deviate in e_{11} than it was in e_1 in the previous state diagram (for P=50, we get $\delta_1 \geq 0.82$, for P=90, $\delta_1 \geq 0.98$). This is due to the fact that the good payoff 100 comes back later, so that it is more interesting to try to benefit from the deviating payoff 98+P.

At last, player 1 is less induced to deviate in e_{12} than in e_{11} (the condition is $99-P+100\delta_1+100\delta_1^2+100\delta_1^3 \geq 98+P+2\delta_1+(99-P)\delta_1^2+(99-P)\delta_1^3$), player 2 is never induced to deviate in e_{11} and e_{12}, and nobody is induced to deviate in e_3.

And e_{21} and e_{22} are symmetric to e_{11} and e_{12}.

So it derives that the strategy profile in Figure A4.10g is a SPNE provided that $\delta_i \geq [-1+((5P-3)/(P+1))^{0.5}]/2$ and $(1+P)(\delta_i^2+\delta_i^3) \geq (2P-1)-(97-P)\delta_i$, i=1,2.

Comments: With the strategy profile in Figure A4.10g, a player more than recovers the loss due to a deviation from the other player. For example, if player 1 deviates to 99 in e_0, player 2 successively gets 99–P, 99+P, 99+P before turning back to 100. So this profile much benefits the player who doesn't deviate from (100, 100).

We observe that the more P is small, respectively large, the more the conditions on the discount factors are easy, respectively difficult, to fulfil. Clearly, it is more easy to avoid the deviation from (100,100) when P is small. So it is not astonishing that in real life, people spontaneously adopt a different behaviour when facing small and large values of P.

ANSWERS

♣ EXERCISE 15 INFINITE REPETITION OF THE GIFT EXCHANGE GAME

To get the state transition diagram, we make some observations:

- Given that the employer plays before the employee, it is not difficult to punish the employer if she deviates from the expected wage w_3: it is enough for the employee to offer the lowest effort.
- If, after the offer w_3, the employee deviates by offering the lowest effort E_1, it is not sufficient to punish him by playing one time the couple (w_1, E_1) before coming back to (w_3, E_3).

As a matter of fact, if the couple (w_1, E_1) is only played one time after his deviation, the employee gets, by deviating, $5.5 + 2\delta_2 + 3.5\delta_2^2$, which is higher than the payoff obtained by not deviating, $3.5 + 3.5\delta_2 + 3.5\delta_2^2$ regardless of the employee's discount factor δ_2. So at least two periods of punishment are necessary.

That is why we propose the following state transition diagram (Figure A4.11):

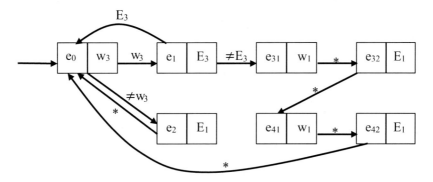

Figure A4.11 * means any action

Let us comment on this diagram. Each state corresponds to one player because the game is a perfect information game. If both players play the expected actions, they switch from e_0 to e_1 and vice versa forever, which means that they always play the couple (w_3, E_3). And if the employer deviates in e_0, we only punish her by the response E_1: that is enough to deter her from deviating, because the employer prefers offering w_3 followed by E_3 (she gets 4), rather than offering a lower wage followed by a low effort (she gets 2 with w_1 and 1 with w_2). If the employee deviates in e_1, we just punish him two times with the couple (w_1, E_1) before coming back to (w_3, E_3). This impedes any deviation for sufficiently large discount factors.

Let us prove that the strategy profile in the state transition diagram in Figure A4.11 is a SPNE. It is obvious that the employer is not incited to deviate in e_0, regardless of her discount factor. We now study the employee's behaviour in e_1: we get the evolution Table A4.4.

ANSWERS

Table A4.4

	t	t+1	t+1	t+2	t+2	t+3	t+3
Effort E_3	e_1 E_3	e_0 w_3	e_1 E_3	e_0 w_3	e_1 E_3	e_0 w_3	e_1 E_3
Payoff	3.5		$3.5\delta_2$		$3.5\delta_2^2$		$3.5\delta_2^3$
Effort$\neq E_3$	e_1 $\neq E_3$	e_{31} w_1	e_{32} E_1	e_{41} w_1	e_{42} E_1	e_0 w_3	e_1 E_3
Payoff	At best 5.5		$2\delta_2$		$2\delta_2^2$		$3.5\delta_2^3$

The employee doesn't deviate from the expected effort E_3 if

$$3.5+3.5\delta_2+3.5\delta_2^2 \geq 5.5+2\delta_2+2\delta_2^2, \text{ i.e. } \delta_2 \geq 0.76.$$

The employer is not incited to deviate in e_{31}, because the future actions do not depend on her offered wage, and each wage is followed by a low effort. A similar remark holds for e_{41}.

And the employee doesn't deviate in e_{32} given that the future actions do not depend on his present behaviour and given that E_1 is her best response to any wage in the stage game. A similar remark holds for e_{42} and e_2.

> So the proposed profile is a SPNE, provided that the employee values the future enough ($\delta_2 \geq 0.76$), in order to prefer getting 3.5 with the highest effort, and thereafter two times 3.5, rather than 5.5 with the lowest effort and thereafter two times the payoff 2.

Comments

If δ_2 doesn't fit the condition $\delta_2 \geq 0.76$, we can add a third punishment period to get a weaker condition on δ_2 (with two additional states e_{51} and e_{52} that lead to the couple of payoffs (2,2)). So it is really not difficult, in a context of infinite repetition, to lead both the employer and the employee to offer high salaries and high efforts.

In real life, employers and employees are usually in a repeated game context. The real game is not infinitely repeated, but the number of repetitions is usually unknown by the players. And it has been shown that small uncertainties on the number of repetitions can lead to the same strategies as an infinite repetition. So it is not astonishing that in real life, high wages and high effort levels are much more frequent than in the (stage game) gift exchange game.

NOTES

1. Rubinstein, A., 1982. Perfect equilibrium in a bargaining model, *Econometrica*, 50, 97–110.

Chapter 5

Answers to Exercises

♣ EXERCISE 1 PERFECT AND PROPER EQUILIBRIUM

1. Player 3's action A_3 at the subgame starting at x_4 is optimal (1=1>0).

   (A_1, A_2) is a Nash equilibrium of the reduced game where the node x_4 is replaced by the payoffs (3,3,1), because A_1 is the best response to A_2 (3>0), and A_2 is a best response to A_1 (3≥3). Hence $E=(A_1,A_2,A_3)$ is a SPNE.

   We call ε_1^k the minimal probability assigned to B_1, ε_2^k the minimal probability assigned to B_2, ε_3^k the minimal probability assigned to B_3, e_3^k the minimal probability assigned to C_3, and we fix the profile π^k as follows:

   $\pi_1^k(A_1)=1-\varepsilon_1^k$, $\pi_1^k(B_1)=\varepsilon_1^k$, $\pi_2^k(A_2)=1-\varepsilon_2^k$, $\pi_2^k(B_2)=\varepsilon_2^k$, $\pi_3^k(A_3)=1-\varepsilon_3^k-e_3^k$, $\pi_3^k(B_3)=\varepsilon_3^k$, $\pi_3^k(C_3)=e_3^k$.

   $\pi^k \to \pi^*$ where π^* is the profile of behavioural strategies corresponding to E (we often say $\pi^k \to$ E) when $k \to \infty$, because the minimal probabilities go to 0 when k goes to ∞.

   We now prove that π^k is a Nash equilibrium of the perturbed game.

Player 3
A_3 leads to the payoff $(1-\varepsilon_1^k)(1-\varepsilon_2^k)1$. B_3 leads to the payoff 0 and C_3 leads to the payoff $(1-\varepsilon_1^k)(1-\varepsilon_2^k)1$. So A_3 is one of his best responses, and a best strategy in the perturbed game is to assign the minimal probabilities to B_3 and C_3 and the remaining probability $1-\varepsilon_3^k-e_3^k$ to A_3.

Player 2
A_2 leads to the payoff $(1-\varepsilon_1^k)[3(1-\varepsilon_3^k-e_3^k)+4\varepsilon_3^k]+2\varepsilon_1^k=(1-\varepsilon_1^k)[3+\varepsilon_3^k-3e_3^k]+2\varepsilon_1^k$, B_2 leads to the payoff $3(1-\varepsilon_1^k)+2\varepsilon_1^k$. We want A_2 giving a higher payoff than B_2, so we need $\varepsilon_3^k \geq 3e_3^k$, which is possible given that $e_3^k \to 0$ when $k \to \infty$. So, by choosing minimal probabilities such that $\varepsilon_3^k \geq 3e_3^k$, A_2 is player 2's best response and assigning the minimal probability ε_2^k to B_2 and the remaining probability $1-\varepsilon_2^k$ to A_2 is his best strategy in the perturbed game.

Player 1
A_1 leads to the payoff $(1-\varepsilon_2^k)3+\varepsilon_2^k$. B_1 leads to the payoff $2\varepsilon_2^k$. $(1-\varepsilon_2^k)3+\varepsilon_2^k \to 3$ and $2\varepsilon_2^k \to 0$ when $k \to \infty$. So A_1 is her best response and assigning the minimal probability ε_1^k to B_1 and the remaining probability $1-\varepsilon_1^k$ to A_1 is her best strategy in the perturbed game.

Conclusion: E is a perfect equilibrium given that there exist minimal probabilities that sustain this equilibrium; these minimal probabilities check $\varepsilon_3^k \geq 3e_3^k$.

ANSWERS

2. If one eliminates the weakly dominated strategy B_3, it is no longer possible to check $\varepsilon_3^k \geq 3e_3^k$ (ε_3^k is now necessarily equal to 0 given that B_3 has disappeared). A_2 necessarily leads to a lower payoff than B_2 ($3\pi_1^k(A_1)\pi_3^k(A_3)+2\pi_1^k(B_1)<3\pi_1^k(A_1)+2\pi_1^k(B_1)$, because $\pi_3^k(A_3)<1$ by definition).

   It follows that player 2 has to assign the minimal probability to A_2, that goes to 0: so, given that player 2 plays A_2 with probability 1 in E, E is not a perfect equilibrium in the new game without B_3. This leads to two remarks:
   - Given that E is a perfect equilibrium in the original game and is no more a perfect equilibrium once the weakly dominated strategy B_3 is removed from the game, we observe that *the suppression of weakly dominated strategies may eliminate perfect equilibria*.
   - We can also observe that, once B_3 is eliminated, A_2 becomes weakly dominated by B_2. So, in the new game, (A_1,A_2,A_3) contains a weakly dominated strategy and it is not perfect because of this weakly dominated strategy. Yet (A_1,A_2,A_3) is still a SPNE. Hence a SPNE that contains a weakly dominated strategy often does not pass the perfect equilibrium test.

3. To prove that (A_1,A_2,A_3) is not a proper equilibrium, we write the game's normal form in the matrices A5.1.

   According to Myerson's proper equilibrium concept, the probabilities in a perturbed game check $p_3^k(B_3) \leq \varepsilon_3^k p_3^k(C_3)$ and $p_3^k(B_3) \leq \varepsilon_3^k p_3^k(A_3)$, given that B_3 is weakly dominated by A_3 and C_3.

   A_2 leads to the payoff $p_1^k(A_1) [3p_3^k(A_3)+4p_3^k(B_3)]+2p_1^k(B_1)$ whereas B_2 leads to the payoff $p_1^k(A_1) [3p_3^k(A_3)+3p_3^k(B_3)+3p_3^k(C_3)]+2p_1^k(B_1)$. Yet $4p_3^k(B_3)<3p_3^k(B_3)+3p_3^k(C_3)$ because $p_3^k(B_3) \leq \varepsilon_3^k p_3^k(C_3)$. Hence we necessarily have $p_2^k(A_2) \leq \varepsilon_2^k p_2^k(B_2)$. So it is not possible to build a perturbed equilibrium p^k that goes to E, because $p_2^k(A_2) \to 0$ instead of 1 when $k \to \infty$. E isn't a proper equilibrium.

	2 A_2	B_2
1 A_1	(3,3,1)	(1,3,1)
B_1	(0,2,1)	(2,2,1)

A_3

	2 A_2	B_2
1 A_1	(3,4,0)	(1,3,1)
B_1	(0,2,1)	(2,2,1)

B_3

	2 A_2	B_2
1 A_1	(3,0,1)	(1,3,1)
B_1	(0,2,1)	(2,2,1

C_3

3   Matrices A5.1

♣ EXERCISE 2 PERFECT EQUILIBRIUM WITH WEAKLY DOMINATED STRATEGIES

The game has only one proper subgame starting at x_6 and a_1 is player 1's optimal strategy in this subgame (3>2). $(B_1/x_2, B_2/x_3, A_3/h)$ is a Nash equilibrium of the reduced game where the node x_6 is replaced by the payoffs (3,0,0): B_1 is a player 1's best response because B_1 and A_1 yield the payoff 3, B_2 is player 2's best response because A_2 and B_2 respectively yield the payoffs 1 and 2, and player 3's strategy is optimal because he is not called on to play. So $E=((B_1/x_2, a_1/x_6), B_2/x_3, A_3/h)$ is a SPNE.

$(B_1/x_2, a_1/x_6)$ is weakly dominated by $(A_1/x_2, a_1/x_6)$, because, if Nature plays A, $(B_1/x_2, a_1/x_6)$ only yields the payoff 3 whereas $(A_1/x_2, a_1/x_6)$ yields the payoffs 3 and/or 5 (when Nature chooses B, both strategies lead to the same payoff).

We call ε_1^k the minimal probability assigned to A_1, e_1^k the minimal probability assigned to b_1, ε_2^k the minimal probability assigned to A_2, ε_3^k the minimal probability assigned to B_3, and we define the profile π^k as follows.

$$\pi_{1x2}^k(A_1)=\varepsilon_1^k,\ \pi_{1x2}^k(B_1)=1-\varepsilon_1^k,\ \pi_{1x6}^k(a_1)=1-e_1^k,\ \pi_{1x6}^k(b_1)=e_1^k,\ \pi_2^k(A_2)=\varepsilon_2^k,$$
$$\pi_2^k(B_2)=1-\varepsilon_2^k,\ \pi_3^k(A_3)=1-\varepsilon_3^k,\ \pi_3^k(B_3)=\varepsilon_3^k.$$

$\pi^k \to \pi^*$ where π^* is the profile of behavioural strategies corresponding to E (we say $\pi^k \to$ E) when $k \to \infty$, because the minimal probabilities go to 0 when k goes to ∞.

Now we show that π^k is a Nash equilibrium of the perturbed game. Given that we can test the optimality locally, we proceed in this way.

Player 1 at x_6

a_1 is her best response because a_1 gives the payoff ¼ $\varepsilon_1^k(1-\varepsilon_3^k)3$ and b_1 gives the lower payoff ¼ $\varepsilon_1^k(1-\varepsilon_3^k)2$. So assigning the minimal probability e_1^k to b_1 and the remaining probability $1-e_1^k$ to a_1 is player 1's best local strategy at x_6 in the perturbed game.

Player 3

A_3 leads to the payoff ¼ $\varepsilon_1^k e_1^k 3 + ¾ \varepsilon_2^k 3$, B_3 leads to the payoff ¼ $\varepsilon_1^k 2 + ¾ \varepsilon_2^k 2$. We want A_3 yielding a higher payoff than B_3, so we need ¾ $\varepsilon_1^k e_1^k + 9\varepsilon_2^k/4 \geq 2\varepsilon_1^k/4 + 6\varepsilon_2^k/4$, i.e. $\varepsilon_2^k \geq \varepsilon_1^k(2-3e_1^k)/3$, which is possible given that $\varepsilon_1^k(2-3e_1^k) \to 0$ when $k \to \infty$. In literary terms, player 3 needs player 2 trembling enough (in comparison with player 1), because A_3 does better than B_3 at x_5 but worse at x_4 (given player 1's behaviour at x_6). So, provided that $\varepsilon_2^k \geq \varepsilon_1^k(2-3e_1^k)/3$, A_3 is player 3's best response, and assigning the minimal probability ε_3^k to B_3 and the remaining probability $1-\varepsilon_3^k$ to A_3 is his best strategy in the perturbed game.

Player 2

A_2 yields the payoff ¾ $[(1-\varepsilon_3^k)+4\varepsilon_3^k]$, B_2 yields the payoff ¾. $2=6/4$. ¾ $[(1-\varepsilon_3^k)+4\varepsilon_3^k] \to 3/4$ when $k \to \infty$, so B_2 is player 2's best response, and assigning the minimal probability ε_2^k to A_2 and the remaining probability $1-\varepsilon_2^k$ to B_2, is his best strategy in the perturbed game.

Player 1 at x_2

A_1 leads to the payoff ¼ $(1-\varepsilon_3^k)[3(1-e_1^k)+2e_1^k]+¼ \times 5\varepsilon_3^k = (3+2\varepsilon_3^k-e_1^k+\varepsilon_3^k e_1^k)/4$. B_1 leads to the payoff ¼. $3=¾$. We want B_1 yielding a higher payoff than A_1, so we need $3 \geq 3+2\varepsilon_3^k-e_1^k+\varepsilon_3^k e_1^k$ i.e. $e_1^k \geq 2\varepsilon_3^k/(1-\varepsilon_3^k)$, which is possible given that $2\varepsilon_3^k/(1-\varepsilon_3^k) \to 0$ when $k \to \infty$. So, provided that $e_1^k \geq 2\varepsilon_3^k/(1-\varepsilon_3^k)$, B_1 is her best response, and assigning the minimal probability ε_1^k to A_1 and the remaining probability $1-\varepsilon_1^k$ to B_1 is her best local strategy at x_2 in the perturbed game.

Conclusion: E is a perfect equilibrium given that there exist minimal probabilities that sustain this equilibrium; these minimal probabilities check $e_1^k \geq 2\varepsilon_3^k/(1-\varepsilon_3^k)$ and $\varepsilon_2^k \geq \varepsilon_1^k(2-3e_1^k)/3$.

In literary terms, $e_1^k \geq 2\varepsilon_3^k/(1-\varepsilon_3^k)$ means that, to play B_1, player 1 has to expect that she will deviate more later, at x_6, than player 3 will deviate at x_4. Providing she trembles enough at x_6, the payoff 2 she gets with b_1 outweighs the payoff 5 she gets when player 3 trembles toward B_3. This explains why, despite $(B_1/x_2,a_1/x_6)$ being weakly dominated by $(A_1/x_2,a_1/x_6)$, it can be played in a perfect equilibrium. In other words, dominated strategies can appear in a perfect equilibrium when

ANSWERS

they consist of several local strategies, because you can't be sure today that you play your best local strategy tomorrow (player 1, at node x_2, is not sure to play a_1 at x_6).

♣♣ EXERCISE 3 PERFECT EQUILIBRIUM AND INCOMPATIBLE PERTURBATIONS

A_3 is player 3's optimal action at the subgame starting at x_4 (2>0) and a_3 is player 3's optimal action at the subgame starting at x_5 (1>0). (A_1, A_2) is a Nash equilibrium of the reduced game with the payoffs (4,2,2) and (4,2,1) at the nodes x_4 and x_5 (A_1 is a best response to A_2 (4≥4), and A_2 is a best response to A_1 (2≥2)). So $E=(A_1, A_2, (A_3/x_4, a_3/x_5))$ is a SPNE.

We introduce minimal probabilities to try to show that the profile is perfect. We call ε_1^k the minimal probability assigned to B_1, ε_2^k the minimal probability assigned to B_2, ε_3^k the minimal probability assigned to B_3, e_3^k the minimal probability assigned to b_3, and we set the profile π^k as follows:

$$\pi_1^k(A_1)=1-\varepsilon_1^k, \pi_1^k(B_1)=\varepsilon_1^k, \pi_2^k(A_2)=1-\varepsilon_2^k, \pi_2^k(B_2)=\varepsilon_2^k, \pi_{3x4}^k(A_3)=1-\varepsilon_3^k$$
$$\pi_{3x4}^k(B_3)=\varepsilon_3^k, \pi_{3x5}^k(a_3)=1-e_3^k, \pi_{3x5}^k(b_3)=e_3^k.$$

$\pi^k \to \pi^*$ where π^* is the profile of behavioural strategies corresponding to E, when $k \to \infty$, because the minimal probabilities go to 0 when k goes to ∞.

We now try to show that π^k is a Nash equilibrium of the perturbed game.

Player 3 at x_4

A_3 leads to the payoff $(1-\varepsilon_1^k)(1-\varepsilon_2^k)2$, B_3 leads to the payoff 0, so A_3 is his best response, and assigning the minimal probability ε_3^k to B_3 and the remaining probability $1-\varepsilon_3^k$ to A_3 is his best local strategy at x_4 in the perturbed game.

Player 3 at x_5

a_3 leads to the payoff $(1-\varepsilon_1^k)\varepsilon_2^k 1$, b_3 leads to the payoff 0, so a_3 is his best response. It follows that assigning the minimal probability e_3^k to b_3 and the remaining probability $1-e_3^k$ to a_3 is his best local strategy at x_5 in the perturbed game.

Player 2

A_2 leads to the payoff $(1-\varepsilon_1^k)[2(1-\varepsilon_3^k)+3\varepsilon_3^k]+2\varepsilon_1^k$, B_2 leads to the payoff $(1-\varepsilon_1^k)[2(1-e_3^k)+3e_3^k]+2\varepsilon_1^k$. We want A_2 yielding a higher payoff than B_2 so we need $2(1-\varepsilon_3^k)+3\varepsilon_3^k \geq 2(1-e_3^k)+3e_3^k$ i.e. $\varepsilon_3^k \geq e_3^k$, which is possible given that $e_3^k \to 0$ when $k \to \infty$. In literary terms player 2 needs that player 3 more often trembles at x_4 than at x_5, in order to more often get the payoff 3 at x_4 than at x_5, which justifies her choice A_2 leading to x_4. So, provided that $\varepsilon_3^k \geq e_3^k$, A_2 is his best response, and assigning the minimal probability ε_2^k to B_2 and the remaining probability $1-\varepsilon_2^k$ to A_2 is his best strategy in the perturbed game.

Player 1

A_1 leads to the payoff $(1-\varepsilon_2^k)4(1-\varepsilon_3^k)+\varepsilon_2^k[4(1-e_3^k)+5e_3^k]=4-4\varepsilon_3^k(1-\varepsilon_2^k)+\varepsilon_2^k e_3^k$, B_1 leads to the payoff 4. We want A_1 yielding a higher payoff than B_1 so we need $4-4\varepsilon_3^k(1-\varepsilon_2^k)+\varepsilon_2^k e_3^k \geq 4$, i.e. $\varepsilon_3^k \leq \varepsilon_2^k e_3^k/4(1-\varepsilon_2^k)$.

■ 398

Yet, given that $\varepsilon_2^k/4(1-\varepsilon_2^k) \to 0$ when $k \to \infty$, we can't have simultaneously $\varepsilon_3^k \leq e_3^k \varepsilon_2^k/4(1-\varepsilon_2^k)$ and $\varepsilon_3^k \geq e_3^k$ (the condition required for player 2).

Conclusion: the conditions on the minimal probabilities required for player 1 and player 2 are incompatible. So E is not a perfect equilibrium.

In literary terms, player 1, to justify A_1, needs the payoff 5 she can get when player 3 trembles at x_5, and fears the 0 she can get when player 3 trembles at x_4. By contrast, player 2, to justify A_2, needs the payoff 3 he can get when player 3 trembles at x_4 and fears the 3 he can get when player 3 trembles at x_5. So player 1 needs that player 3 more often trembles at x_5, whereas player 2 needs that player 3 more often trembles at x_4, so they have incompatible needs, that prevent E from becoming a perfect equilibrium.

♣ EXERCISE 4 STRICT SEQUENTIAL EQUILIBRIUM

The beliefs assigned to π^*, the profile of behavioural strategies assigned to E=(A_1, A_2, (A_3/x_4, a_3/x_5)), given that all the information sets are singletons or reached on the equilibrium path, are $\mu(x_1)=1=\mu(x_2)=\mu(x_4)=\mu(x_5)$ and $\mu(x_3)=0$ ($\mu(x_2)$ and $\mu(x_3)$ are obtained by applying Bayes's rule to the equilibrium strategies).

(μ, π^*) is sequentially rational. As a matter of fact A_1 is optimal given the other players' strategies (4≥4). A_2 is optimal given $\mu(x_2)=1$ and player 3's strategies (2≥2). Player 3's action is optimal at x_4, given $\mu(x_4)=1$ and 2>0. And player 3's action is optimal at x_5, given $\mu(x_5)=1$ and 1>0.

(μ, π^*) is consistent. Just set $\pi_1^k(A_1)=1-\varepsilon^k$, $\pi_1^k(B_1)=\varepsilon^k$, $\pi_2^k(A_2)=1-\varepsilon^k$, $\pi_2^k(B_2)=\varepsilon^k$, $\pi_{3x4}^k(A_3)=1-\varepsilon^k \pi_{3x4}^k(B_3)=\varepsilon^k$, $\pi_{3x5}^k(a_3)=1-\varepsilon^k$, $\pi_{3x5}^k(b_3)=\varepsilon^k$.

We get: $\mu^k(x_1)=1=\mu^k(x_4)=\mu^k(x_5)$, $\mu^k(x_2)=1-\varepsilon^k$ and $\mu^k(x_3)=\varepsilon^k$.

So $\mu^k \to \mu$ and $\pi^k \to \pi^*$ when $k \to \infty$ (the ε^k are supposed to go to 0 when $k \to \infty$). So (μ, π^*) is consistent.

Given that it is also sequentially rational, (μ, π^*) is a sequential equilibrium. Yet it is not a strict sequential equilibrium because, for example, B_1 yields the same payoff than A_1 (4=4) but is not in the equilibrium's support. So (μ, π^*) is only a weak sequential equilibrium.

We already know that E is not a perfect equilibrium. So we get an often observed fact: a sequential equilibrium that is not a perfect equilibrium is generally a weak sequential equilibrium.

♣ EXERCISE 5 SEQUENTIAL EQUILIBRIUM, INERTIA IN THE PLAYERS' BELIEFS

1. The game has no proper subgame, so E=(C_1, A_2, (A_3/h_1, A_3/h_2)) is a SPNE in that it is a Nash equilibrium (players 2 and 3 are not called on to play, so their strategies are optimal and player 1 gets her highest possible payoff with C_1).

    We call π^* the profile of behavioural strategies assigned to E. For (μ,π^*) to be sequentially rational, player 2 has to play rationally at h given his beliefs $\mu(x_2)$ and $\mu(x_3)$, player 3 has to play rationally at h_1 given his beliefs $\mu(y_1)$ and $\mu(y_2)$ and player 3 has to play rationally at h_2 given his beliefs $\mu(y_3)$ and $\mu(y_4)$.

ANSWERS

Player 2

A_2 leads to the payoff $1\mu(x_2)+2\mu(x_3)$. B_2 leads to the payoff $0\mu(x_2)+3\mu(x_3)$. C_2 leads to the payoff $0\mu(x_2)+3\mu(x_3)$. So A_2 is player 2's best strategy provided that $\mu(x_2)+2\mu(x_3) \geq 3\mu(x_3)$, i.e. $\mu(x_2) \geq 0.5$ ($\mu(x_3)=1-\mu(x_2)$). For example we can set $\mu(x_2)=0.6$ and $\mu(x_3)=0.4$.

Player 3 at h_1

A_3 leads to the payoff $0\mu(y_1)+1\mu(y_2)$, B_3 leads to the payoff $1\mu(y_1)+0\mu(y_2)$. A_3 is his best response if $\mu(y_2) \geq \mu(y_1)$, i.e. $\mu(y_2) \geq 0.5$ For example we can set $\mu(y_1)=0.4$ and $\mu(y_2)=0.6$.

Player 3 at h_2

A_3 leads to the payoff $1\mu(y_3)+0\mu(y_4)$, B_3 leads to the payoff $0\mu(y_3)+5\mu(y_4)$. A_3 is his best response if $\mu(y_3) \geq 5\mu(y_4)$, i.e. $\mu(y_3) \geq 5/6$. For example we can set $\mu(y_3)=0.9$ and $\mu(y_4)=0.1$.

2. The problem lies in player 3's beliefs at h_1 and h_2. We have:

$$\mu^k(y_1)=\pi_1^k(A_1)\pi_2^k(B_2)/(\pi_1^k(A_1)\pi_2^k(B_2)+\pi_1^k(A_1)\pi_2^k(C_2))=\pi_2^k(B_2)/(\pi_2^k(B_2)+\pi_2^k(C_2))$$

and

$$\mu^k(y_3)=\pi_1^k(B_1)\pi_2^k(B_2)/(\pi_1^k(B_1)\pi_2^k(B_2)+\pi_1^k(B_1)\pi_2^k(C_2))=\pi_2^k(B_2)/(\pi_2^k(B_2)+\pi_2^k(C_2))$$

where $\pi^k \to \pi^*$.

$\mu^k(y_1)=\mu^k(y_3)$ and so both beliefs converge to a same value when $k \to \infty$. Given that $\mu^k(y_1) \to \mu(y_1)$ and $\mu^k(y_3) \to \mu(y_3)$, it follows that $\mu(y_1)=\mu(y_3)$ so that it is not possible to have $\mu(y_1) \leq 0.5$ and $\mu(y_3) \geq 5/6$. In other words, if player 3, at h_1, believes that player 2 more often deviates to C_2 than to B_2, he has to keep the same beliefs at h_2.

So we can't build a couple (μ,π^*) that is sequentially rational and consistent, that is to say we can't build a sequential equilibrium with $(C_1, A_2, (A_3/h_1, A_3/h_2))$.

3. If we interchange the bold underlined payoffs 1 and 5, the sequential rationality condition for player 3 at h_2 becomes:

A_3 leads to the payoff $5\mu(y_3)+0\mu(y_4)$, B_3 leads to the payoff $0\mu(y_3)+1\mu(y_4)$. A_3 is his best response if $5\mu(y_3) \geq \mu(y_4)$, i.e. $\mu(y_3) \geq 1/6$.

It follows that we can find $\mu(y_1)$ and $\mu(y_3)$ that satisfy player 3's requirements at h_1 and at h_2. For example we can set $\mu(y_1)=\mu(y_3)=0.4$, $\mu(y_2)=\mu(y_4)=0.6$, these beliefs both check $\mu(y_1)=\mu(y_3) \geq 1/6$ and $\mu(y_1)=\mu(y_3) \leq 1/2$.

And we keep $\mu(x_2)=0.6$, $\mu(x_3)=0.4$, and $\mu(x_1)=1$

So we can build a couple (μ,π^*) which is sequentially rational and consistent.

We can set:

$\pi_1^k(A_1)=6\varepsilon^k$, $\pi_1^k(B_1)=4\varepsilon^k$, $\pi_1^k(C_1)=1-10\varepsilon^k$, $\pi_2^k(A_2)=1-10\varepsilon^k$, $\pi_2^k(B_2)=4\varepsilon^k$, $\pi_2^k(C_2)=6\varepsilon^k$, $\pi_{3h1}^k(A_3)=1-\varepsilon^k$, $\pi_{3h1}^k(B_3)=\varepsilon^k$, $\pi_{3h2}^k(A_3)=1-\varepsilon^k$, $\pi_{3h2}^k(B_3)=\varepsilon^k$

with $\varepsilon^k \to 0$ when $k \to \infty$.

We check:

$\pi^k \to \pi^*$ when $k \to \infty$.

$\mu^k(x_1)=\mu(x_1)=1$

$\mu^k(x_2)=6\varepsilon^k/(6\varepsilon^k+4\varepsilon^k)=0.6=\mu(x_2)$, $\mu^k(x_3)=0.4=\mu(x_3)$

$\mu^k(y_1)=6\varepsilon^k 4\varepsilon^k/(6\varepsilon^k 4\varepsilon^k+6\varepsilon^k 6\varepsilon^k)=0.4=\mu(y_1)$, $\mu^k(y_2)=0.6=\mu(y_2)$

$\mu^k(y_3)=4\varepsilon^k 4\varepsilon^k/(4\varepsilon^k 4\varepsilon^k+4\varepsilon^k 6\varepsilon^k)=0.4=\mu(y_3)$, $\mu^k(y_4)=0.6=\mu(y_4)$

Hence $\mu^k \to \mu$ when $k \to \infty$. So $(\mu^k, \pi^k) \to (\mu, \pi^*)$ when $k \to \infty$, (μ, π^*) is consistent. Given that it is also sequentially rational, it is a sequential equilibrium.

♣ ♣ EXERCISE 6 CONSTRUCT THE SET OF SEQUENTIAL EQUILIBRIA

1. $(A_1, (A_2/x_2, b_2/x_3), A_3)$ is a Nash equilibrium and consequently a SPNE given that the game has no proper subgame (A_1 is optimal for player 1 (2>0), player 2's action at x_3 is optimal because it has no impact and A_2 is optimal at x_2 (2>0), and A_3 is optimal for player 3 (3>0).

    If player 1 plays A_1, player 3 reaches his information set so his beliefs have to follow Bayes's rule. Given that player 1 plays A_1, $\mu(x_4)+\mu(x_5)=1$, hence $\mu(x_6)=\mu(x_7)=0$. It follows that player 3's sequentially rational strategy consists in playing A_3 because A_3 is his optimal strategy, both at x_4 (3>0) and at x_5 (3>0). It follows that player 2's sequentially rational action at x_3, given player 3's strategy and $\mu(x_3)=1$, is a_2 because 3>2. It follows that player 1's sequentially rational strategy is B_1, given that B_1 is followed by a_2 and A_3, hence leads to the payoff 3, whereas A_1 can only lead to the payoff 2. So it is not possible to build a sequential equilibrium such that player 1 plays A_1.

2. Let us try to build a sequential equilibrium where player 1 plays B_1. In that case, player 3 reaches his information set, so his beliefs have to follow Bayes's rule.

    It follows that $\mu(x_6)+\mu(x_7)=1$, hence $\mu(x_4)=\mu(x_5)=0$. So player 3's sequentially rational strategy is B_3 because B_3 is optimal, both at x_6 (3>0) and at x_7 (3>0). It follows that player 1's sequentially rational strategy is A_1, given that B_1 is followed by B_3, hence only leads to the payoff 0, whereas A_1 always leads to the payoff 2. So it is not possible to build a sequential equilibrium where player 1 plays B_1.

3. Given that the game is finite, a sequential equilibrium exists.

    Given the previous results, player 1 necessarily plays A_1 with probability p and B_1 with probability 1–p, with 0<p<1.

    Necessarily player 3 plays his both actions with positive probability because we know from above that, if he only plays A_3, player 1 only plays B_1, and if he only plays B_3, player 1 only plays A_1. So player 3 plays A_3 with probability r and B_3 with probability 1–r, with 0<r<1. Given that 3r>2r, player 2 necessarily plays a_2 at x_3. For player 1 playing her both actions, we need that the payoff obtained with A_1, 2, is equal to the payoff obtained with B_1, i.e. 3r. Hence r=2/3. It follows that player 2 plays A_2 at x_2, given that 2r>2(1–r). Bayes' rule leads to $\mu(x_4)$= p/(p+(1–p))=p, $\mu(x_6)$=1–p and $\mu(x_5)=\mu(x_7)=0$. So player 3 is indifferent between A_3 and B_3 only if $3\mu(x_4)$ (=3p) is equal to $3\mu(x_6)$ (=3(1–p)), hence p=½.

    Let us check that (μ, π^*) with:

    $\pi_1^*(A_1)=\pi_1^*(B_1)=½$, $\pi_{2x2}^*(A_2)=\pi_{2x3}^*(a_2)=1$, $\pi_{2x2}^*(B_2)=\pi_{2x3}^*(b_2)=0$,
    $\pi_3^*(A_3)=2/3$ $\pi_3^*(B_3)=1/3$, $\mu(x_1)=\mu(x_2)=\mu(x_3)=1$, $\mu(x_4)=\mu(x_6)=1/2$,
    $\mu(x_5)=\mu(x_7)=0$ is a sequential equilibrium.

ANSWERS

We already know that (μ,π^*) is sequentially rational (by construction). Let us show that it is consistent.

Given that player 1 and player 3 play a completely mixed strategy, we can set $\pi_1^k = \pi_1^*$ and $\pi_3^k = \pi_3^*$ and let us set $\pi_{2x2}^k(A_2) = \pi_{2x3}^k(a_2) = 1-\varepsilon^k$, $\pi_{2x2}^k(B_2) = \pi_{2x3}^k(b_2) = \varepsilon^k$, with $\varepsilon^k \to 0$ when $k \to \infty$

$\pi^k \to \pi^*$ when k goes to ∞.

And we have
$\mu^k(x_1) = \mu^k(x_2) = \mu^k(x_3) = 1 = \mu(x_1) = \mu(x_2) = \mu(x_3)$
$\mu^k(x_4) = \frac{1}{2}(1-\varepsilon^k)/[\frac{1}{2}(1-\varepsilon^k)+\frac{1}{2}\varepsilon^k+\frac{1}{2}(1-\varepsilon^k)+\frac{1}{2}\varepsilon^k] = \frac{1}{2}(1-\varepsilon^k) \to \frac{1}{2} = \mu(x_4)$ when $k \to \infty$
$\mu^k(x_5) = \frac{1}{2}\varepsilon^k/[\frac{1}{2}(1-\varepsilon^k)+\frac{1}{2}\varepsilon^k+\frac{1}{2}(1-\varepsilon^k)+\frac{1}{2}\varepsilon^k] = \frac{1}{2}\varepsilon^k \to 0 = \mu(x_5)$ when $k \to \infty$
$\mu^k(x_6) = \frac{1}{2}(1-\varepsilon^k)/[\frac{1}{2}(1-\varepsilon^k)+\frac{1}{2}\varepsilon^k+\frac{1}{2}(1-\varepsilon^k)+\frac{1}{2}\varepsilon^k] = \frac{1}{2}(1-\varepsilon^k) \to \frac{1}{2} = \mu(x_6)$ when $k \to \infty$
$\mu^k(x_7) = \frac{1}{2}\varepsilon^k/[\frac{1}{2}(1-\varepsilon^k)+\frac{1}{2}\varepsilon^k+\frac{1}{2}(1-\varepsilon^k)+\frac{1}{2}\varepsilon^k] = \frac{1}{2}\varepsilon^k \to 0 = \mu(x_7)$ when $k \to \infty$

Hence $\mu^k \to \mu$ when $k \to \infty$.

So (μ,π^*) is consistent. Given that it is also sequentially rational, it is a sequential equilibrium. There is no other sequential equilibrium (by construction).

♣♣ EXERCISE 7 PERFECT EQUILIBRIUM, WHY THE NORMAL FORM IS INADEQUATE, A LINK TO THE TREMBLING*-HAND EQUILIBRIUM

1. $E = (A_1a_1, A_2a_2)$ is not a perfect equilibrium because it is not a SPNE (player 2 should play B_2 in the reduced game starting at x_2 (1<4)).

   The normal form of the game is given in Matrix A5.2a:

	A_2a_2	A_2b_2	B_2a_2	B_2b_2
A_1a_1	(5,2)	(5,2)	(5,2)	(5,2)
A_1b_1	(5,2)	(5,2)	(5,2)	(5,2)
B_1a_1	(0,1)	(0,1)	(4,4)	(0,0)
B_1b_1	(0,1)	(0,1)	(0,0)	(1,1)

   Player 2 across top; Player 1 down side.

   Matrix A5.2a

   If it were possible to transpose the perfect equilibrium concept to the game in normal form, we would get a concept that goes as follows:

ANSWERS

> *Take a normal form game, G. Call $G^k=(G,\eta^k)$, a perturbed game, where η^k is a function that assigns a minimal positive probability to each pure strategy. p^k is an equilibrium in G^k if it is a Nash equilibrium in the original game where the set of strategies is reduced to the set of perturbed ones. A profile p^* in G is a normal form perfect equilibrium, if it is possible to find a sequence of perturbed games G^k, with a sequence of equilibria p^k in G^k, such that the minimal probabilities go to 0 when $k\to\infty$ and such that $p^k\to p^*$ when $k\to\infty$.*

Let us apply this criterion to the game in Matrix A5.2a. We show that $E=(A_1a_1, A_2a_2)$ is a normal form perfect equilibrium!

We set $\varepsilon_1^k, \varepsilon_2^k, \varepsilon_3^k, e_1^k, e_2^k, e_3^k$ the minimal probabilities respectively assigned to A_1b_1, B_1a_1, B_1b_1, A_2b_2, B_2a_2 and B_2b_2. Consider the perturbed strategy:

$p_1^k(A_1a_1)=1-\varepsilon_1^k-\varepsilon_2^k-\varepsilon_3^k$, $p_1^k(A_1b_1)=\varepsilon_1^k$, $p_1^k(B_1a_1)=\varepsilon_2^k$, $p_1^k(B_1b_1)=\varepsilon_3^k$

$p_2^k(A_2a_2)=1-e_1^k-e_2^k-e_3^k$, $p_2^k(A_2b_2)=e_1^k$, $p_2^k(B_2a_2)=e_2^k$, $p_2^k(B_2b_2)=e_3^k$

We have $p^k\to p^*$ where p^* is the profile of mixed strategies corresponding to E.

We now show that p^k is an equilibrium of the perturbed game.

Player 1

A_1a_1 yields the payoff 5.

A_1B_1 yields the payoff 5.

B_1a_1 yields the payoff $4e_2^k$.

B_1b_1 yields the payoff e_3^k.

So A_1a_1 yields the highest payoff and player 1 can assign the minimal probabilities to the three other strategies and the remaining probability $1-\varepsilon_1^k-\varepsilon_2^k-\varepsilon_3^k$ to A_1a_1. By doing so, player 1 plays her best strategy in the perturbed game.

Player 2

A_2a_2 leads to the payoff $2(1-\varepsilon_1^k-\varepsilon_2^k-\varepsilon_3^k)+2\varepsilon_1^k+\varepsilon_2^k+\varepsilon_3^k=2-\varepsilon_2^k-\varepsilon_3^k$.

A_2b_2 leads also to the payoff $2-\varepsilon_2^k-\varepsilon_3^k$.

B_2a_2 leads to the payoff $2(1-\varepsilon_1^k-\varepsilon_2^k-\varepsilon_3^k)+2\varepsilon_1^k+4\varepsilon_2^k=2+2\varepsilon_2^k-2\varepsilon_3^k$.

B_2b_2 leads to the payoff $2(1-\varepsilon_1^k-\varepsilon_2^k-\varepsilon_3^k)+2\varepsilon_1^k+\varepsilon_3^k=2-2\varepsilon_2^k-\varepsilon_3^k$.

Clearly, B_2b_2 leads to a lower payoff than A_2a_2. B_2a_2 leads to a lower payoff than A_2a_2 if $2-\varepsilon_2^k-\varepsilon_3^k\geq 2+2\varepsilon_2^k-2\varepsilon_3^k$, i.e. if $\varepsilon_2^k\leq\varepsilon_3^k/3$, which is possible given that $\varepsilon_2^k\to 0$ when $k\to\infty$. So, with $\varepsilon_2^k\leq\varepsilon_3^k/3$, player 2 plays his best strategy in the perturbed game by assigning the probability $1-e_1^k-e_2^k-e_3^k$ to A_2a_2 and the minimal probabilities to the other strategies.

Conclusion: *If we transpose the perfect equilibrium concept to normal form games, it is possible to find minimal perturbations ($\varepsilon_2^k\leq\varepsilon_3^k/3$) so that ($A_1a_1$, A_2a_2) is normal form perfect, despite its not even being a SPNE.*

ANSWERS

This fact seems strange but it is easy to understand.

In the game tree, when player 2 is at x_2, *he knows that player 1 has played B_1* and so he no longer takes the action A_1 into account. *Hence he compares the bold underlined 1 and 4*. It is as if he only focuses on the two lower rows of Matrix A5.2a, i.e. Matrix A5.2b:

			A_2a_2	A_2b_2	B_2a_2	B_2b_2
1	$1-\eta^k$	B_1a_1	(0,<u>1</u>)	(0,1)	(4,<u>4</u>)	(0,0)
	η^k	B_1b_1	(0,<u>1</u>)	(0,1)	(0,0)	(1,1)

Matrix A5.2b

And given that, locally, player 1 plays a_1 and not b_1, we have to put a probability that goes to 1 on B_1a_1 and the minimal probability η^k on B_1b_1. It immediately follows that, in that case, B_2a_2 leads to a higher payoff than A_2a_2 ($4(1-\eta^k)>1$ when $\eta^k \to 0$).

In the normal form game in Matrix A5.2a, player 2 does not compare 1 and 4 but $\varepsilon_2^k+\varepsilon_3^k$ and $4\varepsilon_2^k$, because the probability is mostly focused on A_1a_1. It is impossible to have 1>4, but it is possible to have $\varepsilon_2^k+\varepsilon_3^k>4\varepsilon_2^k$!

2. Jackson et al.'s trembling*-hand perfect equilibrium also accepts (A_1a_1,A_2a_2). Let us focus on extensive form perturbations, hence on the perturbed strategies $\pi_{1x1}^k(A_1)=1-\varepsilon_1^k$, $\pi_{1x1}^k(B_1)=\varepsilon_1^k$, $\pi_{1x3}^k(a_1)=1-e_1^k$, $\pi_{1x3}^k(b_1)=e_1^k$, $\pi_{2x2}^k(A_2)=1-\varepsilon_2^k$, $\pi_{2x2}^k(B_2)=\varepsilon_2^k$, $\pi_{2h}^k(a_2)=1-e_2^k$, $\pi_{2h}^k(b_2)=e_2^k$.

By playing A_2 at x_2 player 2 gets $1\varepsilon_1^k$. By playing B_2 he gets $\varepsilon_1^k[4(1-e_1^k)(1-e_2^k)+e_1^ke_2^k]$. So, by playing A_2, player 2 loses $\varepsilon_1^k[4(1-e_1^k)(1-e_2^k)+e_1^ke_2^k]-\varepsilon_1^k=\varepsilon_1^k[3-4e_1^k-4e_2^k+5e_1^ke_2^k]$. Hence player 2 makes a loss by playing A_2 instead of B_2 but this loss is very small because it is multiplied by the very small probability that player 1 switches to B_1. So, by adding small perturbations on the payoffs, it is possible to outweigh this loss. For example, by fixing $\varepsilon^k=\varepsilon_1^k[3-4e_1^k-4e_2^k+5e_1^ke_2^k]$, which goes to 0 when k goes to infinity, and by adding this ε^k to player 2's payoff when he plays A_2, A_2 becomes again optimal, and (A_1a_1,A_2a_2) is a trembling*-hand perfect equilibrium.

So we find a similar flaw as in the previous approach. By weighting player 2's actions at x_2 by the very small probability that player 1 switches to B_1, we give almost no weight to player 2's actions at x_2.

♣ EXERCISE 8 PERFECT BAYESIAN EQUILIBRIUM

1. Let us first look for separating PBE.

Case 1: t_1 (i.e. player 1 of type t_1) plays m_1 and t_2 (player 1 of type t_2) plays m_2.
If so, $\mu(y_2)=1$ (and $\mu(y_1)=0$) and player 2 plays r_1 after m_2 (because 3>1), which induces t_1 to deviate to m_2 (because 7 is higher than what she can get with m_1).

Case 2: t_1 plays m_2 and t_2 plays m_1.
If so, $\mu(y_1)=1$ (and $\mu(y_2)=0$) and player 2 plays r_2 after m_2 (2>1). $\mu(x_2)=1$ (and $\mu(x_1)=0$) and player 2 plays r_1 after m_1 (2>1). t_1 is best off playing m_1 (1>0), and t_2 is best off playing m_1 (4>1). So each player plays optimally given the behaviour of the other and given the beliefs,

ANSWERS

and the beliefs satisfy Bayes's rule. Hence we have a separating PBE defined by: (m_2/t_1, m_1/t_2, r_1/m_1, r_2/m_2, $\mu(x_1)=\mu(y_2)=0$, $\mu(x_2)=\mu(y_1)=1$).
We now look for pooling PBE.

Case 3: t_1 plays m_1 and t_2 plays m_1.

In that case $\mu(x_1)=0.4$ and $\mu(x_2)=0.6$. So, at h_1, r_1 leads to the payoff $1\times0.4+2\times0.6=1.6$ and r_2 leads to the payoff $3\times0.4+1\times0.6=1.8$: player 2 plays r_2 after observing m_1. He can believe what he wants after m_2 (we can't apply Bayes's rule because the information set is not on the equilibrium path), so we can set $\mu(y_1)=1$ (and $\mu(y_2)=0$), which induces player 2 to play r_2 after m_2 (2>1). It follows that t_1 is best off playing m_1 (5>1) and t_2 is best off playing m_1 (2>1). So each player plays optimally given the behaviour of the other and given the beliefs, and the beliefs satisfy Bayes's rule when it applies. So we have a pooling PBE defined by: (m_1/t_1, m_1/t_2, r_2/m_1, r_2/m_2, $\mu(x_1)=0.4$, $\mu(x_2)=0.6$, $\mu(y_1)=1$, $\mu(y_2)=0$).

Case 4: t_1 plays m_2 and t_2 plays m_2.

In that case $\mu(y_1)=0.4$ and $\mu(y_2)=0.6$. So, at h_2, r_1 leads to the payoff $1\times0.4+3\times0.6=2.2$ and r_2 leads to the payoff $2\times0.4+1\times0.6=1.4$. Hence player 2 plays r_1 after observing m_2. He can believe what he wants after m_1 (we can't apply Bayes's rule because the information set is not on the equilibrium path). Yet, whatever player 2 plays after m_1, neither t_1 nor t_2 will deviate from m_2, given that they get their highest payoff with m_2 (7>5>0, 5>4>2).

Hence we have 2 different pure strategy PBE. In the first one, we set: $\mu(x_1)=1$ (and $\mu(x_2)=0$), which induces player 2 to play r_2 after m_1 (3>1). In the second, we set $\mu(x_1)=0$ (and $\mu(x_2)=1$), which induces player 2 to play r_1 after m_1 (2>1).

The 2 pooling pure strategy PBE are defined by:

(m_2/t_1, m_2/t_2, r_2/m_1, r_1/m_2, $\mu(x_1)=1$, $\mu(x_2)=0$, $\mu(y_1)=0.4$, $\mu(y_2)=0.6$) and

(m_2/t_1, m_2/t_2, r_1/m_1, r_1/m_2, $\mu(x_1)=0$, $\mu(x_2)=1$, $\mu(y_1)=0.4$, $\mu(y_2)=0.6$)

2. Only t_2 plays m_1 so $\mu(x_2)=1$ (and $\mu(x_1)=0$), which induces player 2 to play r_1 after m_1(2>1). We immediately observe that t_1 is best off only playing m_2 (0<1<7).

t_2 gets 4 by playing m_1. Given that she plays m_1 and m_2, she gets the same payoff with m_2.
Call q the probability assigned to r_1 after m_2. We need $4=5q+1(1-q)$, hence q=¾. But player 2 can play r_1 and r_2 with positive probability after m_2 only if he gets the same payoff with both responses. So we need: $\mu(y_1)+3\mu(y_2)=2\mu(y_1)+\mu(y_2)$, hence $\mu(y_1)=2/3$. Call p the probability assigned by t_2 to m_2. We get $0.4/(0.4+0.6p)=2/3$ hence p=1/3.

Each player plays optimally given the behaviour of the other and given the beliefs, and the beliefs satisfy Bayes rule. So we have a semi separating PBE defined by: (m_2/t_1, $\pi_1(m_1/t_2)=2/3$, $\pi_1(m_2/t_2)=1/3$, r_1/m_1, $\pi_2(r_1/m_2)=¾$, $\pi_2(r_2/m_2)=¼$, $\mu(x_1)=0$, $\mu(x_2)=1$, $\mu(y_1)=2/3$, $\mu(y_2)=1/3$).

ANSWERS

♣♣ EXERCISE 9 PERFECT BAYESIAN EQUILIBRIUM, A COMPLEX SEMI SEPARATING EQUILIBRIUM

Let us first look for separating PBE.

Case 1: t_1 plays m_1 and t_2 plays m_2.
If so, $\mu(x_1)=1$ (and $\mu(x_2)=0$) and player 2 plays r_1 after m_1 (3>1). This induces t_2 to deviate to m_1 (because 4>3>2).

Case 2: t_1 plays m_2 and t_2 plays m_1.
If so, $\mu(y_1)=1$ (and $\mu(y_2)=0$) and player 2 plays r_1 after m_2 (2>0). $\mu(x_2)=1$ (and $\mu(x_1)=0$) and player 2 plays r_2 after m_1 (1>0). This induces t_2 to deviate to m_2 (3>1).
So there is no separating PBE.

We now look for pooling PBE.

Case 3: t_1 plays m_1 and t_2 plays m_1.
In that case $\mu(x_1)=0.5$ and $\mu(x_2)=0.5$. So, at h_1, r_1 yields the payoff $3\times 0.5=1.5$ and r_2 yields the payoff $1\times 0.5+1\times 0.5=1$. So player 2 plays r_1 after observing m_1. He can believe what he wants after m_2 (because the information set is not on the equilibrium path), so we can set $\mu(y_2)=1$ (and $\mu(y_1)=0$), which induces player 2 to play r_2 after m_2 (3>0). So t_1 is best off playing m_1 (2>0) and t_2 is best off playing m_1 (4>2). Each player plays optimally given the behaviour of the other and given the beliefs, and the beliefs satisfy Bayes's rule when it applies. So we have a pooling PBE defined by: (m_1/t_1, m_1/t_2, r_1/m_1, r_2/m_2, $\mu(x_1)=0.5$, $\mu(x_2)=0.5$, $\mu(y_1)=0$, $\mu(y_2)=1$).

Observe that we can't build a pooling equilibrium where both types of player 1 play m_1 and where player 2 plays r_1 after m_2 (this would induce t_1 to deviate from m_1 to m_2).

Case 4: t_1 plays m_2 and t_2 plays m_2.
In that case $\mu(y_1)=\mu(y_2)=0.5$. So, at h_2, r_1 leads to the payoff $2\times 0.5=1$ and r_2 leads to the payoff $3\times 0.5=1.5$. So player 2 plays r_2 after observing m_2. This induces t_1 to deviate to m_1 (0<1.5<2).
Let us now look for the semi separating PBE.

Case 5: t_1 only plays m_1 and t_2 plays m_1 and m_2 with positive probability.
In that case, given that t_1 only plays m_1, $\mu(x_1)>0.5$, so player 2 plays r_1 after m_1 (because $3\mu(x_1)>1$). And $\mu(y_2)=1$ (because only t_2 plays m_2), so player 2 plays r_2 after m_2 (3>0). So t_2 can't play m_1 and m_2 with positive probability because she gets 4 with m_1 and only 2 with m_2.

Case 6: t_1 only plays m_2 and t_2 plays m_1 and m_2 with positive probability.
In that case, $\mu(x_2)=1$ (because only t_2 plays m_1), hence player 2 plays r_2 after m_1 (1>0) and t_2 can't play m_1 and m_2 with positive probability (because she gets 1 with m_1 and at least 2 with m_2).

Case 7: t_1 plays m_1 and m_2 with positive probability and t_2 only plays m_1.
In that case, $\mu(y_1)=1$ (because only t_1 plays m_2), so player 2 plays r_1 after m_2 (2>0). It follows that t_1 can't play m_1 and m_2 with positive probability (because she gets 3 with m_2 and at most 2 with m_1).

Case 8: t_1 plays m_1 and m_2 with positive probability and t_2 only plays m_2.
In that case, $\mu(x_1)=1$ (because only t_1 plays m_1), hence player 2 plays r_1 (3>1) after m_1 and t_2 deviates to m_1 (because she gets 4 with m_1 and at most 3 with m_2).
There is only one last case to examine:

Case 9: t_1 and t_2 play m_1 and m_2 with positive probability.
Call q the probability assigned by player 2 to r_1 at h_1 and s the probability assigned by player 2 to r_1 at h_2. t_1 plays m_1 and m_2 with positive probabilities only if the two messages yield the same payoff. So we need: $2q+1.5(1-q)=3s$, hence $0.5q+1.5=3s$. t_2 plays m_1 and m_2 with positive probabilities only if the two messages yield the same payoff. So we need: $4q+(1-q)=3s+2(1-s)$, hence $3q=s+1$. The system $0.5q+1.5=3s$ and $3q=s+1$ has a unique solution, $q=9/17$ and $s=10/17$. Given that q and s are positive, player 2 has to get the same payoff with r_1 and r_2 at h_1, so we need $3\mu(x_1)=\mu(x_1)+\mu(x_2)=1$, i.e. $\mu(x_1)=1/3$. And he has to get the same payoff with r_1 and r_2 at h_2, so we need $2\mu(y_1)=3\mu(y_2)$, hence $\mu(y_1)=3/5$. Call p_1 the probability assigned to m_1 by t_1 and p_2 the probability assigned to m_1 by t_2. We need:

$0.5p_1/(0.5p_1+0.5p_2)=1/3$ hence $p_2=2p_1$ and $0.5(1-p_1)/(0.5(1-p_1)+0.5(1-p_2))=3/5$ hence $3p_2=1+2p_1$. The system $p_2=2p_1$ and $3p_2=1+2p_1$ has a unique solution $p_1=¼$ and $p_2=½$.

So we have a semi separating PBE characterized by:

$(\pi_1(m_1/t_1)=¼, \pi_1(m_2/t_1)=3/4, \pi_1(m_1/t_2)=1/2, \pi_1(m_2/t_2)=1/2, \pi_2(r_1/m_1)=9/17, \pi_2(r_2/m_1)=8/17, \pi_2(r_1/m_2)=10/17, \pi_2(r_2/m_2)=7/17, \mu(x_1)=1/3, \mu(x_2)=2/3, \mu(y_1)=3/5, \mu(y_2)=2/5)$.

Given that we have studied all player 1's possible strategies, there doesn't exist any other PBE.

♣ ♣ EXERCISE 10 LEMON CARS, EXPERIENCE GOOD MARKET WITH TWO QUALITIES

1. Given that the price 60 is only proposed by the low quality seller, the buyer is indifferent between buying and not buying. So he can buy with probability 1 and a low quality seller gets (60-40) with the price 60. It follows that if she also plays another price P, with 100<P<120, she has to get the same payoff with P. So we need $60-40=(P-40)q$, where q is the probability that the consumer buys the good (a car in Akerlof's model) at price P. We get $0<q=20/(P-40)<1$: the consumer buys the good at price P with a positive probability lower than 1. This is only possible if he is indifferent between buying and not buying. So we need $\mu(y_1)(60-P)+\mu(y_2)(120-P)=0$, i.e. $\mu(y_1)=(120-P)/60$ (which is lower than 1 given that P is between 100 and 120).

    Call p the probability assigned by the low quality seller (t_1) to P. Given that the high quality seller (t_2) only plays P, we get: $\mu(y_1)=0.5p/(0.5p+0.5)=(120-P)/60$, so $p=(120-P)/(P-60)$, which is possible, given that 100<P<120.

    So t_1 is indifferent between 60 and P and gets a positive payoff by playing these prices and t_2 gets a positive payoff with P (and has no interest to play 60 which would lead to a negative payoff). Now suppose that the consumer assigns any price P' different from P to t_1, so that he buys with probability 1 if P'<60, and refuses to buy if P'>60. It immediately follows that the low quality seller is not induced to deviate from 60 and P, and that the high quality seller is not induced to deviate from P.

407

ANSWERS

So we have constructed a semi separating PBE characterized by:

($\pi_S(60/t_1)$=(2P–180)/(P–60), $\pi_S(P/t_1)$=(120–P)/(P–60), $\pi_S(P/t_2)$=1, $\pi_B(A/60)$=1, $\pi_B(A/P)$=20/(P–40), $\pi_B(A/P')$=1 if P'<60, $\pi_B(A/P')$=0 if P'>60 and ≠P, $\mu(y_1)$=$\mu(t_1/P)$=(120–P)/60, $\mu(y_2)$=$\mu(t_2/P)$=(P–60)/60, $\mu(t_1/P')$=1 for any P'≠P).

2. We now show that it is possible to build a separating PBE where the low quality seller plays H_1=60, the high quality seller plays H_2=120, and the consumer buys the good at price 60 with probability 1.

   Given that only t_1 plays 60, the consumer is indifferent between buying and not buying, so he can buy with probability 1. Given that only t_2 plays 120, the consumer is indifferent between buying and not buying, so he can buy with probability q. But, to not incite t_1 to deviate to 120, it is necessary that t_1 gets more (or the same amount) with 60 than with 120, i.e: 60–40≥(120–40)q, i.e. q≤1/4. Observe that t_2 gets the positive payoff (120–100)q. So, to avoid that t_1 or t_2 deviate to another price P, it is sufficient that the consumer assigns P to a low quality seller, so that he accepts P if it is lower than 60 and refuses it if it is higher than 60.

   So we get a whole family of separating PBE characterized by:

($\pi_S(60/t_1)$=1, $\pi_S(120/t_2)$=1, $\pi_B(A/120)$=q with q≤1/4, $\pi_B(A/P)$=1 if P≤60, $\pi_B(A/P)$=0 if P>60 and ≠120, $\mu(t_2/120)$=1, $\mu(t_1/P)$=1 for any P≠120).

Remarks

Among these equilibria the most profitable to the high quality seller is the one such that q=¼. The high quality seller gets (120–100)1/4=5. This payoff is higher than the one obtained in the semi separating equilibria in question 1, where the high quality seller gets the payoff (P–100)×20/(P–40)<5 for P<120.

In both question 1 and question 2, we are far from Akerlof's remark (who claimed that only the low quality is sold), given that the high quality is also sold, even at the buyer's reservation price H_2, i.e. the highest possible price in a complete information context. This is possible because the consumer does not buy the good with probability 1.

♣ ♣ ♣ EXERCISE 11 LEMON CARS, EXPERIENCE GOOD MARKET WITH n QUALITIES

First of all, many results directly derive from the single crossing property proved in chapter 3 (see Exercise 27, chapter 3).

1. We focus on a PBE path in which each type of seller gets a positive payoff.

   We first prove that if t_i plays three prices p, p' and p", with p<p'<p", then p'=H_i.

   We need $(p–h_i)q=(p'–h_i)q'=(p"–h_i)q"$ where q, q' and q" are the probabilities of buying at prices p, p' and p": it derives q>q'>q">0 (given the positive payoff of the seller). It follows by single crossing that, for each type t_j with j<i, $(p–h_j)q>(p'–h_j)q'>(p"–h_j)q"$ and for each type t_j with j>i, $(p–h_j)q<(p'–h_j)q'<(p"–h_j)q"$. Therefore p' and p" cannot be played by any type lower than t_i and p and p' cannot be played by any type higher than t_i. It derives that p' is only played by t_i. Given that q' is different from 0 and 1, the consumer is indifferent between buying and not buying; this is only possible if p'=H_i.

It also follows that if t_i plays four prices p, p', p" and p''', with p<p'<p"<p''', then p'=p"=H_i. So each type of seller sets at most three prices, and, if she sets three prices, the middle price is H_i.

We now show that a played price p different from H_i is necessarily played by at least two types. Suppose the contrary, i.e. that p is only played by t_i. Given that it is accepted with positive probability (the seller makes a profit), it checks p<H_i, so it is accepted with probability 1. It follows that p is the lowest price in the game (because, if not, all types playing lower prices would switch to it). By single crossing it follows that t_i=t_1. But t_1 at least plays H_1 given that all the prices lower than H_1 are accepted, so p=H_1, a contradiction to our assumption. It follows that the assumption is wrong, so that p is necessarily played by at least two types.

Let us be more precise by showing that an adjacent type, t_{i-1} or t_{i+1}, plays p.

If p is played by t_j with j<i–1, then, by single crossing, t_{i-1} prefers p to any lower played price. And, given that t_i plays p, t_{i-1} prefers p to any higher played price. It follows that t_{i-1} only plays p.

Symmetrically, if p is played by a type t_j with j>i+1, then, by single crossing, t_{i+1} prefers p to any higher played price. And, given that t_i plays p, t_{i+1} prefers p to any lower played price. It follows that t_{i+1} only plays p.

So, due to single crossing, if t_i plays p and p'>p, with p and p' different from H_i, p is also played by t_{i-1} and p' is also played by t_{i+1}.

It immediately follows that at most (2n–1) different prices are played in the game. As a matter of fact, given that a type t_i at most plays three different prices, and given that, in this case, the middle price is necessarily H_i, t_1 can only play two prices H_1 and p_1>H_1. Hence p_1 is necessarily played by t_2. It follows that t_2 at most plays the three prices, p_1, H_2 and p_2>H_2. It follows that t_3 plays p_2 and at most the three prices p_2, H_3 and p_3>H_3. And so on, until t_{n-1} who at most plays three prices, p_{n-2},H_{n-1} and p_{n-1}. Hence t_n plays p_{n-1}, so she at most plays two different prices, p_{n-1} and H_n. The number (2n–1) follows.

2. H_1 can be (is by continuity) accepted with probability 1, so t_1 has to be indifferent between p_1*=H_1 and p_2*. So we need H_1–h_1=q(p_2*)(p_2*–h_1). It follows q(p_2*)=(H_1–h_1)/(p_2*–h_1)<1 (given that H_1<p_2*)

More generally, t_i, with i from 1 to n–1, is indifferent between p_i* and p_{i+1}* if:

$$(p_{i+1}^*-h_i)q(p_{i+1}^*)=(p_i^*-h_i)q(p_i^*)$$

So q(p_i*) is obtained in a recursive way, starting with q(p_1*)=q(H_1)=1:

$$q(p_{i+1}^*)=(p_i^*-h_i)q(p_i^*)/(p_{i+1}^*-h_i)$$

Given that p_{i+1}*>p_i*, q(p_i*) is lower than 1 and decreasing in i (and p_i*).

Now we check that t_i is not better off switching to p<p_i* or to p>p_{i+1}*.

Suppose first that p is a price out of the equilibrium path. We know that in that case the buyer refuses p with probability 1, unless it is lower than H_1. Yet no type of seller can be better off switching to p<H_1 or to a price leading to no transaction.

Suppose now that p<p_i* is played by a type j<i. By single crossing, t_i prefers p_i* to p_{i-1}* (because t_{i-1} is indifferent between both prices). By single crossing t_{i-1} prefers p_{i-1}* to p_{i-2}* (because t_{i-2} is indifferent between both prices), hence, by single crossing, t_i also prefers p_{i-1}*

to $p_{i-2}*$, hence prefers p_i* to $p_{i-2}*$, and so on. It derives that t_i prefers p_i* to any p_j*, with j from 1 to i–1.

You can show in a symmetric way that t_i prefers $p_{i+1}*$ to any p_j*, with j from i+2 to n.

Finally, we have to check that the buyer is indifferent between accepting and refusing p_i* (>H_1) (to be able to buy with a positive probability lower than 1). Given that p_i* is only played by t_i and t_{i-1}, the buyer is indifferent between buying and refusing the transaction if:

$$(H_{i-1}-p_i*)\rho_{i-1}\pi_{i-1}(p_i*)+(H_i-p_i*)\rho_i\pi_i(p_i*)=0$$

So $\pi_{i-1}(p_i*)=(H_i-p_i*)\rho_i\pi_i(p_i*)/(p_i*-H_{i-1})\rho_{i-1}$

Given that $\pi_n(p_n*)=1$, $\pi_i(p_i*)$ is obtained in a recursive way, starting with $\pi_n(p_n*)$.

$\pi_{i-1}(p_i*)>0$. Let us check that $\pi_{i-1}(p_i*)<1$.

We have $H_i \rho_i + H_{i-1}\rho_{i-1} < h_i (\rho_i + \rho_{i-1})$ by assumption (a)

and $\pi_n(p_n*)=1$.

Hence $H_i \rho_i + H_{i-1}\rho_{i-1} < p_i*(\rho_i + \rho_{i-1})$ because $p_i* > h_i$, hence $(H_i-p_i*)\rho_i/(p_i*-H_{i-1})\rho_{i-1}<1$, hence $(H_i-p_i*)\rho_i\pi_i(p_i*)/(p_i*-H_{i-1})\rho_{i-1}<1$, if $\pi_i(p_i*)\leq 1$.

Given that $\pi_n(p_n*)=1$, it follows that $0<\pi_{n-1}(p_n*)<1$, then $0<\pi_{n-1}(p_{n-1}*)=1-\pi_{n-1}(p_n*)<1$, then $\pi_{n-2}(p_{n-1}*)<1$, and so on till $\pi_1(p_2*)$.

♣ ♣ EXERCISE 12 PERFECT BAYESIAN EQULIBRIA IN ALTERNATE NEGOTIATION WITH INCOMPLETE INFORMATION

We know that the general structure of the game is given in Figure A5.1a, because we know that player 1 plays at most three prices, 2, 7 and 8 and that player 2, in period 2, only proposes 1 or 7.

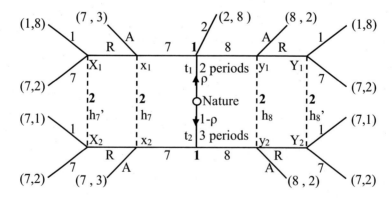

■ Figure A5.1a

There are four types of equilibria, a pooling equilibrium where both types of player 1 play 8, a pooling equilibrium where both types of player 1 play 7 (studied in chapter 5, section 3.3), a semi separating equilibrium where player 1 of type t_1 plays 2 and 8, player 1 of type t_2 playing 8, and a semi separating equilibrium where t_1 plays 2 and 7 and t_2 plays 7.

ANSWERS

Yet, before outlining these equilibria, we simplify the game tree by observing that h_7 is equivalent to $h_{7'}$, in that player 2 doesn't learn anything about player 1 between the moment where he doesn't accept player 1's offer 7, and the moment where he makes the offer 1 or 7. The same observation holds for h_8 and $h_{8'}$. So we get the game tree in Figure A5.1b. We now exclusively work on this (signalling) game tree.

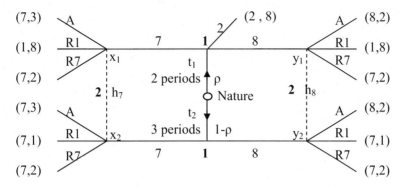

Figure A5.1b R1 and R7 mean that player 2 refuses player 1's offer and proposes 1, respectively 7 at the next period.

Case 1: both types of player 1 play 8

We look for a pooling PBE where t_1 and t_2 play 8 and player 2 accepts 8.

To help this equilibrium to emerge, we suppose that player 2 assigns the offer 7, which is not played, to t_1, so refuses it and proposes the offer 1 (action R1) (8>3>2).

Given that t_1 and t_2 play 8, we get $\mu(y_1)=\rho$, $\mu(y_2)=1-\rho$, so, at h_8, player 2 compares $8\rho+1-\rho$ to 2.

Subcase 1: $\rho \leq 1/7$
Player 2 accepts 8. It follows that player 1 is best off playing 8 regardless of type.

> So, for $\rho \leq 1/7$, we have the pooling equilibrium E_8 defined by:
> $((8/t_1, 8/t_2, A/8, R1/7), \mu(x_1)=1, \mu(x_2)=0, \mu(y_1)=\rho, \mu(y_2)=1-\rho)$

Subcase 2: $\rho > 1/7$
Player 2 refuses the offer 8 and offers 1 (action R1) at h_8. It follows that t_1 is best off deviating to the offer 2 (2>1). So the pooling equilibrium with 8 is not possible.

Case 2: both types of player 1 play 7 (see chapter 5, section 3.3).

> We established, for $\rho \leq 2/7$, the pooling equilibrium E_7 defined by:
> $((7/t_1, 7/t_2, A/7, R1/8), \mu(y_1)=1, \mu(y_2)=0, \mu(x_1)=\rho, \mu(x_2)=1-\rho)$

ANSWERS

Case 3: t_1 plays 8 with probability p and 2 with probability 1–p and t_2 plays 8.

To help this equilibrium to emerge, we suppose again that player 2 assigns the offer 7 to t_1, so he refuses it and offers 1 (action R1) (8>3>2).

Call q_1 the probability to accept 8, q_2 the probability to play R1 and q_3 the probability to play R7 at h_8. Player 2 is indifferent between A and R7, so we suppose that he plays A rather than R7 (we set $q_3=0$) in the spirit of this negotiation game. t_1 can play 2 and 8 if the offer 8 is both followed by A and R1 with positive probability. We need $2=8q_1+q_2$, hence $q_1=1/7$ and $q_2=6/7$. It derives that t_2 only plays 8 because $8/7+7\times 6/7=50/7>7$

To play $q_1=1/7$ and $q_2=6/7$, player 2 needs $8\mu(y_1)+1-\mu(y_1)=2$, i.e. $\mu(y_1)=1/7$. By calling p the probability assigned by t_1 to the offer 8, we get: $\rho p/(\rho p+1-\rho)=1/7$, i.e. $p=(1-\rho)/6\rho$, which is possible if $\rho \geq 1/7$.

> So, for $\rho \geq 1/7$, we get a semi separating equilibrium E_{82} defined by:
>
> $(\pi_{1t1}(2)=(7\rho-1)/(6\rho), \pi_{1t1}(8)=(1-\rho)/(6\rho), 8/t_2, \pi_{2h8}(A)=1/7, \pi_{2h8}(R1)=6/7,$
> $R1/7, \mu(y_1)=1/7, \mu(y_2)=6/7, \mu(x_1)=1, \mu(x_2)=0)$

Let us have a look at the equilibrium payoffs: t_1 gets 2, so does not benefit from incomplete information. t_2 gets: $8/7+7\times 6/7=50/7$. Player 2 gets: $\rho(1-p)8+2(\rho p+1-\rho)=7\rho+1\geq 2$. This payoff can go from 2 to 8 when ρ goes to 1.

Case 4: t_1 plays 7 with probability p and 2 with probability 1–p and t_2 plays 7.

To help this equilibrium to emerge, we suppose that player 2 assigns the offer 8 to t_1, so he refuses it and offers 1 (action R1) (8>2=2).

Each strategy that leads to play R7 at h_7 is weakly dominated by the strategy that, all else being equal, leads to play A at h_7. Call q the probability to accept 7 and (1–q) the probability to refuse it and offer 1. t_1 can play 2 and 7 if she gets the same payoff with both actions, so $2=7q+1-q$, i.e. $q=1/6$. Player 2 accepts playing A and R1 at h_7 if $3=8\mu(x_1)+1-\mu(x_1)$, i.e. $\mu(x_1)=2/7$. By calling p the probability assigned by t_1 to 7, we get: $\rho p/(\rho p+1-\rho)=2/7$, i.e. $p=2(1-\rho)/5\rho$, which is possible if $\rho \geq 2/7$.

t_2 gets 7 with the offer 7 and the offer 8, so he can offer 7 with probability 1.

> So, for $\rho \geq 2/7$, we have a semi separating equilibrium E_{72} defined by:
>
> $(\pi_{1t1}(2)=(7\rho-2)/(5\rho), \pi_{1t1}(7)=2(1-\rho)/(5\rho), 7/t_2, \pi_{2h7}(A)=1/6,$
> $\pi_{2h7}(R1)=5/6, R1/8, \mu(x_1)=2/7, \mu(x_2)=5/7, \mu(y_1)=1, \mu(y_2)=0)$

Let us have a look at the equilibrium payoffs: t_1 gets 2, t_2 gets 7 and player 2 gets $\rho(1-p)8+3(\rho p+1-\rho)=1+7\rho$, which goes from 3 to 8 when ρ goes to 1.

We summarize these equilibria in Table A5.1 and Table A5.2.

ANSWERS

Table A5.1

ρ	0	1/7	2/7	1
equilibria	E_8	E_{82}	E_{82}	
	E_7	E_7	E_{72}	

Table A5.2

equilibria	E_8	E_7	E_{82}	E_{72}
payoffs	$t_1=8$	$t_1=7$	$t_1=2$	$t_1=2$
	$t_2=8$	$t_2=7$	$t_2=50/7$	$t_2=7$
	player 2=2	player 2=3	player 2=from 2 to 8 (when ρ→1)	player 2=from 3 to 8 (when ρ→1)

To conclude, we see that player 1 can benefit from incomplete information. As long as ρ<2/7, he can get 7 and even 8 (if ρ<1/7), even if there are only two periods of negotiation. But if ρ>2/7, then player 1, if there are three periods of negotiation, suffers from incomplete information (7<50/7<8). Player 2 can also benefit from incomplete information. For example, he can get 3 even if there are three periods of negotiation when ρ<2/7. And, for ρ>2/7, his payoff, equal to 1+7ρ, can go to 8, when ρ→1, which means that player 2 can get a high payoff even if there are three periods, provided that the probability of two periods is large. But he suffers from incomplete information if there are only two periods of negotiation, especially in the pooling equilibria.

Finally observe that if we suppress t_2's weakly dominated strategy 7, then we only keep the equilibria E_8 and E_{82}.

Chapter 6

Answers to Exercises

♣ EXERCISE 1 BEST REPLY MATCHING IN A NORMAL FORM GAME

1. Matrix A6.1a is the best reply matrix.

		Player 2		
		q_1 A_2	q_2 B_2	q_3 C_2
Player 1	p_1 A_1	b_1b_2		b_2
	p_2 B_1		b_1b_2	b_1b_2
	p_3 C_1	b_1b_2		b_1b_2

Matrix A6.1a

The BRM concept leads to the equations:

$p_1=q_1/2$ $q_1=p_1/2+p_3/2$
$p_2=q_2+q_3/2$ and $q_2=p_2/2$ and $p_1+p_2+p_3=q_1+q_2+q_3=1$
$p_3=q_1/2+q_3/2$ $q_3=p_1/2+p_2/2+p_3/2$

The solution is: $p_1=1/8$, $p_2=1/2$, $p_3=3/8$, $q_1=q_2=0.25$ $q_3=0.5$.

B_1 is the action most played by player 1 (because it is the only best response to B_2 and one of the two best responses to player 2's most played action C_2). C_2 is the action most played by player 2 (because it is a best response to any action of player 1, and it weakly dominates A_2 and B_2). But the other actions are also played with a significant probability. This derives from the fact that in this game each action is at least one time, and often several times, a best response (remember that there are five pure strategy Nash equilibria in this game).

2. With the GBRM concept, we get more equilibria, and we can get couples of equilibria that resist iterative elimination of dominated strategies.

    For example, after suppressing b_1 in the box corresponding to (C_1,C_2) (we can suppress it because there is another best response), we get:

$p_1=q_1/2$ $q_1=p_1/2+p_3/2$
$p_2=q_2+q_3$ and $q_2=p_2/2$ and $p_1+p_2+p_3=q_1+q_2+q_3=1$
$p_3=q_1/2$ $q_3=p_1/2+p_2/2+p_3/2$

414

The solution is: $p_1=0$, $p_2=1$, $p_3=0$, $q_1=0$, $q_2=q_3=0.5$: so we get the set $\{(B_1,B_2),(B_1,C_2)\}$, which is one of the set of couples of strategies resisting iterative elimination of weakly dominated strategies (each couple is played with probability ½).

We can also get the sets of couples resisting the other orders of elimination of dominated strategies. For example, by keeping the best responses in Matrix A6.1b, we get the couple (B_1,C_2) ($p_2=q_3=1$, $p_1=p_3=q_1=q_2=0$); by keeping the best responses in Matrix A6.1c, we get the set $\{(B_1,C_2),(C_1,C_2)\}$ ($q_3=1$, $p_2=p_3=1/2$, $p_1=q_1=q_2=0$), each couple being played with probability ½. And by keeping the best responses in Matrix A6.1d, we get the set $\{(C_1,A_2),(C_1,C_2)\}$ ($p_3=1$, $q_1=q_3=1/2$, $p_1=p_2=q_2=0$), each couple being played with probability ½.

		P2		
		A_2	B_2	C_2
P1	A_1	b_1		b_2
	B_1		b_1	b_1b_2
	C_1	b_2		

Matrix A6.1b

		P2		
		A_2	B_2	C_2
P1	A_1			b_2
	B_1		b_1	b_1b_2
	C_1	b_1		b_1b_2

Matrix A6.1c

		P2		
		A_2	B_2	C_2
P1	A_1			b_2
	B_1		b_1	b_2
	C_1	b_1b_2		b_1b_2

Matrix A6.1d

		P2		
		A_2	B_2	C_2
P1	A_1	b_1b_2		
	B_1		b_1b_2	b_1
	C_1	b_2		

Matrix A6.1e

But we also get other sets of couples of strategies. For example, by keeping the best responses in Matrix A6.1e, we get the set $\{(A_1,A_2),(B_1,B_2),(A_1,B_2),(B_1,A_2)\}$ with $p_1=q_1$, $p_2=q_2$, $p_1+p_2=q_1+q_2=1$.

The many possible outcomes derive from the fact that the game has many best responses; in such a game, the GBRM (and the BRM) equilibrium concept does not make a strong selection.

We now take the values of the payoffs into account (as in section 1.4 in chapter 6).
We get:

$p_1=q_1(7/14)+q_2(3/10)$ $q_1=p_1(8/19)+p_3/2$
$p_2=q_2(4/10)+q_3(1/2)$ and $q_2=p_1(3/19)+p_2(4/8)$
$p_3=q_1(7/14)+q_2(3/10)+q_3(1/2)$ $q_3=p_1(8/19)+p_2(4/8)+p_3/2$
and $p_1+p_2+p_3=q_1+q_2+q_3=1$

The solution is: $p_1=0.22>1/8$, $p_2=0.32<0.5$, $p_3=0.46>3/8$, $q_1=0.32>0.25$, $q_2=0.2<0.25$, $q_3=0.48<0.5$.

By comparison with the BRM equilibrium, the probabilities assigned to A_1, C_1 and A_2 rise, because these actions can ensure high payoffs to player 1 and player 2. The probability assigned to C_2 is almost the same, and the probabilities assigned to B_1 and B_2 decrease because they lead to lower payoffs than those possible with the actions A and C.

Yet, once again, all the probabilities are rather large: taking into account cardinal payoffs and best replies in this game doesn't help in finding a unique couple of actions to play.

ANSWERS

♣♣ EXERCISE 2 BEST-REPLY MATCHING IN THE TRAVELLER'S DILEMMA

1. The normal form of the game and the best reply matrix are given in Matrix A6.2a and Matrix A6.2b:

		\multicolumn{7}{c}{P2}
		2	3	4	5	6	7	8
	2	(2,2)	(4,0)	(4,0)	(4,0)	(4,0)	(4,0)	(4,0)
	3	(0,4)	(3,3)	(5,1)	(5,1)	(5,1)	(5,1)	(5,1)
	4	(0,4)	(1,5)	(4,4)	(6,2)	(6,2)	(6,2)	(6,2)
P1	5	(0,4)	(1,5)	(2,6)	(5,5)	(7,3)	(7,3)	(7,3)
	6	(0,4)	(1,5)	(2,6)	(3,7)	(6,6)	(8,4)	(8,4)
	7	(0,4)	(1,5)	(2,6)	(3,7)	(4,8)	(7,7)	(9,5)
	8	(0,4)	(1,5)	(2,6)	(3,7)	(4,8)	(5,9)	(8,8)

Matrix A6.2a

		\multicolumn{7}{c}{P2}
		q_2 2	q_3 3	q_4 4	q_5 5	q_6 6	q_7 7	q_8 8
	p_2 2	$b_1 b_2$	b_1					
	p_3 3	b_2		b_1				
	p_4 4		b_2		b_1			
P1	p_5 5			b_2		b_1		
	p_6 6				b_2		b_1	
	p_7 7					b_2		b_1
	p_8 8						b_2	

Matrix A6.2b

Observe the structure of the best replies: the b_1 and the b_2 are above and below the diagonal and this structure holds for any M and any P>1 (given that the best answer to x is x–1, when x>2, and the best answer to x=2 is 2). We get:

$p_2=q_2+q_3$ and $q_2=p_2+p_3$ and $p_2+p_3+p_4+p_5+p_6+p_7+p_8=1$
$p_3=q_4$ $q_3=p_4$ and $q_2+q_3+q_4+q_5+q_6+q_7+q_8=1$
$p_4=q_5$ $q_4=p_5$
$p_5=q_6$ $q_5=p_6$
$p_6=q_7$ $q_6=p_7$
$p_7=q_8$ $q_7=p_8$
$p_8=0$ $q_8=0$

The solution is: $p_2=q_2=1$ and $p_i=q_i=0$ for i from 3 to 8.

So we get the only pure strategy Nash equilibrium of this game, which is also the unique couple that resists iterative elimination of dominated strategies.

The result generalizes for any value of M and any P>1. Keeping the same notations, with i from 2 to M, we get:

$p_2=q_2+q_3$ and $q_2=p_2+p_3$ and $\sum_{i=2}^{M} p_i = 1$ and $\sum_{i=2}^{M} q_i = 1$

ANSWERS

$p_i=q_{i+1}$ $q_i=p_{i+1}$ for i from 3 to M–1
$p_M=0$ $q_M=0$
The solution is $p_2=q_2=1, p_i=q_i=0$ for i from 3 to M.

So BRM cannot do more than stability to unilateral deviations: regardless of the values of M and P (>1), both players play 2 at the unique BRM equilibrium.

2. The normal form of the game and the best response matrix are given in Matrix A6.2c and Matrix A6.2d:

P2

		2	3	4	5	6	7	8
	2	(2,2)	(4,1)	(4,2)	(4,3)	(4,4)	(4,5)	(4,6)
	3	(1,4)	(3,3)	(5,2)	(5,3)	(5,4)	(5,5)	(5,6)
	4	(2,4)	(2,5)	(4,4)	(6,3)	(6,4)	(6,5)	(6,6)
P1	5	(3,4)	(3,5)	(3,6)	(5,5)	(7,4)	(7,5)	(7,6)
	6	(4,4)	(4,5)	(4,6)	(4,7)	(6,6)	(8,5)	(8,6)
	7	(5,4)	(5,5)	(5,6)	(5,7)	(5,8)	(7,7)	(9,6)
	8	(6,4)	(6,5)	(6,6)	(6,7)	(6,8)	(6,9)	(8,8)

Matrix A6.2c

P2

		q_2	q_3	q_4	q_5	q_6	q_7	q_8
		2	3	4	5	6	7	8
	p_2 2							b_2
	p_3 3							b_2
	p_4 4				(b_1)			b_2
P1	p_5 5			(b_2)			b_1	b_2
	p_6 6					b_2		b_1
	p_7 7						b_2	b_1
	p_8 8	b_1	b_1	b_1	b_1		b_2	

Matrix A6.2d

Observe the particular structure of the best reply matrix: for large values x of the opponent (x>5), the best is to play x–1 in order to get x–1+2, but, for low values (x<5), the best is to play 8 because 8–2 is higher than x–1+2. For x=5, both 4 and 8 are best responses (4+2=8–2)

We get:

$p_2=p_3=0$ and $q_2=q_3=0$ and $\sum_{i=2}^{8} p_i = 1$ and $\sum_{i=2}^{8} q_i = 1$
$p_4=q_5/2$ $q_4=p_5/2$
$p_5=q_6$ $q_5=p_6$
$p_6=q_7$ $q_6=p_7$
$p_7=q_8$ $q_7=p_8$
$p_8=q_2+q_3+q_4+q_5/2$ $q_8=p_2+p_3+p_4+p_5/2$

The solution is: $p_2=p_3=q_2=q_3=0$, $p_4=q_4=1/9$, and $p_5=p_6=p_7=p_8=q_5=q_6=q_7=q_8=2/9$.

ANSWERS

This result is close to the Nash equilibrium which assigns probability 0.25 to each amount 5, 6, 7 and 8, and it is also close to the Nash equilibrium of the continuous version of this game, which is the uniform distribution on [4,8] (see Exercise 16 in chapter 3).

Observe that GBRM allows to get the Nash equilibrium, by eliminating the two circled best responses in Matrix A6.2d (player 1 can choose to best reply with the amount 8 to player 2's amount 5 and symmetrically).

The equations become:

$p_2=p_3=p_4=0$ and $q_2=q_3=q_4=0$ and $\sum_{i=2}^{8} p_i = 1$ and $\sum_{i=2}^{8} q_i = 1$
$p_5=q_6$ $q_5=p_6$
$p_6=q_7$ $q_6=p_7$
$p_7=q_8$ $q_7=p_8$
$p_8=q_2+q_3+q_4+q_5$ $q_8=p_2+p_3+p_4+p_5$

The solution is $p_2=p_3=p_4=q_2=q_3=q_4=0$, $p_i=q_i=0.25$ for i from 5 to 8.

Let us insist on the intuitive content of this equilibrium:

7 is the best reply to 8, 6 is the best reply to 7, 5 is the best reply to 6, 8 is the best reply to 5. Given that each amount is the best reply to exactly one other amount, it is really intuitive to ask for each of them with the same probability. GBRM equilibrium is more intuitive than the Nash equilibrium which requires the equality of payoffs, i.e.

$5q_5+7q_6+7q_7+7q_8=4q_5+6q_6+8q_7+8q_8=5q_5+5q_6+7q_7+9q_8=6q_5+6q_6+6q_7+8q_8$
(for player 1)

$5p_5+7p_6+7p_7+7p_8=4p_5+6p_6+8p_7+8p_8=5p_5+5p_6+7p_7+9p_8=6p_5+6p_6+6p_7+8p_8$
(for player 2)

3. The above result generalizes for any M and $1<P<(M-2)/2$.

The BRM equations are:

$p_2=p_3=\ldots=p_{M-2P-1}=0$ and $q_2=q_3=\ldots=q_{M-2P-1}=0$
$p_{M-2P}=q_{M-2P+1}/2$ $q_{M-2P}=p_{M-2P+1}/2$
$p_i=q_{i+1}$ for i from $M-2P+1$ to $M-1$ $q_i=p_{i+1}$ for i from $M-2P+1$ to $M-1$
$p_M=q_2+q_3+\ldots+q_{M-2P}+q_{M-2P+1}/2$ $q_M=p_2+p_3+\ldots+p_{M-2P}+p_{M-2P+1}/2$

and $\sum_{i=2}^{M} p_i = 1$ and $\sum_{i=2}^{M} q_i = 1$

The solution is $p_{M-2P}=q_{M-2P}=1/(1+4P)$, $p_i=q_i=2/(1+4P)$ for i from $M-2P+1$ to M.

ANSWERS

♣♣ EXERCISE 3 BEST-REPLY MATCHING IN THE PRE-EMPTION GAME

1. Matrix A6.3a is the best reply matrix:

		Player 2			
		q_1 $(S_2/x_2 S_2/x_4)$	q_2 $(S_2/x_2 C_2/x_4)$	q_3 $(C_2/x_2 S_2/x_4)$	q_4 $(C_2/x_2 C_2/x_4)$
Player 1	p_1 $(S_1/x_1 S_1/x_3 S_1/x_5)$	$b_1 b_2$	$b_1 \underline{b_2}$	b_2	b_2
	p_2 $(S_1/x_1 S_1/x_3 C_1/x_5)$	$b_1\ b_2$	$b_1 \underline{b_2}$	b_2	b_2
	p_3 $(S_1/x_1 C_1/x_3 S_1/x_5)$	$b_1\ b_2$	$b_1 \underline{b_2}$	b_2	b_2
	p_4 $(S_1/x_1 C_1/x_3 C_1/x_5)$	$b_1\ b_2$	$b_1 \underline{b_2}$	b_2	b_2
	p_5 $(C_1/x_1 S_1/x_3 S_1/x_5)$	b_2	b_2	b_1	
	p_6 $(C_1/x_1 S_1/x_3 C_1/x_5)$	b_2	b_2	b_1	
	p_7 $(C_1/x_1 C_1/x_3 S_1/x_5)$			b_2	b_1
	p_8 $(C_1/x_1 C_1/x_3 C_1/x_5)$				b_2

Matrix A6.3a

We get the equations:

$p_1 = q_1/4 + q_2/4$ and $q_1 = (p_1+p_2+p_3+p_4)/4 + p_5/2 + p_6/2$
$p_2 = q_1/4 + q_2/4$ $q_2 = (p_1+p_2+p_3+p_4)/4 + p_5/2 + p_6/2$
$p_3 = q_1/4 + q_2/4$ $q_3 = (p_1+p_2+p_3+p_4)/4 + p_7$
$p_4 = q_1/4 + q_2/4$ $q_4 = (p_1+p_2+p_3+p_4)/4 + p_8$
$p_5 = q_3/2$
$p_6 = q_3/2$
$p_7 = q_4$
$p_8 = 0$ and $\sum_{i=1}^{8} p_i = 1$ and $\sum_{i=1}^{4} q_i = 1$

The solution is: $p_1 = p_2 = p_3 = p_4 = p_5 = p_6 = p_7 = 1/7$, $p_8 = 0$, $q_1 = q_2 = q_3 = 2/7$ and $q_4 = 1/7$

So player 1 stops at x_1 with probability 4/7, she goes on at x_1 but stops at x_3 with probability 2/7 and she goes on till x_5 but stops at this node with probability 1/7. She never goes on at x_5. Player 2 stops at x_2 with probability 4/7, goes on at x_2 but stops at x_4 with probability 2/7, and goes on at all decision nodes with probability 1/7.

So BRM allows the game to go on till the last decision node, even if the probability to stop at the first decision node is high (4/7); observe that player 1 stops at her last decision node with probability 1, but that player 2 goes on at x_4 with positive probability.

The associated behavioural strategies are those given in *Exercise 8 in chapter 1*:

$\pi_{1x1}(S_1) = 4/7$, $\pi_{2x2}(S_2) = 4/7$, $\pi_{1x3}(S_1) = 2/3$, $\pi_{2x4}(S_2) = 2/3$, $\pi_{1x5}(S_1) = 1$

They nicely express that the probability to stop grows over time (4/7<2/3<1), which seems in accordance with real behaviours: people know that it is good to go on but fear the impact of the logical stopping decision at the end of the game.

ANSWERS

2. The reduced normal form leads to the best reply Matrix A6.3b:

		Player 2		
		q_1 S_2/x_2	q_2 $(C_2/x_2 S_2/x_4)$	q_3 $(C_2/x_2 C_2/x_4)$
	p_1 S_1/x_1	b_1b_2	b_2	b_2
Player 1	p_2 $(C_1/x_1 S_1/x_3)$	b_2	b_1	
	p_3 $(C_1/x_1 C_1/x_3 S_1/x_5)$		b_2	b_1
	p_4 $(C_1/x_1 C_1/x_3 C_1/x_5)$			b_2

Matrix A6.3b

The diagonal structure of best replies becomes obvious (if a player stops at x_i, it is best for the opponent to stop at x_{i-1}). The first line of b_2 expresses that player 2 is indifferent between all his actions if player 1 stops immediately (given that he is not called on to play).

The equations become:

$p_1 = q_1$ $q_1 = p_1/3 + p_2$
$p_2 = q_2$ $q_2 = p_1/3 + p_3$
$p_3 = q_3$ $q_3 = p_1/3 + p_4$
$p_4 = 0$ and $\sum_{i=1}^{4} p_i = 1$ and $\sum_{i=1}^{3} q_i = 1$

The solution is: $p_1 = q_1 = 1/2$, $p_2 = q_2 = 1/3$, $p_3 = q_3 = 1/6$, $p_4 = 0$.

So player 1 stops at x_1 with probability 1/2 (<4/7), she goes on at x_1 but stops at x_3 with probability 1/3 (>2/7), she goes on till x_5 but stops at this node with probability 1/6 (>1/7). She never goes on at x_5. Player 2 stops at x_2 with probability 1/2 (<4/7), goes on at x_2 but stops at x_4 with probability 1/3 (>2/7), goes on at all decision nodes with probability 1/6 (>1/7). As expected, the reduced normal form and the normal form lead to similar BRM equilibria. Yet *they are not identical, so the BRM equilibria may differ depending on whether we work on the normal or the reduced normal form.* The associated behavioural strategies are also partially different.

These strategies are given in Exercise 8 in chapter 1.

$\pi_{1x1}(S_1) = p_1/1 = 1/2$, $\pi_{2x2}(S_2) = q_1/(q_1+q_2+q_3) = 1/2$,
$\pi_{1x3}(S_1) = p_2/(p_2+p_3+p_4) = 2/3$, $\pi_{2x4}(S_2) = q_2/(q_2+q_3) = 2/3$,
$\pi_{1x5}(S_1) = p_3/(p_3+p_4) = 1$

We observe again that the probability of stopping at each node increases over time (1/2<2/3<1), the probability of stopping at the first node being lower than in the normal form game.

Yet these slight differences are not so important: thanks to GBRM, you can get both equilibria in both representative forms (see Umbhauer 2007).[1]

3. GBRM, applied to the reduced normal form, allows all the behaviours given by the equations:

$p_1 = q_1$ $q_1 = p_1\delta_1 + p_2$
$p_2 = q_2$ $q_2 = p_1\delta_2 + p_3$
$p_3 = q_3$ $q_3 = p_1\delta_3 + p_4$
$p_4 = 0$ and $\sum_{i=1}^{4} p_i = 1$ and $\sum_{i=1}^{3} q_i = 1$

ANSWERS

with $\delta_1 \geq 0, \delta_2 \geq 0, \delta_3 \geq 0$ and $\delta_1 + \delta_2 + \delta_3 = 1$. It follows that:
$p_1 = q_1 = 1/(1+\delta_2+2\delta_3)$, $p_2 = q_2 = (\delta_2+\delta_3)/(1+\delta_2+2\delta_3)$, $p_3 = q_3 = \delta_3/(1+\delta_2+2\delta_3)$ and $p_4 = 0$

Let us find the two extreme behaviours, where player 2 chooses to stop immediately when player 1 stops at x_1, and the one where he chooses to always go on.

In the first case, we have $\delta_1 = 1$, $\delta_2 = \delta_3 = 0$, and so

$p_1 = q_1 = 1$ $p_2 = p_3 = p_4 = 0$, $q_2 = q_3 = 0$, *so we get the unique subgame perfect Nash equilibrium path where both players stop the game as soon as possible.*

In the second case, we have $\delta_3 = 1$, $\delta_1 = \delta_2 = 0$, and we get:

$p_1 = p_2 = p_3 = q_1 = q_2 = q_3 = 1/3$ $p_4 = 0$

The associated behavioural strategies are:

$\pi_{1x1}(S_1) = p_1/1 = 1/3$, $\pi_{2x2}(S_2) = q_1/(q_1+q_2+q_3) = 1/3$,
$\pi_{1x3}(S_1) = p_2/(p_2+p_3+p_4) = 1/2$, $\pi_{2x4}(S_2) = q_2/(q_2+q_3) = 1/2$,
$\pi_{1x5}(S_1) = p_3/(p_3+p_4) = 1$

The probability to stop at each node increases with time, but it is much smaller than in the BRM equilibrium (1/3<½, ½<2/3).

So many things are possible: at best both players go on at the first decision node with probability 2/3, at worst, they stop immediately. But the probability to stop always increases with time and player 1 always stops at the last decision node. All these results are quite intuitive, *yet not always observed in reality. My L3 students on a longer game did not behave in this regular way. Many of them went to the end of the game without stopping, even at the last decision node (see Exercise 3 in chapter 4 and Exercise 6 in this chapter).*

♣ EXERCISE 4 MINIMIZING REGRET IN A NORMAL FORM GAME

For player 1:

- the maximal regret associated with A_1 is 1 because, when player 2 plays B_2, she gets 3 with A_1 instead of 4 with B_1, and when player 2 plays C_2, she gets 0 with A_1 instead of 1 with B_1 or C_1.
- the maximal regret associated with B_1 is 7, because player 1 could get 7 (with A_1 or C_1) instead of 0 when player 2 plays A_2.
- the maximal regret associated with C_1 is 1 because player 1 could get 4 (with B_1) instead of 3 when player 2 plays B_2.

So A_1 and C_1 minimize player 1's regret.

For player 2:

- the maximal regret associated with A_2 is 4 because player 2 could get 4 (with B_2 or C_2) instead of 0 when player 1 plays B_1.
- the maximal regret associated with B_2 is 5 because player 2 could get 8 (with A_2 or C_2) instead of 3 when player 1 plays A_1.

ANSWERS

- the maximal regret associated with C_2 is 0 because C_2 is a best response to any action of player 1.

So C_2 minimizes player 2's regret.

To my mind, this way of playing may be a little dangerous, at least with regard to player 1's actions. Given that player 2's actions are not Nature moves but rational decisions, player 1 should not give too much weight to the payoff 7 she gets with A_1 or C_1, because she gets 7 only if player 2 plays A_2: but player 2 is more induced to play C_2 than A_2.

♣♣ EXERCISE 5 MINIMIZING REGRET IN BERTRAND'S DUOPOLY

Suppose that player 1 plays $p_1>c$. She will never play $p_1>(K+ac)/2a$ (her monopoly price).

If player 2 plays $p_2<p_1$, p_1 is a best player 1's answer if $p_2=c$ (in this case there is no regret), and $p_2-\varepsilon$ is her best answer if $p_2>c$. In that case her regret is $(K-ap_2+a\varepsilon)(p_2-\varepsilon-c)-0$ and it is maximal when $p_2=p_1-\varepsilon$: it is $(K-ap_1+2a\varepsilon)(p_1-2\varepsilon-c)$ and goes to $(K-ap_1)(p_1-c)$ when ε goes to 0.

If player 2 plays $p_2>p_1$, player 1's best answer is min $((K+ac)/2a, p_2-\varepsilon)$, and the maximal regret is obtained when $p_2>(K+ac)/2a$. This regret is $(K-ac)^2/4a-(K-ap_1)(p_1-c)$.

If $p_2=p_1$, player 1's best answer is $p_1-\varepsilon$, and the regret is $(K-ap_1+a\varepsilon)(p_1-\varepsilon-c)-(K-ap_1)(p_1-c)/2$; it goes to $(K-ap_1)(p_1-c)/2$ when ε goes to 0 and is clearly lower than $(K-ap_1)(p_1-c)$.

So the maximal regret, M, is given by:

$$M = \max((K-ap_1)(p_1-c), (K-ac)^2/4a-(K-ap_1)(p_1-c))$$

It derives: $M=(K-ap_1)(p_1-c)$ when $2(K-ap_1)(p_1-c) \geq (K-ac)^2/4a$ (case 1)

$= (K-ac)^2/4a-(K-ap_1)(p_1-c))$ when $2(K-ap_1)(p_1-c) \leq (K-ac)^2/4a$ (case 2)

Case 1: in this case, the minimal regret is obtained for $2(K-ap_1)(p_1-c)=(K-ac)^2/4a$, i.e. $p_1=(2-2^{1/2})K/4a+c(2^{1/2}+2)/4$ (lower than $(K+ac)/2a$). The regret is equal to $(K-ac)^2/8a$.

Case 2: in this case, the minimal regret is again obtained for $2(K-ap_1)(p_1-c)=(K-ac)^2/4a$ and is again equal to $(K-ac)^2/8a$ for $p_1=(2-2^{1/2})K/4a+c(2^{1/2}+2)/4$.

So $p_1^*=(2-2^{1/2})K/4a+c(2^{1/2}+2)/4$ minimizes the firm's regret; if both firms set this price, each gets $u_1(p_1^*,p_1^*)=(K-ap_1^*)(p_1^*-c)/2=(K-ac)^2/16a$, i.e. the monopoly profit divided by 4.

So, for example, when $K=100$, $a=2$, $c=10$, we get $p^*=15.86$ and $u_1(p_1^*,p_1^*)=u_2(p_1^*,p_1^*)=200$. The monopoly price is 30 and the monopoly profit is 800, the Bertrand price is $c=10$ and the Bertrand profit is null. So the sum of the profits of both firms is half the monopoly profit, so it is just in between the Bertrand profit and the monopoly profit.

To my mind, Halpern and Pass's minimizing criterion well applies to Bertrand's duopoly. As in the traveller's dilemma, each price is justifiable (x is the best answer to x+ε) so it is difficult to guess the price played by the other, and as in the traveller's dilemma, you may have regrets with two different behaviours: if your price is higher than your opponent's one, you regret to not have played a lower price, but if it is lower, you regret to not have played a higher one.

ANSWERS

♣ ♣ EXERCISE 6 MINIMIZING REGRET IN THE PRE-EMPTION GAME

1. The maximal regret associated with S_1/x_1, respectively to $(C_1/x_1 S_1/x_3)$, $(C_1/x_1 C_1/x_3 S_1/x_5)$, $(C_1/x_1 C_1/x_3 C_1/x_5 S_1/x_7)$, $(C_1/x_1, C_1/x_3, C_1/x_5, C_1/x_7, S_1/x_9)$ and $(C_1/x_1, C_1/x_3, C_1/x_5, C_1/x_7, C_1/x_9)$ is 395, respectively 350, 300, 250, 150 and 225 (see the bold italic terms). So the strategy that minimizes regret is $(C_1/x_1, C_1/x_3, C_1/x_5, C_1/x_7, S_1/x_9)$. Player 1 should go on till x_9 and stop there.

    The maximal regret associated with S_2/x_2, respectively to $(C_2/x_2 S_2/x_4)$, $(C_2/x_2 C_2/x_4 S_2/x_6)$, $(C_2/x_2 C_2/x_4 C_2/x_6 S_2/x_8)$, $(C_2/x_2, C_2/x_4, C_2/x_6, C_2/x_8, S_2/x_{10})$ and $(C_2/x_2, C_2/x_4, C_2/x_6, C_2/x_8, C_2/x_{10})$ is 245, respectively 200, 150, 100, 125 and 210 (see the underlined numbers). So the strategy that minimizes regret is $(C_2/x_2 C_2/x_4 C_2/x_6 S_2/x_8)$. So player 2 should go on till x_8 and stop there.

    First observe that this result doesn't really fit with the students' behaviour: 29.5% of the L3 students admittedly advise player 1 to stop at x_9 but 35.2% advise her to never stop (a similar contrast, 23.1% against 27%, holds for the M1 students), and only 13.1% of the L3 students (15.4% of the M1 students) advise player 2 to stop at x_8. Yet Halpern and Pass's result looks rather convincing. We could argue that players apply backward induction till x_8 but refuse to apply it further because they regret the payoff losses due to stopping too early. So *we have a kind of mix between the strength of backward induction and the power of the increasing payoffs*. Yet this nice result is highly linked to the values of the payoffs, as we will see in question 3.

2. Let us adapt regret minimization to the extensive form game. We do local comparisons of maximal regrets. Stopping at x_1 leads to the maximal regret 395 because player 1 can get 400 by going on; going on at x_1 leads to the maximal regret 5 because player 1 gets 0 if player 2 stops at x_2. So going on minimizes her local regret because 5<395. In the same way, player 1, at x_3 compares 350 to 49 and goes on. At x_5 she compares 300 to 90 and goes on, at x_7 she compares 250 to 135 and goes on, at x_9 she compares 150 to 225 and stops. So player 1 stops at x_9, as in the normal form game.

    We proceed in the same way for player 2. Stopping at x_2 leads to the maximal regret 245 because player 2 can get 250 by going on, whereas going on leads to the maximal regret 4, because he gets 1 if player 1 stops at x_3. So going on minimizes his local regret because 4<245. In the same way, at x_4, he compares 200 to 40 and goes on, at x_6 he compares 150 to 85 and goes on, at x_8 he compares 100 to 125 and stops, at x_{10} he compares 0 to 210 and stops. So player 2 stops at x_8 as in the normal form game.

    So we get the same behaviour *but we do not compare the same numbers*. For player 1, we do not compare 395, 350, 300, 250, 150 and 225, but we compare the amounts linked to each pair of same named arrows in Matrix E6.3 (395 to 5, 350 to 49, 300 to 90, 250 to 135, 150 to 225); nevertheless the main comparison, i.e. the one that leads to the minimizing regret strategy is in both approaches: in both approaches we compare 150 to 225, which leads to stop at x_9. Similar comments hold for player 2.

3. If we replace 400 by 500, nothing changes for player 2, but the regrets change for player 1. In the reduced normal form game, the maximal regret associated with S_1/x_1, respectively to $(C_1/x_1 S_1/x_3)$, $(C_1/x_1 C_1/x_3 S_1/x_5)$, $(C_1/x_1 C_1/x_3 C_1/x_5 S_1/x_7)$, $(C_1/x_1, C_1/x_3, C_1/x_5, C_1/x_7, S_1/x_9)$ and $(C_1/x_1, C_1/x_3, C_1/x_5, C_1/x_7, C_1/x_9)$ becomes 495, respectively 450, 400, 350, 250 and 225, and the strategy that minimizes regret consists in never stopping. And the same is true in the extensive form game. Player 1 now compares 495 to 5, 450 to 49, 400 to 90, 350 to 135, 250 to 225 and

ANSWERS

never stops. Well, *we here observe that Halpern and Pass's concept has nothing to do with backward induction:* why should player 2 stop at x_8 if player 1 never stops? And why does player 1 always go on if player 2 stops at x_8 and x_{10}? Clearly *the strategy that minimizes the regret of a player is established without taking into account the strategies played by the other players:* in some way, there is no strategic thinking. To better highlight this point, suppose that the players only play the reduced game starting at node x_8. The reduced normal form is given in Matrix A6.4:

		Player 2 S_2/x_8	$(C_2/x_8 S_2/x_{10})$	$(C_2/x_8 C_2/x_{10})$
Player 1	S_1/x_9	(15,150)	(250,25)	(250,25)
	C_1/x_9	(15,150)	(25,250)	(500,40)

Matrix A6.4

The maximal regret associated with S_1/x_9 is 250 because player 1 may lose 250 (=500−250) by stopping if player 2 plays $(C_2/x_8 C_2/x_{10})$. Given that the maximal regret associated with C_1/x_9 is only 225 (250−25), player 1 plays C_1/x_9. Yet $(C_2/x_8 C_2/x_{10})$ is strictly dominated by S_2/x_8, so player 1 should not fear that player 2 plays $(C_2/x_8 C_2/x_{10})$, so she should not attach much significance to the regret 250 and rather play S_1/x_9.

♣ ♣ ♣ EXERCISE 7 MINIMIZING REGRET IN THE TRAVELLER'S DILEMMA

This game is more complicated than Basu's one, studied in chapter 6.

Suppose that player 1 plays $x_1 > 2$

- If player 2 plays $x_2 > x_1$, player 1 regrets to not have played $x_2 − 1$ and her maximal regret is $x_2 − 1 + P − x_1 − P = x_2 − 1 − x_1$ which is maximal for $x_2 = M$, and equal to $M − 1 − x_1$.
- If player 2 plays $x_2 = x_1$ (>2), player 1 regrets to not have played $x_2 − 1$ and her max regret is $x_1 − 1 + P − x_1 = P − 1$
- If player 2 plays $x_2 < x_1$ we have to consider two cases: $x_1 > P$ and $x_1 \leq P$

If $x_1 > P$, player 2 can play x_2 higher or lower than P, so player 1 regrets to not have played max(2, $x_2 − 1$) and the regret is $x_2 − 1 + P − \max(0, x_2 − P)$ if $x_2 > 2$ and $2 − \max(0, 2 − P) = 2$ if $x_2 = 2$. The maximal regret is obtained for $x_2 \geq P$ and equal to $2P − 1$.

If $x_1 \leq P$, x_2 is necessarily lower than P, player 1 regrets to not have played max(2, $x_2 − 1$) and the regret is $x_2 − 1 + P$ if $x_2 > 2$ and 2 if $x_2 = 2$.

$x_2 − 1 + P$ is maximal for $x_2 = x_1 − 1$ and the maximal regret is max(2, $x_1 − 2 + P$)=$x_1 − 2 + P$.

Now suppose that player 1 plays $x_1 = 2$. It is easy to check that her maximal regret is $M − 1 − x_1 = M − 3$.

So we have three cases to study:

Case 1: $x_1 = 2$. The maximal regret is $M − 3$.

Case 2: $x_1 > P$. The maximal regret is max($M − 1 − x_1$, $2P − 1$)

ANSWERS

$= M-1-x_1$ if $x_1 \leq M-2P$
$= 2P-1$ if $x_1 \geq M-2P$

There are two possibilities: either M−2P>P, i.e. P<M/3, in which case all the integers in [M−2P, M] minimize the regret, equal to **2P−1**. Or P≥M/3 and x_1>P≥M−2P, in which case all the integers in [P+1, M] minimize the regret, which is again **2P−1**.

Case 3: 2<x_1≤P (which supposes that P>2)
The maximal regret is max(M−1−x_1, x_1−2+P)

$= M-1-x_1$ if $x_1 \leq (M+1-P)/2$
$= x_1-2+P$ if $x_1 \geq (M+1-P)/2$.

There are two possibilities. Either (M+1−P)/2≤P, i.e. P≥(M+1)/3, in which case x_1=(M+1−P)/2 minimizes the regret (for x_1≤P) and this regret is **(M−3+P)/2**.
Or P<(M+1)/3 and x_1≤P<(M+1−P)/2, in which case x_1=P minimizes the regret, which is equal to **M−1−P**.

Now observe that 2P−1>(M−3+P)/2 if P>(M−1)/3 and 2P−1<M−1−P for any P<M/3
So we get:

If M−3>P≥(M+1)/3, then x_1=(M+1−P)/2 minimizes the regret, which is equal to (M−3+P)/2. (This regret is lower than M−3).
If P≤(M−1)/3, all the integers in [M−2P, M] minimize the regret, which is equal to 2P−1. (This regret is lower than M−3).

If (M−1)/3 or (M+1)/3 is an integer we have studied all the cases. If not, P can be equal to **M/3**. In that case, all the integers in [P,M] minimize the regret, which is equal to 2P−1.
So we get the same result than in Basu's version when P is small with regard to M, but we get a quite different solution as soon as P is higher than (M+1)/3.

For M=100 and P=4, 4<99/3, the minimal regret is obtained for all the integers in [92,100], and it is equal to 7.
For M=100 and P=40, 40>101/3, the minimal regret is obtained for x_1=(61)/2 and it is equal to (137)/2. More exactly, the minimal regret is obtained for x_1=30 and x_1=31 and it is equal to 69. *Observe in Table 6.1 in chapter 6 that the mean amount asked by my L3 students for P=40 and M=100 is 33.3 and that the median value is 30. These values are very close to the ones that minimize regret.*
Let us also recall that in Basu's version (studied in chapter 6), the minimal regret, for P=40 and M=100, was obtained for any integer in [20, 100] and it was equal to 2P−1=79. So switching from x_j−P to max (0, x_j−P) strongly changes the nature of the results for high values of P, which is not surprising: for high values of P, x_j−P is often negative while max(0, x_j−P) is never negative.

ANSWERS

♣ EXERCISE 8 LEVEL–K REASONING IN AN ASYMMETRIC NORMAL FORM GAME

Level–0 players:
Player 1 plays each action with probability 1/3.
Player 2 plays each action with probability 1/3.

Level–1 players:
Player 1 best responds to level–0 players 2, so plays C_1, because 11/3>10/3>5/3.
Player 2 best responds to level–0 players 1, so plays C_2, because 13/3>9/3>7/3.

Level–2 players:
Player 1 best responds to level–1 players 2, so plays B_1 and C_1 with probability ½ each, because 1=1>0.
Player 2 best responds to level–1 players 1, so plays A_2 and C_2 with probability ½ each, because 1=1>0.

Level–3 players:
Player 1 best responds to level–2 players 2, so plays C_1, because 8/2>7/2>1/2.
Player 2 best responds to level–2 players 1, so plays C_2, because 5/2>4/2>1/2.

So level–3 players behave as level–1 players and a new cycle begins.

Clearly, this is not the philosophy of level–k reasoning; usually, the level–k philosophy leads each higher lever player to a new strategy that converges to a given behaviour; there is no cycle.

♣♣ EXERCISE 9 LEVEL–1 AND LEVEL–K REASONING IN BASU'S TRAVELLER'S DILEMMA

1. For M=12 and P=5, player 1's payoffs are given in Matrix A6.5:

	P2										
Player 1		2	3	4	5	6	7	8	9	10	11	12	
	2	2	7	7	7	7	7	7	7	7	7	7	72/11
	3	0	3	8	8	8	8	8	8	8	8	8	75/11
	4	0	0	4	9	9	9	9	9	9	9	9	76/11
	5	0	0	0	5	10	10	10	10	10	10	10	75/11
	6	0	0	0	0	6	11	11	11	11	11	11	72/11
	7	0	0	0	0	1	7	12	12	12	12	12	68/11
	8	0	0	0	0	1	2	8	13	13	13	13	63/11
	9	0	0	0	0	1	2	3	9	14	14	14	57/11
	10	0	0	0	0	1	2	3	4	10	15	15	50/11
	11	0	0	0	0	1	2	3	4	5	11	16	42/11
	12	0	0	0	0	1	2	3	4	5	6	12	33/11

Matrix A6.5

In the last column, we write player 1's mean payoff associated with each demand, when facing a level–0 player 2 who plays each amount with the same probability 1/11. It derives that a level–1 player plays 4!

So, as already claimed by Breitmoser, assuming that the level–0 players play the uniform distribution on [2,M] does not mean that level–1 players have to best react to (M+2)/2. So a level–1 player does not best respond to 7, the mean value of [2,12], he does not play 6, but he asks for the lower amount 4.

A level–2 player best responds to the demand 4, so plays 3, a level–3 player best responds to the demand 3, so plays 2. And the model sticks into 2.

For M=12 and P=2, player 1's payoffs are given in Matrix A6.6.

		2	3	4	5	6	7	8	9	10	11	12	
	2	2	4	4	4	4	4	4	4	4	4	4	42/11
	3	0	3	5	5	5	5	5	5	5	5	5	48/11
	4	0	1	4	6	6	6	6	6	6	6	6	53/11
	5	0	1	2	5	7	7	7	7	7	7	7	57/11
	6	0	1	2	3	6	8	8	8	8	8	8	60/11
Player 1	7	0	1	2	3	4	7	9	9	9	9	9	62/11
	8	0	1	2	3	4	5	8	10	10	10	10	63/11
	9	0	1	2	3	4	5	6	9	11	11	11	63/11
	10	0	1	2	3	4	5	6	7	10	12	12	62/11
	11	0	1	2	3	4	5	6	7	8	11	13	60/11
	12	0	1	2	3	4	5	6	7	8	9	12	57/11

Player 2 (top header)

Matrix A6.6

In the last column, we write again player 1's mean payoff associated with each demand, when her opponent is a level–0 player 2 who plays each amount with the same probability 1/11. It derives that the best response of a level–1 player is to play 8 and 9 with probability ½.

Once again, assuming that the level–0 players play the uniform distribution on [2,M] does not mean that level–1 players have to react to (M+2)/2. A level–1 player does not best respond to 7, the mean value of [2,12], he does not play 6, but, this time he plays the higher amounts 8 and 9 with probability ½.

A level–2 player best responds to a level–1 player by playing 7 and 8 with probability ½, a level–3 player best responds to a level–2 player by playing 6 and 7 with probability ½, a level–4 player best responds to a level–3 player by playing 5 and 6 with probability ½, a level–5 player best responds to a level–4 player by playing 4 and 5 with probability ½, a level–6 player best responds to a level–5 player by playing 3 and 4 with probability ½, a level–7 player best responds to a level–6 player by playing 2 and 3 with probability ½, and a level–k player, $k>7$, best responds to a level(k–1) player by playing 2.

2. Let us compare the mean payoff (multiplied by M–1) associated with the demand x and the demand $x+1$, when $x \leq P$ and $x > P+1$

ANSWERS

If x≤P
Payoff with x=x+(x+P)(M−x)=A
Payoff with x+1=x+1+(x+1+P)(M−x−1)=B
B−A=M−P−2x

If x=P+1
Payoff with x=x+(x+P)(M−x)=C
Payoff with x+1=1+x+1+(x+1+P)(M−x−1)=D
D−C=M−P−2x+1=M−3P−1

If x>P+1
Payoff with x=1+…+x−1−P+x+(x+P)(M−x)=E
Payoff with x+1=1+…+x−1−P+x−P+x+1+(x+1+P)(M−x−1)=F
F−E=M−2P−x

So let us consider three cases:

Case P<(M−1)/3
In that case, M−P−2x>0 for x≤P and D−C>0. So the payoff grows till M−2P−x=0, i.e. x=M−2P (and M−2P>P because M>3P). So, for P small relative to M, both M−2P and M−2P+1 ensure the highest payoff. For example, for M=12 and P=2, M−2P=8, and the highest payoff is obtained for 8 and 9.

Case P>(M−1)/3
In that case M−2P−x<0 for x>(P+1), and M−3P−1<0. If P=M/3, the payoff is maximal for x=P and x=P+1. If P>M/3, the payoff stops growing when M−P−2x=0, hence x=(M−P)/2 which is lower than P. So for P large relative to M, if (M−P)/2 is not an integer, the highest payoff is obtained for (M−P+1)/2. If (M−P)/2 is an integer, the highest payoff is obtained for both (M−P)/2 and (M−P)/2+1. Especially, for M=12 and P=5, (M−P)/2=3.5, hence the highest payoff is obtained for x=(M−P+1)/2=4.

Case P=(M−1)/3
In that case the payoff is maximal for x=P+1 and x=P+2

What matters is that both M−2P and (M−P)/2 have no link with M/2, the best answer to the mean demand (2+M)/2. For large values of P (P>M/3), we can add that the demand that maximizes level–1 player's payoff, (M−P)/2 is lower than M/2. By contrast, for P small relative to M (P<(M−1)/3), the demand that maximizes level–1 player's payoff, (M−2P), can be higher or lower than M/2; when M>4P, the demand is higher than M/2. For example, we saw that for M=12 and P=2, a level–1 player plays 8 and 9 (M−2P and M−2P+1) with probability ½, 8 and 9 being higher than 6 (=M/2).

Let us make an additional remark: for P<(M−1)/3, the level −1 best response, which consists of playing M−2P (and M−2P+1), also corresponds to the lowest bound of the interval of values that minimize the player's regret (according to Halpern and Pass, see Exercise 7).

3. In the traveller's dilemma, regardless of the demand of a level–1 player, the process always ends in the Nash equilibrium (2,2). So whether we work with the good level–1 behaviour or a wrong one only modifies the number of levels required to reach the Nash equilibrium.

But in another game, this may have a stronger impact.

ANSWERS

For example, let us come back to our standard game in Matrix E6.1, but let us replace A, B and C by three integers 1, 2 and 3. So, the mean number played by a level–0 player is 2. We get the game in Matrix A6.7.

		Player 2	
	1	2	3
1	(7,8)	(3,3)	(0,8)
Player 1 2	(0,0)	(4,4)	(1,4)
3	(7,1)	(3,0)	(1,1)

Matrix A6.7

Assuming that a level–1 player best reacts to the mean strategy, which is 2 in this game, leads level–1 players 1 to play 2 (because 4>3=3) and level–1 players 2 to play 2 and 3 with probabilities ½ because 4=4>0. This leads level–2 players 1 to also play 2 because $(4+1)/2>(3+1)/2>(3+0)/2$. Level–2 players 2 again play 2 and 3 each with probability ½. So the process ends in the state where all level–k players 1 play 2 (i.e. B_1), and all level–k players 2 play 2 and 3 (i.e. B_2 and C_2) with probability ½. We are very far from the cycle $((B_1½, C_1½), (A_2½, C_2½))(C_1, C_2))$ obtained in Exercise 8!

♣ ♣ EXERCISE 10 LEVEL–1 AND LEVEL–K REASONING IN THE STUDENTS' TRAVELLER'S DILEMMA

1. Player 1's payoffs are given in Matrix A6.8. The last column gives player 1's mean payoff when her opponent is a level–0 player 2 who asks for each amount with the same probability 1/7.

				P2				
		2	3	4	5	6	7	8	
	2	2	4	4	4	4	4	4	26/7
	3	1	3	5	5	5	5	5	29/7
	4	2	2	4	6	6	6	6	32/7
Player 1	5	3	3	3	5	7	7	7	35/7
	6	4	4	4	4	6	8	8	38/7
	7	5	5	5	5	5	7	9	41/7
	8	6	6	6	6	6	6	8	44/7

Matrix A6.8

A level–1 player should play 8, i.e. the highest possible amount. Once again, *the fact that level–0 players play the uniform distribution on [2,M] does not mean that level–1 players have to react to (M+2)/2. A level–1 player does not best respond to 5, the mean value of [2,8], he does not play 4 and 8, each with probability ½, but he plays 8.*

More generally, let us write the mean payoff (multiplied by M–1) of a level–1 player, who reacts to a uniform distribution on [2,M] for P low, so that P<(M–1)/2.

Player 1's payoff when she plays x is $A=(x–P)(x–2)+x+(x+P)(M–x)$

Her payoff when she plays x+1 is $B=(x+1–P)(x–1)+x+1+(x+1+P)(M–x–1)$

429

ANSWERS

B–A=M–1–2P>0. The payoff associated with x is increasing in x up to M, if P<(M–1)/2.

So a level–1 player plays M, for small values of P, which is exactly what we observe with the students in many experimental settings. Many of my students play like level–1 players given that, for P=4 and M=100, 65% play 100, and for P=40 and M=100, still 38% play 100 (see Exercise 16 in chapter 3 for more details on the students' behaviour).

Let us now establish the behaviour of higher level–k players in Matrix A6.8.

A level–2 player plays 7, a level–3 player plays 6, a level–4 player plays 5, a level–5 player plays 4 and 8 with probability ½, a level–6 player plays 7 and 8 with probability ½, a level–7 player plays 6 and 7 with probability ½, a level–8 player plays 5 and 6 with probability ½, a level–9 player plays 4, 5 and 8 with probability 1/3, a level–10 player plays 8, like a level–1 player. So we get again a cycle, something which is not really in the philosophy of level–k reasoning. Observe that the cycle circles on the Nash equilibrium amounts [M–2P,M] ([4,8] in our example).

2. Let us now turn to M=15 and P=8. Player 1's payoffs are given in Matrix A6.9. The last column gives player 1's mean payoff when facing a level–0 player 2 playing each amount with the same probability 1/14.

P2

	2	3	4	5	6	7	8	9	10	11	12	13	14	15	
2	2	10	10	10	10	10	10	10	10	10	10	10	10	10	132/14
3	–5	3	11	11	11	11	11	11	11	11	11	11	11	11	130/14
4	–4	–4	4	12	12	12	12	12	12	12	12	12	12	12	128/14
5	–3	–3	–3	5	13	13	13	13	13	13	13	13	13	13	126/14
6	–2	–2	–2	–2	6	14	14	14	14	14	14	14	14	14	124/14
7	–1	–1	–1	–1	–1	7	15	15	15	15	15	15	15	15	122/14
8	0	0	0	0	0	0	8	16	16	16	16	16	16	16	120/14
9	1	1	1	1	1	1	1	9	17	17	17	17	17	17	118/14
10	2	2	2	2	2	2	2	2	10	18	18	18	18	18	116/14
11	3	3	3	3	3	3	3	3	3	11	19	19	19	19	114/14
12	4	4	4	4	4	4	4	4	4	4	12	20	20	20	112/14
13	5	5	5	5	5	5	5	5	5	5	5	13	21	21	110/14
14	6	6	6	6	6	6	6	6	6	6	6	6	14	22	108/14
15	7	7	7	7	7	7	7	7	7	7	7	7	7	15	106/14

Player 1

Matrix A6.9

This time, a level–1 player plays 2 (*again far from (M+2)/2–1*). A level–2 player plays 15, a level–3 player plays 14, a level–4 player plays 13, a level–i plays 17–i, for i from 4 to 15, and then a new cycle begins: a level–16 player plays 15, a level–i player plays 31–i for i from 17 to 29, and a new cycle begins, which does not make sense.

What is nice observing here is the sharp difference between the level–1 behaviour (playing 2) and the level–1 behaviour in Matrix A6.8, which consisted in playing M. And this switch from M to 2 is a general result.

As a matter of fact, consider P and M with P>(M–1)/2.

ANSWERS

We get again:

Player 1's payoff when she plays x is A=(x–P)(x–2)+x+(x+P)(M–x)=A
Her payoff when she plays x+1 is B=(x+1–P)(x–1)+x+1+(x+1+P) (M–x–1)=B
B–A=M–1–2P<0. So this time the payoff associated with x is always decreasing in x, because P>(M–1)/2, leading the player to play 2.

As regards my L3 players, let me observe that 37.2% played 2 and that 21.2% played 100 for P=70 and M=100. So 58.4% of them played like level–1 and level–2 players (see Exercise 17 in chapter 3 for more details on the students' behaviour).

Anyway, regardless of the starting point – i.e. regardless of the level–1 action, regardless of the values of P and M – we get a cycle which leads from M to max(2, M–2P) and from max(2, M–2P) to M, which is not very intuitive. Cognition hierarchy would not lead to a cycle but would also lead to mix behaviour on all values between max(2, M–2P) and M.

Clearly a level–*k* reasoning better suits for games that slowly converge to a Nash equilibrium.

♣ ♣ EXERCISE 11 STABLE EQUILIBRIUM, PERFECT EQUILIBRIUM AND PERTURBATIONS

1. Call p_1, p_2, p_3, p_4 the probabilities assigned to A_1a_1, A_1b_1, B_1a_1 and B_1b_1. Player 2 is best off playing A_2a_2 rather than B_2a_2 if $2(p_1+p_2)+p_3+p_4 \geq 2(p_1+p_2)+4p_3$, i.e. $p_4 \geq 3p_3$.

   Call ε_3 and ε_4 the minimal probabilities on B_1a_1 and B_1b_1. When $\varepsilon_4 \geq 3\varepsilon_3$, there is no problem to ensure $p_4 \geq 3p_3$. But if $\varepsilon_4 < 3\varepsilon_3$ then $p_4 \geq 3p_3$ leads to $p_4 \geq 3p_3 \geq 3\varepsilon_3 > \varepsilon_4$. So player 1 has to play B_1b_1 with a probability higher than the minimal one, which is impossible, given that she gets 5 with A_1a_1 and at most 1 with B_1b_1. So (A_1a_1, A_2a_2) is not a stable equilibrium.

   This is a rather expected result, given that (A_1a_1, A_2a_2) is not even subgame perfect.

2. Let us set $\pi_{1x1}^k(A_1)=1-\varepsilon_1^k$, $\pi_{1x1}^k(B_1)=\varepsilon_1^k$, $\pi_{1x3}^k(a_1)=1-e_1^k$, $\pi_{1x3}^k(b_1)=e_1^k$, $\pi_{2x2}^k(A_2)=\varepsilon_2^k$, $\pi_{2x2}^k(B_2)=1-\varepsilon_2^k$, $\pi_{2h}^k(a_2)=1-e_2^k$, $\pi_{2h}^k(b_2)=e_2^k$, where ε_1^k, ε_2^k, e_1^k, e_2^k are the minimal probabilities assigned to B_1, A_2, b_1 and b_2. ε_1^k, ε_2^k, e_1^k, $e_2^k \to 0$ when $k \to \infty$.

   $\pi^k \to \pi^*$, where π^* are the behavioural strategies assigned to $E^*=(A_1a_1, B_2a_2)$.

   Player 1 at x_3: a_1 leads to the payoff $4\varepsilon_1^k(1-\varepsilon_2^k)(1-e_2^k)$, b_1 leads to the payoff $\varepsilon_1^k(1-\varepsilon_2^k)e_2^k$. So assigning the minimal probability e_1^k to b_1 and the remaining one to a_1 is player 1's best local strategy in the perturbed game.

   Player 1 at x_1: A_1 leads to the payoff 5, B_1 leads to a payoff lower than 5. So player 1 is right assigning the minimal probability ε_1^k to B_1 and the remaining one to A_1.

   Player 2 at h: a_2 leads to the payoff $4\varepsilon_1^k(1-\varepsilon_2^k)(1-e_1^k)$, b_2 leads to the payoff $\varepsilon_1^k(1-\varepsilon_2^k)e_1^k$. So player 2 is right assigning the minimal probability e_2^k to b_2 and the remaining one to a_2.

   Player 2 at x_2: A_2 leads to the payoff ε_1^k, B_2 leads to the payoff $\varepsilon_1^k[4(1-e_1^k)(1-e_2^k)+e_1^k e_2^k]$ which goes to $4\varepsilon_1^k$ when $k \to \infty$. So player 2 is right assigning the minimal probability ε_2^k to A_2 and the remaining one to B_2.

   So π^k is a Nash equilibrium of the perturbed game and E^* is a perfect equilibrium, without any restriction on the introduced minimal probabilities.

ANSWERS

We would like to conclude that it is also a stable equilibrium given that it resists the introduction of any small probabilities. Yet, strangely enough, this is not the case.

To see why, we turn back to the normal form game and observe that player 2 is best off playing B_2a_2 rather than A_2a_2 if $2(p_1+p_2)+p_3+p_4 \leq 2(p_1+p_2)+4p_3$, i.e. $p_4 \leq 3p_3$.

We call again ε_3 and ε_4 the minimal probabilities on B_1a_1 and B_1b_1. When $\varepsilon_4 \leq 3\varepsilon_3$, there is no problem to ensure $p_4 \leq 3p_3$. But if $\varepsilon_4 > 3\varepsilon_3$ then $3p_3 \geq p_4$ leads to $3p_3 \geq p_4 \geq \varepsilon_4 > 3\varepsilon_3$. So player 1 has to play B_1a_1 with a probability higher than the minimal one, which is clearly impossible, given that she gets 5 with A_1a_1 and at most 4 with B_1a_1. So *(A_1a_1, B_2a_2) is not a stable equilibrium despite it resists any small perturbations on the extensive form game.*

This astonishing result is due to the fact that perfection introduces perturbations in the extensive form, hence on local strategies, whereas Kohlberg and Mertens' stability introduces perturbations on the normal form game. These two kinds of perturbations are completely different. As a matter of fact, to play B_1a_1 in the extensive form game, player 1 has to make only one error, at x_1, because she plays the unexpected action B_1 at x_1 but the expected action a_1 at x_3: the probability of this strategy in the perturbed game is $\varepsilon_1^k(1-e_1^k)$. By contrast, to play B_1b_1, player 1 has to make two errors, because she plays the unexpected actions B_1 at x_1 and b_1 at x_3. The probability of this strategy in the perturbed game is $\varepsilon_1^k e_1^k$. *If we take these probabilities as minimal probabilities in the normal form game, instead of ε_3 and ε_4, $p_4 \leq 3p_3$ is always fulfilled because $\varepsilon_1^k e_1^k < 3\varepsilon_1^k(1-e_1^k)$ regardless of the chosen perturbations!*

So clearly the two logics are different: perfection introduces as many levels of possible errors as information sets, whereas Kohlberg and Mertens' stability introduces one level of error, at the beginning for the whole game.

In literary terms, (A_1a_1, B_2a_2) is not stable because player 2 is induced to switch to A_2a_2 when player 1 more often trembles to B_1b_1 than to B_1a_1 *(something which is impossible with perfection)*. To avoid this switch, player 1 has to play B_1a_1 more often than asked by the trembles, so she has to get at least 5 at equilibrium with B_1a_1, which is not possible.

3. The set $\{(A_1a_1,A_2a_2), (A_1a_1,B_2a_2)\}$, corresponding to the unique outcome (5,2) is stable.

A_1a_1 leads to the highest possible payoff 5 for player 1 regardless on the minimal probabilities. No strategy can lead to a higher payoff, so A_1a_1 resists any perturbations.

A_2a_2 leads to the same payoff than A_2b_2 and to a higher payoff than B_2b_2 in a perturbed environment, regardless of the introduced perturbations.

Call ε_3 and ε_4 the minimal probabilities on B_1a_1 and B_1b_1. When $\varepsilon_4 \leq 3\varepsilon_3$, B_2a_2 is player 2's best strategy. When $3\varepsilon_3 \leq \varepsilon_4$, A_2a_2 is player 2's best strategy. It follows that, regardless of the minimal perturbations introduced, player 2's best response is always in $\{A_2a_2, B_2a_2\}$, which explains why $\{(A_1a_1,A_2a_2), (A_1a_1,B_2a_2)\}$ is a stable set.

… ANSWERS

♣ ♣ EXERCISE 12 FOUR DIFFERENT FORWARD INDUCTION CRITERIA AND SOME DYNAMICS

1. To study Kohlberg and Mertens' stability, call p_1, respectively p_2, the probability assigned to m_2 by t_1, respectively by t_2, and call q the probability assigned by player 2 to r_1 after m_2.

    For E to be a Nash equilibrium, we need $3q \leq 2$ and $3q+2(1-q) \leq 4$ (always true), hence $q \leq 2/3$. So player 2 has to play r_2 with positive probability after m_2 which is only possible if $3\mu(y_2) \geq 2\mu(y_1)$, hence $3p_2 \geq 2p_1$. Now we introduce small perturbations, so that t_1 and t_2 play m_2, respectively with probability ε_1 and ε_2. If $3\varepsilon_2 \geq 2\varepsilon_1$, there is no problem. But if $3\varepsilon_2 < 2\varepsilon_1$, then we need $3p_2 \geq 2p_1 \geq 2\varepsilon_1 > 3\varepsilon_2$ to sustain E. So t_2 has to play m_2 with a probability higher than the minimal one, which requires $3q+2(1-q)=4$, which is impossible. So E is not stable.

    E is also not self-enforcing: $3q+2(1-q)<4$, so m_2 is an inferior strategy for t_2. If one deletes this strategy, then m_2 is assigned to t_1, consequently followed by r_1, and E is destroyed.

    And E does not pass the D criterion. At the first step, we observe that any response after m_2 makes t_1 or nobody better off. So player 2 assigns m_2 to t_1 with probability 1 and plays r_1. It follows that in the second step (and all the subsequent ones till infinity) only t_1 plays m_2 and player 2 only assigns m_2 to t_1 (and plays r_1 after m_2): player 2's beliefs (and associated actions) do not sustain E, which does not pass the D criterion.

    Well the three criteria, and many others, highlight that only t_1 can be better off switching to m_2, and if player 2 recognizes this fact, then he plays r_1, which induces t_1 to deviate to m_2, destroying the tested PBE E.

    Yet E is not dismissed by the CFIEP criterion, because there only exists *(see Exercise 9 in chapter 5)* one other PBE E' in this game, defined by: $(\pi_1(m_1/t_1)=\frac{1}{4}, \pi_1(m_2/t_1)=\frac{3}{4}, \pi_1(m_1/t_2)=1/2, \pi_1(m_2/t_2)=1/2, \pi_2(r_1/m_1)=9/17, \pi_2(r_2/m_1)=8/17, \pi_2(r_1/m_2)=10/17, \pi_2(r_2/m_2)=7/17, \mu(x_1)=1/3, \mu(x_2)=2/3, \mu(y_1)=3/5, \mu(y_2)=2/5)$. But no type of player 1 is better off in E' because t_1's payoff in this equilibrium is $30/17<2$, and t_2's payoff is $44/17<4$.

    Umbhauer's concept does not eliminate E due to Stiglitz's remark according to which, if an out of equilibrium action conveys new information (with regard to the equilibrium one), then *an equilibrium action may also convey new information*. In other words, if m_2 is assigned to t_1 because she is the only type who can be better off with this message, then t_1 plays this message (because followed by r_1). *But then, how should player 2 interpret the equilibrium message m_1?* If t_1 plays m_2, then, faced to m_1, player 2 has to assign it to t_2. So he switches from r_1 to r_2 (because $1>0$). But then t_2 is better off switching to m_2 too (because $3>1$). But if so, player 2 is better off playing r_2 after m_2, which deters any type from deviating from E.

2. Let us observe that, if we interchange player 2's payoffs 2 and 3 after m_2, then Kohlberg and Mertens, Kohlberg and Banks and Sobel still eliminate E for the same reasons as above (as regards Kohlberg and Mertens, the main inequation now becomes $2p_2 \geq 3p_1 \geq 3\varepsilon_1 > 2\varepsilon_2$ but this changes nothing), but now E isn't a CFIEP either. As a matter of fact there exists now a new PBE E'', where both types of player 1 play m_2 followed by r_1 (m_1 being followed by r_2, because assigned to t_2) and which checks:

    D, the set of types playing the out (E's) equilibrium message m_2, and getting more by so doing, is the singleton $\{t_1\}$ (because t_2 also plays m_2 but is worse off by so doing). So D is not empty – there exists somebody that is better off in the new equilibrium – (condition i), and S, the set of best responses in front of D is the singleton $\{r_1\}$. And t_1, the type in D, is best off

433

ANSWERS

deviating when player 2 plays r_1 (condition (iii)): condition (iii) ensures that if player 2 focuses his beliefs on the types who are happy to deviate (and not on those that are compelled to do so, only to not be recognized), they are still happy to deviate. Condition (ii) is also satisfied (t_2, out of D, is worse off playing m_2).

Let us observe here that Stiglitz's remark now leads to *a kind of dynamic reasoning* that destroys E and stabilizes in E". So, let us start in E. If m_2 is assigned to t_1 because she is the only type who can be better off with this message, then t_1 plays this message (because followed by r_1). But then, player 2 assigns m_1 to t_2 and plays r_2 after m_1. This leads t_2 to switch to m_2 too. This doesn't change player 2's action after m_2, which stabilizes the process in E".

♣♣ EXERCISE 13 FORWARD INDUCTION AND THE EXPERIENCE GOOD MARKET

1. With regard to the CFIEP criterion, E* checks all the necessary properties to eliminate an equilibrium E. It contains an out of equilibrium message, 120.

    We have (60–40)=q(P–40) in E and (60–40)=q*(120–40) in E*, where P is the price played by t_2 in E, and q and q* are the probabilities of accepting P in E and 120 in E*.

    The equality q(P–40)=q*(120–40) implies, by single crossing: q(P–100)<q*(120–100). So t_1 gets the same payoff in E* and E, but t_2 gets more in E* than in E. Hence D={t_2} and accepting 120 belongs to S, the set of best responses when the beliefs focus on t_2. Conditions (i) and (iii) are checked (given that t_2 is better off switching to 120 when it is followed by acceptation) and condition (ii) is also satisfied, given that t_1 is not better off in E* than in E (she is indifferent).

    So each equilibrium E is eliminated by E*.

    By contrast E* can't be eliminated by an equilibrium E.

    (60–40)=q(P–40)=q*(120–40) implies by single crossing q(P–100)<q*(120–100), so it follows that D=∅. Nobody is better off switching to E, so condition (i) is not satisfied and E* can't be eliminated by E.

2. With regard to the self-enforcing criterion, start in an equilibrium E (P is the price played in E by t_2 and q the probability of accepting it). The probability q' of accepting the out of equilibrium message 120 has to check:

    q'(120–40)≤q(P–40)=(60–40) and q'(120–100)≤q(P–100)

    By single crossing, q'(120–100)≤q(P–100) ⇒ q'(120–40)<q(P–40).

    So 120 is an inferior strategy for t_1. It derives that 120 is affected to t_2 and therefore accepted, which destroys E (because t_2 deviates to 120). Why do we say that 120 is accepted by player 2 even when player 2 is indifferent between accepting and refusing? The reason is that if we replace 120 by 119.999, player 2 will accept 119.999: by continuity, he accepts 120.

    In fact, instead of 120, we could choose any price p' with 120>p'>P: player 2 would again assign p' to t_2, accept p', and t_2 would deviate to p'.

    By contrast, E* can't be eliminated by Kohlberg's criterion (q* is the probability of accepting 120 in E*). As a matter of fact, the probability q' of accepting an out of equilibrium message P has to check: q'(P–40)≤q*(120–40) and q'(P–100)≤q*(120–100)

Given that P<120, we get by single crossing: q'(P–40)≤q*(120–40) ⇒ q'(P–100)<q*(120–100). Hence P is an inferior strategy for t_2. So P is assigned to t_1, accepted if lower than 60 and refused if not. It follows that nobody is induced to deviate to P, so E* is self-enforcing.

So forward induction leads us far from Akerlof's result. First, both qualities are sold with positive probability at E. Second, each type plays the highest possible price, i.e. the price accepted by the consumer in a context of complete and perfect information. Observe that forward induction helps the good type, in that t_2 never fears to be recognized.*

♣ EXERCISE 14 FORWARD INDUCTION AND ALTERNATE NEGOTIATION

1. In E_7 and E_{72}, t_2 plays a weakly dominated strategy, 7 (dominated by 8). After eliminating this strategy, 7 can only be assigned to t_1, so will be refused and never played. It follows that E_7 and E_{72} are not self-enforcing. It also follows that E_8 and E_{82} are self-enforcing.
2. Let us turn to the CFIEP. For ρ lower than 1/7, E_7 is eliminated by E_8 because D=T (both types are better off deviating from 7 to 8 and if player 2 focuses on T, he accepts the offer 8 which makes both types better off). And E_8 can't be eliminated by E_7 because nobody is better off in this equilibrium (D=∅). For ρ between 1/7 and 2/7, E_7 is eliminated by E_{82}: only t_2 is better off deviating to 8 and if player 2 focuses his beliefs on t_2, he accepts 8 which makes t_2 better off (and condition (ii) is also satisfied because t_1 is worse off with 8 in E_{82}). And E_{82} can't be eliminated by E_7 because only t_1 is better off in E_7, D={t_1}, but he is worse off when player 2 focuses on t_1 (so refuses 7). And for ρ higher than 2/7, E_{72} is eliminated by E_{82}: as in the previous case, only t_2 is better off deviating to 8 and if player 2 focuses his beliefs on t_2, he accepts 8 which makes t_2 better off (and condition (ii) is also satisfied in that t_1 gets the same payoff with 8 than in the starting equilibrium). And E_{82} can't be eliminated by E_{72} because nobody is better off in this equilibrium (D=∅).

   To conclude, for any ρ, there is only one CFIEP, either E_8 or E_{82}.

We can observe that the fact that player 2, when he recognizes t_2, accepts 8, helps this type to always get more than 7. This phenomenon is similar to the one observed in Exercise 13 and highlights that forward induction helps the types who are happy to be recognized.

NOTES

1 Umbhauer, G., 2007. Best-reply matching and the centipede game, BETA working paper n°2007–25, Strasbourg.

Index

Akerlof's lemon cars *see* lemon cars
all pay auction *see* all pay auction with incomplete information; ascending all pay auction; first price/second price sealed bid all pay auction; Stackelberg all pay auction
all pay auction with incomplete information 113–15
almost common value auction 97, 333–5
ascending all pay auction 23–6, 30, 35–40; and backward induction 151–5, 160–4, 201, 371–3; and best-reply matching 257–8; and dominance 67–8, 73–4, 78–9; and Nash equilibrium 107–10, 123; non incremental 26–7, 165–8, 201–2, 373–5 and regret minimization 266–71
ascending all pay auction with incomplete information 43–5, 215–16, 229, 232
auction *see* all pay auction; almost common value auction; Dutch auction; first price/ second price sealed bid auction; wallet game

backward induction: and dominance 154–6; inconsistency 163–4, 174–5; in finite games 153–4, 156–63; and in infinite games 179; *see also* learning; repeated game
bargaining *see* negotiation
battle of the sexes 10–11, 100, 288–9; sequential 176–7, 193–5
bazooka game 57, 140, 301–2, 341

behavioural strategy 35; link with mixed strategy 38–40; *see also* errors in mixed/ behavioural strategies
behavioural strategy Nash equilibrium *see* Nash equilibrium
beliefs *see* perfect Bayesian equilibrium; sequential equilibrium; *see also* consistency of actions and beliefs; forward induction; inconsistency of actions and beliefs
Bertrand duopoly 56, 94, 143, 293, 324–5, 347, 422
best reply matching equilibrium (Kosfeld, Droste and Voorneveld) 250–60; generalized (Umbhauer) 251–2
burning money game 93, 138–9, 288–9, 321–2, 338–9

carrier pigeon game 227–8
cautious behaviour 8, 129–35, 144–5, 163, 351, 358; *see also* risk
cautious equilibrium 130–5
centipede game *see* pre-emption game
centroïd 239
chicken game 7–8, 51, 63–4, 138, 337
children's pie game 57, 301
cognitive hierarchy 271, 275
commitment in negotiation 160; *see also* Li's negotiation game
common knowledge 44
common value auction *see* almost common value auction; wallet game
complete/incomplete information *see* information

437

INDEX

completely mixed strategy *see* mixed strategy
conflict game *see* zero sum game
congestion game 9
consistency of actions and beliefs 217–20; *see also* consistent forward induction equilibrium path; Nash equilibrium; perfect Bayesian equilibrium
consistent forward induction equilibrium path (Umbhauer) 286–8
continuity at infinity *see* payoffs
coordination game 10, 100, 122
correlated equilibrium 253–4
cowboy story *see* duel guessing game
crossed levels of knowledge 44
crossed rationality 77–8; and inconsistency 78–9

D criterion (Banks and Sobel) 284–5, 288
dirty faces game *see* prisoners' disks game
dominance 63–72; *see also* iterated dominance; learning; Nash equilibrium; perfect equilibrium; strictly/weakly dominated strategy
duel game *see* duel guessing game
duel guessing game 58–9, 95–6, 142, 198–9, 305–6, 327–30, 344–5, 367–8
Dutch auction 60–1, 149, 312–15, 363–5

electronic mail game 232–5
embedded game *see* bazooka game
envelope game 33–5, 48; and crossed rationality 77–8; and dominance 75; and level-k reasoning 76–7, 273, 276–7; and Nash equilibrium 127–9; and regret minimization 265–6
ε equilibrium (Jackson et al) 236
equivalent strategy 37
errors in mixed/behavioural strategies 39–40, 246–7, 295–6, 402–4, 431–2; *see also* perturbations
evolutionary criteria 136
evolution table in an infinitely repeated game 181
existence: Nash equilibrium 99; perfect equilibrium 210; sequential equilibrium 217; subgame perfect Nash equilibrium 153
experience good *see* lemon cars
experimental game theory 20, 22, 30–1, 87–8; *see also* students' way to play
ex-post regret (no) 147, 356–8
extensive form game 18–19; story extensive form games 20–8; *see also* link between normal and extensive form games

fairness 49–51
feelings 51–2
final offer arbitration 160
finite game 28
finitely repeated game: extensive form game 175–9; and facts 173–4; and inconsistency 174–5; normal form game 169–73
first price sealed bid all pay auction 12–15, 18–20, 28–9, 49; asymmetric 90–1, 130, 132, 145–6, 317–19, 352–3; and best reply matching 254–5; and cautious behaviour 132; and dominance 64–5; and Nash equilibrium 102–6; and repeated game 174–5, 180–93
first price sealed bid auction 60–1, 115–16, 148–9, 312–15, 361, 363–5
focal point 122–3, 145, 352
folk theorem 183–9
Fort Boyard sticks game 31–2; and backward induction 156–7; and dominance 68–72
forward induction 24, 84, 123, 161, 164, 175, 278–89
French plea bargaining game 46–7, 52–4, 64, 224, 287–8
French rock paper scissors game *see* rock paper scissors game

game of 21 *see* race game
generalized best reply matching equilibrium (Umbhauer) *see* best reply matching equilibrium
gift exchange game 22–3, 30–1, 50–1, 141, 202, 205–6, 342–3, 375–6, 393–4
grab the dollar game 24, 163, 258; *see also* ascending all pay auction

gradualism 160; *see also* Li's negotiation game
guessing game 57, 94, 143, 272, 274–8, 323–4, 348

hawk-dove game 7–8, 17

inconsistency of actions and beliefs in real life 109–10, 128–9, 137, 220
inconsistency of actions and beliefs in theory *see* backward induction; crossed rationality; finitely repeated game
incredible threat 124, 126, 151
individual rational payoff *see* payoff
inferior strategy 282
infinitely repeated game: extensive form game 193–5; in practice 189–93; normal form game 180–9
information: changing incomplete into imperfect 43–5; complete/incomplete 43; perfect/imperfect 40; set 18, 20, 27; value of 91, 319–20
iterated dominance 72–5, 277–8; *see also* crossed rationality; prisoners' disks game

Kuhn's theorem 39, 60, 309–12

learning: backward induction 195–6; a dominated strategy 87–8
lemon cars 59, 148–9, 248–9, 296–7, 307–9, 362, 407–10, 434–5
level-0 reasoning 271, 273–4
level-k reasoning 76–7, 271–6; and iterated dominance 277–8; and risk 276–7
levels of crossed rationality *see* crossed rationality
lexicographic consistency 219–20
Li's negotiation game 160, 205, 384–5
link between normal and extensive form games 36–40; and Nash equilibrium 106, 110–12; and perfect equilibrium 246–7, 402–4; and stable equilibrium set 295–6, 431–2
local strategy 35
logit equilibrium 239
lottery 11, 33, 47–8; *see also* Nature

marginal approach of Nash equilibria *see* second price auction wallet game; second price sealed bid auction
matching pennies game 11, 138, 337–8
meeting game 10, 122
minmax value 182–3
mixed strategy 28–9; completely mixed 28; *see also* behavioural strategy
mixed strategy Nash equilibrium *see* Nash equilibrium; *see also* best reply matching
multiplicity of Nash equilibria *see* Nash equilibrium

Nash equilibrium: behavioural strategy 99, 107–9; and consistency 108–9, 128, 137; and dominance 101–2; existence 99; mixed strategy 99, 103–6, 136; multiplicity of 122; and ordinal differences 136; outcome 99; path 99; pure strategy 99–102; support of 99; *see also* incredible threat; links between normal and extensive form games
Nature 33, 35, 44–5
negotiation 158–60, 200, 370–1; with an unknown number of rounds 229–32, 249, 297–8, 410–13, 435; *see also* commitment; final offer arbitration; gradualism; Li's negotiation game; Rubinstein's bargaining game
nodes 18
normal form game 6; story normal form games 6–12; *see also* link between normal and extensive form games

order of elimination *see* iterated dominance
outcome (equilibrium outcome) *see* Nash equilibrium

paranoid behaviour 51–2, 131
path (equilibrium path) *see* Nash equilibrium
payoff 18; cardinal/ordinal 66–7, 133–6, 259–60; continuous at infinity 179; individually rational 183; *see also* equivalent strategy; utility
perfect Bayesian equilibrium (Harsanyi) 221–5

439

INDEX

perfect equilibrium (Selten) 208–11, 213–14; and dominance 212, 214–16; existence 210; *see also* errors in mixed/behavioural strategies; proper equilibrium; quantal response equilibrium; replicator equations; sequential equilibrium; stable equilibrium set; trembling*-hand perfect equilibrium

perfect/imperfect information *see* information

perfect recall 40

perturbations of payoffs *see* quantal response equilibrium; trembling*-hand perfect equilibrium

perturbations of strategies in extensive form games *see* perfect Bayesian equilibrium; perfect equilibrium; sequential equilibrium

perturbations of strategies in normal form games *see* proper equilibrium; stable equilibrium set; trembling*-hand perfect equilibrium

perturbations of the game structure *see* structure of a game (changes in)

perturbed game 209; test sequence of perturbed games 209; *see also* structure of a game (changes in)

plea bargaining game *see* French plea bargaining game

pooling perfect Bayesian equilibrium 221

posterior probability distribution 45

pre-emption game 26, 58, 93–4, 142–3, 199–200, 292–4, 303–4, 322–3, 345–7, 368–70, 419–21, 423–4

preference relations 6; *see also* normal form game (story games); utility

prior probability distribution 43–5

prisoner's dilemma 6–7, 65–7, 101–2, 174–5

prisoners' disks game 41–2, 52; and dominance 79–84; and Nash equilibrium 118–21; and rule of behaviour 84–6

proper equilibrium (Myerson) 225–7

pure strategy in an extensive form game 30; and completeness 30–1; link with the pure strategy in a normal form game 36–7; number of 31

pure strategy in a normal form game 28; and cleverness 28; number of 37

pure strategy Nash equilibrium *see* Nash equilibrium

quantal response equilibrium (Kelvey and Palfrey) 238–9

race game 87–8, 195–6

rationalizability (Bernheim, Pearce) 251

reciprocity *see* fairness

reduced field of vision 126

reduced game 153, 181

reduced normal form game 37–8, 110

regret minimization (Halpern and Pass) 260–71; *see also* traveller's dilemma

renegotiation (Benoît and Krishna) 174

repeated game *see* finitely repeated game; infinitely repeated game

replicator equations 240–1

representative form of a game *see* extensive form game; normal form game; reduced normal form game

reprisal 169, 181; *see also* repeated game

reward 169, 181; *see also* repeated game

risk 6, 8, 47–8, 128–30, 132, 263; and level-k reasoning 273, 276–7; *see also* cautious behaviour; risk dominance

risk dominance 135

rock paper scissors game 1–2, 11–12, 139–40, 340

Rubinstein's bargaining game: finite 204, 380–3; infinite 204, 383–4

rule of behaviour: and the Fort Boyard sticks game 70–1; and the prisoners' disks game 84–6, 118; and sealed bid auction 115, 117; and the wallet game 146, 355–6

screening game 45–6

sealed bid auction *see* first price/second price sealed bid all pay auction; first price/second price sealed bid auction

second price sealed bid all pay auction 16–18, 29, 49; and best reply matching 255–7, 259; and Nash equilibrium 148, 360–1

INDEX

second price sealed bid auction 96, 116–17, 331–2, 356; and marginal approach 117–18
self-enforcing outcome (Kohlberg) 282–4, 287–8
Selten's horse game 27, 208, 210–11, 213–14, 220–1, 225, 236–41, 279–83
semi-separating perfect Bayesian equilibrium 221
separating perfect Bayesian equilibrium 221
sequential equilibrium (Kreps and Wilson) 216–21
sequential rationality 216; *see also* perfect Bayesian equilibrium
signal *see* forward induction
signalling game 45–7; *see also* perfect Bayesian equilibrium
simultaneous game 18, 38
single crossing 148–9, 361–2, 363–5, 408–9, 434
stable equilibrium set (Kohlberg and Mertens) 279–82, 284; and perfect equilibrium 295–6, 431–2
Stackelberg all pay auction 20–1, 31, 37, 92, 124–6, 133, 177–9, 198, 320–1, 366–7
stage game 169
state 181; state transition diagram 180–1
strategic form game *see* normal form game
strategy *see* behavioural strategy; equivalent strategy; local strategy; mixed strategy; pure strategy
strict sequential equilibrium 219
strictly dominated/dominant strategy 63–8, 72, 88n2; and Nash equilibrium 101; *see also* prisoners' disks game
structure of a game 2, 5, 9, 43; changes in 227–35; and equilibria 54–5, 60–1, 312–15
students' way to play the ascending all pay auction/war of attrition 25–6, 109, 162–3
students' way to play the centipede game/ pre-emption game 199–200, 293–4, 368–70
students' way to play the envelope game 48, 75–7, 129, 265–6, 273
students' way to play the first price sealed bid all pay auction 14–15; the second price sealed bid all pay auction 17–18, 49, 256–7

students' way to play the gift exchange game 50–1
students' way to play the guessing game 272
students' way to play the race game 87–8, 195–6
students' way to play the traveller's dilemma: Basu's version 261–3, 349–50; students' version 144–5, 351
subgame 27–8; proper subgame 27–8
subgame perfect Nash equilibrium (Selten) 151–4, 179; *see also* backward induction
support of an equilibrium *see* Nash equilibrium
syndicate game 9, 57–8, 138, 302–3, 337
system of beliefs 216, 221

talk: before the game 122–3; cheap 122, 228; *see also* forward induction
threshold 24, 33, 54, 114–15, 163, 256
transition state diagram *see* state
traveller's dilemma, Basu's version (and variant) 56, 260; and best reply matching 291–2, 416–18; and dominance 95, 325–6; and focal point 145, 352; and Nash equilibrium 143, 348–50; and level-k reasoning 273–4, 277, 295, 426–9; and regret minimization 260–4, 294, 424–5; and repeated game 205, 385–92
traveller's dilemma, students' version: and best reply matching 291–2, 416–18; and cautious behaviour 144–5, 351; and dominance 95, 326; and focal point 145, 352; and level-k reasoning 295, 429–31; and Nash equilibrium 143–5, 350–1
tree 18
trembles *see* perturbations
trembling hand equilibrium 209
trembling*-hand perfect equilibrium (Jackson et al.) 236–8
type of player 45

ultimatum game 20–1, 49–50, 158, 380
unilateral deviation 99, 108, 172
utility 13–14, 47–54; Von Neumann Morgenstern (VNM) 6, 47–8; *see also* payoff; risk

441

INDEX

wallet game: first price auction 146, 354–6; and marginal approach 147, 359; second price auction 147, 356–9; *see also* almost common value auction

war of attrition game *see* ascending all pay auction

weak sequential equilibrium 219

weakly dominated/dominant strategy 63–8; and learning 87–8; and Nash equilibrium 101; *see also* Fort Boyard sticks game

winner's curse 146, 354

zero sum games 11–12, 131; *see also* duel guessing game